CAMBRIDGE MONOGRAPHS ON APPLIED AND COMPUTATIONAL MATHEMATICS

34 Numerical Bifurcation Analysis of Maps

The *Cambridge Monographs on Applied and Computational Mathematics* series reflects the crucial role of mathematical and computational techniques in contemporary science. The series publishes expositions on all aspects of applicable and numerical mathematics, with an emphasis on new developments in this fast-moving area of research.

State-of-the-art methods and algorithms as well as modern mathematical descriptions of physical and mechanical ideas are presented in a manner suited to graduate research students and professionals alike. Sound pedagogical presentation is a prerequisite. It is intended that books in the series will serve to inform a new generation of researchers.

A complete list of books in the series can be found at
www.cambridge.org/mathematics.
Recent titles include the following:

19. Matrix preconditioning techniques and applications, *Ke Chen*
20. Greedy approximation, *Vladimir Temlyakov*
21. Spectral methods for time-dependent problems, *Jan Hesthaven, Sigal Gottlieb & David Gottlieb*
22. The mathematical foundations of mixing, *Rob Sturman, Julio M. Ottino & Stephen Wiggins*
23. Curve and surface reconstruction, *Tamal K. Dey*
24. Learning theory, *Felipe Cucker & Ding Xuan Zhou*
25. Algebraic geometry and statistical learning theory, *Sumio Watanabe*
26. A practical guide to the invariant calculus, *Elizabeth Louise Mansfield*
27. Difference equations by differential equation methods, *Peter E. Hydon*
28. Multiscale methods for Fredholm integral equations, *Zhongying Chen, Charles A. Micchelli & Yuesheng Xu*
29. Partial differential equation methods for image inpainting, *Carola-Bibiane Schönlieb*
30. Volterra integral equations, *Hermann Brunner*
31. Symmetry, phase modulation and nonlinear waves, *Thomas J. Bridges*
32. Multivariate approximation, *V. Temlyakov*
33. Mathematical modelling of the human cardiovascular system, *Alfio Quarteroni, Luca Dede', Andrea Manzoni & Christian Vergara*

Numerical Bifurcation Analysis of Maps

From Theory to Software

YURI A. KUZNETSOV
Utrecht University and University of Twente

HIL G. E. MEIJER
University of Twente

CAMBRIDGE
UNIVERSITY PRESS

CAMBRIDGE
UNIVERSITY PRESS

University Printing House, Cambridge CB2 8BS, United Kingdom

One Liberty Plaza, 20th Floor, New York, NY 10006, USA

477 Williamstown Road, Port Melbourne, VIC 3207, Australia

314–321, 3rd Floor, Plot 3, Splendor Forum, Jasola District Centre, New Delhi – 110025, India

79 Anson Road, #06–04/06, Singapore 079906

Cambridge University Press is part of the University of Cambridge.

It furthers the University's mission by disseminating knowledge in the pursuit of education, learning, and research at the highest international levels of excellence.

www.cambridge.org
Information on this title: www.cambridge.org/9781108499675
DOI: 10.1017/9781108585804

First published 2019

Printed and bound in Great Britain by Clays Ltd, Elcograf S.p.A.

A catalogue record for this publication is available from the British Library

ISBN 978-1-108-49967-5 Hardback

To our families
for all their support and understanding
while we were writing this book

Contents

Preface

When a researcher is faced with experimental results seeming to obey a deterministic law, usually a specific mathematical model is built, tested, and validated. Many models are indeed formulated as recurrent relations defining iterated maps. These are models describing dynamics of populations with non-overlapping generations, as well as biological, economical, and industrial systems subject to periodic environmental influence. With varying parameters in a model, different behavior can be observed, providing explanations of the experimental results. To actually analyze such a model one needs to draw on theoretical knowledge but also appropriate numerical methods. Preferably these will be available through user-friendly software. With this goal we have worked on developing theory and algorithms for our MATLAB® toolbox MatcontM over the past decade.

This book covers discrete-time dynamical systems generated by iterated nonlinear maps. In particular, it explains how their dynamics change under variation of parameters, which is a subject of bifurcation theory. We present these topics via a systematic treatment of bifurcations of fixed points and cycles up to and including cases in which two system parameters are involved. Theoretical results for two-parameter bifurcations have been obtained during the past 40 years. There are a number of recent developments available to experts in the field through research papers only. This textbook fills this gap by presenting the theory systematically and consistently, from an introductory level up to current research topics.

Through our recent work, the work of collaborators, and other researchers in the field, we have obtained a fairly complete understanding of local bifurcations of maps and can apply these results to concrete models. Local bifurcation theory gives good indicators and descriptions of how a certain model behaves, but in practice global characteristics are used too. Therefore, our treatment also includes several of these complementary methods, such as Lyapunov exponents, invariant manifolds and homoclinic structures, and parts of chaos theory.

The power of the developed theory, methods, and computer algorithms will be illustrated on both elementary and more realistic models. We provide step-by-step tutorials to introduce the reader to MatcontM. Here, we focus on the functionality using rather simple dynamical models defined by one- and two-dimensional maps. These tutorials illustrate how the general numerical methods described in the book and implemented in MatcontM can be used. Even in the simplest situations, this provides useful insight. In addition, we show how to study more complicated models from engineering, ecology, and economics. We provide code to reproduce the numerical results using our free toolbox, MatcontM.

This book is written for those who study discrete-time dynamical models that frequently appear in various scientific disciplines. It is accessible not only for applied mathematicians, but also for researchers with a moderate mathematical background (e.g., basic differential equations and numerical analysis). Researchers from different areas can use it as a reference text for some advanced topics. Some results will be new even to experts. Active support for the software has given us valuable feedback about where users experience difficulties. Moreover, our teaching experience has shown that parts of this book can be used in regular and advanced (post-)graduate courses on nonlinear dynamics and mathematical modeling. This book can be used as

- material for systematic study of bifurcation theory of maps, if you read it from beginning to end;
- a theoretical reference book for specific topics, e.g., a particular bifurcation;
- description of numerical bifurcation methods for maps that one could implement her/himself;
- a user-guide for particular software amenable to all theory and methods, including step-by-step tutorials; and
- a source of case studies ranging from elementary to recent research topics.

The famous notion of a Poincaré map intimately relates our exposition also to continuous-time dynamics of ordinary differential equations (ODEs), and thus the material is also useful for applications with limit cycles.

There are a number of books treating local codim 1 and 2 bifurcation of maps theoretically. Such books are either entirely devoted to maps (Neimark (1972); Iooss (1979); Mira (1987); Devaney (2003)), or have many chapters about maps (Guckenheimer and Holmes (1990); Arnold (1983); Arrowsmith and Place (1990); Kuznetsov (2004); Wiggins (2003)). It should be noted that generic two-parameter bifurcations of fixed points and cycles also involve global bifurcations leading to fractal parameter portraits and chaotic dynamics. A detailed treatment of these complications is usually only discussed at a theoretical level. The number of books on numerical analysis of maps is very limited, (e.g., Nusse and Yorke (1998); Abraham, Gardini, and Mira (1997)), but these focus on the visualization of the phase space of planar maps and non-invertibility. To the best of our knowledge, no existing book systematically describes numerical techniques for continuation, normal forms, invariant manifolds, and Lyapunov exponents to study maps depending on several parameters. We aim to fill this gap. This book provides theoretical and practical details to study the dynamics in generic two-parameter families of maps. In particular, we not only describe dynamics of approximating ODEs, but systematically study effects of their nonsymmetric perturbations, including quasi-periodic bifurcations. This will be helpful to elucidate the route to chaos in many models. The book will also teach the reader how to use the MATLAB software toolbox MatcontM that implements the developed numerical algorithms.

In Part One we first introduce analytical techniques that will be used later to study bifurcations. We briefly summarize without proof well-known results on local bifurcations in the one-parameter families following Kuznetsov (2004). The parameter-dependent normal forms on the center manifolds are given. We treat only those global bifurcations that appear near codim 2 bifurcations studied later, i.e., homoclinic tangencies, and some quasi-periodic bifurcations of closed invariant curves and 2D tori. Then we systematically present with proofs results on normal form analysis for all 11 local bifurcations of codim 2. Our exposition is complete, yet brief for the simplest cases, which are also treated by Arrowsmith and Place (1990) and Kuznetsov (2004), i.e., cusp, generalized period-doubling bifurcations, fold-flip, and strong resonances. We provide complete proofs and correct mistakes occurring in the literature. We also treat the most complicated codim 2 cases (flip-NS, fold-NS, and double NS bifurcations), which currently have been studied only in journal articles. In all cases, we derive critical and parameter-dependent normal forms and study their local bifurcations, then we obtain relevant approximating ODEs and analyze their local and global bifurcations, thus gaining insight into the main features of canonical local bifurcation diagrams. Moreover, we include new results on homoclinic and quasi-periodic bifurcations near codim 2 points by considering representative nonsymmetric perturbations of the truncated normal forms. Finally, we derive explicit formulas for the normal form coefficients of the restricted maps to the relevant center manifolds, which are then used to construct efficient predictors for codim 1 local bifurcation curves from codim 2 points.

Part Two is devoted to various algorithms for numerical bifurcation analysis of smooth maps, combining continuation techniques with normal form computations and constructing of Lyapunov charts. While modern methods for numerical bifurcation analysis of ODEs are systematically presented in several texts (e.g., Kuznetsov (2004); Govaerts (2000)), no single book is available with such methods for maps. These algorithms are scattered across journal publications, including ours, and we collect them here using uniform notation. We also discuss methods to compute the necessary partial derivatives, including automatic differentiation, and also for maps obtained via numerical integration. Then we describe the functionality of MatcontM and provide detailed step-by-step tutorials on how to use this toolbox. Here, we use simple models, e.g., the Ricker map and the delayed logistic map.

In Part Three we demonstrate the effectiveness of the developed methods and software MatcontM on more complicated models that range from the generalized Hénon map (which plays an important role in theoretical analysis of codim 2 homoclinic bifurcations of maps) to models from engineering (adaptive control map) and economics (duopoly model of Kopel). Practically all results – some of which are novel – are obtained using MatcontM. The last example (SEIR epidemic model) is a periodically forced ODE system. In this case, we apply the developed techniques and software to the numerically computed Poincaré return map.

While working on the topics included in this book, we collaborated with many colleagues and friends. First of all, we want to thank Willy Govaerts (Ghent University, Belgium) for long-term collaboration on developing numerical methods and interactive software for the analysis of continuous- and discrete-time dynamical systems, and for developing the MATLAB bifurcation toolboxes Matcont and MatcontM. We thank Stephan van Gils (University of Twente, Enschede, the Netherlands) for supporting this project. We acknowledge Odo Diekmann and Ferdinand Verhulst (Utrecht University, the Netherlands) for discussions on numerous topics. We are also thankful to Eusebius Doedel, Bernd Krauskopf, Hinke Osinga, and Renato Vitolo for stimulating discussions of various aspects of numerical bifurcation analysis of maps and ODEs. We acknowledge contributions of the (post-)graduate students we supervised, namely Reza Khoshsiar Ghaziani, Niels Neirynck, and Matthias Aengenheyster.

Part One

Theory

1
Analytical Methods

1.1 Setting and basic terminology

We will deal with maps

$$x \mapsto f(x), \quad x \in \mathbb{R}^n, \tag{1.1}$$

where $f\colon \mathbb{R}^n \to \mathbb{R}^n$ is sufficiently smooth, i.e., has all required continuous partial derivatives with respect to its arguments.[1] To simplify our presentation, we assume that f is a diffeomorphism $f\colon \mathbb{R}^n \to \mathbb{R}^n$, so that its inverse $f^{-1}\colon \mathbb{R}^n \to \mathbb{R}^n$ is globally defined and smooth. A sequence of points $x_n \in \mathbb{R}^n$ is called an *orbit* of (1.1) if

$$x_{k+1} = f(x_k), \qquad k \in \mathbb{Z}.$$

One says that $x_0 \in \mathbb{R}^n$ is a starting point of the orbit. In general, an orbit can be finite, i.e., undefined starting from some (positive or negative) k. The part of an orbit with $k \geq 0$ is called the *forward orbit*. If f is invertible, the *backward orbit* is uniquely defined.

A *fixed point* x_0 satisfies $f(x_0) = x_0$. The orbit starting at a fixed point x_0 is constant:

$$\ldots, x_0, x_0, x_0, \ldots .$$

A nonconstant K-periodic orbit $\{x_k\}$, i.e., such that

$$x_K = x_0,$$

where $K > 1$ is the minimal integer possible, is called a *cycle* with *period K* or *K-periodic orbit*. A cycle with period K defines a set of K distinct points,

$$C = \left\{x_0, f(x_0), f^{(2)}(x_0), \ldots, f^{(K-1)}(x_0)\right\},$$

[1] If f is only defined on an open region $U \subset \mathbb{R}^n$ and one is interested in studying dynamics generated by (1.1), then, usually, it is possible to extend f to the whole state space and study a smooth map $f\colon \mathbb{R}^n \to \mathbb{R}^n$ and restrict to U.

with $x_0 = f^{(K)}(x_0)$. Here, $f^{(k)}$ denotes the composition of k copies of f, also called the kth *iterate* of f. Each point in C is a fixed point of $f^{(K)}$.

A subset $S \subset \mathbb{R}^n$ is said to be *invariant* if any orbit starting at $x_0 \in S$ is located in S, i.e., $f^{(k)}(x_0) \in S$ for all $k \in \mathbb{Z}$. Fixed points and cycles are the simplest invariant sets, but more complicated ones exist, e.g., *invariant manifolds* (closed curves, tori) and *fractal invariant sets.*

Let S be an invariant set of a diffeomorphism $f\colon \mathbb{R}^n \to \mathbb{R}^n$. The set

$$W^s(S) := \{x \in \mathbb{R}^n : f^{(k)}(x) \to S \text{ as } k \to \infty\}$$

is called the *stable set* of S. It is composed of all points converging to S under iteration of f. Similarly,

$$W^u(S) := \{x \in \mathbb{R}^n : f^{(-k)}(x) \to S \text{ as } k \to \infty\}$$

is called the *unstable set* of S.

A fixed point x_0 of (1.1) is called *hyperbolic* if the Jacobian matrix $A = f_x(x_0) := Df(x_0)$ is nonsingular and has no eigenvalues with $|\lambda| = 1$. If x_0 is hyperbolic, A has n_s *stable eigenvalues* with $|\lambda| < 1$ and n_u *unstable eigenvalues* with $|\lambda| > 1$ with $n_s + n_u = n$. Denote by E^s (E^u) the generalized invariant eigenspace of A corresponding to the union of its stable (unstable) eigenvalues.

Theorem 1.1 (Local Stable and Unstable Invariant Manifolds (Palis and de Melo, 1982)) *Near a hyperbolic fixed point x_0, the map* (1.1) *has two smooth embedded invariant manifolds $W^s(x_0)$ and $W^u(x_0)$ that are tangent at x_0 to the eigenspaces E^s and E^u, respectively.*

The next key notion is that of the *equivalence* of maps. We introduce another map

$$x \mapsto g(x), \quad x \in \mathbb{R}^n, \tag{1.2}$$

where $g\colon \mathbb{R}^n \to \mathbb{R}^n$ is sufficiently smooth. The maps (1.1) and (1.2) are *topologically equivalent* if there is a homeomorphism $h\colon \mathbb{R}^n \to \mathbb{R}^n$ that maps orbits of (1.1) onto orbits of (1.2). Analytically, this means that

$$f(x) = h^{-1}(g(h(x)), \quad x \in \mathbb{R}^n,$$

or, equivalently, but easier in practice,

$$h(f(x)) = g(h(x)), \quad x \in \mathbb{R}^n.$$

The number and stability of invariant sets are the same for both maps. If the homeomorphism h is a diffeomorphism, we call the two maps *smoothly equivalent.* One can consider two smoothly equivalent maps as one map written in two different coordinate systems. If we restrict our attention to an open neighborhood U of a fixed point or a cycle, we say that the corresponding equivalence is *local*.

Theorem 1.2 (Grobman–Hartman) *Consider a smooth map*

$$x \mapsto Ax + F(x), \quad x \in \mathbb{R}^n, \tag{1.3}$$

where A is an $n \times n$ matrix and $F(x) = O(\|x\|^2)$. If $x = 0$ is a hyperbolic fixed point of (1.3), *then* (1.3) *is locally topologically equivalent near this point to its linearization*

$$x \mapsto Ax, \quad x \in \mathbb{R}^n.$$

Consider now a family of maps

$$x \mapsto f(x, \alpha), \quad x \in \mathbb{R}^n, \ \alpha \in \mathbb{R}^p, \tag{1.4}$$

where $f\colon \mathbb{R}^n \times \mathbb{R}^p \to \mathbb{R}^n$ is smooth. The parameter point $\alpha_0 \in \mathbb{R}^p$ is called a *bifurcation point* if arbitrarily close to it there is $\alpha \in \mathbb{R}^p$ such that (1.4) is not topologically equivalent to

$$x \mapsto f(x, \alpha_0), \quad x \in \mathbb{R}^n,$$

in some domain $U \subset \mathbb{R}^n$. The appearance of a topologically nonequivalent map under a variation of parameters is called a *bifurcation*. Our main goal in this book is to classify and study local bifurcations occurring in generic one- and two-parameter families of smooth maps, and to provide the necessary analytical and numerical tools to analyze these bifurcations in concrete maps. Here, "local" means happening in a small but fixed neighborhood of a fixed point. The minimal number of parameters required to meet a particular bifurcation in a generic family (1.4) is called the *codimension* of the bifurcation. Hence, we focus on a systematic study of local codim 1 and 2 bifurcations. It must be noted immediately that global bifurcations of codim 1 involving cycles and more complicated invariant sets may occur near local codim 2 bifurcation points. We treat the most important aspects of these global bifurcations.

It should also be clear that hyperbolic fixed points do not bifurcate. Indeed, in a smooth family (1.4), a hyperbolic fixed point can only move slightly under small parameter variations, and the local orbit structure near this point remains unchanged due to the Grobman–Hartman Theorem 1.2. Thus, only non-hyperbolic fixed points require further analysis.

1.2 Center manifold reduction

Consider a smooth map

$$x \mapsto Ax + F(x), \quad x \in \mathbb{R}^n, \tag{1.5}$$

where A is a nonsingular $n \times n$ matrix and $F(x) = O(\|x\|^2)$. This map has a fixed point $x = 0$ and we would like to study the orbit structure near the origin. Now, suppose that $x = 0$ is a nonhyperbolic fixed point, so that there are in general $n_c > 0$ *critical* eigenvalues of A satisfying $|\lambda| = 1$, n_s *stable* eigenvalues with $|\lambda| < 1$, and n_u *unstable* eigenvalues with $|\lambda| > 1$. Counting these eigenvalues with their algebraic multiplicities, we have $n_c + n_s + n_u = n$. Let E^c, E^s and E^u be the generalized invariant eigenspaces of A corresponding to the critical, stable, and unstable eigenvalues. The following direct-sum decomposition holds: $\mathbb{R}^n = E^c \oplus E^s \oplus E^u$.

It turns out that the map (1.5) possesses an invariant manifold near $x = 0$.

Theorem 1.3 (Center Manifold) *There exists an invariant manifold W_0^c locally defined near $x = 0$ for* (1.5) *with* $\dim W_0^c = n_c$ *that is tangent to E^c at $x = 0$ and has the same (finite) smoothness as F.*

The manifold W_0^c is called the *center manifold*. In general, it is not unique. The map (1.5) is smoothly (linearly) equivalent to the map

$$\begin{pmatrix} \xi \\ u \\ v \end{pmatrix} \mapsto \begin{pmatrix} A_0\xi + F_0(\xi, u, v) \\ A_1 u + F_1(\xi, u, v) \\ A_2 v + F_2(\xi, u, v) \end{pmatrix}, \tag{1.6}$$

where the components of $\xi \in \mathbb{R}^{n_c}$ are coordinates in E^c, the components of $u \in \mathbb{R}^{n_s}$ are coordinates in E^s, and the components of $v \in \mathbb{R}^{n_u}$ are coordinates in E^u. According to Theorem 1.3, the center manifold W_0^c can be represented locally by a graph of a smooth mapping

$$H\colon \mathbb{R}^{n_c} \to \mathbb{R}^{n_s} \times \mathbb{R}^{n_u}, \quad H(0) = 0, H_\xi(0) := DH(0) = 0$$

(see Figure 1.1). In this setting, we have the following theorem.

Theorem 1.4 (Reduction Principle) *The map* (1.6) *is locally topologically equivalent near the origin to*

$$\begin{pmatrix} \xi \\ u \\ v \end{pmatrix} \mapsto \begin{pmatrix} A_0\xi + F_0(\xi, H(\xi)) \\ A_1 u \\ A_2 v \end{pmatrix}. \tag{1.7}$$

This theorem states that dynamics along the stable and unstable subspaces are separated and are determined by the linear maps $u \mapsto A_1 u$ and $v \mapsto A_2 v$,

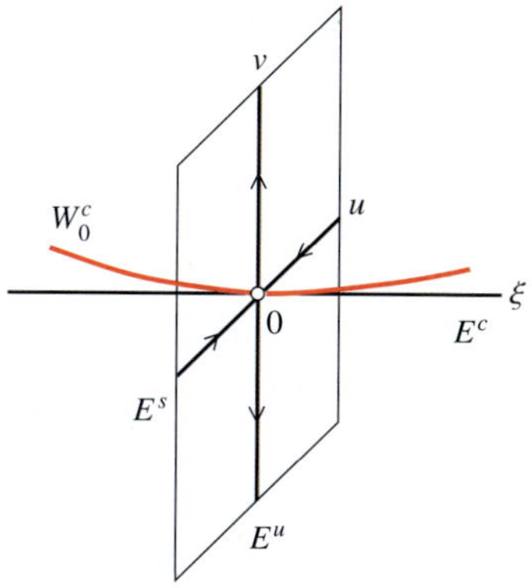

Figure 1.1 Critical center manifold W_0^c for $n_c = n_s = n_u = 1$.

so that the center manifold is *normally hyperbolic*. These dynamics are trivial since all eigenvalues of A_1 satisfy $|\lambda| < 1$, while for those of A_2 we have $|\lambda| > 1$. The dynamics on the center manifold is governed by the nonlinear n_c-dimensional map $\xi \mapsto A_0\xi + f_0(\xi, H(\xi))$, where the linear part has all its n_c eigenvalues on the unit circle. This map is called the *restriction* of (1.6) to its center manifold W_0^c. While the center manifold may not be unique, all such manifolds are represented by functions H having coinciding Taylor expansions. This leads to restricted equations, which can only differ by "flat" functions.

Thus, the analysis of the map (1.5) reduces to that of its restriction to the center manifold. Since the number of critical eigenvalues is usually small, we achieve a considerable simplification.

For a smooth family of smooth maps

$$x \mapsto f(x, \alpha), \quad x \in \mathbb{R}^n, \; \alpha \in \mathbb{R}^p, \tag{1.8}$$

where $f(x, 0) = Ax + F(x)$ as in (1.5), there exists a smooth continuation of W_0^c for small $|\alpha|$, i.e., a family of locally defined invariant normally hyperbolic manifolds $W_\alpha^c \subset \mathbb{R}^n$, carrying all interesting local dynamics of $x \mapsto f(x, \alpha)$. This can be shown by considering the *extended* map

$$\begin{pmatrix} x \\ \alpha \end{pmatrix} \mapsto \begin{pmatrix} f(x, \alpha) \\ \alpha \end{pmatrix}, \quad (x, \alpha) \in \mathbb{R}^n \times \mathbb{R}^p, \tag{1.9}$$

and applying Theorem 1.3 to this map. Indeed, for this map, the point $(x, \alpha) = (0, 0)$ is nonhyperbolic with $n_c + p$ eigenvalues on the unit circle. It has therefore a $(n_c + p)$-dimensional center manifold with n_c-dimensional α-slices defining W_α^c.

1.3 Normal forms

A smooth map near a fixed point, e.g., the restriction of some map to a center manifold, can be simplified by nonlinear transformations. There is a systematic method to remove as many terms as possible from the Taylor expansion of the map. This method is called *Poincaré normalization*.

Let H_k be the linear space of vector-valued functions whose components are homogeneous polynomials of order k. Consider a smooth map

$$x \mapsto Ax + f^{(2)}(x) + f^{(3)}(x) + \cdots, \quad x \in \mathbb{R}^n, \tag{1.10}$$

where $f^{(k)} \in H_k$ for $k \geq 2$. Introduce new coordinates $y \in \mathbb{R}^n$ by the substitution

$$x = y + h^{(m)}(y), \tag{1.11}$$

where $h^{(m)} \in H_m$ for some fixed $m \geq 2$. At this moment, $h^{(m)}$ is an arbitrary function from H_m. Notice that the substitution (1.11) is close to the identity near the origin and thus invertible there, and the inverse transformation

$$y = x - h^{(m)}(x) + O(\|x\|^{m+1}) \tag{1.12}$$

is also smooth. In the new coordinates y, the map (1.10) has the form

$$y \mapsto Ay + \sum_{k=2}^{m-1} f^{(k)}(y) + \left[f^{(m)}(y) - (M_A h^{(m)})(y)\right] + O(\|y\|^{m+1}), \tag{1.13}$$

where the linear operator M_A is defined by the formula

$$(M_A h)(y) := h(Ay) - Ah(y). \tag{1.14}$$

If $h \in H_m$, then $M_A h \in H_m$ for all $m \geq 2$.

Notice that all terms of order less than m in (1.13) are the same as in (1.10), while the terms of order m have changed and differ from $f^{(m)}(y)$ by $-(M_A h^{(m)})(y)$. Now, we define the *linear homological equation* in H_m:

$$M_A h^{(m)} = f^{(m)}. \tag{1.15}$$

If $f^{(m)}$ belongs to the *range* $M_A(H_m)$ of M_A, then there is a solution $h^{(m)}$ to (1.15), meaning that there is a transformation (1.11) that eliminates all homogeneous terms of order m in (1.10). In general, however, $f^{(m)} = g^{(m)} + r^{(m)}$, where $g^{(m)} \in M_A(H_m)$, while $r^{(m)}$ belongs to a *complement* $\widetilde{H}_m$ to $M_A(H_m)$ in H_m. Therefore, only the $g^{(m)}$ part of $f^{(m)}$ can be eliminated from (1.10) by a transformation (1.11). The remaining $r^{(m)}$ terms are called the *resonant terms* of order m. Since $\widetilde{H}_m$ is not uniquely defined, the same is true for the resonant terms.

Applying the above elimination procedure recursively for $m = 2, 3, 4, \ldots$, one proves the following theorem going back to Poincaré.

Theorem 1.5 (Poincaré Normal Form) *There is a polynomial change of coordinates*

$$x = y + h^{(2)}(y) + h^{(3)}(y) + \cdots + h^{(m)}(y), \ h^{(k)} \in H_k,$$

that transforms a smooth map

$$x \mapsto Ax + f(x), \quad x \in \mathbb{R}^n, \tag{1.16}$$

with $f(x) = O(\|x\|^2)$ into

$$y \mapsto Ay + r^{(2)}(y) + r^{(3)}(y) + \cdots + r^{(m)}(y) + O(\|y\|^{m+1}), \tag{1.17}$$

where each $r^{(k)}$ contains only resonant terms of order k, i.e., $r^{(k)} \in \widetilde{H}_k$ for $k = 2, 3, \ldots, m$.

If all eigenvalues $\lambda_1, \lambda_2, \ldots, \lambda_n$ of A are real and different, one can assume that A is diagonal, while the standard unit vectors $\{e_j\}_{j=1,2,\ldots,n}$ are the corresponding eigenvectors. In the space H_m, the operator M_A then has eigenvalues $(\lambda_1^{m_1} \lambda_2^{m_2} \cdots \lambda_n^{m_n} - \lambda_j)$, where $m_1 + m_2 + \cdots + m_n = m$. In this case, the homogeneous vector-monomials

$$x_1^{m_1} x_2^{m_2} \cdots x_n^{m_n} \, e_j$$

are the eigenvectors of M_A in H_m. If a *resonance* occurs, i.e.,

$$\lambda_j = \lambda_1^{m_1} \lambda_2^{m_2} \cdots \lambda_n^{m_n}$$

with $m_j \geq 0, \ m \geq 2$, the corresponding vector-monomial is not in the range of M_A and thus defines a resonant term. This allows determining resonant terms without long computations.

Note that all formulated results are also valid in the complex case, when $x, y \in \mathbb{C}^n$ and the complex matrix A has n different eigenvalues.

System (1.17) is called the *Poincaré normal form* of (1.16). In Chapter 4 we will give an efficient method to find coefficients of the normal forms of maps restricted to center manifolds, that combines the Poincaré normalization with the computation of the center manifold.

When considering a family of maps (1.8) depending on parameters, two approaches to its parameter-dependent normal forms are possible. One can try to find a normalizing transformation in $\mathbb{R}^n$ with coefficients that smoothly depend on parameters. Alternatively, one can consider the extended map (1.9) in the (x, α)-space and apply a normalization there. The former approach works well if the critical fixed point has a smooth continuation for nearby parameter values, i.e., there is no eigenvalue 1. The latter approach is necessary if such an eigenvalue is present.

1.4 Approximating ODEs

When dealing with local codim 2 bifurcations, we will repeatedly use the approximation of maps near their fixed points by shifts along orbits of certain systems of autonomous ordinary differential equations (ODEs). This allows us to predict *global* bifurcations of closed invariant curves and tori happening in the maps near cyclic, homo-, and heteroclinic bifurcations of the approximating ODEs. Although the exact bifurcation structure is *different* for maps and approximating ODEs, they provide information that is hardly available by analysis of the maps alone.

Consider a map having a fixed point $x = 0$:

$$x \mapsto f(x) = Ax + f^{(2)}(x) + f^{(3)}(x) + \cdots, \quad x \in \mathbb{R}^n, \tag{1.18}$$

where A is the Jacobian matrix of f at $x = 0$, while each component of $f^{(k)} \in H_k$ is a homogeneous polynomial of order k, $f^{(k)}(x) = O(\|x\|^k)$:

$$f_i^{(k)}(x) = \sum_{j_1+j_2+\cdots+j_n=k} b^{(k)}_{i,j_1j_2\cdots j_n} x_1^{j_1} x_2^{j_2} \cdots x_n^{j_n}.$$

In addition, consider a system of differential equations of the *same* dimension as the map (1.18) having an equilibrium at the point $x = 0$:

$$\dot{x} = F(x) = \Lambda x + F^{(2)}(x) + F^{(3)}(x) + \cdots, \quad x \in \mathbb{R}^n, \tag{1.19}$$

where Λ is a matrix and the terms $F^{(k)}$ have the same properties as the corresponding $f^{(k)}$ above. Denote by $\varphi^t(x)$ the (local) flow associated with (1.19). An interesting question is whether it is possible to construct a system (1.19), whose *unit-time shift* φ^1 along orbits coincides with (or at least approximates) the map f given by (1.18).

The map (1.18) is said to be *approximated up to order* k by system (1.19) if its Taylor expansion coincides with that of the unit-time shift φ^1 along the orbits of (1.19) up to and including terms of order k:

$$f(x) = \varphi^1(x) + O(\|x\|^{k+1}).$$

System (1.19) is then called an *approximating ODE system.*

We can construct the Taylor expansion of $\varphi^t(x)$ with respect to x at $x = 0$ as follows using *Picard iterations*. Namely, set

$$x^{(1)}(t) = e^{\Lambda t} x.$$

So, $x^{(1)}$ is the solution of the linear equation $\dot{x} = \Lambda x$ with initial condition x, and define the Picard iteration

$$x^{(k+1)}(t) = e^{\Lambda t}x + \int_0^t e^{\Lambda(t-\tau)} \left(F^{(2)}(x^{(k)}(\tau)) + \cdots + F^{(k+1)}(x^{(k)}(\tau)) \right) d\tau. \quad (1.20)$$

Clearly, the $(k + 1)$ iteration does not change $O(\|x\|^l)$ terms for any $l \leq k$. Substituting $t = 1$ into $x^{(k)}(t)$ provides the correct Taylor expansion of $\varphi^1(x)$ up to and including terms of order k:

$$\varphi^1(x) = e^{\Lambda}x + g^{(2)}(x) + g^{(3)}(x) + \cdots + g^{(k)}(x) + O(\|x\|^{k+1}). \quad (1.21)$$

Next we require that the corresponding terms in (1.21) and (1.18) coincide:

$$e^{\Lambda} = A, \quad (1.22)$$

and

$$g^{(k)}(x) = f^{(k)}(x), \ \ k = 2, 3, \ldots, \quad (1.23)$$

and then try to find Λ and the coefficients of $g^{(k)}$ (and, eventually, the coefficients of $F^{(k)}$) in terms of those of $f^{(k)}$, i.e., $b^{(k)}_{i,j_1 j_2 \cdots j_n}$. This is *not* always possible.

First of all, (1.22) does not always have a *real* solution matrix Λ, even if A is nonsingular. A sufficient condition for the solvability is that all eigenvalues of A are positive. Moreover, not all equations (1.23) may be solvable for the coefficients $b^{(k)}_{i,j_1 j_2 \cdots j_n}$ with $|j| := j_1 + j_2 + \cdots + j_n = k$. The corresponding conditions could be formulated explicitly in a rather general form. We will not do this, since in our cases we will verify the solvability explicitly. Actually, these conditions are always satisfied if the map (1.18) is close to identity. More results on the existence of the approximating vector field g can be found in Gramchev and Walcher (2005) and Takens (1974).

In the parameter-dependent case, one approximates the extended map by an extended flow, thus obtaining a parameter-dependent ODE system.

1.5 Simplest bifurcations of planar ODEs

In our analysis of bifurcations of maps, we will often encounter auxiliary smooth planar autonomous ODEs depending on one or two parameters. While the main purpose of these auxiliary vector fields is the study of global bifurcations, their local bifurcations are also useful. Therefore, for further reference, we summarize all necessary results without proof about bifurcations of such systems. Of course, this overview is not a substitute for a systematic study of this classical part of bifurcation theory.

At fixed parameter values, one defines for such an ODE system its *orbits* (oriented by the advance of time) and *phase portrait*. Two such systems are considered as *topologically equivalent* (in some domains of $\mathbb{R}^2$) if their phase portraits are homeomorphic, i.e., one can be obtained from the other by a continuous invertible deformation. Note that such a transformation maps orbits into orbits, but not necessarily solutions into solutions. An appearance of a topologically nonequivalent phase portrait is called a *bifurcation*. Since the number of equilibria, the number of periodic orbits, and their stability, as well as the presence of connecting orbits, are topological invariants, a bifurcation of the 2D system means a change of (some of) these properties.

All bifurcations can be divided into *local*, i.e., occurring in an arbitrary small fixed neighborhood of an equilibrium, and *global*. Each bifurcation is characterized by a number of *bifurcation conditions*. Similarly as for maps, this number is called *codimension* and is equal to the number of independent parameters needed to unfold this bifurcation in a generic system, i.e., systems without symmetries or integrals of motion. Bifurcation theory studies canonical unfoldings (*normal forms*) of bifurcations (if they exist) and provides techniques to find out which of the possible unfoldings actually occurs in the particular ODE system. One describes unfoldings by means of *bifurcation diagrams*, i.e., stratifications of the parameter space near a bifurcation point induced by the topological equivalence of phase portraits.

1.5.1 Generic one-parameter local bifurcations in 2D ODEs

Consider a smooth one-parameter planar ODE

$$\dot{u} = f(u, \alpha), \quad u \in \mathbb{R}^2, \alpha \in \mathbb{R}. \tag{1.24}$$

Suppose $u_0 \in \mathbb{R}^2$ is an *equilibrium* of (1.24) at $\alpha_0 \in \mathbb{R}$, i.e., $f(u_0, \alpha_0) = 0$. An equilibrium u_0 is called *hyperbolic* if $\Re(\lambda) \neq 0$ for any eigenvalue $\lambda \in \mathbb{C}$ of its Jacobian matrix $A = f_u(u_0, \alpha_0)$. A hyperbolic equilibrium can be smoothly continued with respect to α near α_0, and the Grobman–Hartman Theorem for ODEs ensures that it does not exhibit any local bifurcations. Indeed, the equilibrium remains hyperbolic for parameter values close to α_0 and has a local phase portrait that is topologically equivalent to that of the linearized ODE. Thus, a local bifurcation can happen only for a nonhyperbolic equilibrium with $\Re(\lambda) = 0$.

In generic one-parameter planar ODEs, one can encounter only two types of nonhyperbolic equilibria, i.e., with either

(1) two real eigenvalues, with one eigenvalue $\lambda_1 = 0$; or

(2) two purely imaginary eigenvalues $\lambda_{1,2} = \pm i\omega_0$ with $\omega_0 > 0$.

Each condition indeed defines a codim 1 local bifurcation of generic planar ODEs. Case (1) leads to a *fold* (or *saddle-node*) *bifurcation*. Case (2) implies a *Hopf* (or *Andronov–Hopf*) *bifurcation*. To describe their canonical unfoldings, assume that $\alpha_0 = 0$ and $u_0 = 0$.

Fold (saddle-node) bifurcation in the plane

By a linear invertible change of variables, the critical system $\dot{u} = f(u, 0)$ can be transformed near $u = 0$ into

$$\begin{cases} \dot{x} &= ax^2 + bxy + cy^2 + O(\|(x, y)\|^3), \\ \dot{y} &= \lambda_2 y + O(\|(x, y)\|^2), \end{cases}$$

where $(x, y) \in \mathbb{R}^2$, $a, b, c \in \mathbb{R}$ and $\lambda_2 \neq 0$ is the second (real) eigenvalue of A. The variables (x, y) are coordinates in the directions of the eigenvectors of the Jacobian matrix $A = f_u(0, 0)$ corresponding to eigenvalues $\lambda_1 = 0$ and $\lambda_2 \neq 0$. Let $q, p \in \mathbb{R}^2$ be nonzero vectors satisfying

$$Aq = A^T p = 0$$

and normalized such that $\langle p, q \rangle = 1$. Then a can be computed as the quadratic coefficient in the Taylor expansion

$$\langle p, f(\xi q, 0) \rangle = a\xi^2 + O(\xi^3),$$

i.e.,

$$a = \frac{1}{2} \frac{d^2}{d\xi^2} \langle p, f(\xi q, 0) \rangle \bigg|_{\xi=0}.$$

Theorem 1.6 *If $a \neq 0$ and $\lambda_2 \neq 0$, then the system* (1.24) *is locally topologically equivalent near the fold bifurcation to*

$$\begin{cases} \dot{x} &= \beta(\alpha) + ax^2, \\ \dot{y} &= \lambda_2 y, \end{cases}$$

where $\beta = \beta(\alpha)$ is a smooth function with $\beta(0) = 0$.

If $\beta'(0) \neq 0$, we can use β as the new unfolding parameter and visualize the bifurcation diagram of the canonical unfolding

$$\begin{cases} \dot{x} &= \beta + ax^2, \\ \dot{y} &= \lambda_2 y \end{cases} \tag{1.25}$$

(see Figure 1.2). In this *topological normal form*, two equilibrium points

$$O_{1,2} = \left(\mp \sqrt{-\frac{\beta}{a}}, 0 \right)$$

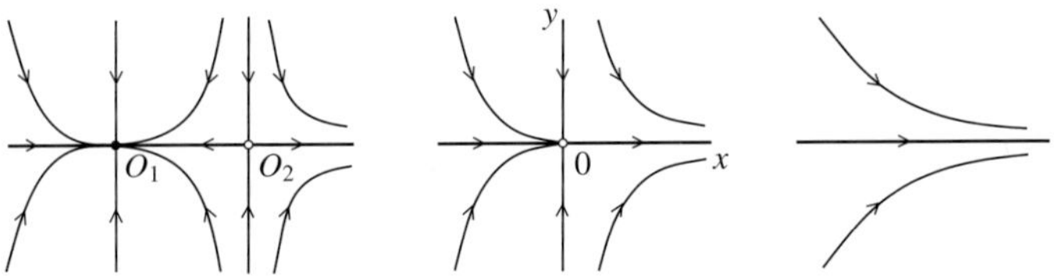

Figure 1.2 Planar fold bifurcation in the topological normal form (1.25): $a > 0, \lambda_2 < 0$.

collide and disappear when β changes sign. This is called a *fold* (or *saddle-node*) bifurcation. In the original coordinates (u_1, u_2), the same topological transition happens in system (1.24), with deformed phase portraits.

Remark 1.7 Notice that all essential rearrangements in system (1.25) occur on the line $y = 0$ that is exponentially stable or unstable, depending on the sign of λ_2. In the original system, this line becomes a smooth (parameter-dependent) curve W^c_α, which is a local *center manifold* of (1.24) near the fold bifurcation.

(Andronov–)Hopf bifurcation in the plane

By a linear invertible change of variables, the critical system $\dot{u} = f(u, 0)$ can be transformed near $u = 0$ into

$$\begin{cases} \dot{x} &= -\omega_0 y + R(x, y), \\ \dot{y} &= \omega_0 x + S(x, y), \end{cases}$$

where $R(x, y) = O(\|(x, y)\|^2)$ and $S(x, y) = O(\|(x, y)\|^2)$ are smooth functions. Introducing $z = x + iy \in \mathbb{C}$ and $\bar{z} = x - iy$, this system can be written as one complex ODE

$$\dot{z} = i\omega_0 z + g(z, \bar{z}), \tag{1.26}$$

where

$$g(z, \bar{z}) = R\left(\frac{z + \bar{z}}{2}, \frac{z - \bar{z}}{2i}\right) + iS\left(\frac{z + \bar{z}}{2}, \frac{z - \bar{z}}{2i}\right) = \sum_{j+k \geq 2} \frac{1}{j!k!} g_{jk} z^j \bar{z}^k .$$

One can directly compute the function $g(z, \bar{z})$ in (1.26) using the original coordinates (u_1, u_2). Let $A = f_u(0, 0)$ be the Jacobian matrix of (1.24) at $(u_0, \alpha_0) = (0, 0)$. Introduce $q, p \in \mathbb{C}^2$, such that

$$Aq = i\omega_0 q, \quad A^T p = -i\omega_0 p,$$

and $\langle p, q\rangle = \bar{p}^T q = 1$. Then

$$g(z,\bar{z}) = \langle p, f(zq + \bar{z}\bar{q})\rangle,$$

so that

$$g_{jk} = \left.\frac{\partial^{j+k}}{\partial z^j \partial \bar{z}^k}\right|_{z=\bar{z}=0} \langle p, f(zq + \bar{z}\bar{q})\rangle,$$

where z and $\bar{z}$ should be considered as independent variables.

There exists a polynomial change of variable

$$z = w + \frac{1}{2}h_{20}w^2 + h_{11}w\bar{w} + \frac{1}{2}h_{02}\bar{w}^2 + \frac{1}{6}h_{30}w^3 + \frac{1}{2}h_{12}w\bar{w}^2 + \frac{1}{6}h_{03}\bar{w}^3,$$

such that (1.26) will take the *Poincaré normal form*

$$\dot{w} = i\omega_0 w + c_1 w|w|^2 + O(|w|^4),$$

where $c_1 \in \mathbb{C}$. Define the *first Lyapunov coefficient*

$$l_1 := \frac{1}{\omega_0}\Re(c_1).$$

One can show that

$$l_1 = \frac{1}{2\omega_0^2}\Re(ig_{20}g_{11} + \omega_0 g_{21}). \tag{1.27}$$

Theorem 1.8 *If $l_1 \neq 0$ and $\omega_0 > 0$, then* (1.24) *is locally topologically equivalent near the Hopf bifurcation to the following system in polar coordinates*

$$\begin{cases} \dot{\rho} &= \rho(\beta(\alpha) + l_1\rho^2), \\ \dot{\varphi} &= 1, \end{cases}$$

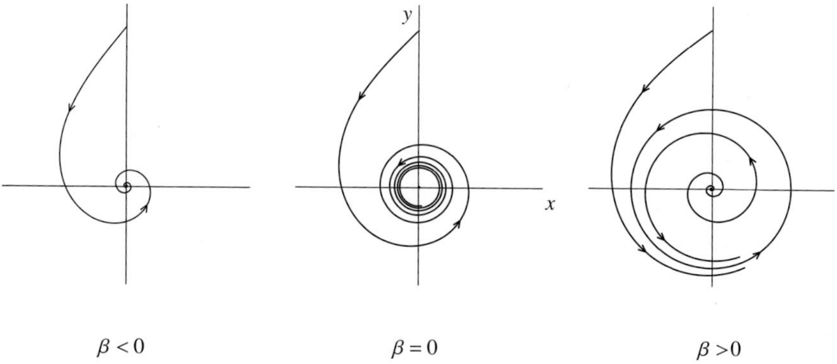

Figure 1.3 Supercritical Hopf bifurcation: $l_1 < 0$.

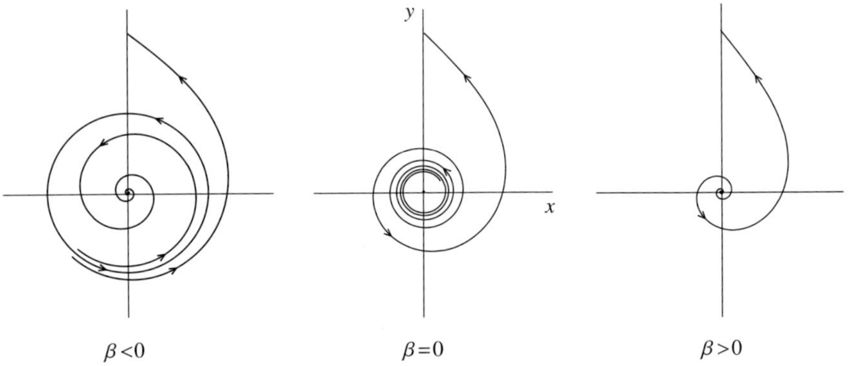

Figure 1.4 Subcritical Hopf bifurcation: $l_1 > 0$.

where $\beta = \beta(\alpha)$ is a smooth function with $\beta(0) = 0$.

If $\beta'(0) \neq 0$, we can use β as the new unfolding parameter and consider the bifurcation diagram of the topological normal form

$$\begin{cases} \dot{\rho} &= \rho(\beta + l_1\rho^2), \\ \dot{\varphi} &= 1. \end{cases} \tag{1.28}$$

A *limit cycle* of radius $\rho_0 = \sqrt{-\frac{\beta}{l_1}} > 0$ appears or disappears, while the focus at the origin changes stability, (Figures 1.3 and 1.4). This phenomenon is called the planar *(Andronov–)Hopf bifurcation*. In the original system (1.24), a deformed limit cycle bifurcates (with the period approaching $2\pi/\omega_0$).

The direction of the cycle bifurcation is determined by the sign of the first Lyapunov coefficient l_1. Notice that the cycle stability is the same as that of the critical equilibrium ("weak focus").

Remark 1.9 The saddle-node and Hopf bifurcations occur also in smooth parameter-dependent n-dimensional ODEs

$$\dot{u} = f(u, \alpha), \quad u \in \mathbb{R}^n, \alpha \in \mathbb{R}.$$

Without loss of generality, we assume that the critical equilibrium is $u = 0$ and the bifurcation takes place at $\alpha = 0$.

At the *fold bifurcation*, the Jacobian matrix $A = f_u(0,0)$ has a simple zero eigenvalue $\lambda_1 = 0$ and no other eigenvalues with $\Re(\lambda) = 0$. In this case, there exists a smooth parameter-dependent invariant curve W^c_α on which the system is locally topologically equivalent to the x-equation in (1.25) with $\beta = \beta(\alpha)$, i.e.,

$$\dot{x} = \beta + ax^2.$$

Thus, generically, two equilibrium points in W^c_α collide and disappear.

The normal form coefficient a can be computed as

$$a = \frac{1}{2}\langle p, B(q,q)\rangle, \tag{1.29}$$

where $q, p \in \mathbb{R}^2$ satisfy $Aq = A^T p = 0$, $\langle q, q\rangle = \langle p, q\rangle = 1$ and

$$B_i(q,r) = \sum_{j,k\in\{1,2,\ldots,n\}} \frac{\partial^2 f_i(0,0)}{\partial u_j \partial u_k} q_j r_k, \quad i = 1,2,\ldots,n. \tag{1.30}$$

At the *(Andronov–)Hopf bifurcation*, the Jacobian matrix $A = f_u(0,0)$ has a pair of simple purely imaginary eigenvalues $\lambda_{1,2} = \pm i\omega_0$ and no other eigenvalues with $\Re(\lambda) = 0$. In this case, there exists a smooth parameter-dependent invariant surface W^c_α on which the system is locally topologically equivalent to (1.28). Hence, a limit cycle bifurcates in W^c_α from a focus that changes stability.

The first Lyapunov coefficient can be computed by the following formula

$$l_1 = \frac{1}{2\omega_0}\Re\left[\langle p, C(q,q,\bar{q}) - 2B(q, A^{-1}B(q,\bar{q})) + B(\bar{q}, (2i\omega_0 I_n - A)^{-1}B(q,q))\rangle\right], \tag{1.31}$$

where $p, q \in \mathbb{C}^n$ satisfy $Aq = i\omega_0 q$, $A^T p = -i\omega_0 p$ and $\langle q, q\rangle = \langle p, q\rangle = 1$ with $\langle p, q\rangle := \bar{p}^T q$. The components of the multilinear form B have been defined above, while those of C are given by

$$C_i(q,r,s) = \sum_{j,k,l\in\{1,2,\ldots,n\}} \frac{\partial^3 f_i(0,0)}{\partial u_j \partial u_k \partial u_l} q_j r_k s_l, \quad i = 1,2,\ldots,n. \tag{1.32}$$

Note that (1.31) is valid for $n \geq 2$. However, for $n = 2$ one may prefer to use formula (1.27), as that does not involve solving any linear system or inverting a matrix.

1.5.2 Generic two-parameter local bifurcations in 2D ODEs

Consider a smooth two-parameter planar ODE

$$\dot{u} = f(u,\alpha), \quad u \in \mathbb{R}^2, \alpha \in \mathbb{R}^2. \tag{1.33}$$

In such planar ODEs, only three types of doubly degenerate equilibrium points can be encountered generically, i.e., either with

(1) one simple eigenvalue $\lambda_1 = 0$ and $a = 0$, i.e., the normal form coefficient (1.29) of the fold vanishes;

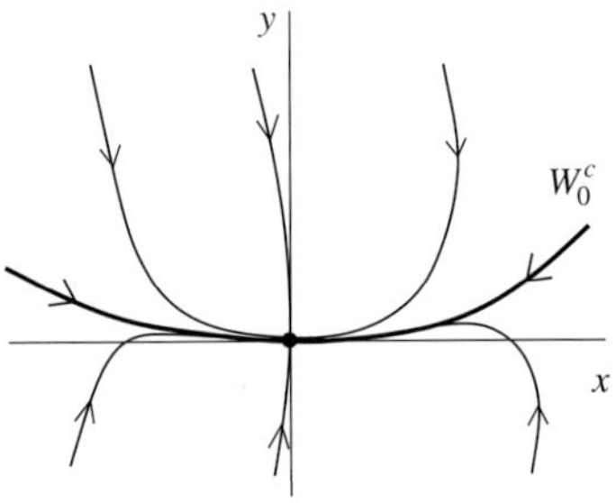

Figure 1.5 Local critical center manifold at cusp bifurcation.

(2) a double zero non-semisimple eigenvalue $\lambda_{1,2} = 0$; or
(3) two purely imaginary eigenvalues $\lambda_{1,2} = \pm i\omega_0$ and $l_1 = 0$, i.e., the first Lyapunov coefficient (1.31) vanishes.

Each condition indeed defines a codim 2 local bifurcation of generic planar ODEs. Case (1) corresponds to a *cusp* bifurcation. Case (2) implies a *Bogdanov–Takens* bifurcation. Case (3) leads to a *generalized Hopf* (or *Bautin*) bifurcation. We now describe their canonical unfoldings. We will assume that the bifurcation occurs at $\alpha_0 = 0$ and the corresponding critical equilibrium is $u_0 = 0$.

Cusp bifurcation

By a linear invertible change of variables, the critical system $\dot{u} = f(u, 0)$ at the cusp bifurcation can be transformed into

$$\begin{cases} \dot{x} = p_{11}xy + \frac{1}{2}p_{02}y^2 + \frac{1}{6}p_{30}x^3 + \frac{1}{2}p_{21}x^2y + \frac{1}{2}p_{12}xy^2 + \frac{1}{6}p_{03}y^3 + O(\|(x,y)\|^4), \\ \dot{y} = \lambda_2 y + \frac{1}{2}q_{20}x^2 + q_{11}xy + \frac{1}{2}q_{02}y^2 + O(\|(x,y)\|^3). \end{cases}$$

As in the fold case, the variables (x, y) are coordinates in the directions of the eigenvectors of the Jacobian matrix $A = f_u(0, 0)$ corresponding to eigenvalues $\lambda_1 = 0$ and $\lambda_2 \neq 0$. It has an invariant center manifold W_0^c that is locally given by the graph of the smooth function

$$y = \frac{1}{2}w_2x^2 + O(x^3), \quad w_2 = -\frac{q_{20}}{\lambda_2},$$

so that the restriction of the critical ODE to W_0^c can be written as

$$\dot{x} = cx^3 + O(x^4),$$

where

$$c = \frac{1}{6}\left(p_{30} - \frac{3}{\lambda_2}q_{20}p_{11}\right).$$

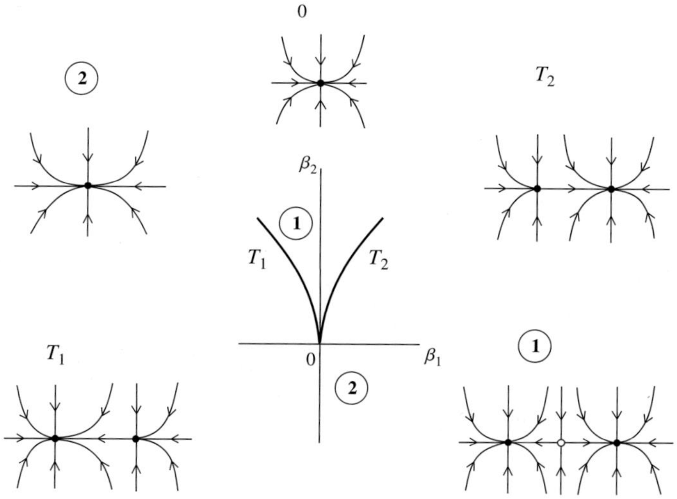

Figure 1.6 Bifurcation diagram of the topological normal form for cusp bifurcation: $s = \sigma = -1$.

Theorem 1.10 *If $c \neq 0$, then (1.33) is locally topologically equivalent near the cusp bifurcation to the system*

$$\begin{cases} \dot{x} &= \beta_1(\alpha) + \beta_2(\alpha)x + sx^3, \\ \dot{y} &= \sigma y, \end{cases}$$

where $\beta = \beta(\alpha)$ is a smooth vector-valued function with $\beta_1(0) = \beta_2(0) = 0$, while $s = \text{sign}(c) = \pm 1$ and $\sigma = \text{sign}(\lambda_2) = \pm 1$.

If the 2D mapping $\alpha \mapsto \beta(\alpha)$ is regular at $\alpha = 0$, i.e., its Jacobian matrix $\beta_\alpha(0)$ is nonsingular, then (β_1, β_2) can be used as the new unfolding parameters. The bifurcation diagram of the topological normal form

$$\begin{cases} \dot{x} &= \beta_1 + \beta_2 x + sx^3, \\ \dot{y} &= \sigma y, \end{cases} \tag{1.34}$$

contains a fold bifurcation curve $T = T_1 \cup T_2$ that delimits a narrow wedge. For parameter values chosen inside the wedge three equilibrium points exist, while outside the wedge only one equilibrium exists.

Remark 1.11 As in the fold case, all essential rearrangements in system (1.34) occur on the line $y = 0$ that is exponentially stable or unstable, depending on the sign of λ_2. In the original system, this line becomes a smooth (parameter-dependent) curve W^c_α which is a local *center manifold* of (1.33) near the cusp bifurcation.

Bogdanov–Takens bifurcation

By a linear invertible change of variables, the critical system $\dot{u} = f(u, 0)$ at the Bogdanov–Takens (BT) bifurcation can be transformed to

$$\begin{cases} \dot{x} &= y + \frac{1}{2}p_{20}x^2 + p_{11}xy + \frac{1}{2}p_{02}y^2 + O(\|(x,y)\|^3) \ =: \ P(x,y), \\ \dot{y} &= \frac{1}{2}q_{20}x^2 + q_{11}xy + \frac{1}{2}q_{02}x^2 + O(\|(x,y)\|^3). \end{cases}$$

The variables (x, y) are coordinates in the directions of the eigenvector and the generalized eigenvector of the Jacobian matrix $A = f_u(0, 0)$ corresponding to its double non-semisimple eigenvalue $\lambda_1 = 0$. The local smooth invertible change of variables

$$\begin{cases} \xi &= x, \\ \eta &= P(x,y) \end{cases}$$

reduces this system near the origin to

$$\begin{cases} \dot{\xi} &= \eta, \\ \dot{\eta} &= a\xi^2 + b\xi\eta + c\eta^2 + O(\|(\xi,\eta)\|^3), \end{cases}$$

where

$$a = \frac{1}{2}q_{20}, \quad b = p_{20} + q_{11}.$$

Theorem 1.12 *If $ab \neq 0$, then* (1.33) *is locally topologically equivalent near the Bogdanov–Takens bifurcation to*

$$\begin{cases} \dot{x} &= y, \\ \dot{y} &= \beta_1(\alpha) + \beta_2(\alpha)x + x^2 + sxy, \end{cases}$$

where $\beta = \beta(\alpha)$ is a smooth vector-valued function with $\beta_1(0) = \beta_2(0) = 0$ and $s = \operatorname{sign}(ab) = \pm 1$.

As in the cusp case, if the 2D mapping $\alpha \mapsto \beta(\alpha)$ is regular at $\alpha = 0$, then (β_1, β_2) can be used as the new unfolding parameters. The bifurcation diagram of the topological normal form

$$\begin{cases} \dot{x} &= y, \\ \dot{y} &= \beta_1 + \beta_2 x + x^2 + sxy \end{cases} \tag{1.35}$$

is presented in Figure 1.7 for $s < 0$. It includes several bifurcation curves near the origin:

- *fold* $T = T_- \cup T_+ : \beta_1 = \frac{1}{4}\beta_2^2$;
- *Andronov–Hopf* $H : \beta_1 = 0,\ \beta_2 < 0$;
- *saddle homoclinic* $P : \beta_1 = -\frac{6}{25}\beta_2^2 + o(\beta_2^2),\ \beta_2 < 0$.

A unique limit cycle appears at the Andronov–Hopf bifurcation curve H and disappears via the saddle homoclinic bifurcation at curve P. The last bifurcation is global. Crossing the curve P, the limit cycle approaches a *homoclinic orbit* that connects a saddle point with itself, and its period tends to infinity. Having located and analyzed the Bogdanov–Takens bifurcation, it is also possible to predict saddle homoclinic orbits by purely algebraic tools.

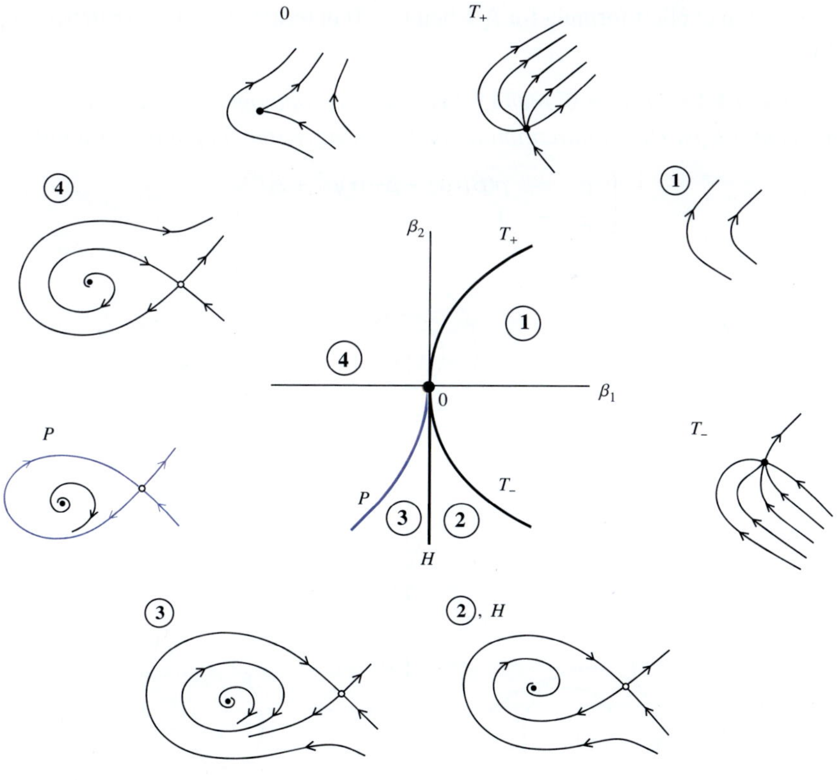

Figure 1.7 Bifurcation diagram of the topological normal form for Bogdanov–Takens bifurcation: $s = -1$.

Generalized Hopf bifurcation

As in the codim 1 Hopf case, the critical system $\dot{u} = f(u, 0)$ at the generalized Hopf bifurcation can be transformed to the complex form

$$\dot{z} = i\omega_0 z + \sum_{2 \leq j+k \leq 5} \frac{1}{j!k!} g_{jk} z^k \bar{z}^j + O(|z|^6),$$

which can locally be reduced by a polynomial change of variables to the *Poincaré normal form*

$$\dot{w} = i\omega w + c_1 w|w|^2 + c_2 w|w|^4 + O(|w|^6),$$

where the first Lyapunov coefficient vanishes: $l_1 = \frac{1}{\omega_0}\Re(c_1) = 0$. Now, we define the *second Lyapunov coefficient*

$$l_2 := \frac{1}{\omega_0}\Re(c_2).$$

There is an explicit formula for l_2 when $l_1 = 0$ in terms of g_{jk} (see (Kuznetsov, 2004)).

Theorem 1.13 *If $l_2 \neq 0$, then* (1.33) *is locally topologically equivalent near the generalized Hopf bifurcation to the following system in polar coordinates:*

$$\begin{cases} \dot{\rho} &= \rho(\beta_1(\alpha) + \beta_2(\alpha)\rho^2 + s\rho^4), \\ \dot{\varphi} &= 1, \end{cases}$$

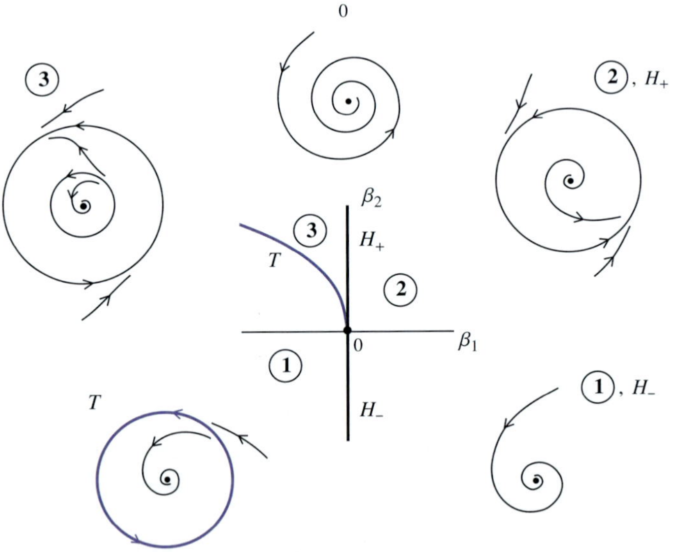

Figure 1.8 Bifurcation diagram of the topological normal form for generalized Hopf bifurcation: $s = -1$.

where $\beta = \beta(\alpha)$ *is a smooth vector-valued function with* $\beta_1(0) = \beta_2(0) = 0$ *and* $s = \text{sign}(l_2) = \pm 1$.

As usual, if the 2D mapping $\alpha \mapsto \beta(\alpha)$ is regular at $\alpha = 0$, then (β_1, β_2) can be used as the new unfolding parameters. The bifurcation diagram of the topological normal form

$$\begin{cases} \dot{\rho} = \rho(\beta_1 + \beta_2\rho^2 + s\rho^4), \\ \dot{\varphi} = 1 \end{cases} \tag{1.36}$$

is presented in Figure 1.8 for $s = -1$. It includes two bifurcation curves near the origin:

- *Andronov–Hopf* H: $\beta_1 = 0$;
- *fold of limit cycles* T: $\beta_1 = -\frac{1}{4}\beta_2^2$, $\beta_2 > 0$.

At the branch H^- of H with $\beta_2 < 0$, the supercritical Hopf bifurcation happens that generates a *stable* limit cycle. On the contrary, at the branch H^+ of H with $\beta_2 > 0$, an *unstable* limit cycle bifurcates via the subcritical Hopf bifurcation. These two cycles collide and disappear at the global bifurcation curve T. For parameter values at this curve, the normal form has a degenerate (nonhyperbolic) limit cycle that is stable from one side and unstable from the other. This cycle has a nontrivial multiplier $+1$.

Remark 1.14 The cusp, Bogdanov–Takens and generalized Hopf bifurcations occur also in smooth n-dimensional ODEs, depending on two parameters

$$\dot{u} = f(u, \alpha), \quad u \in \mathbb{R}^n, \alpha \in \mathbb{R}^2.$$

As usual, assume that the critical equilibrium is $u = 0$ and the bifurcation takes place at $\alpha = 0$.

At the *cusp bifurcation*, the Jacobian matrix $A = f_u(0,0)$ has a simple zero eigenvalue $\lambda_1 = 0$ and no other eigenvalues with $\Re(\lambda) = 0$. In this case, generically, there exists a smooth parameter-dependent invariant curve W^c_α on which the system is locally topologically equivalent to

$$\dot{x} = \beta_1 + \beta_2 x + s x^3,$$

where $s = \text{sign}(c)$ and $\beta = \beta(\alpha)$. Thus, generically, the n-dimensional ODE system has a parametric portrait that is locally equivalent to that of system (1.34). The normal form coefficient c is given by

$$c = \frac{1}{6}\langle p, C(q,q,q) + 3B(q,h_2)\rangle, \tag{1.37}$$

where $q, p \in \mathbb{R}^2$ satisfy $Aq = A^T p = 0,\ \langle q, q\rangle = \langle p, q\rangle = 1$ and $h_2 \in \mathbb{R}^n$ can be computed by solving the nonsingular linear system

$$\begin{pmatrix} A & q \\ p^T & 0 \end{pmatrix}\begin{pmatrix} h_2 \\ r \end{pmatrix} = \begin{pmatrix} -B(q,q) \\ 0 \end{pmatrix}.$$

The expressions for the multilinear forms B and C were given in (1.30) and (1.32).

At the *Bogdanov–Takens bifurcation*, the Jacobian matrix $A = f_u(0,0)$ has a double non-semisimple zero eigenvalue $\lambda_{1,2} = 0$ and no other eigenvalues with $\Re(\lambda) = 0$. In this case, generically, there exists a smooth parameter-dependent invariant surface W^c_α on which the system is locally topologically equivalent to (1.35). The normal form coefficients a and b can be computed as

$$a = \frac{1}{2}\langle p_1, B(q_0, q_0)\rangle, \quad b = \langle p_0, B(q_0, q_0)\rangle + \langle p_1, B(q_0, q_1)\rangle,$$

where $q_{0,1}, p_{0,1} \in \mathbb{R}^n$ satisfy

$$Aq_0 = 0,\ Aq_1 = q_0,\ A^T p_1 = 0,\ A^T p_0 = p_1$$

and are normalized such that $\langle q_0, q_0\rangle = \langle p_0, q_0\rangle = \langle p_1, p_1\rangle = \langle p_1, q_1\rangle = 1$ and $\langle p_0, q_1\rangle = \langle p_1, q_0\rangle = 0$.

At the *generalized Hopf bifurcation*, the Jacobian matrix $A = f_u(0,0)$ has a pair of simple purely imaginary eigenvalues $\lambda_{1,2} = \pm i\omega_0$ and no other eigenvalues with $\Re(\lambda) = 0$. In this case, there exists a smooth parameter-dependent invariant surface W^c_α on which the system is locally topologically equivalent to (1.36). The second Lyapunov coefficient l_2 can be computed by the formula

$$l_2 = \frac{1}{\omega_0}\Re(c_2),$$

where

$$\begin{aligned} c_2 = \frac{1}{12}\langle p, & E(q,q,q,\bar{q},\bar{q}) \\ & + D(q,q,q,\bar{h}_{20}) + 3D(q,\bar{q},\bar{q},h_{20}) + 6D(q,q,\bar{q},h_{11}) \\ & + C(\bar{q},\bar{q},h_{30}) + 3C(q,q,\bar{h}_{21}) + 6C(q,\bar{q},h_{21}) + 3C(q,\bar{h}_{20},h_{20}) \\ & + 6C(q,h_{11},h_{11}) + 6C(\bar{q},h_{20},h_{11}) \\ & + 2B(\bar{q},h_{31}) + 3B(q,h_{22}) + B(\bar{h}_{20},h_{30}) + 3B(\bar{h}_{21},h_{20}) + 6B(h_{11},h_{21})\rangle. \end{aligned}$$

Here, $q, p \in \mathbb{C}^n$ satisfy $Aq = i\omega_0 q,\ A^T p = -i\omega_0 p$ and are normalized according to $\langle q, q\rangle = \langle p, q\rangle = 1$.

The vectors $h_{20}, h_{11}, h_{30} \in \mathbb{C}^n$ are given by

$$\begin{aligned} h_{20} &= (2i\omega_0 I_n - A)^{-1}B(q,q), \\ h_{11} &= -A^{-1}B(q,\bar{q}), \\ h_{30} &= (3i\omega_0 I_n - A)^{-1}[C(q,q,q) + 3B(q,h_{20})], \end{aligned}$$

while $h_{21} \in \mathbb{C}^n$ can be found by solving the nonsingular linear system

$$\begin{pmatrix} i\omega_0 I_n - A & q \\ \bar{p}^T & 0 \end{pmatrix} \begin{pmatrix} h_{21} \\ r \end{pmatrix} = \begin{pmatrix} C(q,q,\bar{q}) + B(\bar{q},h_{20}) + 2B(q,h_{11}) - 2c_1 q \\ 0 \end{pmatrix},$$

where

$$c_1 = \frac{1}{2}\langle p, C(q,q,\bar{q}) + B(\bar{q},(2i\omega_0 I_n - A)^{-1}B(q,q)) - 2B(q,A^{-1}B(q,\bar{q}))\rangle.$$

Recall that c_1 is purely imaginary at the generalized Hopf point.

Finally, we have

$$\begin{aligned} h_{31} &= (2i\omega_0 I_n - A)^{-1}[D(q,q,q,\bar{q}) + 3C(q,q,h_{11}) + 3C(q,\bar{q},h_{20}) \\ &\qquad + 3B(h_{20},h_{11}) + B(\bar{q},h_{30}) + 3B(q,h_{21}) - 6c_1 h_{20}], \\ h_{22} &= -A^{-1}[D(q,q,\bar{q},\bar{q}) + 4C(q,\bar{q},h_{11}) + C(\bar{q},\bar{q},h_{20}) + C(q,q,\bar{h}_{20}) \\ &\qquad + 2B(h_{11},h_{11}) + 2B(q,\bar{h}_{21}) + 2B(\bar{q},h_{21}) + B(\bar{h}_{20},h_{20})]. \end{aligned}$$

In the above formulas, the multilinear forms B and C should be computed via (1.30) and (1.32), while

$$\begin{aligned} D_i(q,r,z,v) &= \sum_{j,k,l,m\in\{1,2,\ldots,n\}} \frac{\partial^4 f_i(0,0)}{\partial u_j \partial u_k \partial u_l \partial u_m} q_j r_k z_l v_m, \\ E_i(q,r,z,v,w) &= \sum_{j,k,l,m,s\in\{1,2,\ldots,n\}} \frac{\partial^5 f_i(0,0)}{\partial u_j \partial u_k \partial u_l \partial u_m \partial u_s} q_j r_k z_l v_m w_s, \end{aligned}$$

for $i = 1, 2, \ldots, n$.

1.6 Pontryagin–Melnikov theory

Consider a planar Hamiltonian system

$$\dot{x} = J\nabla H(x), \quad x = (x_1, x_2) \in \mathbb{R}^2, \tag{1.38}$$

where the *Hamiltonian function* $H: \mathbb{R}^2 \to \mathbb{R}$ is smooth and

$$J = \begin{pmatrix} 0 & 1 \\ -1 & 0 \end{pmatrix}, \quad \nabla H(x) = \begin{pmatrix} H_{x_1}(x) \\ H_{x_2}(x) \end{pmatrix},$$

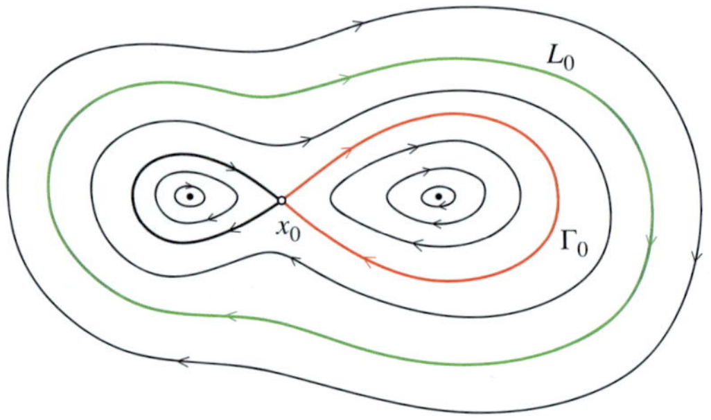

Figure 1.9 Phase portrait of a planar Hamiltonian system.

so that

$$J\nabla H(x) = \begin{pmatrix} H_{x_2}(x) \\ -H_{x_1}(x) \end{pmatrix}.$$

It is well known that periodic orbits in Hamiltonian systems appear in continuous families as closed level curves of H. Generically, such families approach either a center or an orbit homoclinic to a hyperbolic saddle or extend to infinity (see Figure 1.9). We want to study limit cycles and homoclinic orbits in one- and two-parameter generic smooth perturbations of (1.38).

First consider the following one-parameter planar ODE:

$$\dot{x} = J\nabla H(x) + \varepsilon f(x), \quad x = (x_1, x_2) \in \mathbb{R}^2, \tag{1.39}$$

where $\varepsilon \in \mathbb{R}$ is a small parameter, and $f\colon \mathbb{R}^2 \to \mathbb{R}^2$ is a smooth function. For $\varepsilon = 0$, the system (1.39) reduces to the Hamiltonian system (1.38). Since we are interested in non-Hamiltonian perturbations, we assume that $\operatorname{div} f$ does not vanish.

We want to study hyperbolic limit cycles of the perturbed system (1.39). Such cycles branch off from special cycles of the unperturbed system (1.38), as the following theorem ensures (see, e.g., (Andronov et al., 1973; Guckenheimer and Holmes, 1990)).

Theorem 1.15 (Pontryagin, 1934) *Let L_0 be a clockwise-oriented cycle of* (1.39) *for $\varepsilon = 0$ corresponding to a periodic solution $\varphi(t)$ with (minimal) period T_0. If*

$$M_0 := \int_{L_0} f_2(x)\, dx_1 - f_1(x)\, dx_2 = 0,$$

while

$$M_1 := \int_0^{T_0} \operatorname{div} f(\varphi(t))\, dt \neq 0,$$

then

(1) *there exists an annulus around* L_0 *in which the system* (1.39) *has, for all sufficiently small* $\varepsilon > 0$, *a unique hyperbolic limit cycle* L_ε, *such that* $L_\varepsilon \to L_0$ *as* $\varepsilon \to 0$;

(2) *this cycle* L_ε *is stable for* $\varepsilon M_1 < 0$ *and unstable for* $\varepsilon M_1 > 0$.

The theorem is illustrated in Figure 1.10(a), where a stable cycle L_ε is shown.

Notice that Green's Theorem implies

$$M_0 = \int_{\Omega_0} \operatorname{div} f(x)\, dx,$$

where $\Omega_0 \subset \mathbb{R}^2$ is the domain inside the cycle L_0.

Let us now consider perturbations of a saddle homoclinic orbit. Suppose that the Hamiltonian system (1.38) has an orbit Γ_0 that is homoclinic to a hyperbolic saddle point x_0 (see Figure 1.9). Let $H(x_0) = h_0$, so that $\Gamma_0 \subset \{x \in \mathbb{R}^2 : H(x) = h_0\}$.

Introduce now the following two-parameter perturbation of (1.38):

$$\dot{x} = J\nabla H(x) + \varepsilon f(x, \mu), \quad x = (x_1, x_2) \in \mathbb{R}^2, \tag{1.40}$$

where $\varepsilon, \mu \in \mathbb{R}$ are parameters, and $f : \mathbb{R}^2 \times \mathbb{R} \to \mathbb{R}^2$ is a smooth function with nonvanishing $\operatorname{div} f$. The reason for introducing the second parameter will become clear later. Finally, suppose for simplicity that $f(x_0, \mu) = 0$ for $\mu \in \mathbb{R}$. This assumption implies that x_0 is an equilibrium for all values of both parameters. Then the following result holds (see e.g., (Guckenheimer and Holmes, 1990; Sanders and Verhulst, 1985)).

Theorem 1.16 (Melnikov, 1963) *Let* Γ_0 *be an orbit homoclinic to a saddle equilibrium of* (1.40) *for* $\varepsilon = 0$. *Suppose that for some* $\mu = \mu_0$

$$\int_{\Gamma_0} f_2(x, \mu_0)\, dx_1 - f_1(x, \mu_0)\, dx_2 = 0,$$

while

$$\int_{\Gamma_0} \frac{\partial f_2}{\partial \mu}(x, \mu_0)\, dx_1 - \frac{\partial f_1}{\partial \mu}(x, \mu_0)\, dx_2 \neq 0.$$

Then there exists a unique function $\mu_H(\varepsilon)$ *with* $\mu_H(0) = \mu_0$, *and an annulus around* Γ_0 *in which the system* (1.40) *has, for all sufficiently small* ε *and* $\mu = \mu_H(\varepsilon)$, *a homoclinic to* x_0 *orbit* $\Gamma_\varepsilon \to \Gamma_0$ *as* $\varepsilon \to 0$.

The theorem is illustrated in Figure 1.10(b), where a perturbed phase portrait with homoclinic orbit Γ_ε existing when $\mu = \mu_H(\varepsilon)$ is shown.

For some combination of parameters (ε, μ), the system (1.40) can also have nonhyperbolic cycles with multiplier +1. Such degenerate cycles bifurcate

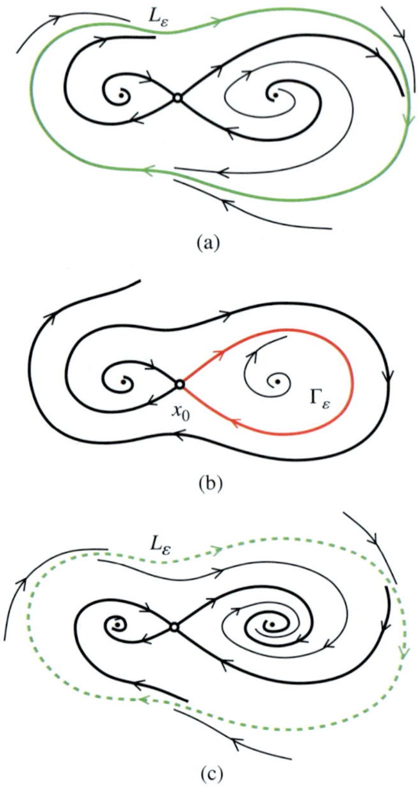

Figure 1.10 Phase portraits of a perturbed Hamiltonian system with small $\varepsilon > 0$: (a) L_ε is a stable limit cycle in (1.39); (b) Γ_ε is a homoclinic orbit to x_0 in (1.40) for $\mu = \mu_H(\varepsilon)$; (c) L_ε is a nonhyperbolic limit cycle in (1.40) for $\mu = \mu_C(\varepsilon)$.

from those cycles of the unperturbed system (1.38), for which M_1 defined in Theorem 1.15 vanishes. Namely, the following result holds.

Theorem 1.17 *Let L_0 be a clockwise-oriented cycle of* (1.40) *for* $\varepsilon = 0$ *corresponding to a periodic solution $\varphi(t)$ with the (minimal) period T_0 and let $\Omega_0 \subset \mathbb{R}^2$ denote the domain inside the cycle L_0. Suppose that for some $\mu = \mu_0$*

$$\int_{\Omega_0} \operatorname{div} f(x, \mu_0)\, dx = 0$$

and

$$\int_0^{T_0} \operatorname{div} f(\varphi(t), \mu_0)\, dt = 0.$$

Then generically there exists a unique function $\mu_C(\varepsilon)$ *with* $\mu_C(0) = \mu_0$, *and an annulus around* L_0 *in which the system* (1.40) *has, for all sufficiently small* $\varepsilon > 0$ *and* $\mu = \mu_C(\varepsilon)$, *a nonhyperbolic cycle* $L_\varepsilon \to L_0$ *as* $\varepsilon \to 0$.

The theorem is illustrated in Figure 1.10(c), where a perturbed phase portrait with a nonhyperbolic limit cycle L_ε existing when $\mu = \mu_C(\varepsilon)$ is shown.

The genericity conditions include

$$\int_{L_0} \frac{\partial f_2}{\partial \mu}(x, \mu_0)\, dx_1 - \frac{\partial f_1}{\partial \mu}(x, \mu_0)\, dx_2 \neq 0,$$

as well as one more integral condition ensuring the nonvanishing of the quadratic part of the Poincaré map of the nonhyperbolic cycle for small $\varepsilon > 0$.

2

One-Parameter Bifurcations of Maps

In this chapter we consider smooth one-parameter families of smooth maps

$$x \mapsto f(x, \alpha), \quad x \in \mathbb{R}^n, \alpha \in \mathbb{R}, \tag{2.1}$$

or more generally for integer $K \geq 1$

$$x \mapsto g(x, \alpha) = f^{(K)}(x, \alpha), \tag{2.2}$$

which is the Kth iterate of (2.1).

This chapter is organized as follows. First, we review codim 1 local bifurcations of fixed points and cycles. This material is standard and is included in many textbooks (e.g., (Arnold, 1983; Arrowsmith and Place, 1990; Arnold et al., 1994; Kuznetsov, 2004)), to which we refer for proofs. Then we consider some global codim 1 bifurcations that occur near codim 2 local bifurcations treated in the next chapter.

2.1 Codim 1 bifurcations of fixed points and cycles

Assume that (2.2) has a fixed point $x = x_0$, that is not a fixed point of $f^{(J)}(\cdot, \alpha_0)$ for $1 \leq J < K$ when $K > 1$. In other words, x_0 is a fixed point or a cycle with minimal period K of the map $f(\cdot, \alpha_0)$. If the Jacobian matrix $A = g_x(x_0, \alpha_0)$ is invertible and has no eigenvalue λ with $|\lambda| = 1$, then x_0 is a hyperbolic fixed point. It follows that the dynamics near x_0 is topologically equivalent to that of the linear map $x \mapsto Ax$ (Grobman–Hartman Theorem 1.2). If eigenvalues with $|\lambda| = 1$ are present, the Center Manifold Theorem 1.3 guarantees the existence of local stable, unstable and center invariant manifolds near the fixed point for parameter values close to α_0. On the stable and unstable manifolds, the local dynamics is still determined by the linear part of the map. In contrast, the dynamics on the center manifold depends on both linear and nonlinear terms.

Table 2.1 Codim 1 bifurcations of cycles

Label	Bifurcation condition	Name
LP	$\lambda_1 = 1$	Fold
PD	$\lambda_1 = -1$	Period-doubling
NS	$\lambda_{1,2} = \mathrm{e}^{\pm i\theta_0}, 0 < \theta_0 < \pi$	Neimark–Sacker

Not all nonlinear terms are equally important, since some of them can be eliminated by a smooth coordinate transformation, also depending on parameters, coordinate transformation that puts the map restricted to the center manifold into a normal form, at least up to some order. Fixed points with eigenvalues satisfying $|\lambda| = 1$ bifurcate, i.e., the dynamics near such points changes topologically under parameter variations. The birth of extra invariant objects, such as cycles or tori, is described by a parameter-dependent normal form of the restriction of g to a center manifold. Even though neither the center manifold nor the normal form on it are unique, the qualitative conclusions do not depend on the choices that are made to compute the center manifold.

Assuming sufficient smoothness of g, we write its Taylor expansion about x_0 at $\alpha = \alpha_0$ as

$$g(x_0 + x, \alpha_0) = x_0 + Ax + \frac{1}{2}B(x, x) + \frac{1}{6}C(x, x, x) + \cdots , \tag{2.3}$$

where all functions are multilinear forms of their arguments and the dots denote higher-order terms in x. The components of the multilinear functions B and C are given by

$$B_i(x, y) = \sum_{j,k=1}^{n} \frac{\partial^2 g_i(x_0, \alpha_0)}{\partial \xi_j \partial \xi_k} x_j y_k, \quad C_i(x, y, z) = \sum_{j,k,l=1}^{n} \frac{\partial^3 g_i(x_0, \alpha_0)}{\partial \xi_j \partial \xi_k \partial \xi_l} x_j y_k z_l, \tag{2.4}$$

for $i = 1, 2, \ldots, n$. From now on, I_n is the unit $n \times n$ matrix and $||x|| = \sqrt{\langle x, x \rangle}$, where $\langle u, v \rangle = \bar{u}^T v$ is the standard inner product in $\mathbb{C}^n$ (or $\mathbb{R}^n$).

It is well known that in generic maps only three codim 1 bifurcations of cycles occur (see Table 2.1). It is assumed that the critical eigenvalues are simple and no other eigenvalue of A with $|\lambda| = 1$ exists. In all cases, $\beta = \beta(\alpha)$ is a new real control parameter with critical value 0, and the left (p) and right (q) eigenvectors are normalized such that

$$\langle q, q \rangle = \langle p, q \rangle = 1. \tag{2.5}$$

Below we briefly describe what happens in the state space near the bifurcation.

2.1.1 Fold bifurcation

Lemma 2.1 *If the fixed point x_0 at the critical parameter value α_0 has a simple eigenvalue $\lambda_1 = 1$ and no other eigenvalues on the unit circle, then the restriction of* (2.2) *to a one-dimensional center manifold at α_0 is locally smoothly equivalent to*

$$\xi \mapsto \xi + b_0\xi^2 + O(\xi^3), \tag{2.6}$$

where

$$b_0 = \frac{1}{2}\langle p, B(q,q)\rangle \tag{2.7}$$

with $Aq = q$ and $A^T p = p$ satisfying (2.5).

Proof See Kuznetsov (2004, section 5.4.2, pp. 183–185). □

Here ξ is an appropriate local coordinate in the critical center manifold, e.g., a coordinate in the direction of vector q. Thus

$$b_0 = \frac{1}{2}\left.\frac{d^2}{d\xi^2}\langle p, g(x_0 + \xi q, \alpha_0)\rangle\right|_{\xi=0}.$$

Theorem 2.2 *If $b_0 \neq 0$, then for parameter values α close to α_0, the restriction of* (2.2) *to a parameter-dependent center manifold is locally smoothly equivalent to*

$$\xi \mapsto \beta(\alpha) + \xi + b(\alpha)\xi^2 + O(\xi^3), \tag{2.8}$$

where $\beta(\alpha_0) = 0, b(\alpha_0) = b_0$, and the O-terms may also depend on α.

Proof See Kuznetsov (2004, theorem 4.1, pp. 123–124). □

Suppose that $\beta'(\alpha_0) \neq 0$ so that β can be used as the new unfolding parameter. Also assume that $b_0 \neq 0$. When the parameter β crosses its critical value $\beta = 0$ corresponding to the fold bifurcation at α_0, two fixed points of g collide and disappear. This implies the collision of two period-K cycles of the original map f. This bifurcation is often called the *saddle-node bifurcation* or *Limit Point* (LP).

Actually, the O-terms in (2.8) are irrelevant provided $b_0 \neq 0$.

Theorem 2.3 *The smooth family of smooth maps*

$$\xi \mapsto \beta + \xi + b(\beta)\xi^2 + O(\xi^3), \tag{2.9}$$

where $b(0) = b_0 \neq 0$, is locally topologically equivalent to

$$\xi \mapsto \beta + \xi + \sigma\xi^2, \tag{2.10}$$

where $\sigma = \text{sign}(b_0) = \pm 1$.

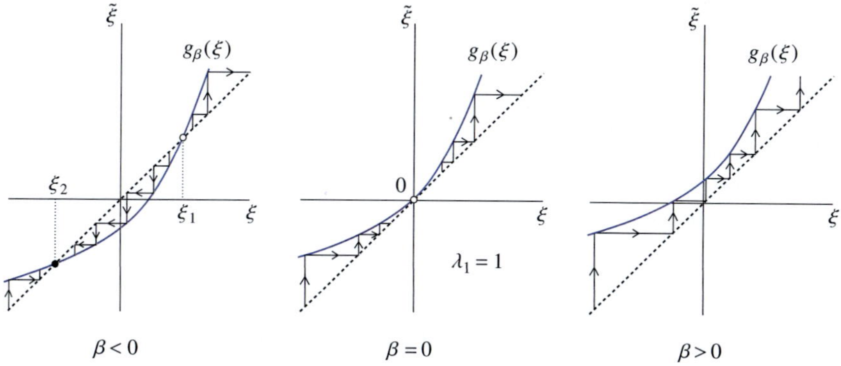

Figure 2.1 Bifurcation diagram of $\xi \mapsto g_\beta(\xi) = \beta + \xi + \xi^2$. Two fixed points ($\xi_1$ and ξ_2) of g_β collide and disappear at $\beta = 0$.

Proof A recent elementary proof is given by Balibrea, Oliveira, and Valverde (2017), see also Katok and Hasselblatt (1995, proposition 7.3.3, p. 300). □

The map (2.10) is called the *topological normal form* for the LP-bifurcation. Its bifurcation diagram is universal and is presented in Figure 2.1 for the case $\sigma = 1$.

2.1.2 Period-doubling (flip) bifurcation

Lemma 2.4 *If the fixed point x_0 at the critical parameter value α_0 has a simple eigenvalue $\lambda_1 = -1$ and no other eigenvalues on the unit circle, then the restriction of* (2.2) *to a one-dimensional center manifold at α_0 is locally smoothly equivalent to the form*

$$\xi \mapsto -\xi + c_0\xi^3 + O(\xi^4), \qquad (2.11)$$

where

$$c_0 = \frac{1}{6}\langle p, C(q,q,q) + 3B(q,(I_n - A)^{-1}B(q,q))\rangle \qquad (2.12)$$

with $Aq = -q$ and $A^T p = -p$ satisfying (2.5).

Proof Since the lowest resonance here is $\lambda_1 = \lambda_1(\lambda_1^2) = \lambda_1^3$, Theorem 1.5 from Section 1.3 immediately implies that the Poincaré normal form of the restriction to the critical center manifold is given by (2.11). Alternatively, one can apply the first part of the proof of Lemma 3.4 in Appendix 3.A.

The formula (2.12) can be found in Kuznetsov (2004, section 5.4.2, pp. 183–185). □

Here ξ is a local coordinate in the critical center manifold selected so that the restriction of (2.2) at α_0 to this manifold has no quadratic term, which is always possible in the considered case.

Theorem 2.5 *For parameter values α close to α_0 defined in* Lemma 2.4, *the restriction of* (2.2) *to a parameter-dependent center manifold is locally smoothly equivalent to*

$$\xi \mapsto -(1+\beta(\alpha))\xi + c(\alpha)\xi^3 + O(\xi^4), \tag{2.13}$$

where $\beta(\alpha_0) = 0, c(\alpha_0) = c_0$, and the O-terms may also depend on α.

Proof The formula (2.13) appears if one keeps the resonant cubic term not only at α_0 but for all α close to α_0. Alternatively, one can apply the first part of the proof of Theorem 3.5 in Appendix 3.A. □

Suppose that $\beta'(\alpha_0) \neq 0$ so that β can be used as the new unfolding parameter. Also assume that $c_0 \neq 0$. When in this case the control parameter β crosses its critical value $\beta = 0$ corresponding to a flip bifurcation at α_0, a cycle of period 2 for g bifurcates from the fixed point, i.e., a cycle of period $2K$ for map f. This phenomenon is often called the *period-doubling bifurcation* (PD) or *flip*. If c_0 is positive, the bifurcation is supercritical and the period-2 cycle is stable. If c_0 is negative, it is subcritical and the period-2 cycle is unstable and exists for $\beta < 0$.

Similarly to the LP-case, the O-terms in (2.13) are irrelevant if $c_0 \neq 0$.

Theorem 2.6 *The smooth family of smooth maps*

$$\xi \mapsto -(1+\beta)\xi + c(\beta)\xi^3 + O(\xi^4), \tag{2.14}$$

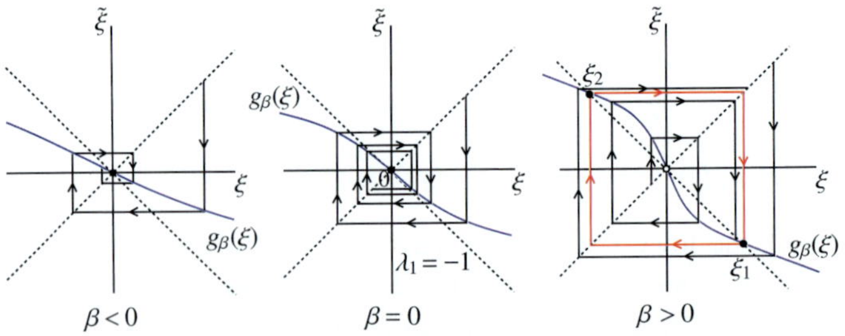

Figure 2.2 Bifurcation diagram of $\xi \mapsto g_\beta(\xi) = -(1+\beta)\xi + \xi^3$. A 2-cycle $\{\xi_1, \xi_2\}$ appears at $\beta = 0$ and exists for $\beta > 0$, while the trivial fixed point $\xi = 0$ becomes unstable.

where $c(0) = c_0 \neq 0$, *is locally topologically equivalent to*

$$\xi \mapsto -(1+\beta)\xi + \sigma\xi^3, \tag{2.15}$$

where $\sigma = \operatorname{sign}(c_0) = \pm 1$.

Proof A recent elementary proof is given by Balibrea, Oliveira, and Valverde (2017); see also Peckham and Kevrekidis (1991) and Katok and Hasselblatt (1995, pp. 301–302). □

The map (2.15) is called the *topological normal form* for the PD-bifurcation. Its bifurcation diagram is universal and is presented in Figure 2.2 for the supercritical case $\sigma = 1$.

2.1.3 Neimark–Sacker bifurcation

Generic NS bifurcation

Lemma 2.7 *If the fixed point* x_0 *at the critical parameter value* α_0 *has simple critical eigenvalues* $\lambda_{1,2} = e^{\pm i\theta_0}$ *and no other eigenvalues on the unit circle, while*

$$e^{ik\theta_0} - 1 \neq 0, \quad k = 1, 2, 3, 4 \text{ (no strong resonances)}, \tag{2.16}$$

then the restriction of (2.2) *to a two-dimensional critical center manifold at the critical parameter value* α_0 *is locally smoothly equivalent to the form*

$$w \mapsto e^{i\theta_0} w + c_1^0 w|w|^2 + O(|w|^4), \tag{2.17}$$

where w *is a complex variable and* c_1^0 *is a complex number given by*

$$c_1^0 = \frac{1}{2}\langle p, C(q,q,\bar{q}) + 2B(q,(I_n - A)^{-1}B(q,\bar{q})) + B(\bar{q},(e^{2i\theta_0}I_n - A)^{-1}B(q,q))\rangle \tag{2.18}$$

with $Aq = e^{i\theta_0}q$ *and* $A^T p = e^{-i\theta_0}p$ *satisfying* (2.5).

Proof Write the restriction of (2.2) to the critical center manifold at α_0 in the complex form

$$z \mapsto e^{i\theta_0} z + \sum_{2\leq j+k\leq 3} \frac{1}{j!k!} g_{jk} z^j \bar{z}^k + O(|z|^4)$$

and introduce the map from $\mathbb{C}^2$ to $\mathbb{C}^2$ by the formula

$$\begin{pmatrix} z \\ \bar{z} \end{pmatrix} \mapsto \begin{pmatrix} \lambda_1 z + \displaystyle\sum_{2\leq j+k\leq 3} \frac{1}{j!k!} g_{jk} z^j \bar{z}^k \\ \lambda_2 \bar{z} + \displaystyle\sum_{2\leq j+k\leq 3} \frac{1}{j!k!} \bar{g}_{kj} z^j \bar{z}^k \end{pmatrix} + O(\|(z,\bar{z})\|^4),$$

where the second component is the complex-conjugate of the first one, while z and $\bar{z}$ are treated as independent complex variables. We have $\lambda_1\lambda_2 = 1$, and we see that the only possible resonances of order less than 4 are

$$\lambda_1 = \lambda_1(\lambda_1\lambda_2) \quad \text{and} \quad \lambda_2 = \lambda_2(\lambda_1\lambda_2),$$

while all others are excluded by (2.16). Then, Theorem 1.5 from Section 1.3 immediately implies that the Poincaré normal form of the restriction to the critical center manifold is given by (2.17).

The derivation of formula (2.18) can be found by Kuznetsov (2004, section 5.4.2, pp. 185–187). □

In this case, w is a local coordinate in the two-dimensional critical center manifold selected so that the restriction of (2.2) at α_0 to this manifold has no quadratic terms and only the displayed cubic term.

Theorem 2.8 *In the absence of strong resonances, for parameter values α close to α_0, the restriction of (2.2) to a parameter-dependent center manifold is locally smoothly equivalent to*

$$w \mapsto e^{i\theta(\alpha)}(1+\beta(\alpha))w + c_1(\alpha)w|w|^2 + O(|w|^4), \tag{2.19}$$

where $\beta(\alpha_0) = 0, \theta(\alpha_0) = \theta_0$ and $c_1(0) = c_1^0$.

Proof The formula (2.19) emerges if one keeps the resonant cubic term not only at α_0 but for all α close to α_0. Alternatively, see Kuznetsov (2004, theorem 4.5, pp. 140–141). □

As usual, suppose that $\beta'(\alpha_0) \neq 0$ so that β can be used as the new unfolding parameter. Notice that at $\beta = 0$ the trivial fixed point $w = 0$ of (2.19) loses stability.

In this case, called *Neimark–Sacker bifurcation* (NS), the O-terms in (2.19) *cannot be neglected* (even in the generic situation) and the truncated unfolding

$$w \mapsto e^{i\theta(\beta)}(1+\beta)w + c_1(\beta)w|w|^2 \tag{2.20}$$

is *not* a topological normal form for the NS-bifurcation. Fortunately, generically (2.19) and the truncated map (2.20) share an important property described by the following theorem.

Theorem 2.9 *Consider a smooth family of smooth maps*

$$w \mapsto \Phi_\beta(w) := e^{i\theta(\beta)}(1+\beta)w + c_1(\beta)w|w|^2 + O(|w|^4), \tag{2.21}$$

where $\theta(0) = \theta_0, 0 < \theta_0 < \pi$, and suppose that

$$L_1 := \Re(e^{-i\theta_0}c_1(0)) = \Re(e^{-i\theta_0}c_1^0) \neq 0.$$

If $L_1 < 0$ *then* (2.21) *has a unique stable closed invariant curve around the fixed point* $w = 0$ *that bifurcates at* $\beta = 0$ *from this point and exists for sufficiently small* $\beta > 0$.

If $L_1 > 0$ *then* (2.21) *has a unique unstable closed invariant curve around the trivial fixed point* $w = 0$ *that shrinks at* $\beta = 0$ *to this point and exists for sufficiently small* $\beta < 0$.

Proof See Iooss (1979) and Kuznetsov (2004, lemma 4.8, pp. 149–154). □

Thus, generically, a unique closed invariant curve for g around the fixed point appears on the center manifold, when the control parameter α crosses its critical value α_0 corresponding to the Neimark–Sacker bifurcation. For the original map, this means the appearance of K disjoint curves, cyclically shifted by f.

If L_1 is negative, the bifurcation is supercritical and the invariant curve is stable (see Figure 2.3). For L_1 positive, it is subcritical and the invariant curve is unstable. The quantity L_1 is often called the *first Lyapunov coefficient* for the Neimark–Sacker bifurcation.

As for the Hopf bifurcation in planar ODE systems, in the case of maps (2.2) with $n = 2$, there is an alternative for (2.18) if we want to compute L_1, that requires neither inversion of matrices nor solving linear systems. Introduce the function

$$G(z, \bar{z}) := \langle p, g(x_0 + zq + \bar{z}\bar{q}, \alpha_0) - x_0 \rangle$$

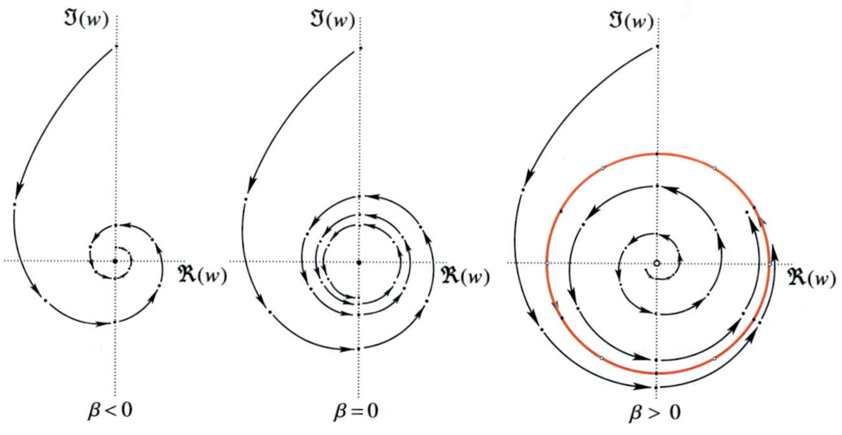

Figure 2.3 Neimark–Sacker bifurcation in (2.20) in the supercritical case $\sigma = \text{sign}(L_1) = -1$. The unique stable closed invariant curve bifurcates for $\beta > 0$ at $\beta = 0$.

and consider its Taylor coefficients

$$g_{jk} = \left. \frac{\partial^{j+k}}{\partial z^j \partial \bar{z}^k} \langle p, g(x_0 + zq + \bar{z}\bar{q}, \alpha_0) - x_0 \rangle \right|_{z=\bar{z}=0}.$$

Then

$$L_1 = \Re\left(\frac{e^{-i\theta_0} g_{21}}{2}\right) - \Re\left(\frac{(1 - 2e^{i\theta_0})e^{-2i\theta_0}}{2(1 - e^{i\theta_0})} g_{20} g_{11}\right) - \frac{1}{2} |g_{11}|^2 - \frac{1}{4} |g_{02}|^2 \quad (2.22)$$

(see, e.g., Kuznetsov (2004)).

Remark 2.10 It should be noted that the bifurcating closed invariant curve has only finite smoothness, that increases as $\beta \to 0$. Moreover, near the Neimark–Sacker bifurcation, stable and unstable cycles generically coexist on the closed invariant curve for parameter values from small open intervals. These cycles collide and disappear via the fold bifurcations at the end points of each interval, so that only dense (quasi-periodic) orbits remain in the invariant curve. Away from the Neimark–Sacker bifurcation, the closed invariant curve usually loses its smoothness and disappears. More details are given in Section 2.1.3.

Remark 2.11 There is an alternative way to understand the appearance of a closed invariant curve for the map (2.21) under the variation of β. Write the map Φ_β defined by (2.21) in the form

$$\Phi_\beta(w) = e^{i\theta(\beta)} \left[(1 + \mu(\beta))w + d_1(\beta)w|w|^2 + O(|w|^4) \right],$$

where

$$d_1(\beta) = e^{-i\theta(\beta)} c_1(\beta).$$

We have $\Re(d_1(0)) = L_1$ with L_1 defined above. Then the map Φ_β can be approximated by the unit-time flow φ_β^1 of the *approximating ODE*

$$\dot{w} = (\mu(\beta) + i\theta(\beta))w + d_1(\beta)w|w|^2, \quad (2.23)$$

where $\mu(\beta) = \ln(1 + \beta)$ (so that $\mu(0) = 0$). More precisely, it holds that

$$\Phi_\beta(w) = \varphi_\beta^1(w) + O(|w|^4).$$

The ODE (2.23) is the normal form for the *Andronov–Hopf bifurcation*[1] of ODEs that describes the appearance of a unique limit cycle, provided that the first Lyapunov coefficient for ODE (2.23)

$$l_1 = \frac{\Re(d_1(0))}{\theta_0} = \frac{L_1}{\theta_0} \neq 0.$$

[1] This explains why the Neimark–Sacker bifurcation is sometimes unfortunately called "Hopf bifurcation for maps."

Therefore, one can naturally expect the appearance of a closed invariant curve of the map (2.21) that is approximated by the unit-time flow of (2.23). Note that the stability of the bifurcating limit cycle is determined by the sign of l_1 (or L_1). Of course, there is no topological equivalence between Φ_β and φ^1_β, even locally.

NS bifurcation at weak resonances

The flow approximation technique is very useful for a more detailed study of the Neimark–Sacker bifurcation at *weak resonances*, where $e^{iq\theta_0} = 1$ for some $q \geq 5$. This is the second equality-type condition defining the critical case. Thus, to unfold it properly, we have to consider a *two-parameter* family of smooth maps

$$x \mapsto g(x, \alpha), \quad x \in \mathbb{R}^n, \alpha \in \mathbb{R}^2. \tag{2.24}$$

Lemma 2.12 *If the fixed point x_0 at the critical parameter value α_0 has only simple critical eigenvalues*

$$\lambda_{1,2} = e^{\pm i\theta_0}, \quad \theta_0 = \frac{2\pi p}{q},$$

where $p/q < 1/2$ is irreducible and $q \geq 5$, then the restriction of (2.24) to a parameter-dependent center manifold is locally smoothly equivalent to the map

$$z \mapsto \lambda(\alpha)z + \sum_{m=1}^{[(q-1)/2]} A_m(\alpha)z|z|^{2m} + B(\alpha)\bar{z}^{q-1} + O(|z|^{q+1}), \tag{2.25}$$

where $w \in \mathbb{C}$, $\lambda(0) = \lambda_1 = e^{i\theta_0}$, and A_m, B are smooth complex-valued functions.

Proof See Iooss (1979, chapter III.3). □

If $\lambda(\alpha)$ crosses the unit circle under variation of a component of α, the map (2.25) satisfies the conditions of Theorem 2.9, provided

$$L_1 = \Re(e^{-i\theta_0}A_1(0)) \neq 0.$$

If this condition is satisfied, a single closed invariant curve of (2.25) bifurcates from the origin.

To study the dynamics of (2.25) restricted to the closed invariant curve, approximate it by a flow. Since $\lambda^q(0) = 1$, we can write

$$\lambda^q(\alpha) = e^{\beta_1(\alpha)+i\beta_2(\alpha)},$$

where $\beta_1(0) = \beta_2(0) = 0$. If we assume that the map $\alpha \mapsto \beta(\alpha) = (\beta_1(\alpha), \beta_2(\alpha))$

is regular at α_0, then the components of β can be used as the new unfolding parameters.

Now consider the *qth iterate* of the map (2.25)

$$z \mapsto \Psi_\beta(z) := e^{\beta_1 + i\beta_2} z + \sum_{m=1}^{[(q-1)/2]} C_m(\beta) z|z|^{2m} + D(\beta)\bar{z}^{q-1} + O(|z|^{q+1}), \quad (2.26)$$

where C_m and D are smooth functions of β. The map Ψ_β can be written as

$$\Psi_\beta(w) = \varphi_\beta^1(w) + O(|w|^{q+1}),$$

where φ^t is the flow corresponding to the approximating ODE

$$\dot{w} = (\beta_1 + i\beta_2)w + \sum_{m=1}^{[(q-1)/2]} a_m(\beta) z|z|^{2m} + b(\beta)\bar{z}^{q-1}. \quad (2.27)$$

Recall that $q \geq 5$, while a_m and b here are smooth complex-valued functions of $\beta = (\beta_1, \beta_2)$. One can show that $\Re(a_1(0))$ differs only by a positive factor from L_1.

Notice that equation (2.27) is invariant with respect to the rotation through $2\pi/q$, i.e., the transformation

$$z \mapsto R_q z := e^{-2\pi i/q} z$$

does not change (2.27). This $\mathbb{Z}_q$ symmetry will be visible in the phase portraits.

Using $w = \rho e^{i\varphi}$, we can rewrite (2.27) in polar coordinates (ρ, φ) as

$$\begin{cases} \dot{\rho} &= \rho\left(\beta_1 + \sum_{m=1}^{[(q-1)/2]} \Re(a_m(\beta))\rho^{2m+1} + \rho^{q-2}\Re(b(\beta)e^{-iq\varphi})\right), \\ \dot{\varphi} &= \beta_2 + \sum_{m=1}^{[(q-1)/2]} \Im(a_m(\beta))\rho^{2m} + \rho^{q-2}\Im(b(\beta)e^{-iq\varphi}). \end{cases}$$

Lemma 2.13 *Suppose that*

$$\Re(a_1(0)) \neq 0 \quad \textit{and} \quad b(0) \neq 0.$$

Then the local bifurcation diagram of the approximating system (2.27) *near* $\beta = 0$ *includes only the following curves:*

(1) *the Andronov–Hopf bifurcation curve*

$$H = \{(\beta_1, \beta_2) \in \mathbb{R}^2 : \beta_1 = 0\},$$

at which a limit cycle branches from the trivial equilibrium;

(2) *two saddle-node bifurcation curves existing when* $\beta_2 \Re(a_1(0)) < 0$

$$T_{1,2} = \left\{(\beta_1, \beta_2) \in \mathbb{R}^2 : \beta_2 = K_1\beta_1 \pm K_2 \left(-\frac{\beta_1}{\Re(a_1(0))}\right)^{(q-2)/2} + o(|\beta_1|^{(q-2)/2})\right\},$$

at which q symmetric pairs of nontrivial equilibria appear or disappear on the limit cycle. Here

$$K_1 := \frac{\Im(a_1(0))}{\Re(a_1(0))}, \quad K_2 := \frac{|a_1(0)|\,|b(0)|}{\Re(a_1(0))}.$$

Moreover, these bifurcations are non-degenerate and are the only bifurcations happening in (2.27) *near* $\beta = 0$ *in a neighborhood of the origin.*

Proof See Iooss and Adelmeyer (1998, section III.3.6, pp. 178–182). □

The bifurcation diagram of the approximating system (2.27) is presented in Figure 2.4. Notice that the open region between two saddle-node bifurcation

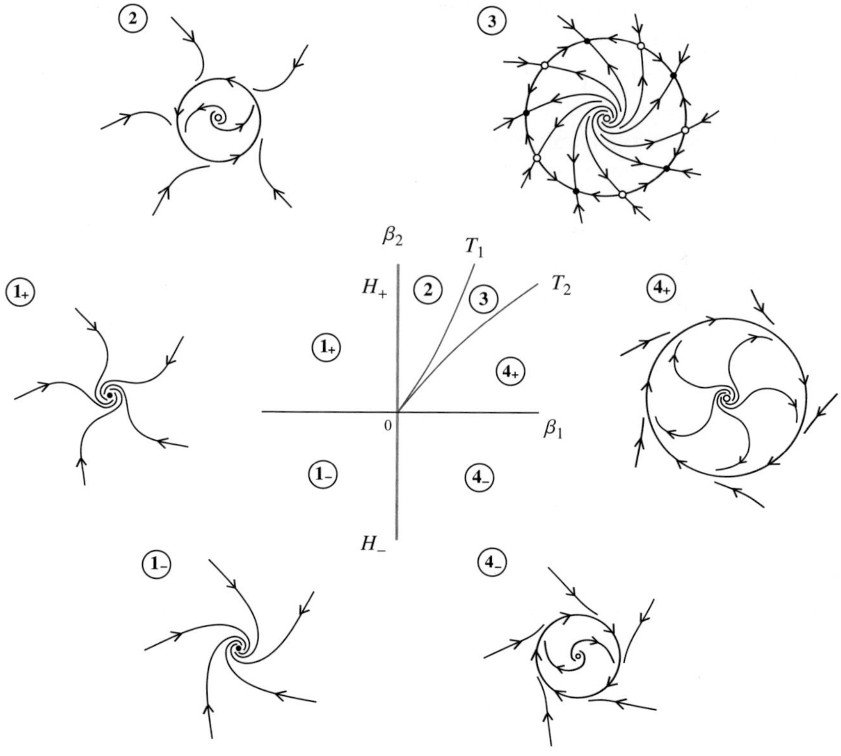

Figure 2.4 Local bifurcation diagram of the approximating system (2.27) with $q = 5$ and $\Re(a_1(0)) < 0$, $\Im(a_1(0)) < 0$.

curves $T_{1,2}$, where the nontrivial equilibria exist, approaches the axis $\beta_1 = 0$ as a narrow tongue with width of order $O(|\beta_1|^{(q-1)/2})$.

What information about the smooth normal form (2.25) of the original map near a weak resonance can be extracted from Lemma 2.13? In general, the map (2.25) is not the unit-time shift of any ODE. Also, it does not have $\mathbb{Z}_q$-symmetry. However, the major features of the approximating ODE (2.27) have direct counterparts in the dynamics generated by (2.25). Recall that the ODE (2.27) approximates the qth iterate Ψ_β of the normal form (2.25). Therefore, while the trivial equilibrium $w = 0$ corresponds to the trivial fixed point of (2.25), the q pairs of nontrivial equilibria shifted by R_q actually correspond to *two period-q cycles* of this map. The Hopf bifurcation line H corresponds to the bifurcation line $NS^{(1)}$ of the map (2.25), at which near $\beta = 0$ the Neimark–Sacker bifurcation occurs that produces the single closed invariant curve. The mentioned q-cycles are all located on this closed invariant curve, and one of them is stable, while the other is unstable. The saddle-node bifurcation curves $T_{1,2}$ approximate correctly two fold bifurcation curves $LP^{(q)}_{1,2}$ at which two period-q cycles collide and disappear. The narrow region delimited by the fold curves $LP^{(q)}_{1,2}$ is called the *Arnold tongue*, or resonance tongue. Inside this region, the rotation number of the restriction of the map (2.25) to its closed invariant curve is rational and equals to p/q.

In the complex $(\Re(\lambda), \Im(\lambda))$-plane, the Neimark–Sacker bifurcation curve $NS^{(1)}$ is defined by $|\lambda| = 1$ and Arnold tongues approach the unit circle at points $\lambda = e^{2\pi i p/q}$ with various rational p/q (see Figure 2.5). Thus, a generic one-parameter family of maps exhibiting a Neimark–Sacker bifurcation crosses an

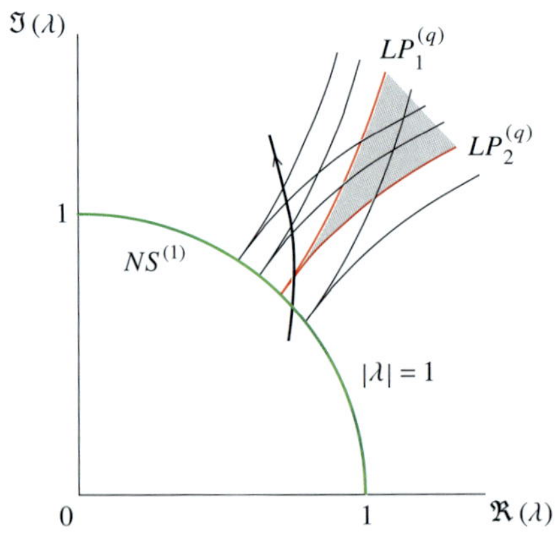

Figure 2.5 Arnold tongues near the Neimark–Sacker bifurcation.

infinite number of Arnold tongues, corresponding to different rotation numbers. In such a family, thus, an infinite number of fold bifurcations of long-periodic cycles occur.

Remark 2.14 While for the analysis of the Neimark–Sacker bifurcation the flow approximation hardly provides any information that cannot be obtained by analyzing the original (normalized) map and its iterates, we shall see in Chapter 3 that this technique is very useful in the study of codim 2 local bifurcations of maps. For such bifurcations, the flow approximation provides information on global bifurcations that is impossible to get otherwise.

2.2 Some global codim 1 bifurcations

Several codim 1 global bifurcations of generic one-parameter maps have been analyzed theoretically. Let us briefly discuss only those appearing near codim 2 local bifurcations, without pretending to give a complete picture.

2.2.1 Homoclinic tangencies in planar maps

Consider a family of planar diffeomorphisms having a hyperbolic fixed point O. Its stable and unstable one-dimensional invariant manifolds $W^s(O)$ and $W^u(O)$ can intersect along homoclinic orbits. Generically, such an intersection is transversal. It has been shown that this implies the presence of an infinite number of saddle limit cycles near each homoclinic orbit. However, at a certain parameter value, say α_0, the manifolds can become *tangent* to each other and then split (see Figure 2.6, where the rearrangement of the manifolds is sketched). This bifurcation phenomenon was first studied by Gavrilov and Shilnikov (1972, 1973) and by Palis and Takens (1993).

Theorem 2.15 *Suppose a smooth family of orientation-preserving planar diffeomorphisms*

$$u \mapsto P_\alpha(u), \quad u \in \mathbb{R}^2, \tag{2.28}$$

has at $\alpha = \alpha_0$ a homoclinic orbit to a hyperbolic fixed point O, and along this orbit the manifolds $W^s(O)$ and $W^u(O)$ have quadratic tangency. Then, generically, there are two infinite sequences of parameter values converging (from one side) to $\alpha = 0$ and corresponding to period-doubling and fold bifurcations of cycles located near the critical homoclinic orbit.

Proof See, e.g., Kuznetsov (2004, section 7.2.1, pp. 262–266). □

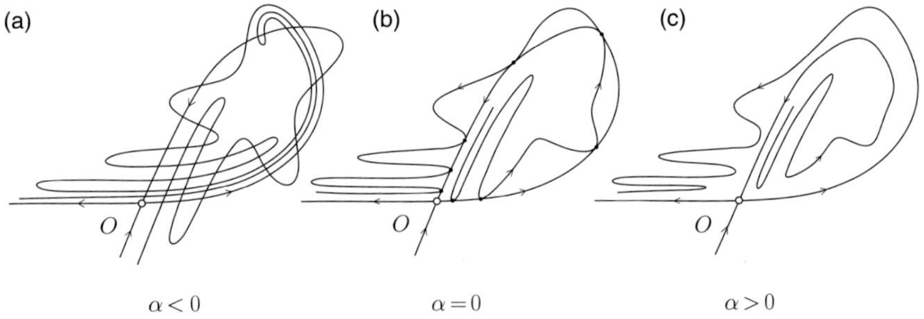

Figure 2.6 Configuration of stable and unstable manifolds of the saddle O: (a) transversal intersection yielding a persistent homoclinic orbit; (b) homoclinic tangency; (c) separated manifolds.

Note that all these bifurcations occur when the invariant manifolds of the saddle point have no intersection (see case $\alpha > 0$ in Figure 2.6).

2.2.2 Quasi-periodic bifurcations of invariant tori

Here we describe some properties of invariant tori in more detail, focusing on implications for dynamics. We do not pretend to give a full exposition of relevant parts of KAM-theory, as it would distract from our more practical approach. To start we discuss invariant curves, and sketch a generalization for higher-dimensional tori.

A closed invariant curve has several characteristics. The first is its geometry so an invariant curve (IC) is topologically equivalent to a torus $\mathbb{T}^1$, i.e., the unit circle. Here we may assume that there exists a (at least continuous) parametrization $x\colon \mathbb{T}^1 \to \mathbb{R}^n$. The second is the rotation number ρ, that is the average angle by which the map f "rotates" $\mathbb{T}^1$. In the case of a *rigid rotation*, the dynamics on the IC can be described by the translation $\theta \mapsto \theta+\rho$ for $\theta \in \mathbb{T}^1$, or in other words

$$f(x(\theta)) = x(\theta + \rho), \tag{2.29}$$

where ρ is the corresponding rotation number. In case of an n-dimensional torus $\mathbb{T}^n$, this quantity becomes an n-dimensional *frequency vector*. The third is the smoothness and normal behavior related to reducibility and stability.

The rotation number ρ is especially important for the persistence of the invariant curve if we vary some parameters as it is a global dynamical object. To make this more concrete we ask ourselves if we can transform the dynamics of the map $x \mapsto f(x)$ restricted to the invariant curve to the translation $\theta \mapsto \theta + \rho$ by a conjugacy Φ. In order for Φ to exist, we need to impose a *Diophantine condition* on the (necessarily irrational) rotation number ρ:

$$\left|\rho - \frac{p}{q}\right| \geq \frac{C}{q^\tau}, \quad p, q \in \mathbb{N}, \tag{2.30}$$

for positive constants C, τ. If ρ satisfies this condition, it is said to be sufficiently irrational, or *Diophantine*. The set of excluded rotation numbers has measure $O(\gamma)$ as $\gamma \to 0$. Even though many values of ρ will obviously not satisfy this condition, the complement is a set of positive measures corresponding to Diophantine ρ. Then it follows by KAM-theory (see, e.g., (Broer et al., 1990; Vitolo, Broer, and Simó, 2011b)), given an invariant curve with Diophantine ρ, that Φ exists and the invariant curve will persist under small variations of the parameters.

The complement contains so-called resonance gaps where the quasi-periodic behavior becomes periodic instead. These are the flat parts of the Devil's staircase. In practice, only when $\rho \approx \frac{p}{q}$ with q a small natural number, below say 20, will it be prominent that dynamics on the invariant curve is qualitatively different from a rotation. If the denominator q is higher, this is easily skipped over during numerical exploration of a given system. This is also visible in Figure 2.7.

To study stability we consider a small perturbation h at a particular point $x(\theta)$ of the invariant curve

$$f(x(\theta) + h) = f(x(\theta)) + D_x f(x(\theta))h + O(|h|^2).$$

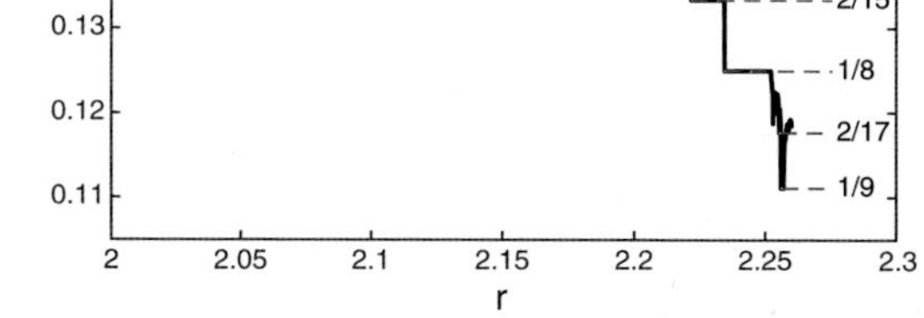

Figure 2.7 The rotation number of the invariant curve for the Delayed Logistic Map (7.1) as a function of r (with $eps = 0$). Flat parts indicate resonances and several are indicated.

It follows that the normal behavior is described by the following linear quasi-periodic *skew product map*

$$\begin{pmatrix} h \\ \theta \end{pmatrix} \mapsto \begin{pmatrix} A(\theta)h \\ \theta + \rho \end{pmatrix}, \tag{2.31}$$

where $A(\theta) = D_x f(x(\theta))$. The local behavior near the invariant curve can depend on θ, but there is a class of systems for which a simpler description is available.

System (2.31) is called *reducible* if there exists a coordinate change $y = C(\theta)x$ such that (2.31) can be transformed to

$$\begin{pmatrix} y \\ \theta \end{pmatrix} \mapsto \begin{pmatrix} By \\ \theta + \rho \end{pmatrix}, \tag{2.32}$$

where $B := C(\theta + \rho)^{-1} A(\theta) C(\theta)$ does not depend on θ.

So if (2.31) is reducible, the eigenvalues of B determine the stability of the invariant curve. This is similar to fixed points and cycles, where eigenvalues within the unit circle correspond to stable directions, and those outside to unstable directions. Note that for autonomous maps, 1 is also an eigenvalue of B corresponding to the tangent direction along the invariant curve. Hence, the remaining eigenvalues determine the normal behavior near the invariant curve. For now, we will assume that the invariant curve is reducible. There are also cases where, due to resonances or loss of smoothness, this is no longer true. Essentially, the normal dynamics (2.31) will depend on θ.

The invariant curve bifurcates when an eigenvalue of B crosses the unit circle as a parameter is varied. As may be expected, this includes quasi-periodic equivalents of the saddle-node, period-doubling and Neimark–Sacker bifurcation of fixed points. However, using an algorithm by Ginelli et al. (2007) to compute the Lyapunov bundles, an additional case was identified (Kamiyama et al., 2014; Komuro et al., 2016). The case of period-doubling can generate a connected set or disconnected set. We remark that, especially for higher-dimensional tori, the bifurcation may occur in the normal direction but may also be due to resonances leading to lower dimensional attractors (Baesens et al., 1991; Komuro et al., 2016). For invariant curves this would be phase-locking.

Lyapunov exponents are also used to study bifurcations of tori as they correspond to the logarithms of the absolute value of the eigenvalues of B. The dimension of a torus is then given by the number of zero Lyapunov exponents. One must take care of the numerical accuracy of the exponents to draw this conclusion. We mention that continuation of invariant curves has been

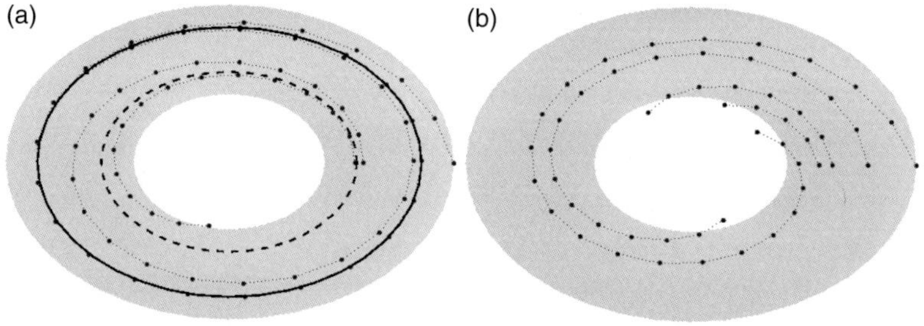

Figure 2.8 The dynamics on the annulus for a quasi-periodic saddle-node bifurcation. (a) for $\mu < 0$ there are two closed invariant curves (black circles), one unstable and one stable. Iterates starting outside the inner circle converge to the outer invariant curve. Here we fixed $a = -1$. (b) for $\mu > 0$ there is no invariant curve, and the iterates pass through the annulus.

proposed by Jorba (2001) and Vitolo, Broer, and Simó (2011b) but do not discuss this here.

Now the bifurcation scenario depends critically on the rotation number. If ρ is irrational, then there is a path in parameter space such that the bifurcation looks simple. Indeed the tangent dynamics will still be a rigid rotation, while the dynamics in the normal direction on a center manifold may be modeled by the normal forms (2.8), (2.13) and (2.19). So for a *quasi-periodic saddle-node bifurcation*, a model map is given by

$$\begin{pmatrix} x \\ \theta \end{pmatrix} \mapsto \begin{pmatrix} x + \mu - x^2 \\ \theta + \rho \end{pmatrix}, \tag{2.33}$$

where μ is the unfolding parameter. This describes the dynamics on an annulus $r = r_0 + x$ and is valid if x is small. Then the typical bifurcation scenario is as shown in Figure 2.8.

Quasi-periodic periodic-doubling and Neimark–Sacker bifurcations exhibit similar bifurcation scenarios, except that their occurrence requires a phase space that is at least three-dimensional. Similar to the period-doubling and Neimark–Sacker bifurcations for cycles, the quasi-periodic counterparts can be sub- or supercritical.

For the *quasi-periodic period-doubling bifurcation*, we have the following model map on $\mathbb{T}^1 \times \mathbb{R}$

$$\begin{pmatrix} x \\ \theta \end{pmatrix} \mapsto \begin{pmatrix} -(1+\mu)x + bx^3 \\ \theta + \rho \end{pmatrix}. \tag{2.34}$$

Embedding this manifold into the original phase space, it may be orientable or not in the normal direction. For the orientable case, Figure 2.9 shows char-

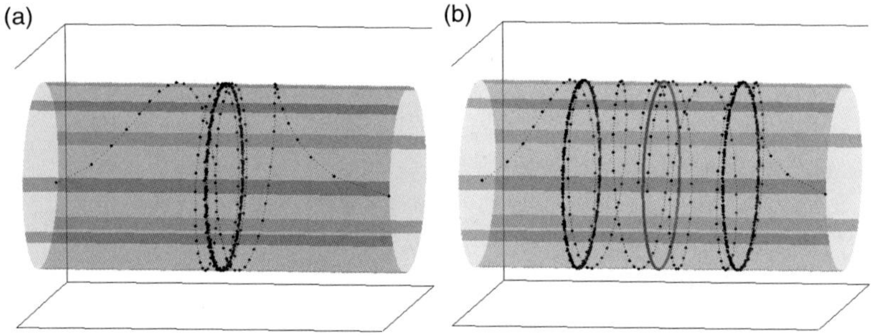

Figure 2.9 A supercritical quasi-periodic doubling bifurcation. (a) for $\mu < 0$, the dynamics restricted to the cylinder converge to the invariant curve. (b) for $\mu > 0$ the original invariant curve is unstable, while a stable "doubled" invariant curve has appeared on the cylinder. Note that iterates actually *hop* from one side of the cylinder to the other and back. For clarity, only the second iterate is shown of every trajectory.

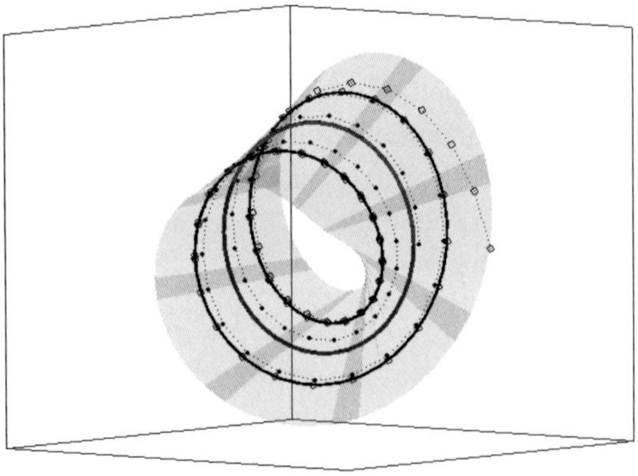

Figure 2.10 A "doubled" invariant curve in case the center manifold is non-orientable. Again, for clarity, only the second iterate is shown of every trajectory.

acteristic phase portraits before and after the bifurcation. Note that the "doubled" invariant curve consists of two disconnected sets. In the case that the center manifold is non-orientable, the bifurcating invariant curve is connected as shown in Figure 2.10. Hence, it is also called double covering bifurcation (Komuro et al., 2016).

In case of a *quasi–periodic Neimark–Sacker bifurcation*, the normal behavior around the invariant curve is of focus type. So orbits spiral around the invariant curve (Figure 2.11). Beyond the bifurcation, there is a 2-torus whose stability is determined by the criticality. A secondary Neimark–Sacker bifur-

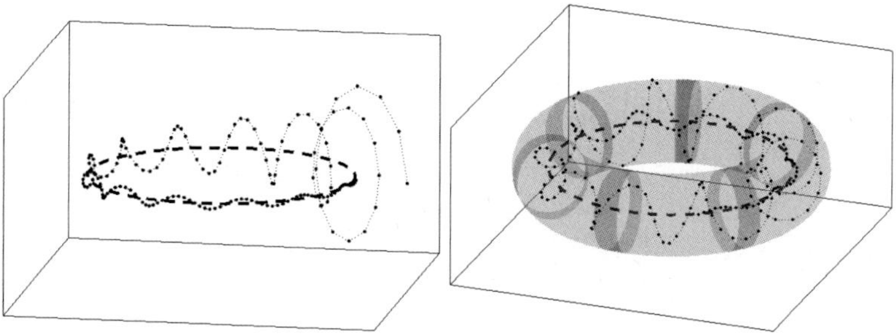

Figure 2.11 (a) Before a supercritical quasi-periodic Neimark–Sacker bifurcation, orbits spiral to an invariant curve. (b) After the bifurcation, the invariant curve has lost stability and orbits converge to a torus.

cation also introduces an additional rotation ρ_2. Provided the ratio ρ_1/ρ_2 is irrational, a trajectory will give a dense covering of the torus. If the ratio is rational, the torus is made up of invariant curves. Only some will survive a generic perturbation. This will exhibit the Arnol'd resonance web, (see (Broer et al., 2008a). In this case, a model map on $\mathbb{R} \times \mathbb{T}^2$ is given by

$$\begin{pmatrix} x \\ \theta_1 \\ \theta_2 \end{pmatrix} \mapsto \begin{pmatrix} (1+\mu)x + cx^3 \\ \theta + \rho_1 \\ \theta + \rho_2 \end{pmatrix}, \qquad (2.35)$$

where it should be noted that $0 \times \mathbb{T}^2 \equiv \mathbb{T}^1$ as it corresponds to the invariant curve.

If ρ is rational, the bifurcation scenario is much more complex, and interesting. This involves various codim 2 bifurcations, which we therefore defer to our later discussion about generic perturbations (unfoldings) of normal forms.

3
Two-Parameter Local Bifurcations of Maps

In this chapter we consider smooth two-parameter families of smooth maps

$$x \mapsto f(x, \alpha), \quad x \in \mathbb{R}^n, \alpha \in \mathbb{R}^2, \tag{3.1}$$

or more generally for integer $K \geq 1$

$$x \mapsto g(x, \alpha) = f^{(K)}(x, \alpha), \tag{3.2}$$

which is the Kth iterate of (3.1). Assuming sufficient smoothness of g, we write its Taylor expansion about (x_0, α_0) as

$$\begin{aligned} g(x_0 + x, \alpha_0 + \alpha) = {} & x_0 + Ax + \tfrac{1}{2}B(x, x) + \tfrac{1}{6}C(x, x, x) + \tfrac{1}{24}D(x, x, x, x) \\ & + \tfrac{1}{120}E(x, x, x, x, x) + \tfrac{1}{720}F(x, x, x, x, x, x) \\ & + J_1\alpha + \tfrac{1}{2}J_2(\alpha, \alpha) \\ & + A_1(x, \alpha) + \tfrac{1}{2}B_1(x, x, \alpha) + \tfrac{1}{6}C_1(x, x, x, \alpha) \\ & + \tfrac{1}{24}D_1(x, x, x, x, \alpha) + \tfrac{1}{120}E_1(x, x, x, x, x, \alpha) \\ & + \tfrac{1}{2}A_2(x, \alpha, \alpha) + \tfrac{1}{4}B_2(x, x, \alpha, \alpha) + \tfrac{1}{12}C_2(x, x, x, \alpha, \alpha) \\ & + \cdots, \end{aligned} \tag{3.3}$$

where all functions are multilinear forms of their arguments and the dots denote higher-order terms in x and α. The components of the multilinear functions B and C were given in (2.4). Likewise, we have

$$J_2(x, y) = \sum_{j,k=1}^{p} \frac{\partial^2 g(x_0, \alpha_0)}{\partial \alpha_j \partial \alpha_k} x_j y_k, \quad B_1(x, y, z) = \sum_{l=1}^{p} \sum_{j,k=1}^{n} \frac{\partial^2 g(x_0, \alpha_0)}{\partial \xi_j \partial \xi_k \partial \alpha_l} x_j y_k z_l \tag{3.4}$$

and so on. In general, with the lower index ν of B_ν we denote the order of differentiation with respect to parameters.

The 11 codim 2 bifurcations of cycles appearing in generic two-parameter families of maps are listed in Table 3.1.

Table 3.1 Generic codim 2 bifurcations of cycles. Abbreviations are used as labels in the figures and in MatcontM. Coefficients b_0, c_0 and L_1 are defined by formulas (2.7), (2.12) and (2.18).

	Label	Name	Bifurcation conditions
1	CP	Cusp	$\lambda_1 = 1,\ b_0 = 0$
2	GPD	Generalized period-doubling	$\lambda_1 = -1,\ c_0 = 0$
3	CH	Chenciner bifurcation	$\lambda_{1,2} = e^{\pm i\theta_0},\ L_1 = 0$
4	R1	1:1 resonance	$\lambda_1 = \lambda_2 = 1$
5	R2	1:2 resonance	$\lambda_1 = \lambda_2 = -1$
6	R3	1:3 resonance	$\lambda_{1,2} = e^{\pm i\theta_0},\ \theta_0 = \frac{2\pi}{3}$
7	R4	1:4 resonance	$\lambda_{1,2} = e^{\pm i\theta_0},\ \theta_0 = \frac{\pi}{2}$
8	LPPD	Fold–flip	$\lambda_1 = 1,\ \lambda_2 = -1$
9	LPNS	Fold–NS	$\lambda_1 = 1,\ \lambda_{2,3} = e^{\pm i\theta_0}$
10	PDNS	Flip–NS	$\lambda_1 = -1, \lambda_{2,3} = e^{\pm i\theta_0}$
11	NSNS	Double NS	$\lambda_{1,2} = e^{\pm i\theta_0},\ \lambda_{3,4} = e^{\pm i\theta_1}$

This chapter is organized as follows. We give normal forms to which the restriction of a generic map $g(x, \alpha) = f^{(K)}(x, \alpha)$ to the parameter-dependent center manifold can be transformed near the corresponding bifurcation by smooth invertible coordinate and parameter transformations. In Chapter 4 we deal with center manifold reduction to obtain explicit expressions for all normal form coefficients. The qualitative picture is determined by the lowest-order terms as in the normal forms. In cases 3, 9, 10 and 11, it is always assumed that $\lambda^k \neq 1$ for at least $k = 1, 2, 3, 4$. A historical overview of the progress is given at the end of this chapter. To facilitate reading, all proofs are confined to the Appendices.

3.1 Cusp and generalized period-doubling bifurcations

3.1.1 CP (cusp)

Lemma 3.1 *The restriction of* (3.2) *to the one-dimensional center manifold at* α_0 *corresponding to a cusp bifurcation can be written in the form*

$$\xi \mapsto \xi + c_0\xi^3 + O(\xi^4), \quad \xi \in \mathbb{R}. \tag{3.5}$$

The lemma follows from the definition; the expression for c_0 will be given in Chapter 4.

Theorem 3.2 *If* $c_0 \neq 0$, *then for parameter values* α *close to* α_0, *the restriction of* (3.2) *to a parameter-dependent center manifold is locally smoothly equivalent to*

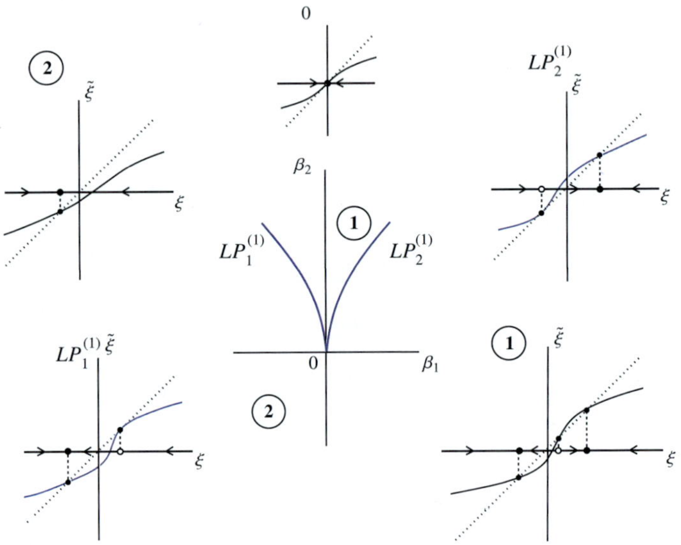

Figure 3.1 Bifurcation diagram of the normal form (3.8) with $\sigma = -1$.

$$\xi \mapsto \beta_1(\alpha) + \xi + \beta_2(\alpha)\xi^2 + c(\alpha)\xi^3 + O(\xi^4), \quad \xi \in \mathbb{R}, \tag{3.6}$$

where $\beta_1(\alpha_0) = \beta_2(\alpha_0) = 0$, $c(\alpha_0) = c_0$, *and the O-terms may also smoothly depend on* α.

Proof See Appendix 3.A. □

Notice that the map (3.6) is monotone near $\xi = 0$ for all sufficiently small $\|\alpha\|$. Next, we suppose that the map $\alpha \mapsto \beta(\alpha) = (\beta_1(\alpha), \beta_2(\alpha))$ is regular at α_0, so that the components of β can be used as the new unfolding parameters.

Theorem 3.3 *The smooth family of smooth maps*

$$\xi \mapsto \beta_1 + \xi + \beta_2\xi^2 + c(\beta)\xi^3 + O(\xi^4), \quad \xi \in \mathbb{R}, \tag{3.7}$$

where $c(0) = c_0 \neq 0$, *is locally topologically equivalent to*

$$\xi \mapsto \beta_1 + \xi + \beta_2\xi^2 + \sigma\xi^3, \quad \xi \in \mathbb{R}, \tag{3.8}$$

where $\sigma = \text{sign}(c_0) = \pm 1$.

Proof See Gheiner (2014). □

The *topological normal form* (3.8) for CP has two fold curves $LP_{1,2}^{(1)}$ in the parameter plane which form a cuspidal wedge

$$LP_1^{(1)} \cup LP_2^{(1)} \cup 0 = \left\{(\beta_1, \beta_2) \in \mathbb{R}^2 : 4\sigma\beta_2^3 + 27\beta_1^2 = 0\right\}.$$

The bifurcation diagram of (3.8) for $\sigma = -1$ is presented in Figure 3.1.

For nearby parameter values, the map g locally has up to three fixed points that pairwise collide along the fold curves, and all merge at the codim 2 point. In the direct product of the state and parameter spaces, the fold curve is smooth and unique, so no branch switching is needed.

3.1.2 GPD (generalized period-doubling)

Lemma 3.4 *The restriction of* (3.2) *to the one-dimensional center manifold at* α_0 *corresponding to a generalized period-doubling bifurcation is locally smoothly equivalent to*

$$\xi \mapsto -\xi + d_0\,\xi^5 + O(\xi^6), \quad \xi \in \mathbb{R}. \tag{3.9}$$

Proof See Appendix 3.A. □

The expression for d_0 in terms of map g from (3.2) will be given in Chapter 4.

Theorem 3.5 *For parameter values* α *close to* α_0 *defined in Lemma 3.4, the restriction of* (3.2) *to a parameter-dependent center manifold is locally smoothly equivalent to*

$$\xi \mapsto -(1 + \beta_1(\alpha))\xi + \beta_2(\alpha)\xi^3 + d(\alpha)\xi^5 + O(\xi^6), \quad \xi \in \mathbb{R}, \tag{3.10}$$

where $\beta_1(\alpha_0) = \beta_2(\alpha_0) = 0$, $d(\alpha_0) = d_0$, *and the O-terms also depend smoothly on* α.

Proof See Appendix 3.A. □

Suppose that the map $\alpha \mapsto \beta(\alpha) = (\beta_1(\alpha), \beta_2(\alpha))$ is regular at α_0, so that the components of β can be used as the new unfolding parameters. Generically, we have $d_0 \neq 0$.

Theorem 3.6 *The smooth family of smooth maps*

$$\xi \mapsto -(1 + \beta_1)\xi + \beta_2\xi^3 + d(\beta)\xi^5 + O(\xi^6), \quad \xi \in \mathbb{R}, \tag{3.11}$$

where $d(0) = d_0 \neq 0$, *is locally topologically equivalent to*

$$\xi \mapsto -(1 + \beta_1)\xi + \beta_2\xi^3 + \sigma\xi^5, \quad \xi \in \mathbb{R}, \tag{3.12}$$

where $\sigma = \operatorname{sign}(d_0) = \pm 1$.

Proof See Peckham and Kevrekidis (1991). □

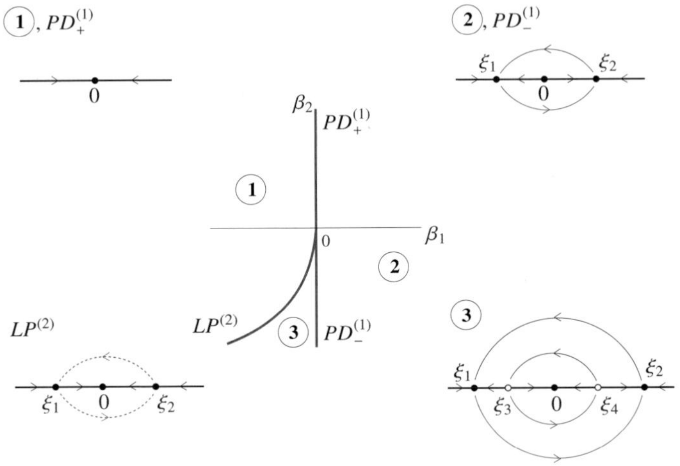

Figure 3.2 Bifurcation diagram of the normal form (3.12) with $\sigma = 1$.

The bifurcation diagram of the *topological normal form* (3.12) for the GPD bifurcation with $\sigma = 1$ is presented in Figure 3.2. The fixed point $w = 0$ of the map (3.12) exhibits a flip bifurcation along the line $PD^{(1)}$ where $\beta_1 = 0$. It is well known that from the point $\beta = 0$, corresponding to the generalized period-doubling bifurcation, a fold curve of double-period cycles emanates. In the parameter plane it is given by

$$LP^{(2)} = \left\{ \beta = (\beta_1, \beta_2) \in \mathbb{R}^2 : \beta_1 = -\frac{1}{4}\beta_2^2, \ \ \sigma\beta_2 < 0 \right\},$$

while an asymptotic expression for this curve in (3.11) is

$$(\xi, \beta_1, \beta_2) = \left(\varepsilon, -c_2^0 \varepsilon^4 + O(\varepsilon^5), -2c_2^0 \varepsilon^2 + O(\varepsilon^3) \right). \tag{3.13}$$

3.2 CH (Chenciner bifurcation)

3.2.1 Normal forms

Lemma 3.7 *If* $e^{ik\theta_0} \neq 1$ *for* $k = 1, 2, \ldots, 6$, *then the restriction of* (3.2) *to the two-dimensional center manifold at* α_0 *corresponding to a Chenciner bifurcation is locally smoothly equivalent to*

$$z \mapsto e^{i\theta_0} z + c_1^0 z|z|^2 + c_2^0 z|z|^4 + O(|z|^6), \quad z \in \mathbb{C}, \tag{3.14}$$

where the first Lyapunov coefficient vanishes, i.e., $L_1 = \Re(e^{-i\theta_0} c_1^0) = 0$.

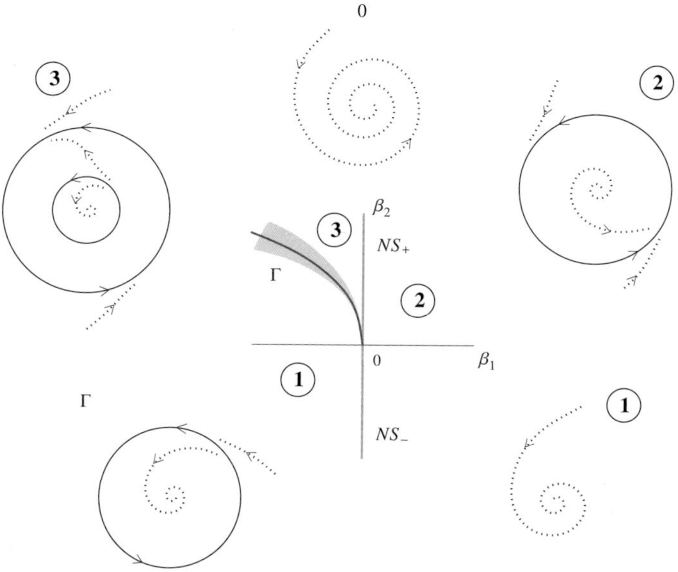

Figure 3.3 The primary bifurcation diagram in the vicinity of a Chenciner bifurcation for $L_2 < 0$.

Proof See Appendix 3.B. □

The expressions for c_1^0 and c_2^0 will be given in Chapter 4. The quantity

$$L_2 := \Re(e^{-i\theta_0}c_2^0) + \frac{1}{2}\left[\Im(e^{-i\theta_0}c_1^0)\right]^2$$

is called the *second Lyapunov coefficient*.

Theorem 3.8 *For parameter values α close to α_0 defined in* Lemma 3.7, *the restriction of* (3.2) *to a parameter-dependent center manifold is locally smoothly equivalent to*

$$z \mapsto e^{i\theta(\alpha)}(1+\beta_1(\alpha))z + (\beta_2(\alpha) + i\Im(c_1(\alpha)))z|z|^2 + c_2(\alpha)z|z|^4 + O(|z|^6), \quad z \in \mathbb{C}, \tag{3.15}$$

where $\beta_1(0) = \beta_2(0) = 0$, $\theta(0) = \theta_0$, and $c_j(0) = c_j^0$, $j = 1, 2$.

Proof The formula (3.15) emerges if one keeps the displayed resonant terms not only at α_0 but for all α close to α_0. □

If the mapping $\alpha \mapsto \beta(\alpha)$ is regular at $\alpha = 0$, we can use $\beta = (\beta_1, \beta_2)$ as the new unfolding parameters. Introducing polar coordinates $z = \rho e^{i\psi}$, we can approximate the radial dynamics by the following map

$$\rho \mapsto \rho(1 + \beta_1 + \beta_2\rho^2 + L_2\rho^4). \tag{3.16}$$

Assuming that $L_2 \neq 0$, this map (3.16) exhibits a fold bifurcation along the curve

$$\Gamma := \left\{\beta \in \mathbb{R}^2 : \beta_1 = \frac{\beta_2^2}{4L_2},\ \beta_2 > 0\right\}.$$

The salient aspects of the whole bifurcation diagram are shown in Figure 3.3, where we assume that $\theta/2\pi$ is sufficiently irrational. For $\beta_1 = 0$, the map (3.15) has a Neimark–Sacker (NS) bifurcation which is supercritical for $\beta_2 < 0$, and subcritical for $\beta_2 > 0$. Combining this with the fold bifurcation of the truncated radial map, one can expect that a generic two-parameter unfolding of this singularity entails a complicated bifurcation set where two closed invariant curves of different stability, born via sub- and supercritical NS bifurcations, respectively, "collide." This phenomenon is also called a *quasi-periodic saddle-node bifurcation* (see Section 2.2.2). There are no small-period cycle bifurcation curves rooted at this bifurcation, although there may be resonance tongues for nearby parameter values. These have an effect on the details of the bifurcation set, see below.

3.2.2 Effect of higher-order terms

Above we have been careful not to make any statement about the bifurcation structure along the curve Γ. To do so requires us to account for the angular dynamics. We assume that the rotation number ρ of the invariant curve along Γ varies, i.e., $\frac{\partial\rho}{\partial\beta_2} \neq 0$. If ρ is irrational it follows immediately that along Γ the collision of the two invariant curves happens according to the model map (2.33). This leads to a string of holes in which the dynamics is more subtle due to resonances. Here we will sketch what can be said about the dynamics for parameter values in the holes using analytical techniques and a numerical investigation. While we follow the steps of Chenciner (1988), we give a derivation for a specific case only.

To start, we adapt (3.15) as follows

$$\begin{aligned} z \quad \mapsto \quad & e^{i\theta}((1 + \beta_1)z + (\beta_2 + id_1)z|z|^2 + c_2 z|z|^4) \\ & + c_3\bar{z}^{q-1} + c_4 z|z|^{q-1} + c_5 z^{q+1} + c_6 z\bar{z}^q + c_7 z^{2q-1}, \end{aligned} \tag{3.17}$$

where $p, q \in \mathbb{N}$, $q > 6$, $\theta = 2\pi p/q + \tau + a\beta_2$ and $d_i = \Im(c_i^0)$ for short. The additional parameter τ controls the detuning of the rotation number, while a, γ are some parameters. The additional higher-order terms signify the presence of the resonance.

For $c_3 = 0$, there is a single curve $R_{p/q}$ in the parameter plane for which $r_0(\beta)$ has rotation number $2\pi/q$. Again switching to polar coordinates, we now find

$$\begin{pmatrix} r \\ \psi \end{pmatrix} \mapsto \begin{pmatrix} f(r,\psi) \\ g(r,\psi) \end{pmatrix} = \begin{pmatrix} r(1+\beta_1+\beta_2 r^2 + L_2 r^4) + O(r^6) \\ \psi + 2\pi p/q + \gamma\tau + a\beta_2 + d_1 r^2 + d_2 r^4 + O(r^5) \end{pmatrix}. \tag{3.18}$$

Then the conditions $(r, 2\pi p/q) = (f(r,\psi), g(r,\psi))$ give the following lowest-order approximation

$$\begin{aligned} rlR_{p/q}\colon\ (\beta_1, r^2) = \Bigg(& -\frac{\gamma\tau}{d_1}\left(\beta_2 + \frac{\gamma\tau}{a}\right) - \frac{a(L_2 d_1 a - d_1^2 + d_2\gamma\tau)}{d_1^3}\left(\beta_2 + \frac{\gamma\tau}{a}\right)^2, \\ & -\frac{a}{d_1}\left(\beta_2 + \frac{\gamma\tau}{a}\right) - \frac{d_2 a^2}{d_1^3}\left(\beta_2 + \frac{\gamma\tau}{a}\right)^2 \Bigg) + O\left(\beta_2 + \frac{\gamma\tau}{a}\right)^3. \end{aligned} \tag{3.19}$$

The first term indicates that the slope depends on the imaginary part similar to Lemma 2.13(2). Now combining $R_{p/q}$ with Γ we find that these curves have a quadratic tangency at the following point:

$$R_{p/q}\colon\quad (\beta_1^*, \beta_2^*, r^*) = \left(L_2\frac{\tau^2}{\gamma^2} + \frac{2L_2 d_2\tau^3}{\gamma^4}, -2L_2\frac{\tau}{\gamma} - \frac{2L_2 d_2 \tau^2}{\gamma^3}, \frac{\tau}{\gamma} + \frac{d_2\tau^2}{\gamma^3} \right), \tag{3.20}$$

where we have chosen $\gamma = (2aL_2 - d_1)$ to simplify expressions. We have plotted these curves and intersection points in Figure 3.4. One can show that this quadratic tangency persists in general when accounting for higher-order terms. Near this intersection point we find an interaction of the weak p:q resonance with the quasi-periodic saddle-node bifurcation. For this we zoom in on a particular resonance tongue fixing $\theta_0 = 2\pi p/q + \gamma\tau$ for $c_3 \neq 0$.

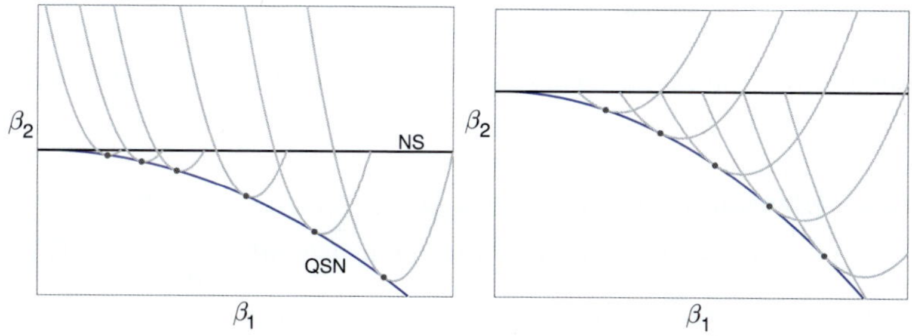

Figure 3.4 The curves R_p in the β_1, β_2-plane. They start from the Neimark–Sacker curve and then have a quadratic tangency with the QSN-curve. Depending on d_1^0, the curves bend to the left or the right.

We will now present a vector field model describing the dynamics within a resonance bubble. We outline a derivation for the case $p = 1, q = 7$ in Appendix 3.B. We can argue, though, that the non-degeneracy conditions show that this is the general model. The approximating vector field describes the dynamics on an annulus near two periodic orbits traversing the entire annulus. These two periodic orbits exhibit a saddle-node bifurcation corresponding to the curve QSN. Due to the resonance, the saddle-node bifurcation is interrupted by a complex diagram. This vector field is given in the following lemma.

Lemma 3.9 *In the vicinity of the parameter values (β_1^*, β_2^*) the dynamics of* (3.17) *on an annulus near r^* is approximated by the time-1 flow of the vector field*

$$\begin{pmatrix} \dot{\psi} \\ \dot{x} \end{pmatrix} = \begin{pmatrix} x \\ \mu_1 + \mu_2 x + \gamma x^2 + \delta_1 \cos(\psi) + x(\delta_2 \cos(\psi) + \delta_3 \sin(\psi) + \delta_4 \sin(2\psi)) \end{pmatrix}. \tag{3.21}$$

Proof See Appendix 3.B. □

The derivation involves ignoring any higher-order term, i.e., setting $\Xi = 0$ and $\tau = 1$ in (3.108), as well as rescaling time and redefining coefficients. This vector field with $\delta_2 = \delta_3 = \delta_4 = 0$ was proposed by Chenciner, Gasull, and Llibre (1987) as a model vector field. The additional harmonics present the lowest-order relevant Fourier expansion of the linear coefficient of x. It is reasonable to assume the boundary of the resonance tongue to be smooth, i.e., δ_1 is nonzero. We have singled out one additional higher-order term, a subharmonic, which is required to prove that the bifurcation diagram of this approximating vector field is generic. For the local bifurcations of (3.21) we have the following lemma.

Lemma 3.10 *The planar ODE* (3.21) *exhibits the following codim 1 bifurcations:*

(1) *fold bifurcations at two curves* $F^+\colon (x, \psi, \mu_1) = (0, 0, -\delta_1)$ *and* $F^-\colon (x, \psi, \mu_1) = (0, \pi, \delta_1)$, *which are non-degenerate if* $\dfrac{\delta_1}{\mu_2 \pm \delta_2} \neq 0$.

(2) *a Hopf bifurcation at the curve*

$$H\colon (x, \mu_1, \mu_2) = (0, -\delta_1 \cos\psi, -\delta_2 \cos\psi - \delta_3 \sin\psi - \delta_4 \sin 2\psi),$$

when $\delta_1 \sin\psi > 0$, *while the other part corresponds to a neutral saddle.*

The first Lyapunov coefficient is

$$\begin{aligned} l_1 = & \frac{2}{\sin\psi}\left((\gamma(\delta_3 \sin 2\psi + \delta_2 \cos 2\psi) - \gamma\delta_2 - \delta_3)\right. \\ & \left. + \delta_4\left(2\gamma(\sin 3\psi - \sin\psi) + (\cos 3\psi - 3\cos\psi)\right)\right). \end{aligned} \tag{3.22}$$

In addition, the ODE (3.21) *has the following local codim 2 bifurcations*:

(3) *Two Bogdanov–Takens points* $BT_\pm$ *at* $(\mu_1, \mu_2) = \pm(\delta_1, \delta_2)$. *The critical normal form coefficients are* $a_0 = \mp\delta_1$ *and* $b_0 = 2\delta_4 \pm \delta_3$.
(4) *One or two generalized Hopf (Bautin) bifurcation(s) GH in the Hopf bifurcation curve whenever* $l_1 = 0$. *Depending on the coefficients* δ_i *and* γ, *the second Lyapunov coefficient may be zero.*

Proof See Appendix 3.B. □

At this point we can make several remarks. First, the Hopf bifurcation curve in the plane is given by the following quartic curve $(\delta_1\delta_3 - 2\delta_4\mu_1)^2(\delta_1^2 - \mu_1) - \delta_1^2(\delta_1\mu_2 - \delta_2\mu_1)^2 = 0$. If $|2\delta_4| > \delta_3$, the corresponding curve is a "figure of eight." Otherwise it is an ellipse. Second, if the coefficient b for one BT point is zero, then the normal form coefficient b for the other BT point is not zero when $\delta_4 \neq 0$. This shows the genericity of the vector field. If $\delta_4 = 0$, there are either zero or two degenerate Hopf bifurcation points. This follows from the first Lyapunov coefficient l_1, which then only has second harmonics in the numerator.

Another issue concerns global bifurcations. One may observe that (3.21) is reminiscent of a pendulum equation. This hints at homoclinic and heteroclinic orbits. Setting $\mu_2 = \delta_2 = \delta_3 = \delta_4 = 0$, one readily checks the vector field (3.21) to be conservative. Indeed, following Chenciner, Gasull, and Llibre (1987), one can derive the following conserved quantity

$$H(\phi, x) = \left(x^2 + \frac{\mu_1}{\gamma} + \frac{\delta_1}{4\gamma^2 + 1}(4\gamma\cos\phi - 2\sin\phi)\right)e^{-2\gamma\phi}.$$

Along $\beta = 0$ we now find several characteristic phase portraits. Decreasing μ_1 we first have two limit cycles. Then for $\mu_1 < 1$ we find two equilibria, one center and one saddle. Decreasing μ_1 further to $\mu_1 \approx -0.38$ (for $\delta_1 = 1, \gamma = 0.2$), we encounter a heteroclinic cycle N. Here, the periodic orbits and the homoclinic orbit disappear into a heteroclinic connection of the saddle with itself. Below N, we have no periodic orbits except the family around the center. And finally for $\mu < -1$, the equilibria have disappeared too, and all invariant sets are gone. The heteroclinic cycle may appear rather degenerate as it comes from the conservative approximation, yet the global bifurcations persist in the dissipative setting.

We have described all local bifurcations, and cannot achieve any more analytically. With the goal of describing global bifurcations, we turn to a numerical inventory of the dynamics. We set $\delta_1 = 1, \delta_2 = 0.2, \delta_4 = -0.5$ and consider δ_3 as an additional deformation parameter. We choose several values for δ_3 such that the bifurcation diagram in the (μ_1, μ_2)-plane exhibits zero, one or two degenerate Hopf bifurcations. The bifurcation diagrams for the vector field approximation are shown in Figure 3.5. The essence is that as μ_1 is increased, the global invariant curve typically disappears in one of the fold of cycles bifurcations. For a small set of parameters a different scenario is encountered as the equilibria may have homoclinic and heteroclinic orbits that the global invariant curve may merge with as parameters are varied. Note that this happens when μ_2 is sufficiently small where the Hopf curve and the global bifurcations occur. Meanwhile, the equilibria corresponding to the q-periodic fixed points exist for $|\mu_1| \leq 1$. As we vary δ_3, we notice that the Hopf part (blue solid) and the neutral saddle part of the curve (blue dashed) exchange position. During this transition we find that the Bogdanov–Takens points become degenerate and two additional Bautin points appear, which eventually merge and disappear.

At this point we have collected all information we can gain from the vector field approximation. To finish our treatment of the bifurcation near the resonance bubble, we interpret the various invariant sets and bifurcation for the original map. Equilibria turn into q-periodic cycles and periodic orbits become closed invariant curves. All local bifurcations are similar too with their discrete time equivalents. Note it is remarkable that within the unfolding of the Chenciner point one or more Chenciner points can reappear. The fold bifurcation of the "big" periodic orbits is the quasi-periodic saddle-node bifurcation of invariant curves. Clearly, it is interrupted by the bubble. Homoclinic and heteroclinic orbits have their counterparts for maps too. They will persist, though under small parameter variations, within wedges of homoclinic and heteroclinic tangencies. Typically they are too small to be observed, but in the vicinity of the heteroclinic cycle there may be chaos if the heteroclinic and homoclinic wedges overlap.

Note that our bifurcation diagrams cover the cases of Baesens and MacKay (2007) and Broer et al. (2011). We have noticed for bubbles in other examples that the point N might lie outside the Hopf-neutral saddle curve. It is unclear which parts of the model map are important for this, but it suggests a few more unfoldings. In our treatment we focused on a single p:q-resonance horn, but in real applications horns of nearby periods may play a role too.

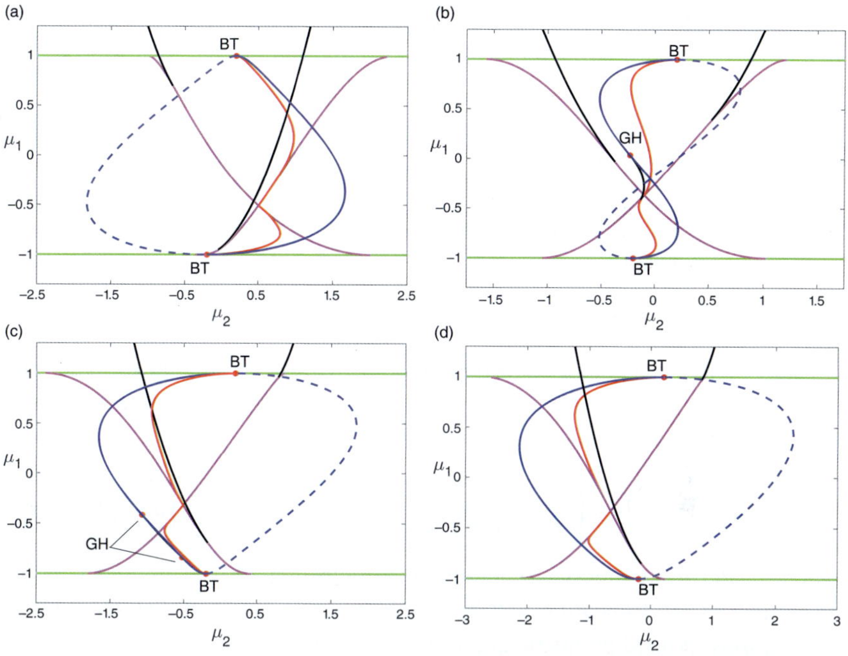

Figure 3.5 Bifurcation diagrams of (3.21) for $\delta_3 = -1.5$ (a), $\delta_3 = 0.2$ (b), $\delta_3 = 1.5$ (c), and $\delta_3 = 2.0$ (d). The resonance tongue is demarcated by the fold curves (green). Hopf/neutral saddle bifurcation curves start at the Bogdanov–Takens (BT) points as well as homoclinic bifurcation curves (red). The latter terminate at the heteroclinic cycle point, given by the intersection of the heteroclinic bifurcation curves (magenta). From a generalized Hopf bifurcation (GH) a limit point of cycles *LPC* curve (black) starts. There are also *LPC*s corresponding to the global periodic orbits coming in from the top of the figure. Theoretically they end at the intersection of the neutral saddle and heteroclinic curves. Above the resonance tongue the global invariant curves do not exist between the two *LPC*s.

3.3 Strong resonances

3.3.1 R1 (resonance 1:1)

Lemma 3.11 *The restriction of the map* (3.2) *to the two-dimensional center manifold at* α_0 *corresponding to a* 1:1 *resonance is locally smoothly equivalent to*

$$\begin{pmatrix} \xi_1 \\ \xi_2 \end{pmatrix} \mapsto \begin{pmatrix} \xi_1 + \xi_2 \\ \xi_2 + a_0\xi_1^2 + b_0\xi_1\xi_2 \end{pmatrix} + O(\|\xi\|^3), \quad \xi \in \mathbb{R}^2. \tag{3.23}$$

Proof See (Kuznetsov, 2004, lemma 9.7, pp. 427–428) for the critical parameter values. □

The expressions for a_0 and b_0 in terms of the map (3.2) will be given in Chapter 4.

Theorem 3.12 *If $a_0 \neq 0$, then for parameter values α close to α_0, the restriction of* (3.2) *to a parameter-dependent center manifold is locally smoothly equivalent near $\xi = 0$ to*

$$\begin{pmatrix} \xi_1 \\ \xi_2 \end{pmatrix} \mapsto \begin{pmatrix} \xi_1 + \xi_2 \\ \xi_2 + \beta_1(\alpha) + \beta_2(\alpha)\xi_2 + A(\alpha)\xi_1^2 + B(\alpha)\xi_1\xi_2 \end{pmatrix} + O(\|\xi\|^3), \quad (3.24)$$

where $\beta_1(\alpha_0) = \beta_2(\alpha_0) = 0$, $A(\alpha_0) = a_0$, $B(\alpha_0) = b_0$, and the O-terms may also smoothly depend on α.

Proof See (Kuznetsov, 2004, lemma 9.7, pp. 427–428). □

Suppose that the map $\alpha \mapsto \beta(\alpha) = (\beta_1(\alpha), \beta_2(\alpha))$ is regular at α_0. Then the components of β can be used as the new unfolding parameters and we can first study the *truncated normal form*

$$N_\beta : \begin{pmatrix} \xi_1 \\ \xi_2 \end{pmatrix} \mapsto \begin{pmatrix} \xi_1 + \xi_2 \\ \xi_2 + \beta_1 + \beta_2\xi_2 + a(\beta)\xi_1^2 + b(\beta)\xi_1\xi_2 \end{pmatrix}, \quad (3.25)$$

where a and b are smooth functions of β such that $a(0) = a_0$, $b(0) = b_0$. The effect of the truncation will be discussed later. One can directly analyze bifurcations of fixed points of (3.25).

Lemma 3.13 *Suppose that $s := 2a_0(b_0 - 2a_0) \neq 0$. Then the truncated map* (3.25) *exhibits a non-degenerate fold bifurcation at the curve*

$$LP^{(1)} = \{(\beta_1, \beta_2) \in \mathbb{R}^2 : \beta_1 = 0,\ \beta_2 \neq 0\}, \quad (3.26)$$

while a non-degenerate NS bifurcation happens at the curve with the following asymptotic expansion

$$NS^{(1)} = \{(\beta_1, \beta_2) \in \mathbb{R}^2 : \beta_1 = -\frac{a_0}{(b_0 - 2a_0)^2}\beta_2^2 + O(\beta_2^3),\ s\beta_2 > 0\}. \quad (3.27)$$

Proof See Appendix 3.C. □

Since global bifurcations are involved, it is convenient to approximate the map N_β by the unit shift along orbits of an auxiliary planar ODE.

Theorem 3.14 *For all sufficiently small $\|\beta\|$, the map* (3.25) *can be represented near $\xi = 0$ as*

$$N_\beta(\xi) = \varphi_\beta^1(\xi) + O\left(\|(\xi, \beta)\|^3\right), \quad (3.28)$$

where φ_β^t is the flow of the planar system

$$\dot{u} = \begin{pmatrix} 0 & 1 \\ 0 & 0 \end{pmatrix} u + \begin{pmatrix} -\frac{1}{2}\beta_1 \\ \beta_1 \end{pmatrix} + F_\beta^{(2)}(u), \quad (3.29)$$

where $u = (u_0, u_1) \in \mathbb{R}^2$ *and*

$$F_\beta^{(2)}(u) = \begin{pmatrix} a_{00}(\beta) \\ b_{00}(\beta) \end{pmatrix} + \begin{pmatrix} a_{10}(\beta)u_0 + a_{01}(\beta)u_1 \\ b_{10}(\beta)u_0 + b_{01}(\beta)u_1 \end{pmatrix} + \begin{pmatrix} \frac{1}{2}a_{20}u_0^2 + a_{11}u_0u_1 + \frac{1}{2}a_{02}u_1^2 \\ \frac{1}{2}b_{20}u_0^2 + b_{11}u_0u_1 + \frac{1}{2}b_{02}u_1^2 \end{pmatrix}$$

with

$$\begin{aligned}
a_{00}(\beta) &= \tfrac{1}{20}\,(2b_0 - a_0)\beta_1^2 + \tfrac{1}{3}\beta_1\beta_2, & b_{00}(\beta) &= \left(\tfrac{1}{30}a_0 - \tfrac{1}{12}b_0\right)\beta_1^2 - \tfrac{1}{2}\beta_1\beta_2,\\
a_{10}(\beta) &= \left(\tfrac{1}{3}b_0 - \tfrac{1}{2}a_0\right)\beta_1, & b_{10}(\beta) &= \left(\tfrac{2}{3}a_0 - \tfrac{1}{2}b_0\right)\beta_1,\\
a_{01}(\beta) &= \left(\tfrac{1}{5}a_0 - \tfrac{5}{12}b_0\right)\beta_1 - \tfrac{1}{2}\beta_2, & b_{01}(\beta) &= \left(\tfrac{1}{2}b_0 - \tfrac{1}{6}a_0\right)\beta_1 + \beta_2,\\
a_{20} &= -a_0, & b_{20} &= 2a_0,\\
a_{11} &= \tfrac{2}{3}a_0 - \tfrac{1}{2}b_0, & b_{11} &= b_0 - a_0,\\
a_{02} &= \tfrac{2}{3}b_0 - \tfrac{1}{3}a_0, & b_{02} &= \tfrac{1}{3}a_0 - b_0.
\end{aligned}$$

Proof: See (Kuznetsov, 2004, lemma 9.8, pp. 429–432) for the linear with respect to parameter terms; the quadratic in (β_1, β_2) terms were first obtained by Al-Hdaibat et al. (2018) using the same approach. □

The planar system (3.29) exhibits a Bogdanov–Takens (BT) bifurcation at $\beta = 0$. Indeed, for $\beta = 0$ the system reduces to

$$\dot{u} = \begin{pmatrix} 0 & 1 \\ 0 & 0 \end{pmatrix} u + \begin{pmatrix} \frac{1}{2}a_{20}u_0^2 + a_{11}u_0u_1 + \frac{1}{2}a_{02}u_1^2 \\ \frac{1}{2}b_{20}u_0^2 + b_{11}u_0u_1 + \frac{1}{2}b_{02}u_1^2 \end{pmatrix}$$

and has a non-degenerate BT-singularity if

$$b_{20}(a_{20} + b_{11}) = 2a_0(-a_0 + b_0 - a_0) = 2a_0(b_0 - 2a_0) \neq 0.$$

Lemma 3.15 *Suppose that*

$$s := 2a_0(b_0 - 2a_0) \neq 0.$$

Then the bifurcation diagram of the approximating system (3.29) *near* $\beta = 0$ *includes only three bifurcation curves:*

(1) *the saddle-node curve*

$$S = \{\beta \in \mathbb{R}^2 : \beta_1 = 0\}$$

with two branches S_- *and* S_+ *corresponding to* $\beta_2 < 0$ *and* $\beta_2 > 0$ *respectively*;

(2) *the Andronov–Hopf curve*

$$H = \left\{\beta \in \mathbb{R}^2 : \beta_1 = -\frac{a_0}{(b_0 - 2a_0)^2}\beta_2^2 + O(\beta_2^3),\;\; s\beta_2 > 0\right\};$$

(3) *the saddle homoclinic curve*

$$P = \left\{\beta \in \mathbb{R}^2 : \beta_1 = -\frac{49}{25}\frac{a_0}{(b_0 - 2a_0)^2}\beta_2^2 + O(\beta_2^3),\;\; s\beta_2 > 0\right\}.$$

Moreover, all mentioned bifurcations are non-degenerate.

Proof See Appendix 3.C. □

We see that the equations for S and $LP^{(1)}$ coincide, while those for H and $NS^{(1)}$ agree at leading order (see (3.26) and (3.27) in Lemma 3.13). The local bifurcation diagram of the approximating planar ODE (3.29) is sketched in Figure 3.6.

Remark 3.16 Actually, as shown by Al-Hdaibat et al. (2018), a more accurate approximation of the homoclinic bifurcation curve P than used in the proof of Lemma 3.15 is given by

$$\begin{pmatrix} \beta_1 \\ \beta_2 \end{pmatrix} = \frac{\varepsilon^2}{a_0}\begin{pmatrix} 0 \\ \frac{10}{7}(b_0 - 2a_0) \end{pmatrix} - \frac{\varepsilon^4}{a_0}\begin{pmatrix} 4 \\ \delta \end{pmatrix} + O\left(\varepsilon^5\right), \quad 0 < \varepsilon \ll 1, \quad (3.30)$$

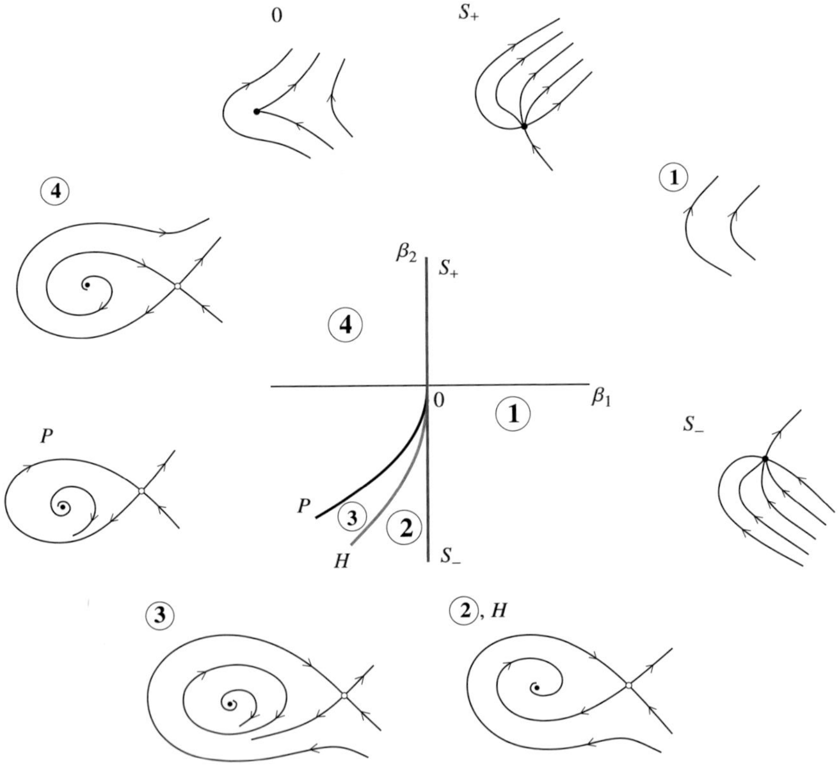

Figure 3.6 Bifurcation diagram of the approximating system (3.29) with $b_0 - 2a_0 < 0, a_0 > 0$.

where

$$\delta = \frac{(b_0 - 2a_0)\left(857a_0^2 - 3650a_0b_0 - 288b_0^2\right)}{2401a_0^2} + \frac{2a_0^2 - 5a_0b_0 + b_0^2}{a_0}. \quad (3.31)$$

The transition from the smooth normal form (3.24) to the approximating ODE (3.29) involves two steps:

(1) truncation of the $O(\|\xi\|^3)$-terms that results in the map N_β;
(2) neglecting the difference between N_β and the time-1 shift along orbits of the approximating planar ODE.

Some features of the approximating ODE have a direct counterpart for the original map, while for others it is not the case. Equilibria of the approximating planar ODE correspond to fixed points of the map N_β, and their bifurcations imply bifurcations of the fixed points. The saddle-node bifurcation corresponds to a fold bifurcation of the map (see Section 2.1.1), while the Andronov–Hopf bifurcation suggests an NS bifurcation (see Section 2.1.3).[1] If $s < 0$, the bifurcating invariant curve will be stable near the Neimark–Sacker bifurcation.

The single homoclinic bifurcation curve P for the ODE (3.29) is replaced by two bifurcation curves HT_1 and HT_2, along which the map N_β has a *homoclinic tangency*, i.e., the stable and unstable manifolds of the saddle have a quadratic contact at (infinitely many) non-transverse homoclinic points (see Figure 3.7). This codim 1 global bifurcation was discussed in Section 2.2.1. The curves HT_1 and HT_2 delimit an exponentially narrow parameter wedge near P, inside which the invariant manifolds of the saddle intersect transversally, forming a *Poincaré homoclinic structure*. As is shown in Section 2.2.1, crossing each of the curves $HT_{1,2}$ is accompanied by an infinite sequence of period-doubling and fold bifurcations. Moreover, the closed invariant curve born via the NS bifurcation loses its smoothness and disappears away from it. Thus, even the full bifurcation diagram of the truncated normal form N_β is complicated and involves global bifurcations near the codim 2 point $\beta = 0$. In the smooth normal form (3.24) with any $O(\|\xi\|^3)$-terms, this complicated bifurcation structure is also present, provided $s \neq 0$ and the map $\alpha \mapsto \beta$ is regular at α_0.

In a generic map g having this codim 2 bifurcation, an NS bifurcation curve of the fixed point meets tangentially the fold bifurcation curve at the 1:1 resonance point.[2] The full bifurcation diagram near the codim 2 point thus involves global bifurcations, i.e., tangencies of stable and unstable invariant manifolds

[1] This explains the choice of notations for the curves in Lemma 3.15.

[2] The local branch switching problem is trivial here, since both curves correspond to fixed points of g.

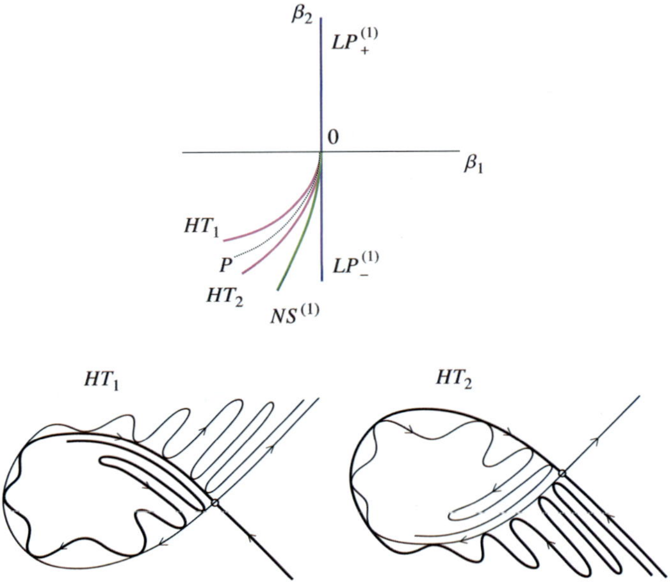

Figure 3.7 Homoclinic tangencies along HT_1 and HT_2 bifurcation curves.

of saddle fixed points of g and destruction of a closed invariant curve born via the NS bifurcation.

To demonstrate the structure we consider the following cubic perturbation of the normal form

$$\begin{pmatrix} x_1 \\ x_2 \end{pmatrix} \mapsto \begin{pmatrix} x_1 + x_2 \\ \beta_1 + x_2 + \beta_2 x_2 + a_0 x_1^2 + b_0 x_1 x_2 + \varepsilon_1 x_1^2 x_2 + \varepsilon_2 x_2^3 \end{pmatrix}, \tag{3.32}$$

where we choose $a_0 = 0.3, b_0 = 0.2, \varepsilon_1 = -0.22, \varepsilon_2 = 0.25$. With these values we only have attractors for parameters near the origin in the third quadrant. Scanning for attractors and their bifurcations results in Figure 3.8, which corroborates our above discussion.

3.3.2 R2 (resonance 1:2)

Theorem 3.17 *Let α_0 correspond to the resonance 1:2. Then for parameter values α close to α_0, the restriction of* (3.2) *to a parameter-dependent center manifold is locally smoothly equivalent near $\xi = 0$ to*

$$\begin{pmatrix} \xi_1 \\ \xi_2 \end{pmatrix} \mapsto \begin{pmatrix} -\xi_1 + \xi_2 \\ -\xi_2 + \beta_1(\alpha)\xi_1 + \beta_2(\alpha)\xi_2 + C(\alpha)\xi_1^3 + D(\alpha)\xi_1^2\xi_2 \end{pmatrix} + O(\|\xi\|^4), \tag{3.33}$$

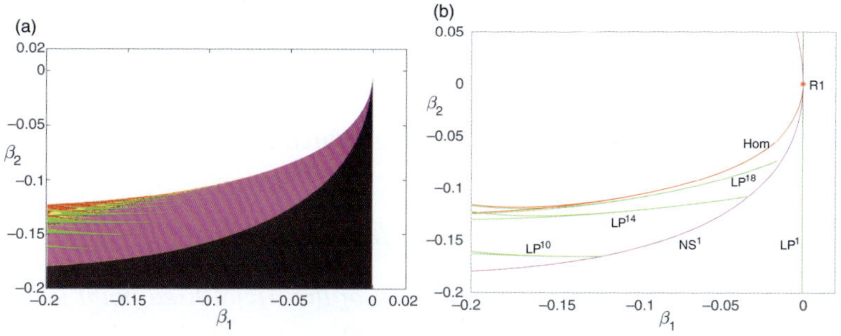

Figure 3.8 Bifurcation diagram of the perturbed normal form (3.32). (a) inventory of attractors based on Lyapunov exponents. Colors indicate stable fixed point (dark-gray), cycles of period up to 18 (green) and higher (yellow), invariant curve (magenta), chaos (red) and no attractor found (white). The latter transition is either via the fold curve LP^1 or the lower branch of the homoclinic tangency. (b) curves found with continuation. The primary fold LP^1 and Neimark–Sacker NS^1 curves meet at $\beta_1 = \beta_2 = 0$ at the 1:1 resonance point. The emerging homoclinic tangency bifurcation curve Hom has been sketched for $\beta_1 > -0.016$ (dashed). For lower values of β_1 it consists of two branches HT_1 and HT_2 of inner and outer tangencies. Resonance tongues of periods 10, 14 and 18 are shown. As β_1 decreases, the tongue boundaries approach either HT_1 or HT_2.

where $\beta_1(\alpha_0) = \beta_2(\alpha_0) = 0$ *and the* O*-terms may also smoothly depend on* α.

Proof: See (Kuznetsov, 2004, lemma 9.9, pp. 437–439). □

The expressions for $c_0 = C(\alpha_0)$ and $d_0 = D(\alpha_0)$ will be given in Chapter 4.

Suppose that the map $\alpha \mapsto \beta(\alpha) = (\beta_1(\alpha), \beta_2(\alpha))$ is regular at α_0. Then the components of β can be used as the new unfolding parameters. We first study the *truncated normal form*

$$N_\beta\colon \begin{pmatrix} \xi_1 \\ \xi_2 \end{pmatrix} \mapsto \begin{pmatrix} -\xi_1 + \xi_2 \\ -\xi_2 + \beta_1\xi_1 + \beta_2\xi_2 + c(\beta)\xi_1^3 + d(\beta)\xi_1^2\xi_2 \end{pmatrix}, \tag{3.34}$$

where c and d are smooth functions of β such that $c(0) = c_0,\ d(0) = d_0$. The effect of the truncation will be discussed later. Note that (3.34) is *invariant* under the reflection

$$x \mapsto Rx, \quad R = \begin{pmatrix} -1 & 0 \\ 0 & -1 \end{pmatrix}, \tag{3.35}$$

for which $R^2 = I_2$. The phase portraits that follow will reflect this $\mathbb{Z}_2$-symmetry.

Lemma 3.18 *Assume that*

$$c_0(d_0 + 3c_0) \neq 0.$$

Then the truncated map (3.34) *has the bifurcation curve*

$$PD^{(1)} = \{(\beta_1, \beta_2) \in \mathbb{R}^2 : \beta_1 = 0,\ \beta_2 \neq 0\},$$

at which a non-degenerate period-doubling bifurcation of the trivial fixed point occurs, as well as the bifurcation curve

$$NS^{(1)} = \{(\beta_1, \beta_2) \in \mathbb{R}^2 : \beta_2 = -\beta_1,\ \beta_1 < 0\},$$

at which a non-degenerate NS bifurcation of the trivial fixed point happens; the bifurcation is supercritical if $d_0 + 3c_0 > 0$.

Moreover, if $c_0 < 0$, *then there is a bifurcation curve*

$$NS^{(2)} = \left\{(\beta_1, \beta_2) \in \mathbb{R}^2 : \beta_2 = \left(2 + \frac{d_0}{c_0}\right)\beta_1 + O(\beta_1^2),\ \beta_1 > 0\right\},$$

at which the 2-cycle exhibits a non-degenerate NS bifurcation; the bifurcation is supercritical if $d_0 + 3c_0 < 0$.

Proof See Appendix 3.C. □

Remark 3.19 Having in mind the branch-switching problem, we point out that the $NS^{(2)}$ branch has the following asymptotic representation in the (ξ, β)-space:

$$(\xi_1^2, \xi_2, \beta_1, \beta_2) = \left(-\frac{1}{c_0}, 0, 1, \left(2 + \frac{d_0}{c_0}\right)\right)\varepsilon + O(\varepsilon^2),\quad \varepsilon > 0. \tag{3.36}$$

To study global bifurcations associated with the destruction of closed invariant curves, it is useful to approximate the map $R \circ N_\beta$ by the time-1 flow of a planar ODE.

Theorem 3.20 *For sufficiently small* $\|\beta\|$, *for the map* (3.34) *holds*

$$R\,N_\beta(\xi) = \varphi_\beta^1(\xi) + O(\|\xi\|^4),$$

where φ_β^t *is the flow of the planar system*

$$\dot{\xi} = \Lambda(\beta)\xi + U(\xi, \beta), \tag{3.37}$$

where

$$\Lambda(\beta) = \begin{pmatrix} -\frac{1}{2}\beta_1 & -1 - \frac{1}{3}\beta_1 - \frac{1}{2}\beta_2 \\ -\beta_1 & -\frac{1}{2}\beta_1 - \beta_2 \end{pmatrix} + O(\|\beta\|^2)$$

and

$$U(\xi, \beta) = \begin{pmatrix} A_{30}(\beta)\xi_1^3 + A_{21}(\beta)\xi_1^2\xi_2 + A_{12}(\beta)\xi_1\xi_2^2 + A_{03}(\beta)\xi_2^3 \\ B_{30}(\beta)\xi_1^3 + B_{21}(\beta)\xi_1^2\xi_2 + B_{12}(\beta)\xi_1\xi_2^2 + B_{03}(\beta)\xi_2^3 \end{pmatrix}$$

with

$$\begin{aligned}
A_{30}(0) &= -\tfrac{1}{2}c_0, & B_{30}(0) &= -c_0,\\
A_{21}(0) &= -c_0 - \tfrac{1}{2}d_0, & B_{21}(0) &= -\tfrac{3}{2}c_0 - d_0,\\
A_{12}(0) &= -\tfrac{1}{2}c_0 - \tfrac{2}{3}d_0, & B_{12}(0) &= -\tfrac{1}{2}c_0 - d_0,\\
A_{03}(0) &= -\tfrac{1}{30}c_0 - \tfrac{1}{6}d_0, & B_{03}(0) &= -\tfrac{1}{6}d_0.
\end{aligned}$$

Proof See Appendix 3.C. □

Notice that the approximating system (3.37) retains the $\mathbb{Z}_2$-symmetry of the truncated normal form (3.34).

Theorem 3.21 *Suppose that* $c_0(d_0 + 3c_0) \neq 0$.

(1) *The bifurcation diagram of the approximating system* (3.37) *near* $\beta = 0$ *includes the following curves related to bifurcations of equilibria, namely*:

(*a*) *the pitchfork bifurcation curve*

$$F = \{\beta \in \mathbb{R}^2 : \beta_1 = 0\}$$

with two branches F_- *and* F_+ *corresponding to* $\beta_2 < 0$ *and* $\beta_2 > 0$ *respectively*;

(*b*) *the Andronov–Hopf bifurcation curve for the trivial equilibrium*

$$H^{(1)} = \{(\beta_1, \beta_2) \in \mathbb{R}^2 : \beta_2 = -\beta_1,\ \beta_1 < 0\};$$

the bifurcation is supercritical if $d_0 + 3c_0 > 0$;

(*c*) *if* $c_0 < 0$, *the Andronov–Hopf bifurcation curve for nontrivial equilibria*

$$H^{(2)} = \left\{(\beta_1, \beta_2) \in \mathbb{R}^2 : \beta_2 = \left(2 + \frac{d_0}{c_0}\right)\beta_1 + O(\beta_1^2),\ \beta_1 > 0\right\};$$

the bifurcation is supercritical if $d_0 + 3c_0 < 0$.

(2) *The bifurcation diagram of* (3.37) *also includes global bifurcation curves*:

(*a*) *if* $c_0 > 0$, *there is a heteroclinic bifurcation curve*

$$C = \left\{(\beta_1, \beta_2) \in \mathbb{R}^2 : \beta_2 = \frac{d_0 - 2c_0}{5c_0}\beta_1 + o(\beta_1)\right\},$$

at which a heteroclinic contour with two saddles exists;

(*b*) *if* $c_0 < 0$, *there exist*

- *the homoclinic figure-of-eight bifurcation curve*

$$P = \left\{(\beta_1, \beta_2) \in \mathbb{R}^2 : \beta_2 = \frac{7c_0 + 4d_0}{5c_0}\beta_1 + o(\beta_1)\right\},$$

at which two orbits homoclinic to the trivial equilibrium exist;

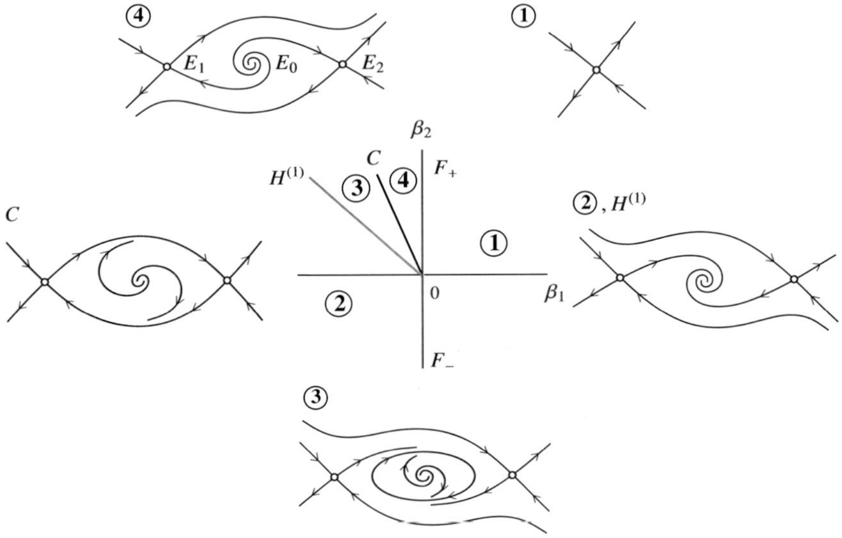

Figure 3.9 Bifurcation diagram of the approximating system (3.37) with $c_0 > 0$ and $d_0 + 3c_0 > 0$.

- *the cyclic fold bifurcation curve*

$$K = \left\{(\beta_1, \beta_2) \in \mathbb{R}^2 : \beta_2 = \left(\kappa_0 \left(3 + \frac{d_0}{c_0}\right) - 1\right)\beta_1 + o(\beta_1)\right\},$$

at which two limit cycles of opposite stability collide and disappear. Here $\kappa_0 = 0.7522\ldots$ *is approximated numerically.*

All mentioned global bifurcations are non-degenerate.

Proof See Appendix 3.C. □

The local bifurcation diagrams of the approximating system are presented in Figures 3.9 and 3.10. Similar to the 1:1 resonance, the transition from the smooth normal form (3.33) for the 1:2 resonance to the approximating ODE (3.37) involves two steps:

(1) truncation of the $O(\|\xi\|^4)$-terms in (3.33) that results in the map N_β given by (3.34);
(2) neglecting the difference between $R \circ N_\beta$ and the time-1 shift along orbits of the planar ODE (3.37).

To apply results about ODE (3.37) to the truncated normal form (3.34), we

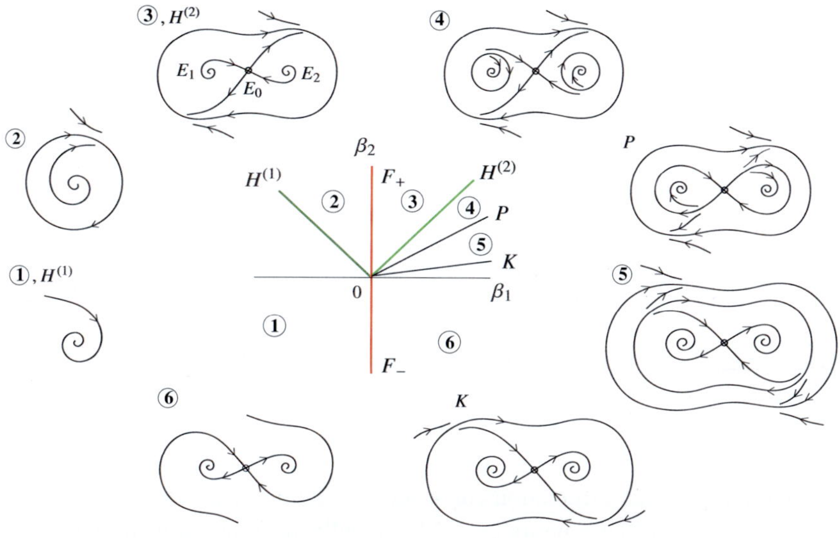

Figure 3.10 Bifurcation diagram of the approximating system (3.37) with $c_0 < 0$ and $d_0 + 3c_0 < 0$.

notice that the time-1 flow of the ODE approximates the map N_β composed with the inversion (3.35).

Thus, the trivial equilibrium of (3.37) corresponds to a fixed point of N_β, while two symmetric nontrivial fixed points actually correspond to a period-2 cycle. Therefore, the pitchfork bifurcation curve F in the ODE approximates the period-doubling bifurcation curve $PD^{(1)}$ of the map N_β where the period-2 cycle bifurcates from the trivial fixed point. The Hopf bifurcation curve $H^{(1)}$ approximates the NS bifurcation curve $NS^{(1)}$ of the trivial fixed point of N_β, while the Hopf bifurcation curve of the nontrivial equilibria approximates the NS bifurcation curve $NS^{(2)}$ of the period-2 cycle. A brief comparison confirms that the leading terms of the expansions for the corresponding curves in Lemma 3.18 and Theorem 3.21 coincide if $3c_0 + d_0 = 0$. Moreover, these expansions are valid for the original (non-truncated) normal form. The corresponding bifurcations in the approximating ODE and in the smooth normal form are non-degenerate if the formulated conditions are satisfied.

Effect of higher-order terms

The situation with global bifurcations in the approximating ODE (3.37) is more involved. In the (β_1,β_2)-plane of the truncated normal form N_β, complicated bifurcation sets exist near the curves C, P and K of the approximating ODE (3.37). Similar to the 1:1 resonance, exponentially narrow wedges appear

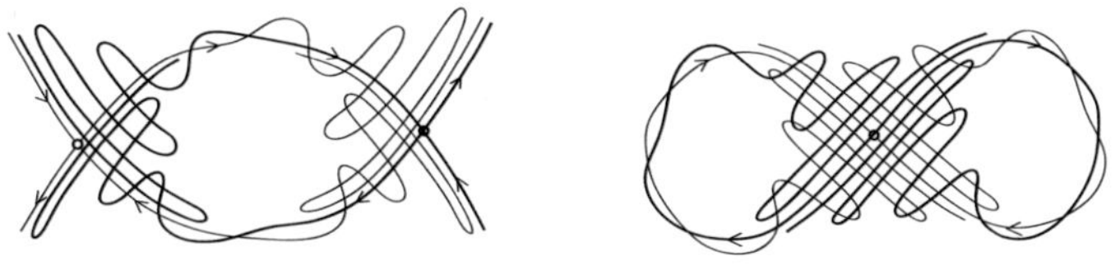

(a) (b)

Figure 3.11 Transversal intersections of stable (a) and unstable (b) manifolds in simple and complex unfoldings of the R2 point.

near curves C and P, inside which transversal homoclinic structures exist (see Figure 3.11) and which are bounded by curves of homoclinic tangencies (see 2.2.1 of Chapter 2). These codim 1 global bifurcations occur also in the full (non-truncated) smooth normal form (3.33) and are accompanied by an infinite number of other bifurcation curves corresponding to period-doubling and fold bifurcations of long-periodic cycles.[3] Finally, near the cyclic fold curve K the truncated (and non-truncated) smooth normal form (3.33) exhibits a quasi-periodic saddle-node bifurcation with the associated complicated bifurcation set (see Sections 2.2.2 and 3.2).

As an illustration, we consider the following representative perturbation of the normal form (3.34)

$$\begin{pmatrix} x \\ y \end{pmatrix} \mapsto \begin{pmatrix} -x+y \\ -y+\beta_1 x+\beta_2 y - x^3 + x^2 y + \gamma_1 y^4 + \gamma_2 x^6 \end{pmatrix}, \tag{3.38}$$

where we choose $c_0 = -1, d_0 = 1$ to have the complex unfolding with $\mathrm{NS}^{(1)}$ subcritical. And we set $\gamma_1 = 0.13, \gamma_2 = 0.02$. We provide two diagrams, one with bifurcation curves and one with a scan of Lyapunov exponents. The bifurcation curves corroborate the interpretation above. Local bifurcation curves remain curves, while the global bifurcations P and K turn into complicated bifurcation sets. Note that the wedge of homoclinic orbits opens as β_2 increases. Here the Lyapunov exponents help to see where an invariant curve exists and where we encounter phase locking. Well before we encounter the homoclinic tangency there are many resonance tongues accumulating onto the homoclinic tangencies. To shed more light on this we look at several resonance tongues emanating from both NS curves. The tongues emanating from NS^2 all have an even period and correspond to saddles and stable nodes. They accumulate immediately onto Hom. The tongues emerging from NS^1 correspond to unstable cycles. Following the tongue we encounter a resonance bubble where the

[3] Notice that both sub-figures of Figure 3.11 illustrate *homoclinic* structures: of the period-2 cycle (a) and of the fixed point (b).

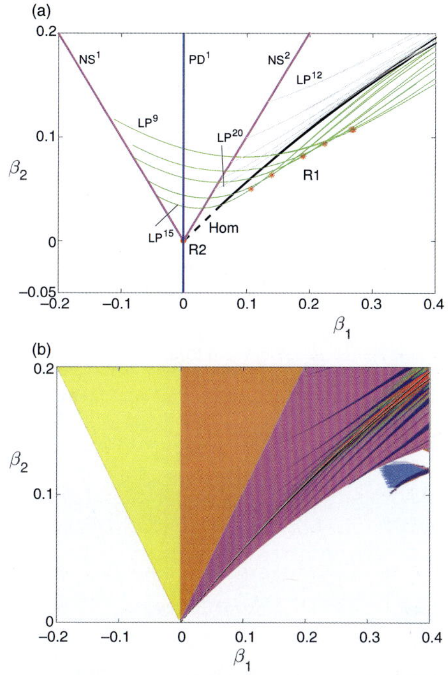

Figure 3.12 Generic bifurcation diagram near 1:2 resonance in the complex case. (a) Neimark–Sacker $\mathsf{NS}^{1,2}$ (magenta) and period-doubling PD^1 curves (blue). Thick transversal intersection of figure-of-eight type occurs between black curves (dashed is theoretical extension). In the upper right, it is visible that it is indeed a wedge. Resonance tongues emerge from both NS curves accumulating on the homoclinic tangency. The period of tongues starting from NS^2 can only be even. The curve K translates to a quasi-periodic saddle-node bifurcation set visible by the R1-points from the resonance bubbles. (b) Scan of Lyapunov exponents; colors indicate fixed point (yellow), period-2 cycle (brown), cycles with higher period (dark blue/green), invariant curve (magenta) and chaos (red). The dark blue region at the middle right corresponds to a different attractor.

cycle source turns into a sink. Next the tongue boundaries approach the curves of homoclinic tangency, but from the other side. The Lyapunov exponents also indicate chaotic dynamics near the homoclinic wedge.

We outlined an unfolding scenario for resonance bubbles near a quasi-periodic saddle-node bifurcation near a Chenciner bifurcation (see Section 3.2). All these scenarios are encountered in this unfolding as well. We zoom in on the period 9 tongue and use γ_1 as an additional deformation parameter. We highlight the NS-curve consisting of an NS bifurcation and a neutral saddle part, omitting the global bifurcations. Starting with $\gamma_1 = 0.10$ the lower

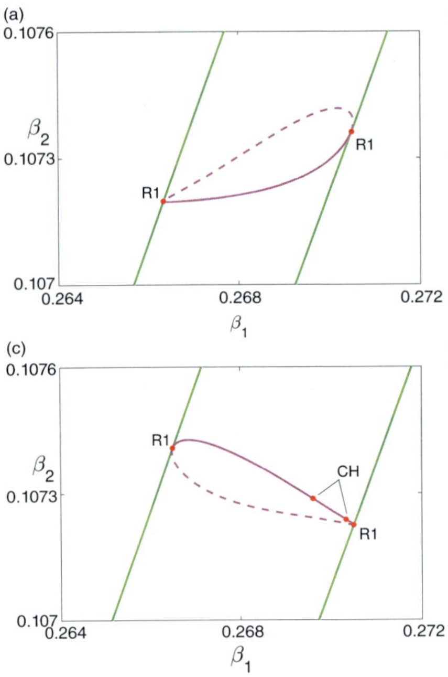

Figure 3.13 Resonance bubbles of period 9 tongue near quasi-periodic saddle-node bifurcation in the unfolding of the 1:2 resonance. The tongue is bounded by two fold curves (green) with R1-points on them. A curve corresponding to a NS bifurcation (solid) and neutral saddles (dashed) connect these two points. (a) $\gamma_1 = 0.10$, (b) $\gamma_1 = 0.12$, (c) $\gamma_1 = 0.13$, (d) $\gamma_1 = 0.14$.

half of the ellipse corresponds to a subcritical NS bifurcation of a cycle of period 9. For $\gamma_1 = 0.12$ the NS-curve intersects itself and has one additional Chenciner point (CH). Moving on to $\gamma_1 = 0.13$ as in Figure 3.12, the left R1-point has also changed type and the NS-curve now has two Chenciner points. These have disappeared for $\gamma_1 = 0.14$. This sequence shows an example of such bubbles in a different model map.

3.3.3 R3 (resonance 1:3)

Theorem 3.22 *Let α_0 correspond to the resonance* 1 : 3. *Then for parameter values α close to α_0, the restriction of* (3.2) *to a parameter-dependent center manifold is locally smoothly equivalent to the complex map*

$$z \mapsto \mu(\alpha)z + B(\alpha)\bar{z}^2 + C(\alpha)z|z|^2 + O(|z|^4), \qquad z \in \mathbb{C}, \tag{3.39}$$

where

$$\mu_0 := \mu(\alpha_0) = e^{2\pi i/3} \tag{3.40}$$

and the O-terms may also smoothly depend on α.

Proof See Kuznetsov (2004, lemma 9.12, pp. 448–449). □

The expressions for $b_0 = B(\alpha_0)$ and $c_0 = C(\alpha_0)$ will be given in Chapter 4.

Since $\mu_0^2 = \bar{\mu}_0 = e^{-2\pi i/3}$, we can write

$$\mu_0^2\mu(\alpha) = e^{\beta_1(\alpha)+i\beta_2(\alpha)},$$

where $\beta_1(0) = \beta_2(0) = 0$. If we assume that the map $\alpha \mapsto \beta(\alpha) = (\beta_1(\alpha), \beta_2(\alpha))$ is regular at α_0, then the components of β can be used as the new unfolding parameters and we can introduce the truncated normal form

$$N_\beta \colon z \mapsto e^{2\pi i/3+\beta_1+i\beta_2}z + b(\beta)\bar{z}^2 + c(\beta)z|z|^2, \tag{3.41}$$

where b and c are smooth functions of β, such that $b(0) = b_0$ and $c(0) = c_0$. Define the linear transformation $R \colon \mathbb{C} \to \mathbb{C}$ by

$$z \mapsto Rz := \overline{\mu}_0 z = e^{-2\pi i/3}z, \tag{3.42}$$

which is the *clockwise* rotation through $2\pi/3$. The truncated normal form (3.41) is invariant with respect to R. Note that $R^3 = \text{id}$. The phase portraits that follow will therefore possess $\mathbb{Z}_3$-symmetry.

Lemma 3.23 *Assume that*

$$\Re(\overline{\mu}_0 c_0) - |b_0|^2 \neq 0.$$

Then the truncated map (3.41) *has the bifurcation curve*

$$NS^{(1)} = \{(\beta_1, \beta_2) \in \mathbb{R}^2 \colon \beta_1 = 0\},$$

at which a non-degenerate NS bifurcation of the trivial fixed point happens, provided $\beta_2 \neq 0$.

Proof See Appendix 3.C. □

Theorem 3.24 *For sufficiently small* $\|\beta\|$, *the map* (3.41) *satisfies*

$$R\,N_\beta(z) = \varphi_\beta^1(z) + O(|z|^4),$$

where φ_β^t *is the flow of the approximating planar system*

$$\dot{z} = (\beta_1 + i\beta_2)z + b_1(\beta)\bar{z}^2 + c_1(\beta)z|z|^2, \qquad z \in \mathbb{C}, \tag{3.43}$$

where b_1 *and* c_1 *are smooth complex-valued functions of* β, *such that*

$$b_1(0) = \overline{\mu}_0 b_0, \quad c_1(0) = \overline{\mu}_0 c_0 - |b_0|^2.$$

Proof See Appendix 3.C. □

If $b_1(0) \neq 0$ (which is equivalent to $b_0 \neq 0$), then we can scale (3.43) by taking $z = \gamma(\beta)\zeta$ with

$$\gamma(\beta) = \frac{1}{|b_1(\beta)|} \exp\left(i\frac{\arg b_1(\beta)}{3}\right).$$

This scaling transforms (3.43) into

$$\dot{\zeta} = (\beta_1 + i\beta_2)\zeta + \bar{\zeta}^2 + c_2(\beta)\zeta|\zeta|^2 \tag{3.44}$$

with

$$c_2(\beta) = \frac{c_1(\beta)}{|b_1(\beta)|^2}.$$

Let $c_2(\beta) = a_2(\beta) + ib_2(\beta)$ with real a_2 and b_2. Writing (3.44) in polar coordinates $\zeta = \rho e^{i\varphi}$, we obtain

$$\begin{cases} \dot{\rho} = \beta_1\rho + \rho^2\cos(3\varphi) + a_2(\beta)\rho^3, \\ \dot{\varphi} = \beta_2 - \rho\sin(3\varphi) + b_2(\beta)\rho^2, \end{cases} \tag{3.45}$$

where

$$a_2(0) = \frac{\Re(\bar{\mu}_0 c_0) - |b_0|^2}{|b_0|^2}, \quad b_2(0) = \frac{\Im(\bar{\mu}_0 c_0)}{|b_0|^2}, \tag{3.46}$$

and μ_0 is defined in (3.40).

Theorem 3.25 *Suppose that*

$$a_2(0) \neq 0.$$

Then the local bifurcation diagram of the approximating system (3.45) *near* $\beta = 0$ *includes two bifurcations curves:*

(1) *the Andronov–Hopf bifurcation curve*

$$H = \{(\beta_1, \beta_2) \in \mathbb{R}^2 : \beta_1 = 0\},$$

at which a limit cycle branches from the trivial equilibrium for small $\beta_2 \neq 0$;

(2) *the heteroclinic bifurcation curve*

$$C = \left\{(\beta_1, \beta_2) \in \mathbb{R}^2 : \beta_1 = -\frac{a_2(0)}{2}\beta_2^2 + o(\beta_2^2)\right\}$$

at which a heteroclinic contour with three saddles exists for small $\beta_2 \neq 0$.

Both bifurcations are non-degenerate when $\beta \neq 0$, *and are the only bifurcations happening in* (3.45) *near* $\beta = 0$ *in a neighborhood of the origin.*

Proof See Appendix 3.C. □

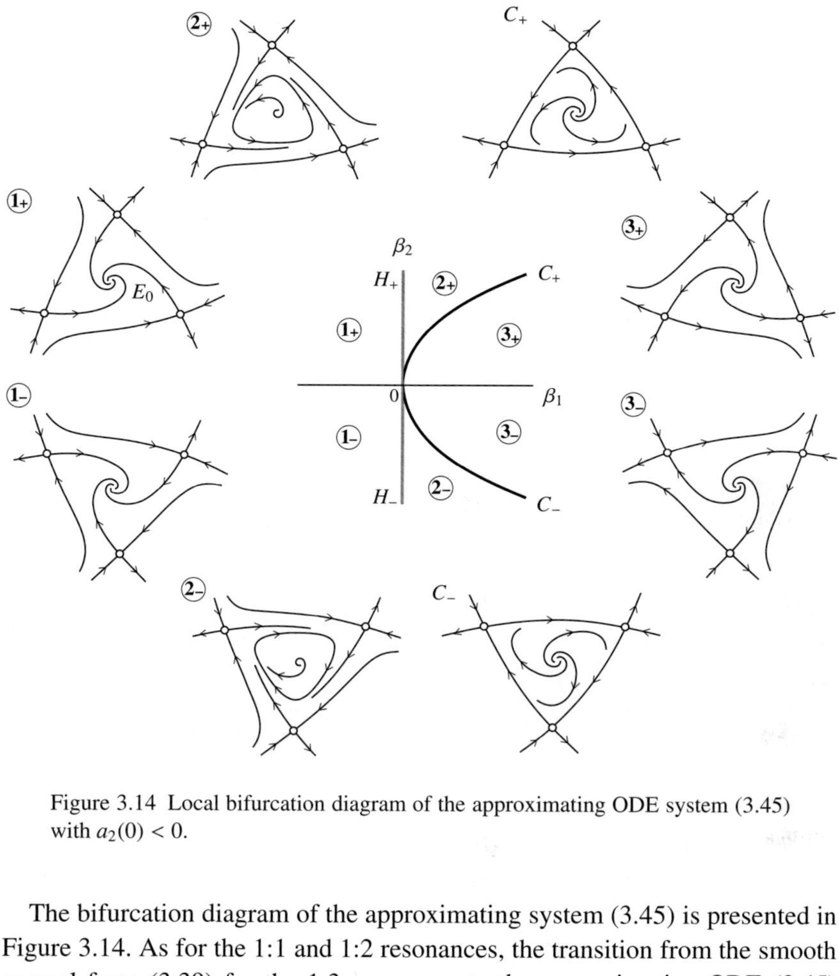

Figure 3.14 Local bifurcation diagram of the approximating ODE system (3.45) with $a_2(0) < 0$.

The bifurcation diagram of the approximating system (3.45) is presented in Figure 3.14. As for the 1:1 and 1:2 resonances, the transition from the smooth normal form (3.39) for the 1:3 resonance to the approximating ODE (3.45) involves two steps:

(1) truncation of the $O(\|z\|^4)$-terms in (3.39) that results in the map N_β given by (3.41);
(2) neglecting the difference between $R \circ N_\beta$ and the time-1 shift along orbits of the planar ODE (3.43) and its rescaled polar form (3.45).

To apply results about (3.45) to the truncated normal form (3.41), we notice that the time-1 flow of the ODE approximates the map N_β composed with the rotation (3.42). Thus, the trivial equilibrium of (3.45) corresponds to a fixed point of N_β, while three symmetric nontrivial equilibrium points actually correspond to one period-3 cycle. Thus, the Hopf bifurcation curve H approximates

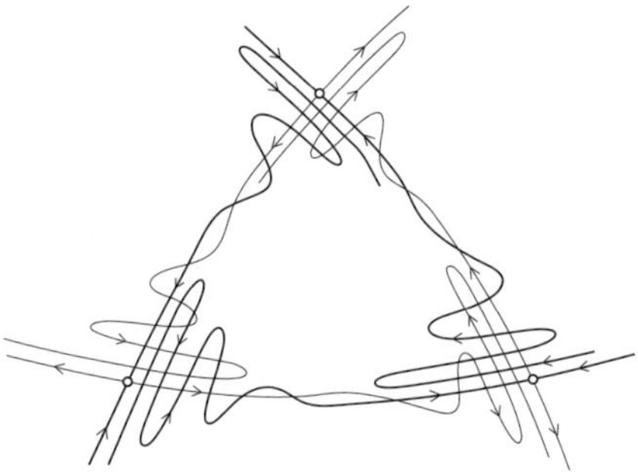

Figure 3.15 Homoclinic structure associated with the period-3 cycle.

the NS bifurcation curve $NS^{(1)}$ of the trivial fixed point of N_β. Indeed, the curve $NS^{(1)}$ in Lemma 3.23 corresponds to the curve H in Theorem 3.25. The same representation is valid for the original (non-truncated) normal form and the corresponding bifurcations in the approximating ODE and in the smooth normal form are simultaneously non-degenerate if the formulated condition is satisfied. The NS bifurcation produces one closed invariant curve surrounding the trivial fixed point. This closed invariant curve is stable if $a_2(0) < 0$ and unstable if $a_2(0) > 0$.

The closed invariant curve disappears near curve C. However, instead of the single heteroclinic bifurcation curve C of (3.45), the map N_β (and its perturbation with arbitrary higher-order terms) possesses a complicated bifurcation set nearby. The stable and unstable invariant manifolds of the period-3 cycle intersect transversally (see Figure 3.15) in an exponentially narrow parameter wedge that approaches $\beta = 0$ and that is bounded by two homoclinic tangency curves. As usual, this homoclinic structure implies existence of infinitely many long-periodic cycles and their fold and period-doubling bifurcations.

Remark 3.26 As we have seen, a generic unfolding of this singularity has a period-3 saddle cycle that does not bifurcate for nearby parameter values, although it merges with the primary fixed point as the parameters approach $\beta = 0$. Note that this period-3 cycle becomes neutral near this bifurcation, i.e., the corresponding fixed point of the third iterate has a pair of real eigenvalues with product 1. This condition is important in analyzing global bifurcations of invariant manifolds of cycles. Moreover, the curve of neutral period-3 saddle cycles may turn into a true NS bifurcation at 1:1 or 1:2 resonances. In the approximating planar system (3.45), the neutral saddle corresponds to a neutral equilibrium, i.e., an equilibrium having two real eigenvalues with zero sum. One can show that the neutral saddle curve has the following asymptotic expression

$$\begin{pmatrix} \rho \\ \varphi \\ \beta_1 \\ \beta_2 \end{pmatrix} = \begin{pmatrix} \varepsilon \\ s(\pi/6 - a_2(0)\varepsilon/3) \\ -2a_2(0)\varepsilon^2 \\ s\varepsilon - b_2(0)\varepsilon^2 \end{pmatrix} + O(\varepsilon^3), \quad \varepsilon > 0, \tag{3.47}$$

where $s = \pm 1$. In the (β_1, β_2)-plane, this is a single smooth curve with the representation $\beta_1 = -2a_2(0)\beta_2^2 + o(\beta_2^2)$.

3.3.4 R4 (resonance 1:4)

Theorem 3.27 *Let α_0 correspond to the resonance* 1 : 4. *Then for parameter values α close to α_0, the restriction of* (3.2) *to a parameter-dependent center manifold is locally smoothly equivalent to the complex map*

$$z \mapsto \mu(\alpha)z + C(\alpha)z|z|^2 + D(\alpha)\bar{z}^3 + O(|z|^4), \qquad z \in \mathbb{C}, \tag{3.48}$$

where

$$\mu_0 := \mu(\alpha_0) = e^{\pi i/2} = i \tag{3.49}$$

and the O-terms may also smoothly depend on α.

Proof See Kuznetsov (2004, lemma 9.14, p. 455). □

The expressions for $c_0 = C(\alpha_0)$ and $d_0 = D(\alpha_0)$ will be given in Chapter 4.

Since $\mu_0^3 = \bar{\mu}_0 = e^{-\pi i/2}$, we can write

$$\mu_0^3\mu(\alpha) = e^{\beta_1(\alpha) + i\beta_2(\alpha)},$$

where $\beta_1(0) = \beta_2(0) = 0$. If we assume that the map $\alpha \mapsto \beta(\alpha) = (\beta_1(\alpha), \beta_2(\alpha))$ is regular at α_0, then the components of β can be used as the new unfolding parameters and we can introduce the truncated normal form

$$N_\beta \colon z \mapsto e^{\pi i/2 + \beta_1 + i\beta_2} z + c(\beta)z|z|^2 + d(\beta)\bar{z}^3, \tag{3.50}$$

where c and d are smooth functions of β, such that $c(0) = c_0$ and $d(0) = d_0$. Define the linear transformation $R \colon \mathbb{C} \to \mathbb{C}$ by

$$z \mapsto Rz := \bar{\mu}_0 z = e^{-\pi i/2} z = -iz, \tag{3.51}$$

which is the *clockwise* rotation through $\pi/2$. The truncated normal form (3.50) is invariant with respect to R. Note that $R^4 = \text{id}$. The phase portraits that follow will therefore possess $\mathbb{Z}_4$-symmetry.

Theorem 3.28 *For sufficiently small* $\|\beta\|$, *the map* (3.41) *satisfies*

$$R\, N_\beta(z) = \varphi_\beta^1(z) + O(|z|^4),$$

where φ_β^t *is the flow of the approximating planar system*

$$\dot{z} = (\beta_1 + i\beta_2)z + c_1(\beta)z|z|^2 + d_1(\beta)\bar{z}^3, \qquad z \in \mathbb{C}, \tag{3.52}$$

where c_1 *and* d_1 *are smooth complex-valued functions of* β, *such that*

$$c_1(0) = \bar{\mu}_0 c_0 = -ic_0, \quad d_1(0) = \bar{\mu}_0 d_0 = -id_0.$$

Proof See Appendix 3.C. □

Using the scaling

$$z = \frac{1}{\sqrt{|d_1(\beta)|}} e^{i\arg(d_1(\beta))/4}\eta$$

and introducing

$$A(\beta) := \frac{c_1(\beta)}{|d_1(\beta)|},$$

we can consider instead of (3.52) the complex ODE

$$\dot{\zeta} = (\beta_1 + i\beta_2)\zeta + A(\beta)\zeta|\zeta|^2 + \bar{\zeta}^3, \qquad z \in \mathbb{C}. \tag{3.53}$$

Writing (3.53) in polar coordinates $\zeta = \rho e^{i\varphi}$, we obtain

$$\begin{cases} \dot{\rho} &= \beta_1\rho + a(\beta)\rho^3 + \rho^3\cos(4\varphi), \\ \dot{\varphi} &= \beta_2 + b(\beta)\rho^2 - \rho^2\sin(4\varphi), \end{cases} \tag{3.54}$$

where $a(\beta) = \Re(A(\beta))$ and $b(\beta) = \Im(A(\beta))$, so that

$$a_0 = \frac{\Re(\bar{\mu}_0 c_0)}{|d_0|}, \quad b_0 = \frac{\Im(\bar{\mu}_0 c_0)}{|d_0|}. \tag{3.55}$$

The bifurcation diagram of the approximating system (3.54) in the (β_1, β_2)-plane depends on $A_0 = a_0 + ib_0 \in \mathbb{C}$. Since the boundaries between the cases in the A_0-plane are symmetric under reflections through the coordinate axes, it is sufficient to consider the quadrant where $a_0 \le 0$ and $b_0 \le 0$. The division of this quadrant of the A_0-plane into regions corresponding to different bifurcation diagrams is shown in Figure 3.16.

The boundaries between the regions correspond to doubly degenerate phase objects that include four nontrivial (with $\rho > 0$) symmetric equilibria of (3.54). Some of these boundaries are known explicitly, namely given by the following conditions

$$a_0 b_0 = 0, \quad a_0^2 + b_0^2 = 1, \quad a_0 = -1,$$

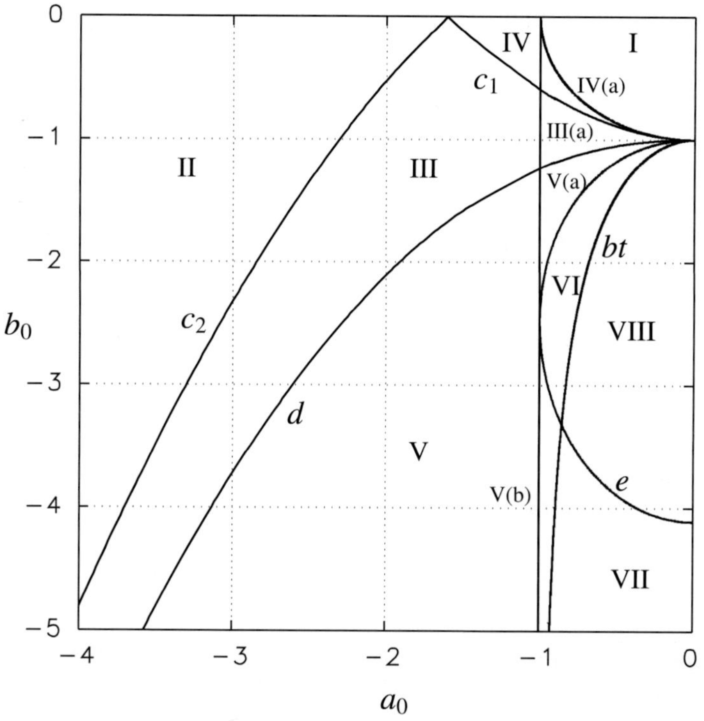

Figure 3.16 Partitioning of the (a_0, b_0)-plane.

as well as

$$|b_0| = \frac{1 + a_0^2}{\sqrt{1 - a_0^2}}. \tag{3.56}$$

Others correspond to codim 2 global bifurcations and are known only approximately, by numerical continuation.

Theorem 3.29 *Suppose that*

$$a_0 b_0 (a_0^2 - 1) \neq 0.$$

Then the local bifurcation diagram of the approximating system (3.45) *near* $\beta = 0$ *includes the following bifurcation curves:*

(1) *the Andronov–Hopf bifurcation curve*

$$N = \{(\beta_1, \beta_2) \in \mathbb{R}^2 : \beta_1 = 0\},$$

at which a limit cycle branches from the trivial equilibrium if $\beta_2 < 0$;

(2) *the saddle-node bifurcation curves*

$$T_{1,2} = \left\{ (\beta_1, \beta_2) \in \mathbb{R}^2 : \beta_2 = \frac{a_0 b_0 \pm \sqrt{\Delta}}{a_0^2 - 1} \beta_1 + O(\beta_1^2) \right\}$$

at which four symmetric pairs of nontrivial equilibrium points appear, provided $\Delta := a_0^2 + b_0^2 - 1 > 0$;

(3) *the Andronov–Hopf bifurcation curve*

$$N' = \left\{ (\beta_1, \beta_2) \in \mathbb{R}^2 : \beta_2 = \frac{b_0 - \operatorname{sign}(b_0)\sqrt{1 - a_0^2}}{2a_0} \beta_1 + O(\beta_1^2) \right\},$$

at which four symmetric limit cycles simultaneously bifurcate from four nontrivial non-saddles, provided $a_0^2 < 1$ *and*

$$|b_0| > \frac{1 + a_0^2}{\sqrt{1 - a_0^2}}.$$

Moreover, all these bifurcations are non-degenerate when $\beta \neq 0$, *and are the only equilibrium bifurcations happening in* (3.54) *near* $\beta = 0$ *in a neighborhood of the origin.*

Besides equilibrium bifurcations described by Theorem 3.29, global bifurcations also occur in the approximating system (3.45). Complete bifurcation diagrams inside regions specified in Figure 3.16 are collected in Figure 3.17.

In order to understand the bifurcation diagrams, we should first mention that (3.54) has at most four nontrivial saddle equilibria if $a_0^2 + b_0^2 < 1$ and up to eight nontrivial equilibria if $a_0^2 + b_0^2 > 1$. In the latter case, the nontrivial equilibria could undergo codim 1 saddle-node $T_{1,2}$ bifurcations (they are labeled by T^{in}, T^{on} and T^{out} in the figure), as well as the Andronov–Hopf (N′) bifurcation. The case T^{on} corresponds to the appearance of a heteroclinic cycle involving four saddle-nodes and can be seen as the appearance of four saddle-nodes precisely on a limit cycle (see Figure 3.18). The other two cases correspond to the appearance of the saddle-nodes inside and outside of the limit cycle.

The trivial equilibrium can also exhibit a Hopf (N) bifurcation that generates a limit cycle. Moreover, there could simultaneously exist four homoclinic orbits (one to each nontrivial saddle point, H^L), as well as two types of heteroclinic contours (H^S and H^C, see Figure 3.19). All these are codim 1 bifurcations generating limit cycles.

The last codim 1 bifurcation is the cyclic fold bifurcation (F), i.e., collision and disappearance of two limit cycles surrounding all equilibria.

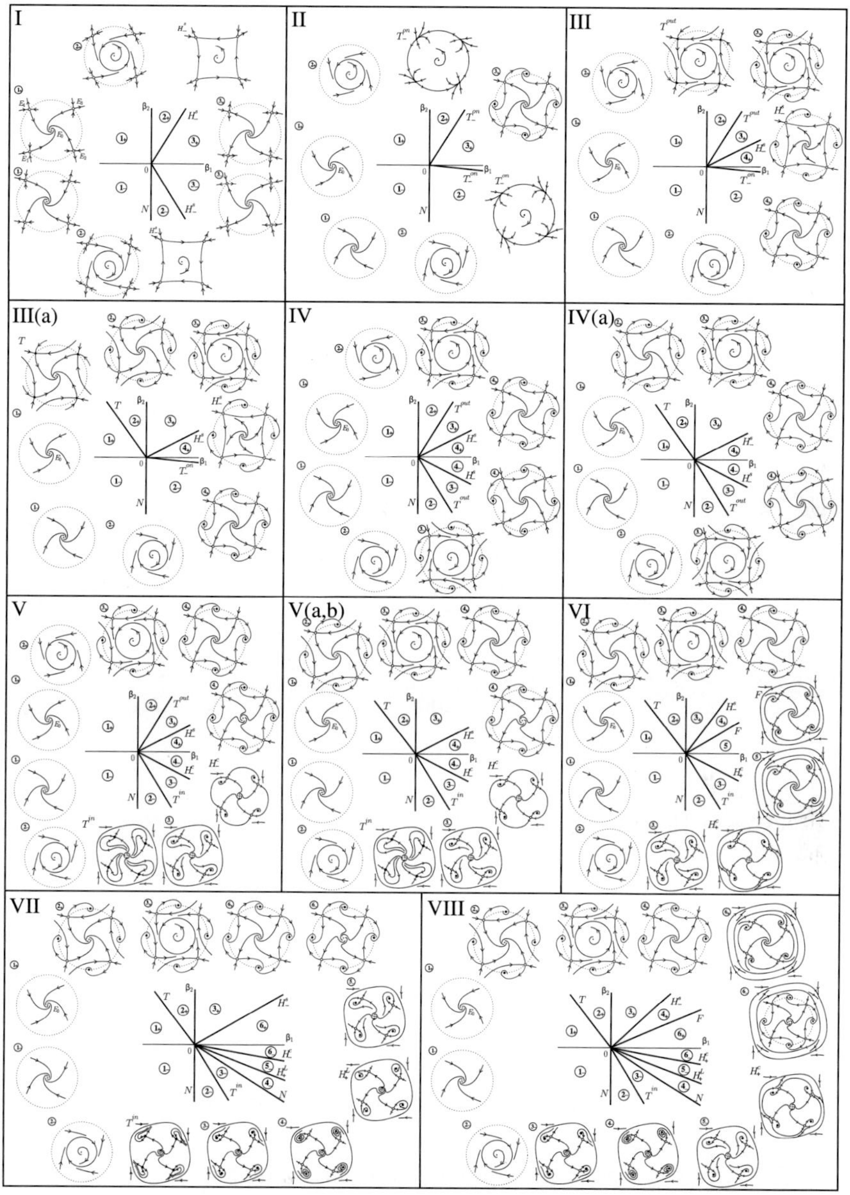

Figure 3.17 Generic bifurcation diagrams of (3.54). The notations for bifurcation curves are explained in the text; ± index indicates two different rotation directions.

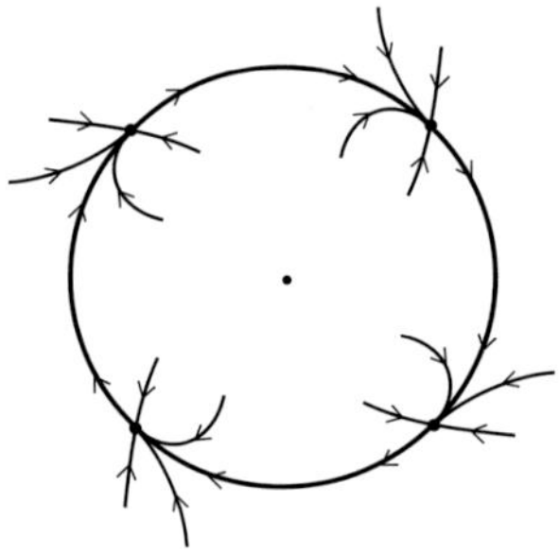

Figure 3.18 Saddle-node heteroclinic bifurcations in system (3.54).

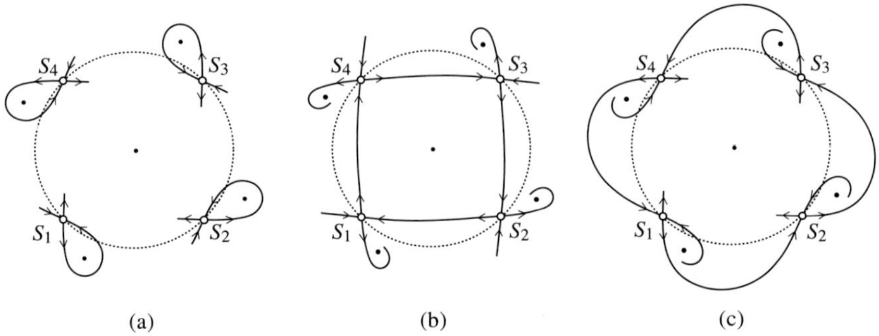

Figure 3.19 Homoclinic and heteroclinic bifurcations in system (3.54) involving saddles S_j, $j = 1, 2, 3, 4$: (a) homoclinic orbits; (b) "square" heteroclinic contour; (c) "clover" heteroclinic contour.

Most boundaries in the A_0-plane separating regions with different bifurcation diagrams are defined by codim 2 bifurcations:

(1) The curve bt defined by (3.56) corresponds to the Bogdanov–Takens (double zero eigenvalue) bifurcation of the nontrivial equilibria.
(2) The curve e corresponds to the neutral "clover" heteroclinic contour, when all four saddles connected by the heteroclinic orbits are neutral (have saddle quantity zero).
(3) The curves $c_{1,2}$ correspond to the non-central saddle-node "square" heteroclinic contour (see Figure 3.20(a)).
(4) The curve d corresponds to the non-central saddle-node "clover" heteroclinic contour (see Figure 3.20(b)).

Only the boundary bt in this list is known analytically, the others have been computed numerically.

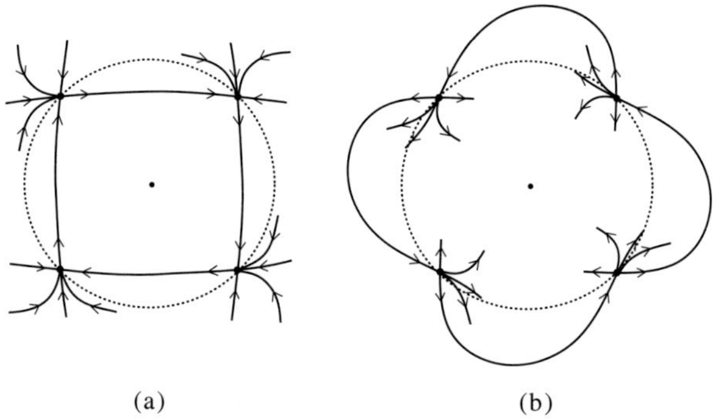

Figure 3.20 Non-central saddle-node heteroclinic contours in system (3.54): (a) "square"; (b) "clover."

What do the obtained results about ODE (3.54) imply for the truncated normal form N_β defined by (3.50) and its perturbation by higher-order terms? As for the other strong resonances, the transition from the smooth normal form (3.48) for the 1:4 resonance to the approximating ODE (3.54) involves two steps:

(1) truncation of the $O(\|z\|^4)$-terms in (3.48) that results in the map N_β given by (3.50);
(2) neglecting the difference between $R \circ N_\beta$ and the time-1 shift along orbits of the planar ODE (3.52) and its rescaled polar form (3.54).

Note that the time-1 flow of the ODE approximates the map N_β composed with the rotation (3.51). Thus, the trivial equilibrium of the approximating system (3.54) corresponds to the trivial fixed point of N_β, while four symmetric nontrivial equilibrium points actually correspond to one period-4 cycle. The Hopf bifurcation line N approximates the NS bifurcation line $NS^{(1)}$ of the trivial fixed point of N_β, while the Hopf bifurcation line N′ approximates the NS bifurcation line $NS^{(4)}$ of the period-4 cycle of N_β. Finally, the saddle-node bifurcation lines approximate the fold bifurcation lines $LP^{(4)}_{1,2}$ at which two period-4 cycles appear or disappear. Moreover, the leading terms of the asymptotic expressions for the corresponding curves coincide.

As usual, homo- and heteroclinic connections in the approximating system (3.54) become homoclinic structures for the map N_β (see Figure 3.21). They are formed by intersections of the stable and the unstable invariant manifolds of the saddle period-4 cycle. These structures imply the existence of an infinite number of periodic orbits. Closed invariant curves corresponding to limit cycles lose their smoothness and are destroyed, almost "colliding" with the

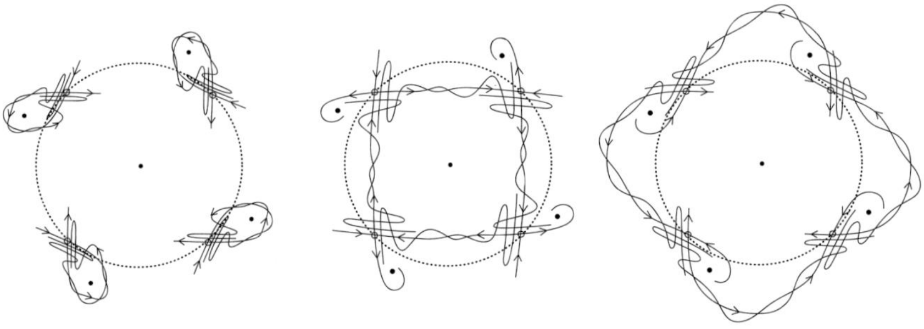

(a) (b) (c)

Figure 3.21 Homoclinic structures near 1:4 resonance: (a) "small"; (b) "square"; and (c) "clover."

saddle period-4 cycle. All these phenomena are also present in the full (non-truncated) normal form (3.48) under the same non-degeneracy conditions.

Remark 3.30 There are three possible local branch switches for this bifurcation. If $\Delta \equiv a_0^2 + b_0^2 - 1 > 0$, then there are two half-lines $LP^{(4)}_{1,2}$ of a fold curve of period-4 cycles, which should be identified with the appropriate lines $\mathrm{T}^{in}, \mathrm{T}^{on}$ and T^{out} in Figure 3.17. If

$$|b_0| > \frac{1 + a_0^2}{\sqrt{1 - a_0^2}},$$

then there is a curve $NS^{(4)}$ along which a period-4 cycles exhibits an NS bifurcation. Using $\zeta = \rho e^{i\varphi}$ we have the following approximations for (3.48)

$$\begin{aligned} LP^{(4)}_{1,2}\colon (\rho^2, \varphi, \beta_1, \beta_2) &= \left(\varepsilon, \frac{1}{4} \arctan\left(\frac{a_0 b_0 \pm \sqrt{\Delta}}{b_0^2 - 1}\right) + O(\varepsilon),\right. \\ &\qquad \left. \frac{-a_0\Delta \mp b_0\sqrt{\Delta}}{a_0^2 + b_0^2}\varepsilon, \frac{-b_0\Delta \pm a_0\sqrt{\Delta}}{a_0^2 + b_0^2}\varepsilon\right) + O(\varepsilon^2), \\ NS^{(4)}\colon (\rho^2, \varphi, \beta_1, \beta_2) &= (\varepsilon + O(\varepsilon^2), \operatorname{sign}(b_0)\arccos(a_0)/4 + O(\varepsilon), \\ &\qquad - 2a_0\varepsilon + O(\varepsilon^2), -(b_0 - \operatorname{sign}(b_0)\sqrt{1 - a_0^2})\varepsilon + O(\varepsilon^2)). \end{aligned} \tag{3.57}$$

If, in the formula for $NS^{(4)}$, we replace $\operatorname{sign}(b_0)$ by $-\operatorname{sign}(b_0)$, then it gives the asymptotic for a neutral period-4 saddle cycle.

3.4 Fold–flip and fold–Neimark–Sacker bifurcations

3.4.1 LPPD (fold–flip bifurcation)

This section is devoted to a detailed analysis of case 8 of Table 3.1, which we call the *fold–flip bifurcation* (LPPD). This bifurcation was first treated by Zholondek (1983) (see also (Gheiner, 1994)). We base our presentation on Kuznetsov, Meijer, and van Veen (2004), where several details were refined. First, we derive the parameter-dependent normal form for a generic fold–flip bifurcation and express explicitly the critical normal form coefficients in two-dimensional systems. Then we present the analysis of local bifurcations for the truncated normal form and study global bifurcations approximating the map by time-shifts along orbits of an auxiliary (approximating) planar ODE. We present (as completely as possible) bifurcation diagrams of the planar ODE, and discuss effects of the truncation and flow approximation. We give the proofs in full detail in Appendix 3.D, as the methods and proofs in the next section are analogous and thus the exposition a little shorter.

Normal form for the fold–flip bifurcation

We start with the derivation of the critical normal form, which will be the same as presented by Gheiner (1994). Moreover, we give explicit expressions for the critical normal form coefficients for planar maps.

Lemma 3.31 (Critical normal form) *Suppose a smooth map $F_0\colon \mathbb{R}^2 \to \mathbb{R}^2$ has the form*

$$\begin{pmatrix} \xi_1 \\ \xi_2 \end{pmatrix} \mapsto \begin{pmatrix} \xi_1 + \displaystyle\sum_{i+j=2,3} \frac{1}{i!j!} g_{ij}\xi_1^i\xi_2^j \\ -\xi_2 + \displaystyle\sum_{i+j=2,3} \frac{1}{i!j!} h_{ij}\xi_1^i\xi_2^j \end{pmatrix} + O(\|\xi\|^4) \tag{3.58}$$

and $h_{11} \neq 0$. Then F_0 is smoothly equivalent near the origin to a map

$$\begin{pmatrix} x_1 \\ x_2 \end{pmatrix} \mapsto \begin{pmatrix} x_1 + a(0)x_1^2 + b(0)x_2^2 + c(0)x_1^3 + d(0)x_1x_2^2 \\ -x_2 + x_1x_2 \end{pmatrix} + O(\|x\|^4), \tag{3.59}$$

where

$$a(0) = \frac{g_{20}}{2h_{11}}, \quad b(0) = \frac{1}{2}g_{02}h_{11}, \quad c(0) = \frac{1}{6h_{11}^2}\left(g_{30} + \frac{3}{2}g_{11}h_{20}\right), \tag{3.60}$$

$$\begin{aligned} d(0) = {} & \frac{1}{2}\left(g_{12} + \frac{1}{2}g_{11}h_{02} - g_{11}^2 - h_{02}^2 - \frac{2}{3}h_{03}\right) \\ & + \frac{3g_{02}\left(h_{21} + \frac{1}{2}h_{20}h_{02} - g_{11}h_{20}\right) - g_{20}\left(h_{03} + \frac{3}{2}h_{02}^2\right)}{6h_{11}}. \end{aligned} \tag{3.61}$$

Proof See Appendix 3.D. □

From the normal forms for the fold and flip bifurcations separately (see (2.8) and (2.13)) one would expect a parameter-dependent normal form

$$\begin{pmatrix} x_1 \\ x_2 \end{pmatrix} \mapsto \begin{pmatrix} \mu_1 + x_1 \\ -(1+\mu_2)x_2 \end{pmatrix} + O(\|x\|^2).$$

For the analysis, however, it turns out to be very convenient to use a form where most of the terms are removed from the second component, unlike in (3.59). We now present this family.

Proposition 3.32 (Parameter-dependent normal form) *Consider a smooth two-parameter family of smooth planar maps*

$$\xi \mapsto F(\xi, \alpha), \quad \xi \in \mathbb{R}^2, \alpha \in \mathbb{R}^2,$$

where $F\colon \mathbb{R}^2 \times \mathbb{R}^2 \to \mathbb{R}^2$ *is smooth and such that:*

1. $F_0\colon \mathbb{R}^2 \to \mathbb{R}^2$, $\xi \mapsto F_0(\xi) = F(\xi, 0)$ *satisfies* Lemma 3.31;
2. *The map* $T\colon \mathbb{R}^2 \times \mathbb{R}^2 \to \mathbb{R}^2 \times \mathbb{R} \times \mathbb{R}$, *defined by*

$$\begin{pmatrix} \xi \\ \alpha \end{pmatrix} \mapsto T(\xi, \alpha) = \begin{pmatrix} F(\xi,\alpha) - \xi \\ \det F_\xi(\xi,\alpha) + 1 \\ Tr\, F_\xi(\xi,\alpha) \end{pmatrix} \tag{3.62}$$

is regular at $(\xi, \alpha) = (0, 0)$.

Then F *is smoothly equivalent near the origin to a family*

$$\begin{pmatrix} x_1 \\ x_2 \end{pmatrix} \mapsto \begin{pmatrix} \mu_1 + (1+\mu_2)x_1 + a(\mu)x_1^2 + b(\mu)x_2^2 + c(\mu)x_1^3 + d(\mu)x_1x_2^2 \\ -x_2 + x_1x_2 \end{pmatrix}$$
$$+O(\|x\|^4), \tag{3.63}$$

where all coefficients are smooth functions of μ *and their values at* $\mu_1 = \mu_2 = 0$ *are given by* (3.60) *and* (3.61).

Proof See Appendix 3.D. □

Remark 3.33 Map (3.62) can be substituted by the map

$$\begin{pmatrix} \xi \\ \alpha \end{pmatrix} \mapsto \begin{pmatrix} F(\xi,\alpha) - \xi \\ \det(F_\xi(\xi,\alpha) + I_2) \\ \det(F_\xi(\xi,\alpha) - I_2) \end{pmatrix}.$$

▷

Bifurcation analysis of the normal form

Here we analyze the codim 1 bifurcations we encounter when unfolding the normal form.

Introduce the *truncated normal form*

$$\begin{pmatrix} x_1 \\ x_2 \end{pmatrix} \mapsto N(x,\mu) = \begin{pmatrix} \mu_1 + (1+\mu_2)x_1 + a(\mu)x_1^2 + b(\mu)x_2^2 + c(\mu)x_1^3 + d(\mu)x_1x_2^2 \\ -x_2 + x_1x_2 \end{pmatrix}. \tag{3.64}$$

Remark 3.34 Note that (3.64) is *invariant* under the reflection in the x_1-axis:

$$x \mapsto Rx, \quad R = \begin{pmatrix} 1 & 0 \\ 0 & -1 \end{pmatrix}, \tag{3.65}$$

for which $R^2 = I_2$. The phase portraits that follow will reflect this $\mathbb{Z}_2$-symmetry.

Denote the critical values of the normal form coefficients by

$$a_0 = a(0),\ b_0 = b(0),\ c_0 = c(0),\ d_0 = d(0).$$

Proposition 3.35 *The family of maps* (3.64) *has the following local codim* 1 *bifurcations in a sufficiently small neighborhood of* $(x,\mu) = (0,0)$.

1. *There is a curve*

$$LP^{(1)}\colon (x_1, x_2, \mu_1) = \left(-\frac{\mu_2}{2a_0} + O(\mu_2^2), 0, \frac{\mu_2^2}{4a_0} + O(\mu_2^3)\right),$$

   on which a non-degenerate fold bifurcation occurs if $a_0 \neq 0$.
2. *There is a curve* $PD^{(1)}\colon (x_1, x_2, \mu_1) = (0,0,0)$ *on which a non-degenerate flip bifurcation occurs if* $b_0 \neq 0$.
3. *If* $b_0 > 0$ *and* $\mu_1 < 0$, *there is a curve*

$$NS^{(2)}\colon (x_1, x_2, \mu_2) = \left(0, \sqrt{-\frac{\mu_1}{b_0}} + O(\mu_1^{3/2}), \frac{(d_0 + 2b_0)\mu_1}{b_0} + O(\mu_1^2)\right),$$

   on which a non-degenerate NS bifurcation of the second iterate of (3.64) *occurs, provided*

$$c_{NS} := 3b_0c_0 - 3a_0b_0 - 2b_0a_0^2 - d_0a_0 \neq 0. \tag{3.66}$$

Proof See Appendix 3.D. □

The invariant curves born from the NS bifurcation cannot exist everywhere. They should disappear through some global bifurcations. First, we note that if the map (3.64) has two saddle fixed points located on the x-axis, it possesses a family of heteroclinic orbits. Indeed, this occurs when $a_0 b_0 > 0$ or $b_0 > 0 > a_0$. If both a_0 and b_0 are positive, then the two saddle fixed points can possess another heteroclinic structure through which the invariant curve is destroyed. To study this global bifurcation phenomenon, we derive a vector field (ODE), such that the unit shift along its orbits approximates (3.64). Bifurcations of this ODE system are easy to analyze, since it is similar to an amplitude system for the zero-Hopf bifurcation (see (Chow, Li, and Wang, 1994; Kuznetsov, 2004)).

Lemma 3.36 (Approximating vector field) *In a small neighborhood of* $(x,\mu) = (0,0)$, *the truncated normal form* (3.64) *satisfies*

$$R\,N(x,\mu) = \varphi^1(x,\mu) + O(\|\mu\|^2) + O(\|x\|^2\|\mu\|) + O(\|x\|^4). \tag{3.67}$$

Here, R is the matrix defined by (3.65), *φ^t is the flow generated by the approximating system*

$$\dot{x} = X(x,\mu), \quad x \in \mathbb{R}^2, \ \nu \in \mathbb{R}^2,$$

where the vector field X is given by

$$X(x,\mu) = \begin{pmatrix} \mu_1 + (-a_0\mu_1 + \mu_2)\,x_1 + a_0 x_1^2 + b_0 x_2^2 + d_1 x_1^3 + d_2 x_1 x_2^2 \\ \frac{1}{2}\mu_1 x_2 - x_1 x_2 + d_3 x_1 x_2^2 + d_4 x_2^3 \end{pmatrix}, \tag{3.68}$$

with

$$d_1 = c_0 - a_0^2, \ d_2 = d_0 + b_0(1-a_0), \ d_3 = \frac{1}{2}(a_0 - 1), \ d_4 = \frac{1}{2}b_0.$$

Proof See Appendix 3.D. □

To explore relationships between the map (3.64) and the vector field (3.68), consider first local bifurcations of the vector field X. One can check that there are two curves,

$$F\colon (x_1, x_2, \mu_1) = \left(-\frac{\mu_2}{2a_0} + O(\mu_2^2), 0, \frac{\mu_2^2}{4a_0} + O(\mu_2^3)\right)$$

and

$$P\colon (x_1, x_2, \mu_1) = (0,0,0),$$

on which equilibria of (3.68) have a zero eigenvalue. These are the same expansions as we computed for the curves $LP^{(1)}$ and $PD^{(1)}$ of the map (3.64). The center manifold reduction shows that in (3.68) a fold (saddle-node) bifurcation occurs on the first curve, while a pitchfork bifurcation happens on the second. Next we compute a Hopf bifurcation curve for (3.68). We indeed get the same expression

$$H\colon (x_1, x_2, \mu_2) = \left(0, \sqrt{-\frac{\mu_1}{b_0}} + O\left(\mu_1^{3/2}\right), \frac{(2b_0 + d_0)\mu_1}{b_0} + O(\mu_1^2)\right)$$

as for the $NS^{(2)}$ curve in Proposition 3.35. Guided by the similarity with the zero-Hopf bifurcation we also find the following result:

Proposition 3.37 *If* $a_0, b_0 > 0$ *and* $\mu_1 < 0$, *then the vector field* (3.68) *has two saddles, which are always connected by a heteroclinic orbit along the* x_1*-axis. There exists another heteroclinic orbit for*

$$J: \mu_2 = \frac{\mu_1}{a_0 b_0(2a_0+3)}\Big(2a_0^2(b_0+d_0) + 2a_0 d_0 + 3b_0(a_0+c_0)\Big) + o(\mu_1). \quad (3.69)$$

Proof See Appendix 3.D. □

Remark 3.38 We obtained the linear approximations to both the Hopf bifurcation curve, i.e.,

$$\mu_2 = \frac{d_0 + 2b_0}{b_0}\mu_1, \quad (3.70)$$

and the heteroclinic bifurcation curve, see (3.69). To analyze their relative position, we compute the difference between their slopes:

$$\frac{1}{a_0 b_0(2a_0+3)}\Big(3b_0c_0 - 3a_0b_0 - 2b_0a_0^2 - d_0a_0\Big).$$

This shows that the curves coincide in the linear approximation if and only if the first Lyapunov coefficient $L_1^{(2)} = c_{NS}$ (3.66) vanishes. Thus, changing the relative position of the two curves changes the stability of the appearing closed invariant curve for the map.

Next we can classify the critical phase portraits of the vector field (3.68). In polar coordinates the vector field at $\mu = 0$ becomes

$$\begin{pmatrix} \dot r \\ \dot\theta \end{pmatrix} = \begin{pmatrix} r^2\left(a_0\cos^2\theta + (b_0-1)\sin^2\theta\right) + O(r^3) \\ -r\sin\theta\left((1+a_0)\cos^2\theta + b_0\sin^2\theta\right) + O(r^2) \end{pmatrix}.$$

We see that there are invariant lines in the critical normal form if $\dot\theta = 0$. This equation is satisfied if $\theta = 0, \pi$, which is expected due to the invariance of the vector field under the map (3.65). Another possibility is that

$$\tan^2\theta = -\frac{1+a_0}{b_0}.$$

Therefore, we find six different critical portraits (see Figures 3.22 and 3.23).

Bifurcation diagrams

Although we had six critical cases, only four bifurcation diagrams will be reported, because the other two differ only at the critical parameter value. Figures 3.24–3.27 show the bifurcation diagrams of the approximating vector field

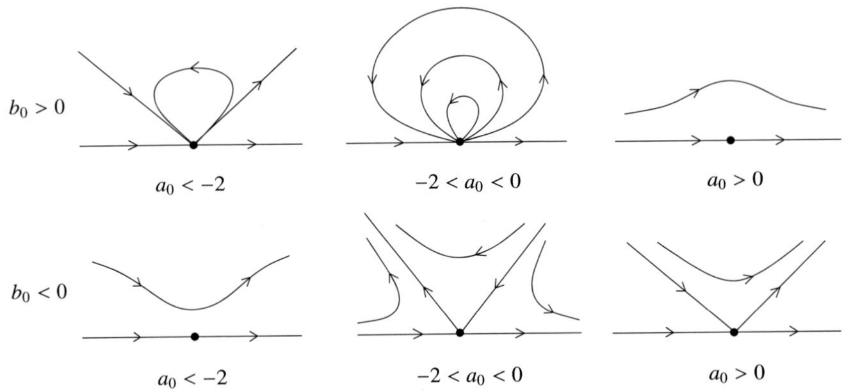

Figure 3.22 Phase portraits of the approximating vector field X at $\mu = 0$.

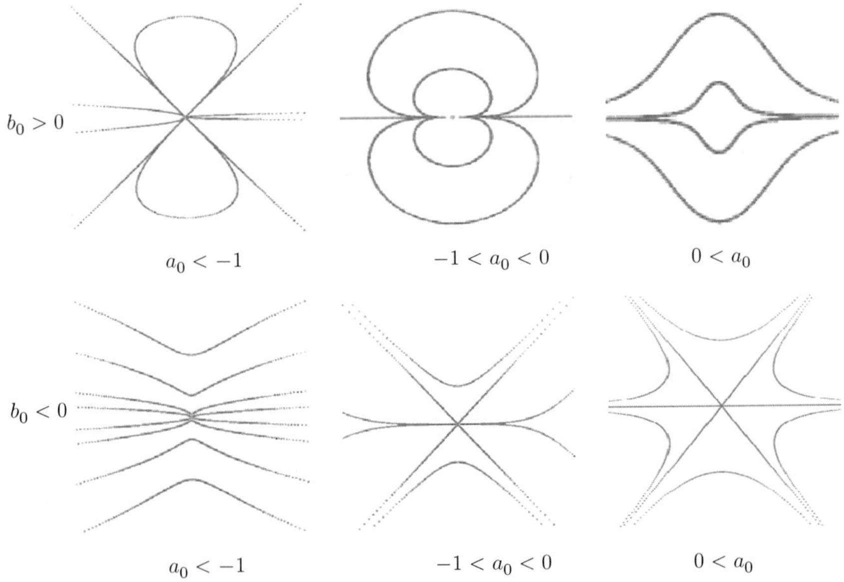

Figure 3.23 Phase portraits of the normal form N at $\mu = 0$. Compare with the orbits of the vector field to see how orbits of the map advance.

(3.68), which are similar to the bifurcation diagrams of the truncated amplitude system for the zero-Hopf bifurcation (see (Guckenheimer and Holmes, 1990; Chow, Li, and Wang, 1994; Kuznetsov, 2004)). In our study, however, we have taken into account explicitly that the Hopf (Neimark–Sacker) bifurcation can

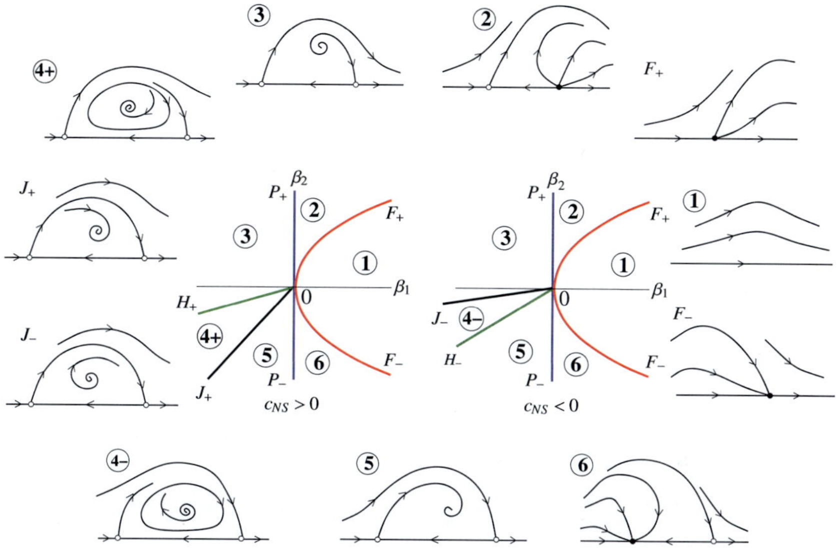

Figure 3.24 Vector field: case 1 ($a_0 > 0, b_0 > 0$).

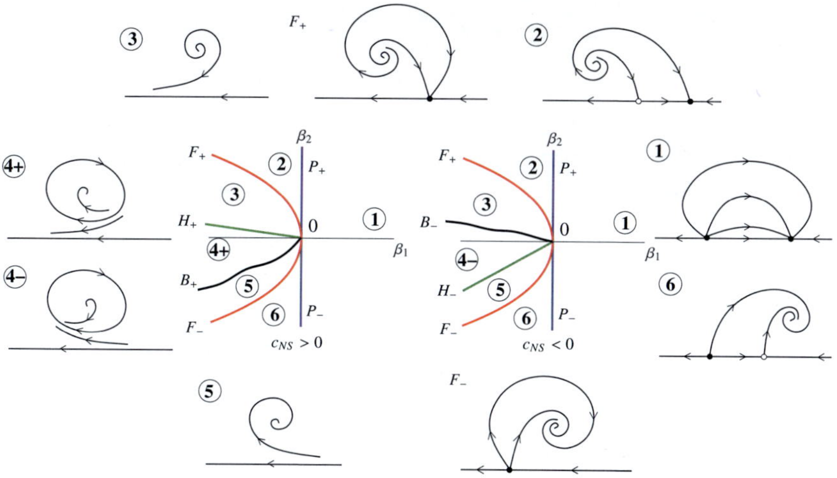

Figure 3.25 Vector field: case 2 ($a_0 < 0, b_0 > 0$).

be either sub- or supercritical, depending on the sign of the first Lyapunov coefficient c_{NS} given by (3.66).

What conclusions about the truncated normal form map (3.59) can be drawn

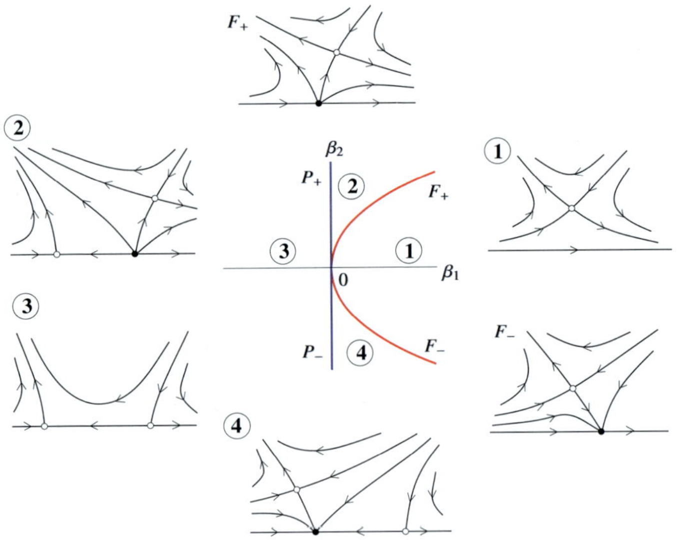

Figure 3.26 Vector field: case 3 ($a_0 > 0, b_0 < 0$).

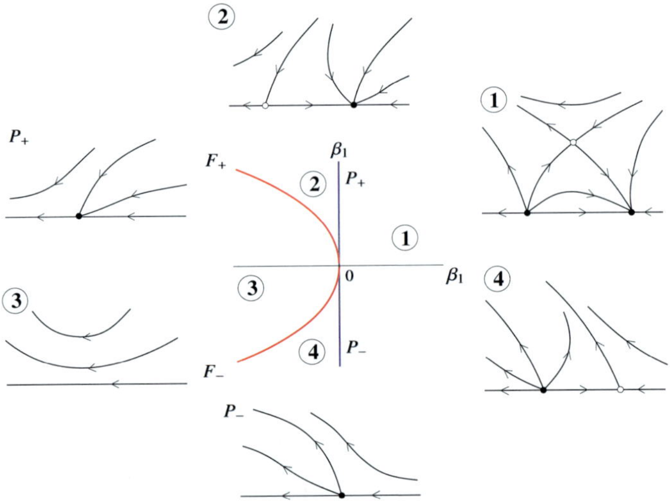

Figure 3.27 Vector field: case 4 ($a_0 < 0, b_0 < 0$).

from the bifurcation diagrams of the approximating ODE (3.68)? The orbits of the map continuously jump from the lower to the upper half-plane and back. This is easily understood from (3.67), which implies that (3.59) can be approximated by the composition of the unit shift along the orbits of X with the reflection $R^{-1} = R$. Thus, any positive equilibrium of (3.68) describes a period-2 cycle of (3.59). We go around the origin in the parameter plane of the approximating ODE and discuss the dynamics of the truncated normal form (3.64). The curve F corresponds to the fold bifurcation curve $LP^{(1)}$ of (3.64), while the pitchfork curve P corresponds to the period-doubling bifurcation curve $PD^{(1)}$ of this map. The positive and negative branches of these curves are denoted by $F_\pm$ and $P_\pm$, respectively. The sub- and supercritical branches of the Hopf bifurcation curve $H_\pm$ in (3.67) correspond to the NS bifurcation curves $NS^{(2)}_\pm$ for the second iterate of (3.59), i.e., the NS bifurcations of the 2-cycle.

Having these interpretations in mind, we can now describe the dynamics of the truncated normal form map (3.59) as follows.

- Case 1. In region **1** orbits merely jump to the right. Crossing F_+ implies the appearance of two fixed points on the horizontal axis. In **2**, one of these fixed points is totally unstable, while the other is a saddle. While crossing curve P_+ from **2** to **3**, the unstable fixed point becomes a saddle and an unstable period-2 cycle appears. If $c_{NS} > 0$, an unstable invariant curve "around" the period-2 cycles appears via the NS bifurcation on H_+, when we go from **3** to **4+**. The invariant curve disappears through a series of bifurcations associated with the heteroclinic bifurcations near J_+, if we come to **5**. The presence of the J-curve for the vector field implies for the map the existence of two curves, along which heteroclinic tangencies occur (see sketches in Figure 3.28). Between these two curves, a heteroclinic structure is present. Figure 3.29 shows unstable and stable manifold configurations computed with MatcontM.

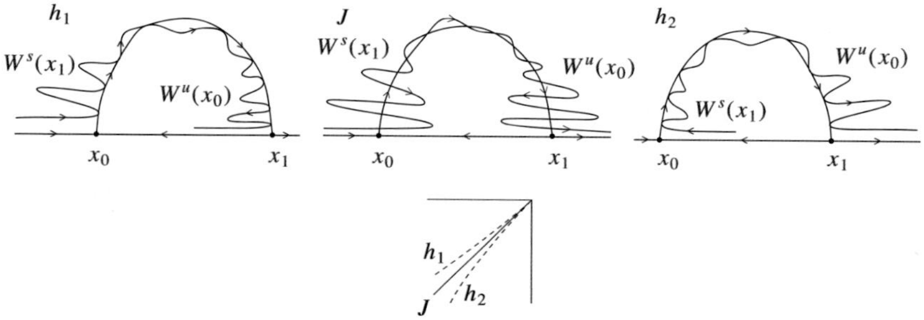

Figure 3.28 Heteroclinic tangencies appearing in the lines $h_{1,2}$ together with a transversal heteroclinic structure between them.

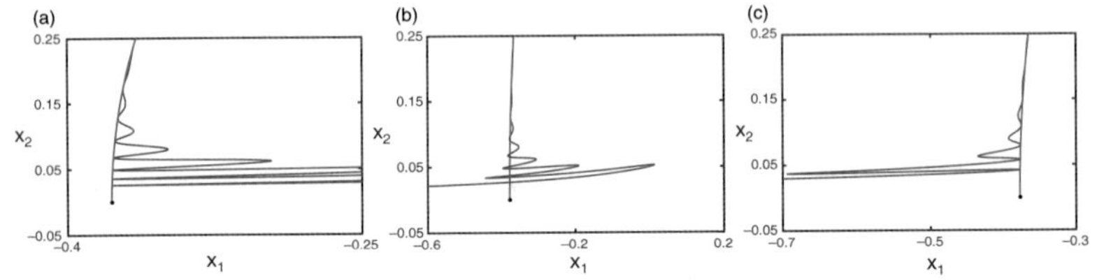

Figure 3.29 Heteroclinic structures in (3.64) near the left fixed point x_0 (black dot) for $a = b = d = 0.5, c = 0.08333, \mu_1 = -0.2$ and $\mu_2 = -0.35346647$ (a); $\mu_2 = -0.35347$ (b); $\mu_2 = -0.3534984$ (c). Red curve denotes $W^u(x_0)$, blue curve is $W^s(x_1)$. Only upper half-plane is shown.

If $c_{NS} < 0$, a stable closed invariant curve emerges in **4**− through a series of bifurcations associated with the heteroclinic structure. This stable invariant curve exists until we cross NS_-, where the stable period-2 cycle become attracting in **5**. Next we cross P_- and the period-2 orbit disappears, leaving us with a stable fixed point and a saddle in **6**. These two collide if we return to **1**.

- Case 2. Fix a phase domain near the origin. Now we start with the two fixed points, one stable and one unstable on the axis in region **1**. Then, crossing the flip curve P_+ to **2**, one fixed point exhibits period-doubling and a period-2 cycle appears. The fixed points on the horizontal axis collide at F_+ separating region **2** from **3**, where a stable period-2 cycle exists. If $c_{NS} > 0$, then an unstable invariant curve appears when we cross the NS bifurcation curve H_+. This invariant curve grows, until it blows up and disappears from the fixed phase domain at some curve B_+. Actually, the invariant curve can lose its smoothness and disappear before touching the boundary of the domain. If $c_{NS} < 0$, then we first encounter the "boundary bifurcation" curve B_-, where a big stable invariant curve appears in our fixed phase domain. The transition from **4**− to **5** destroys the curve via the NS bifurcation. Finally, crossing the fold curve F_- produces two fixed points in **6** and through the flip bifurcation on P_- the period-2 cycle disappears again as we are back in **1**.
- Case 3. We start with a period-2 saddle cycle in **1**. Entering **2** though the fold curve F_+ creates two fixed points on the horizontal axis, a saddle and a repelling one. Then, while crossing the flip curve P_+ to **3**, the period-2 cycle is destroyed and we get two saddles on the x_1-axis. Passing P_- one saddle becomes stable and a period-2 orbit in **4** is created. Finally, the fixed points on the horizontal axis collide on F_- and we are in region **1** again.

- Case 4. Starting in region **1** we have as in case 3, a period-2 saddle cycle, but also a stable and an unstable fixed point on the x_1-axis. The unstable one becomes a saddle when we enter **2** through the P_+ curve. Then nothing special appears but a "saddle-like flow" in **3** after the saddle and the stable point collided on F_+. Going from **3** to **4** we get a saddle and an unstable point through the fold bifurcation on the curve F_-. We are back in **1**, when the flip bifurcation creates the period-two cycle on P_-.

The obtained diagrams give a rather detailed description of bifurcations of the truncated normal form (3.64). However, this description remains incomplete due to the presence of closed invariant curves and heteroclinic tangencies. Indeed, the rotation number on the closed invariant curves can change infinitely many times from rational to irrational and, moreover, the invariant curve can loose smoothness and disappear. Near a heteroclinic tangency, infinite series of bifurcations exist, including cascades of flips and folds (see (Gavrilov and Shilnikov, 1972, 1973; Gonchenko, Shilnikov, and Turaev 1996)).

Effect of higher-order terms

Adding higher order terms to the truncated normal form (3.64), i.e., restoring (3.59), complicates the bifurcation picture further.

Using the Implicit Function Theorem, one can prove that for $||\mu||$ sufficiently small, the map (3.59) has the same *local* bifurcations as (3.64). More precisely, Proposition 3.35 is valid also for the full normal form (3.59) with *any* higher-order terms. Therefore, we know what to expect locally. In particular, in cases 1 and 2 closed invariant curves appear. Moreover, the unit shift along the orbits of the vector field (3.68) composed with the reflection approximates (3.59) as accurately as (3.64). This implies that (3.59) also has two bifurcation curves along which heteroclinic tangencies occur. Between these curves, a heteroclinic structure is present. Higher-order terms in (3.59) do affect these curves, but they both remain tangent to the curve (3.69) from Proposition 3.37.

As mentioned by Gheiner (1994), there are more differences between the phase portraits of (3.64) and a generic (3.59), which are related to other heteroclinic tangencies. Indeed, in the truncated normal form (3.59) the x_1-axis is always invariant. Therefore, in cases 1 and 3 we have the heteroclinic connections between the saddles located on the horizontal axis. However generically, the higher-order terms in (3.59) break the reflection symmetry and the heteroclinic connection along the x_1-axis is lost. This allows for heteroclinic structures caused by intersections of the invariant manifolds of the saddles near the horizontal axis. These intersections can be either transversal (as in Figure 3.30) or tangential. Therefore, in the first three cases, the bifurcation diagrams of (3.64) and a generic (3.59) are not locally topologically equivalent.

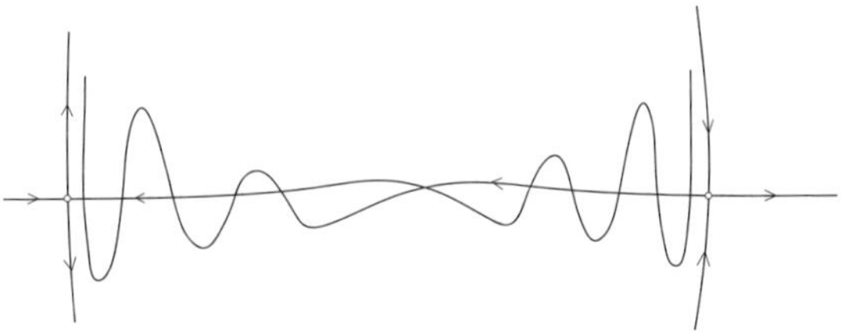

Figure 3.30 A transversal heteroclinic structure near the horizontal axis.

In Gheiner (1994), it is shown that case 3 may have an additional heteroclinic structure: The unstable manifold of the period-2 cycle could intersect tangentially the stable manifold of a saddle fixed point on the x_1-axis. In case 4, Gheiner (1994) gives strong indications that (3.59) is locally topologically equivalent to (3.64), where the cubic terms can be omitted.

3.4.2 LPNS (fold–Neimark–Sacker bifurcation)

This section is devoted to a detailed analysis of case 9 (LPNS) of Table 3.1. This bifurcation occurs for a fixed point having three algebraically simple multipliers $\{1, e^{i\theta_0}, e^{-i\theta_0}\}$ on the unit circle.

Normal form for the fold–Neimark–Sacker bifurcation

Lemma 3.39 *Suppose that* $e^{ik\theta_0} \neq 1$ *for* $k = 1, 2, 3, 4$. *The restriction of the map* (3.2) *to the three-dimensional center manifold at* α_0 *corresponding to the LPNS bifurcation is locally smoothly equivalent near the origin to*

$$\begin{pmatrix} x \\ z \end{pmatrix} \mapsto \begin{pmatrix} x + f^0_{200}x^2 + f^0_{011}z\bar{z} + f^0_{300}x^3 + f^0_{111}xz\bar{z} \\ e^{i\theta_0}z + g^0_{110}xz + g^0_{210}x^2z + g^0_{021}z^2\bar{z} \end{pmatrix} + O(\|(x,z)\|^4), \quad (3.71)$$

where $(x, z) \in \mathbb{R} \times \mathbb{C}$, $f^0_{jkl} \in \mathbb{R}$ *and* $g^0_{jkl} \in \mathbb{C}$.

Proof See Appendix 3.E. □

The expressions for f^0_{jkl} and g^0_{jkl} will be given in Chapter 4.

Theorem 3.40 *If*

$$f^0_{110}f^0_{200} \neq 0, \quad (3.72)$$

then for parameter values α *close to* α_0, *the restriction of* (3.2) *to a parameter-dependent center manifold is locally smoothly equivalent near the origin to the map*

$$\begin{pmatrix} x \\ z \end{pmatrix} \mapsto \begin{pmatrix} x + \beta_1(\alpha) + x^2 + s|z|^2 + C(\alpha)x^3 \\ (1 + \beta_2(\alpha))e^{i\theta(\alpha)}z + A(\alpha)xz + B(\alpha)x^2 z \end{pmatrix} + O(\|(x,z)\|^4), \quad (3.73)$$

where $\beta_1(\alpha_0) = \beta_2(\alpha_0) = 0$ *and* $\theta(\alpha_0) = \theta_0$, *while* $A(\alpha_0) = a_0, B(\alpha_0) = b_0$ *and* $C(\alpha_0) = c_0$ *with*

$$b_0 = \frac{1}{f^0_{011}(f^0_{200})^2}\left(f^0_{011}g^0_{210} + g^0_{110}\left(\frac{1}{2}f^0_{111} + \Re\left(g^0_{021}e^{-i\theta_0}\right)\right) - f^0_{200}g^0_{021}\right),$$

$$a_0 = \frac{g^0_{110}}{f^0_{200}}, \quad c_0 = \frac{f^0_{300}}{(f^0_{200})^2}, \quad s = \text{sign}\left(f^0_{200}f^0_{011}\right) = \pm 1.$$

Proof See Appendix 3.E. □

Suppose that the map $\alpha \mapsto \beta(\alpha) = (\beta_1(\alpha), \beta_2(\alpha))$ is regular at α_0. Then the components of β can be used as the new unfolding parameters.

Bifurcation analysis of the truncated normal form

We can first study the *truncated normal form*

$$\begin{pmatrix} x \\ z \end{pmatrix} \mapsto \begin{pmatrix} x + \beta_1 + x^2 + s|z|^2 + c(\beta)x^3 \\ e^{i\omega(\beta)}z\left(1 + \beta_2 + a(\beta)x + b(\beta)x^2\right) \end{pmatrix}, \quad (3.74)$$

where a, b, c, ω are smooth functions of β such that $a(0) = a_0, b(0) = b_0, c(0) = c_0$, while $\omega(0) = \theta_0$.

Introducing $z = re^{i\varphi}$ we can see that, according to (3.74), $\varphi \mapsto \varphi + \omega(\beta) + \ldots$, which can be seen as a rigid rotation. Meanwhile, the dynamics of (x, r) is independent of φ and is defined by the planar map

$$\begin{pmatrix} x \\ r \end{pmatrix} \mapsto \begin{pmatrix} x + \beta_1 + x^2 + sr^2 + c(\beta)x^3 \\ r\left|1 + \beta_2 + a(\beta)x + b(\beta)x^2\right| \end{pmatrix}. \quad (3.75)$$

This map can be written as

$$(x, r) \mapsto N_\beta(x, r) + O(\|\beta\|^2) + O(\|\beta\| \, \|(x, r)\|^2) + O(\|(x, r)\|^4),$$

where

$$N_\beta(x, r) := \begin{pmatrix} x + \beta_1 + x^2 + sr^2 + c_0 x^3 \\ r + \beta_2 r + d_0 xr + d_1 x^2 r \end{pmatrix} \quad (3.76)$$

with

$$d_0 = \Re(a_0), \; d_1 = \frac{1}{2}[\Im(a_0)]^2 + \Re(b_0).$$

The map (3.76) is $\mathbb{Z}_2$-symmetric, i.e., equivariant with respect to the transformation $r \mapsto -r$. The line $r = 0$ is invariant. Local bifurcations of (3.76) can be analyzed directly. Recall that this map is obtained under the assumption (3.72).

Lemma 3.41 *The family of maps* (3.76) *has the following local codim 1 bifurcations in a sufficiently small neighborhood of* $(x, r, \beta) = (0, 0, 0)$.

1. *There is a curve* $LP^{(1)}: (x, r, \beta_1) = (0, 0, 0)$ *on which a non-degenerate fold bifurcation occurs.*
2. *There is a curve*
$$PF^{(1)}: (x, r, \beta_1) = \left(-\frac{1}{d_0}\beta_2 + O(\beta_2^2), 0, -\frac{1}{d_0^2}\beta_2^2 + O(\beta_2^3)\right)$$
on which a pitchfork bifurcation occurs if $d_0 \neq 0$.
3. *For* $s\beta_1 < 0$, *there is a curve*
$$NS^{(1)}: (x, r^2, \beta_2) = \left(-d_0\beta_1 + O(\beta_1^2), -s\beta_1 + O(\beta_1^2), d_0^2\beta_1 + O(\beta_1^2)\right),$$
on which a non-degenerate NS bifurcation of a nontrivial fixed point occurs, provided
$$C_{NS} := [3d_0(c_0 + d_0) - 2(d_0 + d_1)]d_0 \neq 0. \tag{3.77}$$
Moreover, the NS bifurcation is supercritical if $C_{NS} < 0$ *and subcritical if the opposite inequality holds.*

Proof See Appendix 3.E. □

The closed invariant curve born from the NS bifurcation cannot exist everywhere. It should disappear through some global bifurcation. To study these global bifurcation phenomena, we derive a vector field ("approximating ODE"), such that the unit shift along its orbits approximates (3.76). We have encountered this situation in Section 3.4.1, so our discussion here will be brief.

Theorem 3.42 *For all sufficiently small* $\|\beta\|$, *the map* (3.76) *can be represented near* $(x, r) = (0, 0)$ *as*
$$N_\beta(x, r) = \varphi_\beta^1(x, r) + O(\|\beta\|^2) + O(\|\beta\| \, \|(x, r)\|^2) + O\left(\|(x, r)\|^4\right), \tag{3.78}$$
where φ_β^t *is the flow of the approximating planar system*
$$\begin{cases} \dot{x} = \beta_1 - \beta_1 x + x^2 + sr^2 + (c_0 - 1)x^3 - s(d_0 + 1)xr^2, \\ \dot{r} = \left(\beta_2 - \frac{1}{2}d_0\beta_1\right)r + d_0 xr + \left(d_1 - \frac{1}{2}d_0(d_0 + 1)\right)x^2 r - \frac{1}{2}sd_0 r^3. \end{cases} \tag{3.79}$$

Proof See Appendix 3.E. □

Note that the unit-time flow of (3.79) approximates the map (3.75) with the same accuracy and has the same $\mathbb{Z}_2$-symmetry as (3.76).

One can verify that the planar approximating system (3.79) for small $\|\beta\|$ has two curves

$$F = \left\{(\beta_1,\beta_2) \in \mathbb{R}^2 : \beta_1 = 0\right\}, \quad P = \left\{(\beta_1,\beta_2) \in \mathbb{R}^2 : \beta_1 = -\beta_2^2/d_0^2 + O(\beta_2^3)\right\},$$

on which equilibria with zero eigenvalue exist. These are the same expansions as we computed for the map (3.76); see curves $LP^{(1)}$ and $PF^{(1)}$ in Lemma 3.41. On F, the fold bifurcation happens that generates two fixed points on the x-axis. On P, two symmetric fixed points with $r \neq 0$ branch from a fixed point on the x-axis. The approximating system also exhibits a Hopf bifurcation at the curve

$$H = \left\{(\beta_1,\beta_2) \in \mathbb{R}^2 : \beta_2 = d_0^2\beta_1 + O(\beta_1^2),\ s\beta_1 < 0\right\}.$$

This expansion corresponds to that for $NS^{(1)}$ in Lemma 3.41. Moreover, the first Lyapunov coefficient that determines the stability of the bifurcating limit cycle is given by

$$l_1 = C_+C_{NS} + o(\beta_1),$$

where $C_+ > 0$ is some positive constant and C_{NS} is defined in (3.72). Thus, as one could have expected, the Hopf bifurcation generating a limit cycle corresponds to the NS bifurcation producing a closed invariant curve, and the criticality of these bifurcations near $\beta = 0$ is the same.

Next, we can prove the following result that is analogous to Proposition 3.37 in Section 3.4.1.

Lemma 3.43 *If $s = 1$ and $d_0 < 0$, then for $\beta_1 < 0$ the system* (3.79) *has two saddles, which are always connected by a heteroclinic orbit along the x-axis. Moreover, there exists a heteroclinic orbit with $r > 0$ connecting these saddles for*

$$J : \beta_2 = \frac{3d_0c_0 - 2d_1}{2 - 3d_0}\beta_1 + o(\beta_1). \tag{3.80}$$

Proof See Appendix 3.E. □

We note that if the linear approximations to the Hopf and heteroclinic bifurcation curves collide, this implies $C_{NS} = 0$. So this happens only in a more degenerate case.

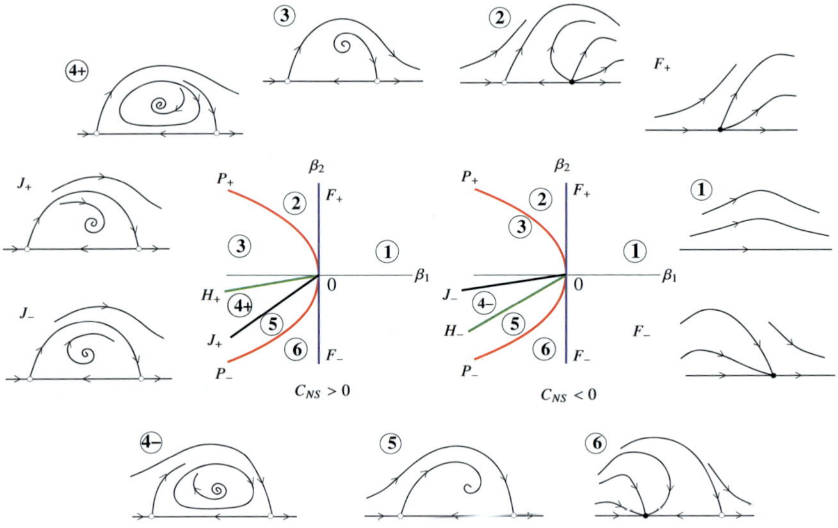

Figure 3.31 Case 1: bifurcation diagram of (3.78) for $s = 1, d_0 < 0$.

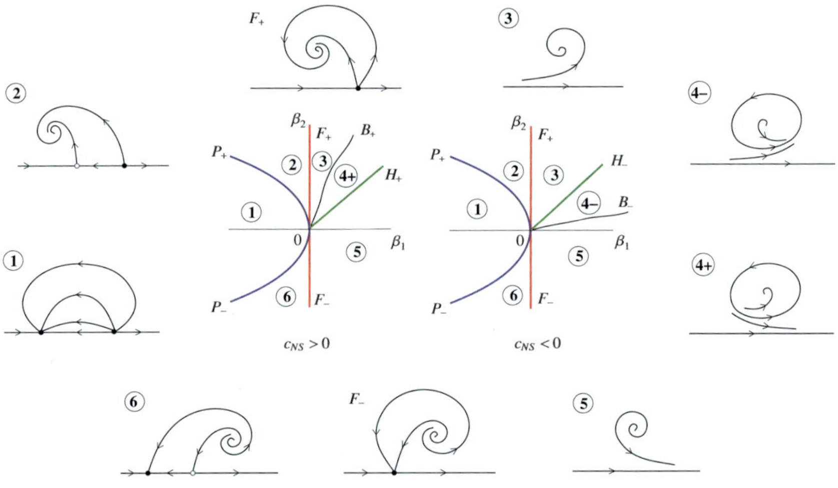

Figure 3.32 Case 2: bifurcation diagram of (3.78) for $s = -1, d_0 > 0$.

Bifurcation diagrams

The bifurcation diagrams of the approximating system (3.78) are very similar to those of the approximating system (3.68) in Section 3.4.1 devoted to the LPPD bifurcation.

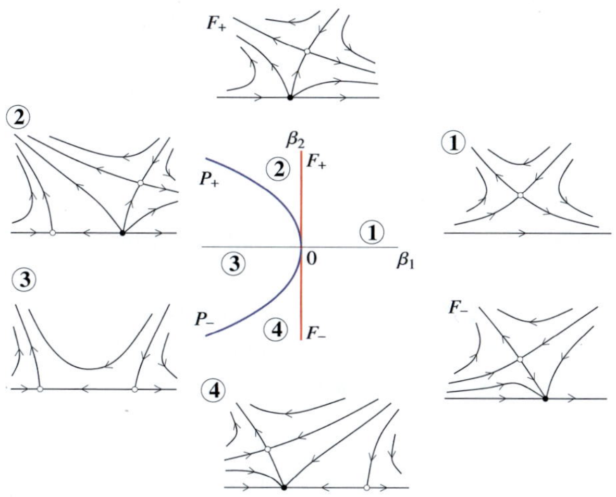

Figure 3.33 Case 3: bifurcation diagram of (3.78) for $s = -1, d_0 < 0$.

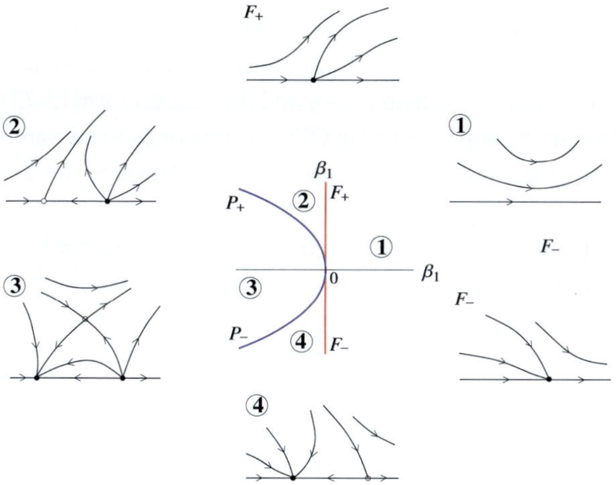

Figure 3.34 Case 4: bifurcation diagram of (3.78) for $s = 1, d_0 > 0$.

There are four cases corresponding to different combinations of signs of s and d_0, which are presented in Figures 3.31–3.34.

In all cases, the planar approximation system (3.78) has at most two trivial equilibria with $r = 0$ and one nontrivial equilibrium with $r > 0$. The trivial equilibria can collide and disappear via the fold bifurcation at curve $F_\pm$, while the nontrivial one branches from a trivial one via a pitchfork bifurcation at curve $P_\pm$. In cases 1 and 2, the nontrivial equilibrium exhibits a (sub- or supercritical) Hopf bifurcation on curve $H_\pm$. The fate of the generated single limit cycle, however, is different in these cases. In case 1, it disappears via a global heteroclinic bifurcation on the corresponding curve $J_\pm$, while in case 2 it leaves a small fixed neighborhood of the origin at some curve $B_\pm$ and is no longer described by the local theory. On the contrary, in cases 3 and 4, the nontrivial equilibrium is a saddle that does not bifurcate.

Effect of higher-order terms

As usual, the transition from the smooth hypernormal form (3.73) to the approximating ODE (3.78) involved two steps:

(1) truncation of the $O(\|(x,z)\|^4)$-terms that results in the 3D map (3.74) and those in the amplitude planar map N_β given by (3.76) as well as assuming rigid rotation around the r-axis;
(2) neglecting the difference between N_β and the time-1 shift along orbits of the approximating planar ODE.

The fold bifurcation of the trivial equilibrium points in the approximating system (3.78) corresponds to a fold (LP) bifurcation of the trivial fixed points in both the planar amplitude map (3.76) and the normal form (3.73). The pitchfork bifurcation in the approximating ODE corresponds to the same bifurcation in the amplitude map, but – taking into account rotation in the φ-coordinate – generates a closed invariant curve in the truncated normal form both in (3.74) and in the full normal form (3.73). This is clearly a Neimark–Sacker bifurcation near the LPNS codim 2 bifurcation in the original map. As usual, the appearing invariant curve is only (finitely) smooth and exists only near the Neimark–Sacker bifurcation, while the orbit structure on it depends on higher-order terms and exhibits an infinite number of LP bifurcations of long-periodic cycles confined to this curve.

The Hopf bifurcation in the approximating system (3.78) implies a NS bifurcation in the planar amplitude map (3.76) and will imply a *quasi-periodic bifurcation* of the map (3.73), in which the closed invariant curve changes stability and a two-dimensional invariant torus bifurcates from it. This complicated phenomenon will be discussed below.

The heteroclinic bifurcation in the approximating system (3.78) gives rise to heteroclinic tangencies of the 1D invariant manifolds of the trivial saddle fixed points in the planar amplitude map (3.76) and splitting of the corresponding 2D invariant manifolds of the saddles in the full map (3.73). Moreover, the line $r = 0$ is no longer invariant for this map, which further complicates its bifurcation diagram.

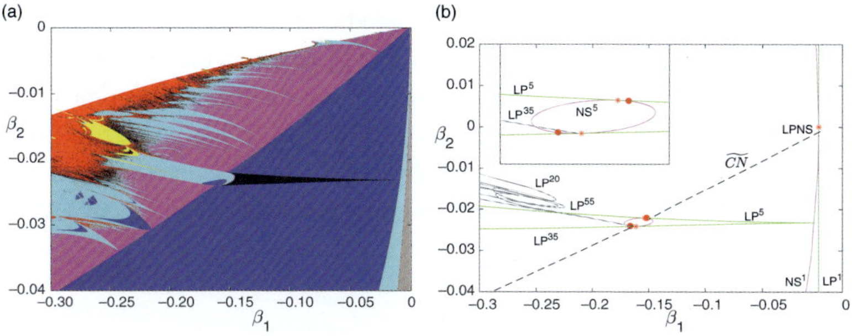

Figure 3.35 Bifurcation diagram of model map (3.81). (a) Inventory using Lyapunov exponents. Colors indicate type of attractor: cycles (gray–yellow–black), invariant curve (light/dark blue), 2-torus (magenta), chaos (red). (b) Local bifurcations including resonances tongues of period 5,20,35,55. The dashed line sketches the location of the quasi-periodic bifurcation set $\widetilde{CN}$.

To have an impression of a generic unfolding involving the case $s = 1, d_0 > 0$, i.e., with the heteroclinic bifurcation, we present results of scanning Lyapunov exponents of a representative perturbation. For the perturbation we start with the normal form (3.74). Next we add fourth-order terms that break the symmetries. Additional fourth-order terms also induce a weak 1:5 resonance on the invariant curve emerging from the NS bifurcation. For this we tune the angle $\theta = 1.29 + .4\beta_2$. This results in the following perturbed normal form

$$\begin{pmatrix} x \\ z \end{pmatrix} \mapsto \begin{pmatrix} x + \beta_1 + x^2 + s|z|^2 + Cx^3 + \varepsilon_1 \left(\Re(z)^4 + \Im(z)^4\right), \\ (1+\beta_2)e^{i\theta(\beta_2)} z \left(1 + Ax + Bx^2\right) + \varepsilon_1(1+i)x^4 + +\varepsilon_2 \left(i\Re(z)^4 + \Im(z)^4\right) \end{pmatrix}. \tag{3.81}$$

We have set $s = 1, A = -0.4 + 0.4i, B = -0.1 - 0.2i, C = -0.6, \varepsilon_1 = 0.1, \varepsilon_2 = 0.2$. For this choice, the NS as well as the quasi-periodic and the heteroclinic bifurcation occur in the third quadrant of the β-plane. Next we computed Lyapunov exponents for a fixed value of β_1 and varied β_2 starting at $\beta_2 = -0.04$ increasing until $\beta_2 = 0$ using the previous state as the initial condition. The result is shown in Figure 3.35.

Starting with $\beta_1 \approx 0$ at the right of Figure 3.35, we start with a fixed point (gray). Moving to the left we encounter a supercritical NS bifurcation. Initially, the invariant curve is of node type (light blue) and then switches to focus type (dark blue) as we decrease β_1 further. For $\beta_2 \approx -0.023$ we observe a 1:5 resonance tongue (black). Next we increase β_2 to encounter the quasi-periodic bifurcation $\widetilde{CN}$ where the invariant curves becomes unstable and a stable 2-torus appears (magenta). On the torus, resonances may occur, visible as the light blue tongues emerging from the quasi-periodic bifurcation. Noteworthy is the resonance bubble where the 1:5 resonance tongue meets the quasi-periodic bifurcation. Increasing β_2 further, the dynamics on the 2-torus becomes chaotic (red). This zone shows also attracting cycles of high period (<100 [yellow] and >100 [black]). Finally, the heteroclinic bifurcation is encountered and no attractors are found anymore except for a few resonant invariant curves.

3.5 Flip–Neimark–Sacker and double Neimark–Sacker bifurcations

We study codim 2 bifurcations of fixed points of dissipative diffeomorphisms with a pair of complex critical eigenvalues together with either an eigenvalue -1 or another such pair. We call them the *flip–Neimark–Sacker* (PDNS) and *double Neimark–Sacker* (NSNS) bifurcations, respectively. This work is not the first contribution on the topic, since such bifurcations have been found in models in Lindtner, Steindl, and Troger (1989); Wen and Xu (2004a,b) but not understood, and in Ding, Xie, and Sun (2004) and Xie and Ding (2005) only a limited analysis is done. Moreover these cases have also been studied by Iooss and Los (1988) and Los (1988; 1989). However those studies were concerned only with interesting but specific details, namely the quasi-periodic bifurcation of a "period-2" invariant circle, which originated via a "doubling" of a "period-1" invariant circle, and a quasi-periodic bifurcation of the invariant circle into a 2-torus ("torification") in the four-dimensional case. Notice that the latter works do not use the term quasi-periodic bifurcation, while in Braaksma and Broer (1987); Broer et al. (1990) and Broer, Huitema, and Sevryuk (1996b) the quasi-periodic bifurcation theory is set up systematically. All these results exclude a region with so-called Chenciner bubbles, where resonances occur. In some unfoldings of these cases, these tori bifurcate once more, into higher-dimensional tori. For such cases, a higher-order normal form is necessary to establish stability results. Here we give a rather complete description of these bifurcations by making the right correspondence with the double Hopf bifurcation of vector fields. Moreover, we study numerically representative perturbations, which break the symmetry of the normal forms. This reveals a peculiar bifurcation structure of the bubbles. Such bubbles were first described by Chenciner (1988) and more recently in the fold-NS case by Vitolo (2003); Broer, Simó, and Vitolo (2008b,a) and Vitolo, Broer, and Simó (2010).

Therefore, the transversality can be checked in the original real coordinates as the regularity of the map G_1 or G_2 at the origin. We refer to Appendix 3.F for some computations related to this remark. A method to verify the transversality as presented by Wen and Xu (2004b) involves the computation of eigenvectors.

Reduction to an amplitude map

As both truncated normal forms (3.85) and (3.86) are equivariant under rotations and reflections, we transform to polar coordinates. Then we truncate higher-order terms and ignore for the moment the dynamics in the angles. Both maps are now reduced to the same truncated amplitude map:

$$H_\mu\colon \begin{pmatrix} x \\ y \end{pmatrix} \mapsto \begin{pmatrix} (1+\mu_1)x + a_{11}x^3 + a_{12}xy^2 + h_{50}x^5 + h_{32}x^3y^2 + h_{14}xy^4 \\ (1+\mu_2)y + a_{21}x^2y + a_{22}y^3 + h_{41}x^4y + h_{23}x^2y^3 + h_{05}y^5 \end{pmatrix}. \tag{3.87}$$

As before, we suppress the parameter dependence of the coefficients a_{ij}. This affects the expressions for the bifurcation curves of the NS and heteroclinic bifurcations in Proposition 3.47, but not the stability of the invariant circle, see also Remark 3.50 in Section 3.5.2. In that sense the suppression is harmless.

For the flip–NS bifurcation we introduce cylindrical coordinates (x, r, ψ), where $z = re^{i\psi}$. Then map (3.85) becomes

$$\begin{pmatrix} x \\ r \\ \psi \end{pmatrix} \mapsto \begin{pmatrix} x\left(-(1+\mu_1) + a_{11}x^2 + a_{12}r^2 + h_{50}x^4 + h_{32}x^2r^2 + h_{14}r^4\right) \\ r\left(1+\mu_2 + a_{21}x^2 + a_{22}r^2 + h_{41}x^4 + h_{23}x^2r^2 + h_{05}r^4\right) \\ \psi + \phi(\mu) + \Im(e^{-i\phi(\mu)}g_{021})r^2 + \Im(e^{-i\phi(\mu)}g_{210})x^2 \end{pmatrix} + O\begin{pmatrix} 0 \\ \|(x,r)\|^6 \\ \|(x,r)\|^3 \end{pmatrix}. \tag{3.88}$$

Note that the first two components of (3.88) are independent of ψ. Ignoring the ψ-equation and the higher-order terms for the moment, we identify (3.85) with (3.87) by composing (3.88) with the reflection $R = \begin{pmatrix} -1 & 0 \\ 0 & 1 \end{pmatrix}$. The correspondence is then given by the formulas a_{ij} and h_{ij} as

$$\begin{pmatrix} a_{11} & a_{12} \\ a_{21} & a_{22} \end{pmatrix} = \begin{pmatrix} -f_{300} & -f_{111} \\ \Re(e^{-i\phi}g_{210}) & \Re(e^{-i\phi}g_{021}) \end{pmatrix}$$

and

$$\begin{aligned} h_{50} &= f_{500}, & h_{41} &= \Re(e^{-i\phi}g_{410}) + \tfrac{1}{2}\Im(e^{-i\phi}g_{210})^2, \\ h_{32} &= f_{311}, & h_{23} &= \Re(e^{-i\phi}g_{311}) + \Im(e^{-i\phi}g_{210})\Im(e^{-i\phi}g_{021}), \\ h_{14} &= f_{122}, & h_{05} &= \Re(e^{-i\phi}g_{032}) + \tfrac{1}{2}\Im(e^{-i\phi}g_{021})^2. \end{aligned}$$

For the NS–NS bifurcation we introduce, as above, polar coordinates $(r_1, r_2, \psi_1, \psi_2)$, where $(w, z) = (r_1 e^{\psi_1}, r_2 e^{\psi_2})$. Then (3.86) transforms into

$$\begin{pmatrix} r_1 \\ r_2 \\ \psi_1 \\ \psi_2 \end{pmatrix} \mapsto \begin{pmatrix} r_1(1 + \mu_1 + a_{11}r_1^2 + a_{12}r_2^2 + h_{50}r_1^4 + h_{32}r_1^2r_2^2 + h_{14}r_2^4) \\ r_2(1 + \mu_2 + a_{21}r_1^2 + a_{22}r_2^2 + h_{41}r_1^4 + h_{23}r_1^2r_2^2 + h_{05}r_2^4) \\ \psi_1 + \phi_1(\mu) + \Im(e^{-i\phi_1(\mu)}g_{2100})r_1^2 + \Im(e^{-i\phi_1(\mu)}g_{1011})r_2^2 \\ \psi_2 + \phi_2(\mu) + \Im(e^{-i\phi_2(\mu)}g_{1110})r_1^2 + \Im(e^{-i\phi_2(\mu)}g_{0021})r_2^2 \end{pmatrix} + O\begin{pmatrix} \|(r_1, r_2)\|^6 \\ \|(r_1, r_2)\|^6 \\ \|(r_1, r_2)\|^3 \\ \|(r_1, r_2)\|^3 \end{pmatrix}, \tag{3.89}$$

where the coefficients are given by

$$\begin{pmatrix} a_{11} & a_{12} \\ a_{21} & a_{22} \end{pmatrix} = \begin{pmatrix} \Re(e^{-i\phi_1} f_{2100}) & \Re(e^{-i\phi_1} f_{1011}) \\ \Re(e^{-i\phi_2} g_{1110}) & \Re(e^{-i\phi_2} g_{0021}) \end{pmatrix}$$

and

$$\begin{aligned} h_{50} &= \Re(e^{-i\phi_1} f_{4100}) + \tfrac{1}{2}\Im(e^{-i\phi_1} f_{2100})^2, \\ h_{41} &= \Re(e^{-i\phi_2} g_{2210}) + \tfrac{1}{2}\Im(e^{-i\phi_2} g_{1110})^2, \\ h_{32} &= \Re(e^{-i\phi_1} f_{2111}) + \Im(e^{-i\phi_1} f_{2100})\Im(e^{-i\phi_1} f_{1011}), \\ h_{23} &= \Re(e^{-i\phi_2} g_{1121}) + \Im(e^{-i\phi_2} g_{1110})\Im(e^{-i\phi_2} g_{0021}), \\ h_{14} &= \Re(e^{-i\phi_1} f_{1022}) + \tfrac{1}{2}\Im(e^{-i\phi_1} f_{1011})^2, \\ h_{05} &= \Re(e^{-i\phi_2} g_{0032}) + \tfrac{1}{2}\Im(e^{-i\phi_2} g_{0021})^2. \end{aligned}$$

As in the flip–NS case, the first two components are independent of ψ_1 and ψ_2, and we obtain the identification of (3.86) with (3.87), if the higher-order terms are truncated.

Hyper-normalization

We see that four cubic terms remain in the amplitude map. We can use these cubic coefficients of the resonant monomials to remove some of the fifth-order terms. In fact, the following statement holds.

Proposition 3.46 *If $a_{11}a_{22} \neq 0$ and $a_{12}a_{21}(a_{12} - a_{22}) \neq 0$, then the family H_μ given by* (3.87) *is locally smoothly equivalent to a family with the 5-jet*

$$F_\mu : \begin{pmatrix} x \\ y \end{pmatrix} \mapsto \begin{pmatrix} x(1 + \mu_1) + s_1x^3 + s_2\theta xy^2 + c_1x^5 \\ y(1 + \mu_2) + s_1\delta x^2y + s_2y^3 + c_4x^4y + c_6y^5 \end{pmatrix}, \tag{3.90}$$

where $s_1 = \operatorname{sign} a_{11}$, $s_2 = \operatorname{sign} a_{22}$,

$$\theta = \frac{a_{12}}{a_{22}},\ \delta = \frac{a_{21}}{a_{11}}, c_1 = \frac{h_{50}}{(a_{11})^2},\ c_6 = \frac{h_{05}}{(a_{22})^2}$$

and

$$c_4 = \frac{h_{41}}{(a_{11})^2} + \frac{a_{21}}{(a_{11})^2}\left(\frac{h_{32}}{a_{12}} - \frac{h_{14}(a_{11}-a_{21})}{a_{12}(a_{12}-a_{22})} - \frac{h_{23}(a_{11}-a_{21})}{a_{21}(a_{12}-a_{22})}\right).$$

Proof See Appendix 3.F. □

3.5.2 Bifurcation analysis of symmetric normal forms

Bifurcations of the hyper-normalized amplitude map

In this section we study bifurcations of the planar map F_μ defined by (3.90). We recall that this map corresponds to the truncated normal forms (3.85) and (3.86). As mentioned before, it appears in other applications involving symmetry-breaking. First, we study fixed points of (3.90) and their stability. Then we consider the correspondence between (3.90) and the full maps (3.85) and (3.86). Due to the symmetries it is enough to consider the positive quadrant in the (x, y)-plane.

Proposition 3.47 *Suppose the coefficients of the map* (3.90) *satisfy the following non-degeneracy conditions*

(D.1) $s_1 s_2 \theta\delta \neq 0$.

(D.2) $\delta\theta - 1 \neq 0$.

(D.3) $\theta, \delta \neq 1$.

(D.4) $L_{NS} = s_1\left(\frac{12(2\delta\theta-\delta-\theta)}{\theta(\theta-1)} + c_1\frac{8(2\delta\theta-\delta-1)}{(\theta-1)(\delta\theta-1)} - c_4\frac{8}{(\delta\theta-1)} + c_6\frac{8\delta(2\delta\theta-\theta-1)}{\theta(\theta-1)(\delta\theta-1)}\right) \neq 0.$

Then, for sufficiently small $\|\mu\|$ *the map* (3.90) *exhibits the following local bifurcations.*

- *The trivial fixed point exhibits a pitchfork bifurcation at* $T_1 = \{\mu \in \mathbb{R}^2 : \mu_1 = 0\}$. *The semitrivial fixed point* $(x, y) = \left(\sqrt{-s_1\mu_1}, 0\right)$ *is stable if* $s_1 < 0$ *and* $\mu_2 - s_1\delta\mu_1 < 0$ *and totally unstable if the inequality signs are reversed. Otherwise the fixed point is a saddle.*
- *The trivial fixed point exhibits another pitchfork bifurcation at* $T_2 = \{\mu \in \mathbb{R}^2 : \mu_2 = 0\}$. *The semitrivial fixed point* $(x, y) = \left(0, \sqrt{-s_2\mu_2}\right)$ *is stable if* $s_2 < 0$ *and* $\mu_1 - s_2\theta\mu_2 < 0$ *and totally unstable if the inequality signs are reversed. Otherwise the fixed point is a saddle.*
- *If both* $\rho_1^2 = \frac{(\mu_1-\theta\mu_2)}{s_1(\delta\theta-1)} > 0$ *and* $\rho_2^2 = \frac{(\mu_2-\delta\mu_1)}{s_2(\delta\theta-1)} > 0$ *hold, then there is a nontrivial fixed point* $(x, r) = (\rho_1, \rho_2) + O(\|\mu\|^2)$. *The fixed point is stable if* $s_1 s_2(\delta\theta - 1) < 0$ *and* $(s_1\rho_1^2 + s_2\rho_2^2) < 0$. *It is unstable if both inequality signs are reversed.*

  The semitrivial fixed points on the x- and y-axes undergo secondary pitchfork bifurcations at $T_3 = \{\mu \in \mathbb{R}^2 : \mu_1 = \theta\mu_2\}$ *when* $\rho_2 > 0$, *or at* $T_4 = \{\mu \in \mathbb{R}^2 : \mu_2 = \delta\mu_1\}$ *when* $\rho_1 > 0$.

- *If* $s_2 = -s_1$ *and* $\delta\theta > 1$, *then the nontrivial fixed point exhibits an NS bifurcation. This happens along the curve* $NS = \{\mu \in \mathbb{R}^2\colon \mu_2 = \mu_{2,NS}(\mu_1)\}$, *where*

$$\mu_{2,NS} = -\tfrac{\delta-1}{\theta-1}\mu_1 + \left(\tfrac{2(\delta\theta-1)^2+(2\delta\theta-\delta-1)c_1-(\theta-1)c_4+(2\delta\theta-\theta-1)c_6}{(\theta-1)^3}\right)\mu_1^2 + O(\mu_1^3),$$

for $\mu_1(\theta-1)s_1 > 0$. *The bifurcating closed invariant curve is stable if* $\mu_1 L_{NS} < 0$ *and unstable if* $\mu_1 L_{NS} > 0$.

Proof See Appendix 3.F. □

Theorem 3.48 *For all sufficiently small* $\|\mu\|$, *the map* (3.90) *can be represented near* $(x,y) = (0,0)$ *as*

$$\begin{aligned} F_\mu(x,y) = {} & \varphi_\mu^1(x,y) \\ & + O(\|\mu\|^3\,\|(x,y)\|) + O(\|\mu\|^2\,\|(x,y)\|^3) + O(\|\mu\|\,\|(x,y)\|^5) \\ & + O\left(\|(x,y)\|^6\right), \end{aligned} \tag{3.91}$$

where φ_μ^t *is the flow of an approximating planar system that is linearly equivalent to*

$$\begin{pmatrix} \dot{x} \\ \dot{y} \end{pmatrix} = \begin{pmatrix} x(\mu_1 - \frac{1}{2}\mu_1^2 + \tilde{a}_{11}x^2 + \tilde{a}_{12}y^2 + \tilde{c}_1x^4 + \tilde{c}_2x^2y^2 + \tilde{c}_3y^4) \\ y(\mu_2 - \frac{1}{2}\mu_2^2 + \tilde{a}_{21}x^2 + \tilde{a}_{22}y^2 + \tilde{c}_4x^4 + \tilde{c}_5x^2y^2 + \tilde{c}_6y^4) \end{pmatrix}, \tag{3.92}$$

where $\tilde{a}_{ij}, \tilde{c}_i$ *are given as*

$$\begin{aligned} \tilde{a}_{11} &= s_1, & \tilde{a}_{12} &= -s_1\theta(1-\mu_1+\mu_2), \\ \tilde{a}_{21} &= s_1\delta(1+\mu_1-\mu_2), & \tilde{a}_{22} &= -s_1, \\ \tilde{c}_1 &= c_1 - \tfrac{3}{2}, & \tilde{c}_4 &= c_4 - \delta(1+\tfrac{1}{2}\delta), \\ \tilde{c}_2 &= \theta(\delta+2), & \tilde{c}_5 &= \delta(\theta+2), \\ \tilde{c}_3 &= -\theta(1+\tfrac{1}{2}\theta), & \tilde{c}_6 &= c_6 - \tfrac{3}{2}. \end{aligned}$$

Proof Performing the standard Picard iterations, we find a vector field containing all terms displayed in (3.92). After applying the scaling

$$(x,y) \to (x/\sqrt{1-2\mu_1},\, y/\sqrt{1-2\mu_2}),$$

we obtain the above expressions. □

Proposition 3.49 *Consider the map* (3.90) *satisfying the conditions listed in Proposition 3.47. If* $s_2 = -s_1$ *and* $\delta\theta > 1$, *and* $\theta < 0$, *then stable and unstable invariant manifolds of the semitrivial fixed points are tangent along two exponentially close bifurcation curves* $HET_{1,2}$, *whose quadratic approximation is given by*

$$\begin{aligned} \mu_{2,HET} = -\frac{\delta-1}{\theta-1}\mu_1 + \Bigg(& \frac{(\delta\theta-1)^2}{2(\theta-1)^3} - \frac{\delta(2\delta\theta-\delta-1)}{(2\delta\theta-\theta-\delta)(\theta-1)}c_1 + \frac{\delta}{(2\delta\theta-\theta-\delta)}c_4 \\ & - \frac{\theta(2\delta\theta-\theta-1)(\delta-1)^2}{(2\delta\theta-\theta-\delta)(\theta-1)^3}c_6\Bigg)\mu_1^2 + O(\mu_1^3). \end{aligned}$$

Proof See Appendix 3.F. □

Remark 3.50 Suppose that the heteroclinic tangencies occur. Then we have

$$\mu_{2,NS} - \mu_{2,HET} = \frac{\theta(\delta\theta - 1)^2}{8(\theta - 1)^2(2\delta\theta - \theta - \delta)}\mu_1^2 s_1 L_{NS} + O(\mu_1^3).$$

Therefore, the quadratic approximations of these curves do not coincide under the imposed non-degeneracy conditions. If we include the parameter dependence of the cubic coefficients, both $\mu_{2,NS}$ and $\mu_{2,HET}$ are affected in the same manner. Their difference is still being proportional to $\mu_1^2 L_{NS}$.

Remark 3.51 If $s_1 s_2 < 0, \delta\theta > 1$ and $\theta > 0$ then the invariant circle of the map (3.90) blows up and disappears through the collision with a fixed boundary of the phase plane (see also case 2 of the fold–flip bifurcation in Section 3.4.1).

The conditions on and signs of the coefficients in the proposition lead to several different unfoldings. We may assume that $\delta \leq \theta$, otherwise we can interchange x and y. We cannot scale time, but since we are working with diffeomorphisms, we can invert the map. So if $s_1 s_2 < 0$, we may assume $\mu_1 L_{NS} < 0$ and there are six different unfoldings. If $s_1 s_2 > 0$, there are five different unfoldings (see Figure 3.36). For each unfolding there is a bifurcation diagram (see Figures 3.37 and 3.39). Then for each diagram we have a sequence of phase portraits as sketched in Figures 3.38 and 3.40. These portraits correspond to the planar system (3.92) approximating (3.90).

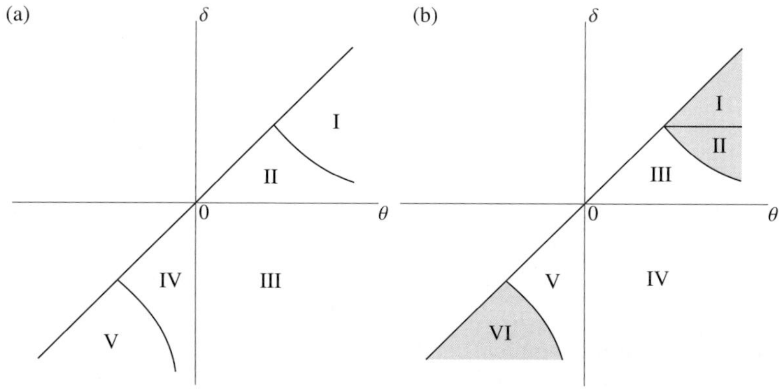

Figure 3.36 There are five or six different unfoldings depending on $s_1 s_2 > 0$ (a) or $s_1 s_2 < 0$ (b), respectively. These are then determined further by θ and δ. The shaded areas indicate the presence of a closed invariant curve in the amplitude map.

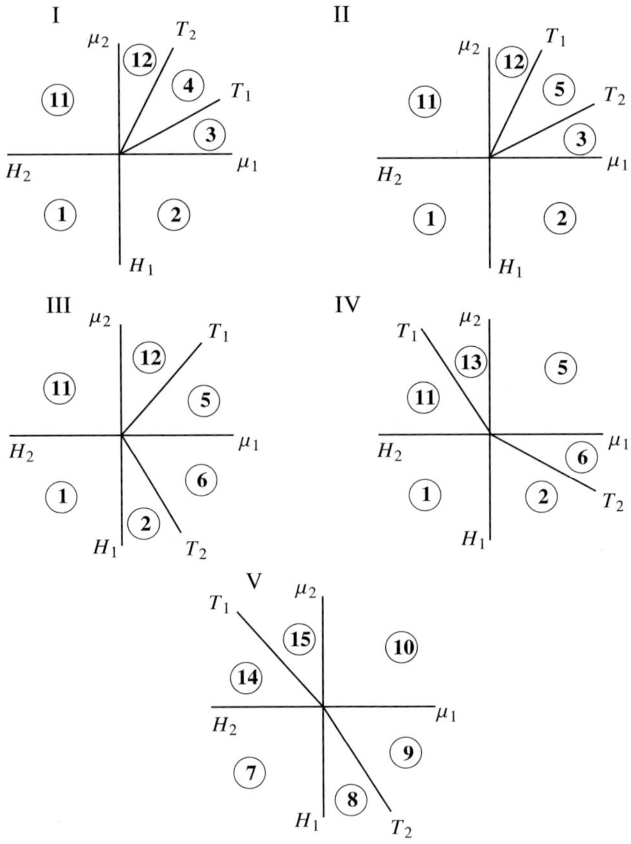

Figure 3.37 Bifurcation diagrams for the simple case $s_1 s_2 > 0$.

Bifurcations of NF_1 and NF_2

In Section 3.5.1 we reduced the normal forms to a planar map by factoring out the $\mathbb{Z}_2$- and $\mathbb{S}_1$-symmetries. In order to study bifurcations of invariant objects of maps (3.85) and (3.86) we must restore these symmetries. Although in an arbitrary map with one of these bifurcations these symmetries are usually broken, the analysis of the truncated normal forms (3.85) and (3.86) still provides a skeleton of the full dynamical catalog.

Symmetric flip–NS

For the flip–NS bifurcation we have a reflection and rotations. The statements of Propositions 3.47 and 3.49 can now be interpreted for the symmetric normal form (3.85) as follows (see also Figure 3.41).

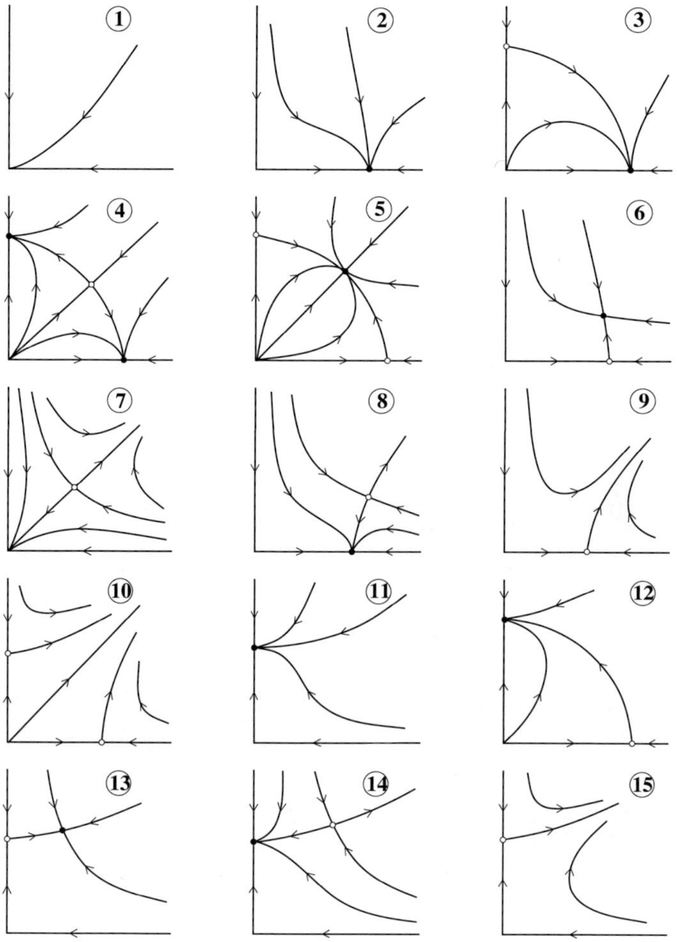

Figure 3.38 Phase portraits for the simple case $s_1 s_2 > 0$.

- The bifurcation at T_1 is the usual flip bifurcation. The origin changes stability in the x-direction and a period-2 orbit (dis)appears if μ_1 crosses zero.
- The bifurcation at T_2 is the usual NS bifurcation. The origin changes stability in the z-direction and a closed invariant curve (dis)appears if μ_2 crosses zero.
- At the bifurcation curve T_3 we encounter a quasi-periodic period-doubling bifurcation, which we will denote by CD. Here, a closed invariant curve consisting of one piece changes stability in the x-direction, which is accompanied by the creation or destruction of the doubled invariant curve.

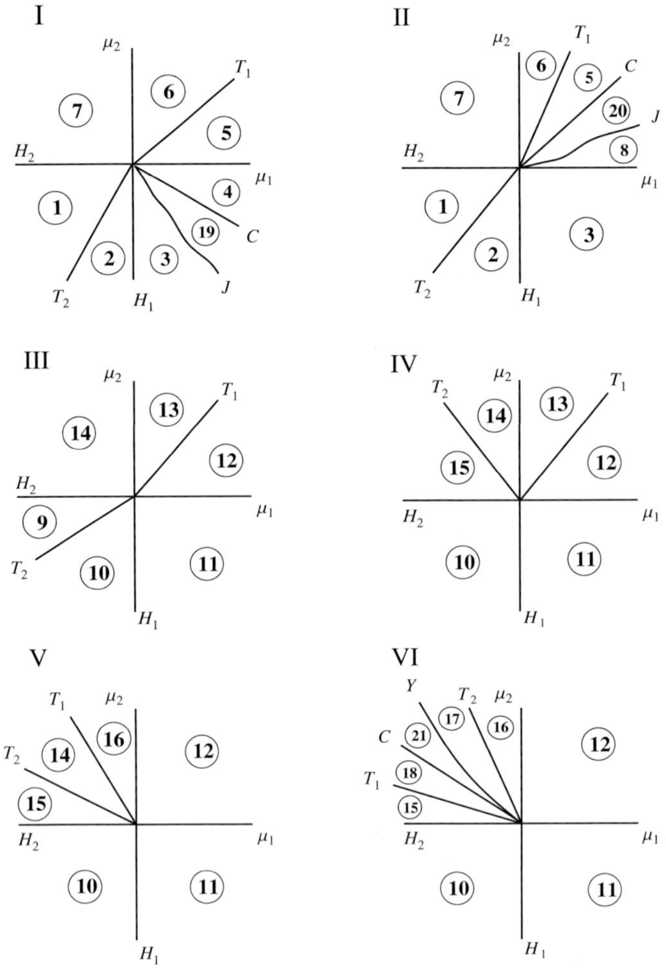

Figure 3.39 Bifurcation diagrams for the difficult case $s_1 s_2 < 0$.

- At the bifurcation curve T_4 the period-2 cycle changes stability and a doubled invariant curve consisting of two disjoint closed curves is created or destroyed via another NS bifurcation.
- If the nontrivial fixed point of (3.90) undergoes the NS bifurcation, then from the doubled invariant curve a 2-torus $\mathbb{T}^2$ bifurcates, which also consists of two disjoint sets. This is also a quasi-periodic Hopf bifurcation. In the following it will be denoted by CN.

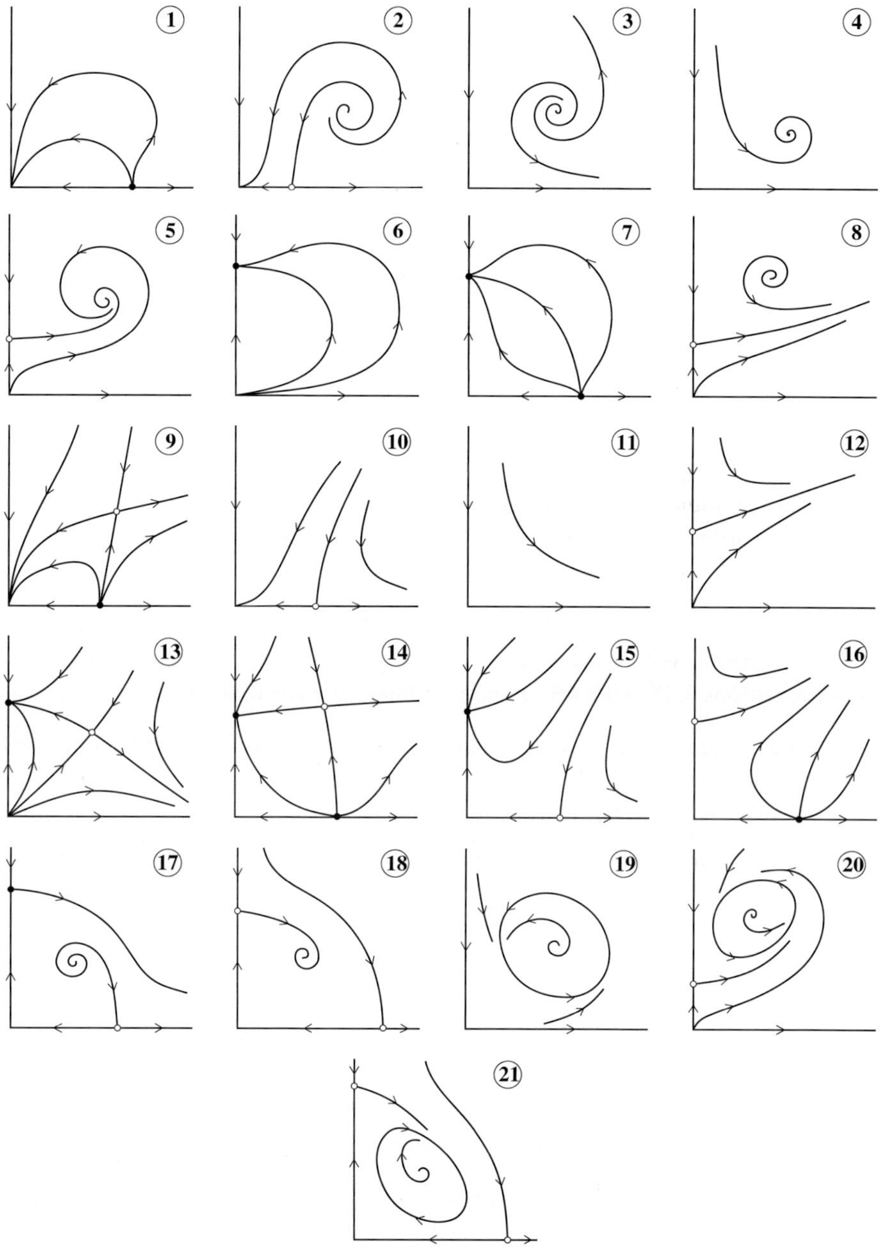

Figure 3.40 Phase portraits for the difficult case $s_1 s_2 < 0$.

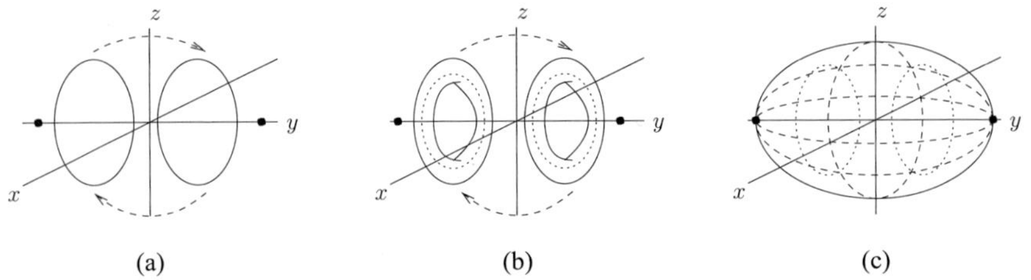

Figure 3.41 Sketches of phase portraits which are specific for the flip–NS bifurcation. The dots represent the period-2 fixed point, which may be present or not. In (a) the doubled invariant curve is stable, in (b) and (c) it is unstable (dotted). In (b) the doubled invariant curve is surrounded by a 2-torus. In (c) the 2-torus has merged with a heteroclinic structure of the manifolds of the period-2 fixed point and the invariant curve. The doubled invariant curve exists inside.

- The $\mathbb{T}^2$ may be destroyed in a heteroclinic bifurcation or by a boundary bifurcation.

Symmetric double Neimark–Sacker

The double NS normal form has two rotational symmetries. The statements of Propositions 3.47 and 3.49 can now be interpreted as follows.

- The bifurcations at T_1 and T_2 correspond to the birth or destruction of closed invariant curves via the standard NS bifurcations.
- A quasi-periodic bifurcation at T_3 or T_4 creates or annihilates an invariant 2-torus $\mathbb{T}^2$ surrounding a closed invariant curve, this is denoted as before by CN.
- The $\mathbb{T}^2$ may change stability such that a $\mathbb{T}^3$ appears in another quasi-periodic bifurcation; it will be labeled TT.
- The $\mathbb{T}^3$ may disappear either through a heteroclinic bifurcation, where the stable and unstable manifolds of both closed invariant curves are connected, or by a collision with a fixed boundary.

3.5.3 Breaking the symmetries

In general, the higher-order terms are not symmetric under the action of $\mathbb{Z}_2$ and $\mathbb{S}_1$. In Broer et al. (1990); Chenciner (1985a,b, 1988); Broer, Huitema, and Sevryuk (1996b); Braaksma and Broer (1987); Iooss and Los (1988); Los (1988, 1989) and Broer, Takens, and Wagener (1999) the question of persistence of the invariant curves and tori under general dissipative perturbations was considered. The doubling and torification curves were the subject of Iooss

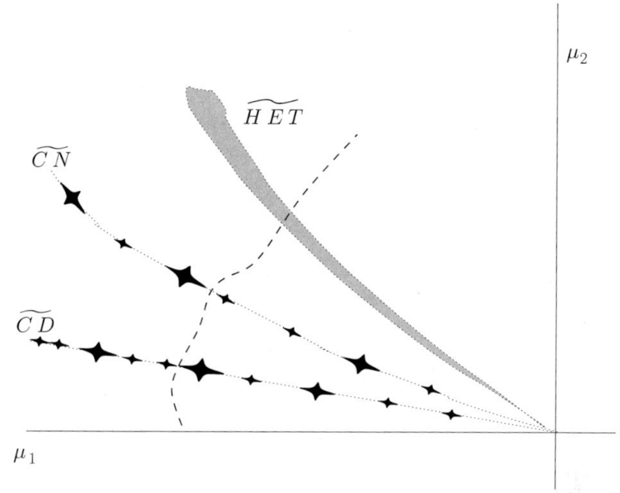

Figure 3.42 A sketch of the (μ_1, μ_2)-plane near a generic flip–NS bifurcation with the resonance bubbles (black diamonds) in the sets $\widetilde{CD}$ and $\widetilde{CN}$. The bubbles have to be avoided by a "good" bifurcation path, like the dashed one, but are relatively small. In the region denoted by $\widetilde{HET}$ the dynamics is complicated by heteroclinic orbits.

and Los (1988); Los (1988, 1989). These works show that there is a set of positive measure in the (μ_1, μ_2)-plane, where the bifurcation path, in general not a straight line, has a "good" bifurcation sequence. Positive measure implies that it occurs with high probability and "good" means the doubling of the invariant curve or its torification without any extra dynamical phenomena. These paths avoid resonance holes and along them the reduction of the original map to the planar map is valid.

In the resonance holes, the bifurcation scenario is different. This question was first considered by Chenciner (1988). Near the curves CD, CN and TT we may encounter resonances on the invariant tori, and these resonance bubbles have to be avoided by a "good" path (see Figure 3.42 for the flip–NS case). Typically, there is a Diophantine condition which decides how large these bubbles are in parameter space.

The bifurcations of fixed points and period-2 cycles have been characterized completely by Proposition 3.47. More interesting are bifurcations of invariant curves and tori in the quasi-periodic sense, as in Braaksma and Broer (1987); Broer et al. (1990); Broer, Huitema, and Sevryuk (1996b). Here we will present numerical studies of the most interesting cases, namely where a torus disappears via a heteroclinic bifurcation from Proposition 3.49. Note that a boundary bifurcation, where the torus moves out of a fixed domain, may be seen as an artifact of the analysis, since in all real applications a specific mechanism destroying the torus will be visible. Phase locking and asymmetry appear if we include the dynamics of the angles ψ in maps (3.85) and (3.86), as well as higher-order nonsymmetric terms. Thus we may encounter resonances on the invariant tori and not follow any "good" bifurcation path. This implies that in a generic nonsymmetric family, complicated bifurcation sets exist near the curves CD and CN for the flip–NS bifurcation and near the curves CN and TT for the NS–NS bifurcation, respectively, hereafter denoted by a tilde.

For a global overview of the dynamical inventory a two-parameter study is rather sufficient. However, near the curves $\widetilde{CD}, \widetilde{CN}$ and $\widetilde{TT}$ a study of three- and four-parameter unfoldings in the spirit of Arnold (1983), where it was first suggested to consider the argument of the multipliers near the NS bifurcation as an additional unfolding parameter, is necessary.

In Takens and Wagener (2000) a theoretical framework for invariant circles with phase locking has been developed, claimed to be representative by the authors. In our numerical case study this provides indeed a skeleton of the bifurcation scenarios. It involves coexistence of attractors and global bifurcations. However, the role of twist-terms (imaginary parts in (3.93)) is subtle in that they lead to a codim 3 bifurcation. So the theoretical framework in Takens and Wagener (2000) for the gaps describes only a part of the possibilities.

To study bifurcations of invariant curves and tori we use several numerical techniques. To study cycles on the invariant tori, i.e., resonances, we use continuation for detection of codim 1 and 2 bifurcations. When dealing with invariant curves and tori, we compute all Lyapunov exponents (see Section 5.7 or (Cincotta, Giordano, and Simó, 2003; Simó, 2005)). We note that Lyapunov exponents are not reliable when we are investigating 2-tori, as the second exponent converges very slowly. Thus, in this case we also analyzed the frequencies, as in Laskar, Froeschlé, and Celletti (1992) and implemented in SDDS Borland et al. (2005). If there were two relevant frequencies we made a continued fraction expansion of the frequency ratio. If the ratio was close to a rational number p/q with $q < 43$, this corresponded to an invariant curve on the 2-torus. In the similar setting of a fold–NS bifurcation, such a study has been done in Vitolo (2003); Broer, Simó, and Vitolo (2008b), and Broer, Simó, and Vitolo (2008a) near a 1:5 resonance bubble.

In our presentation we find the different bifurcation scenarios, the quasi-periodic and the periodic, next to each other. As the precise bifurcation picture depends on the higher-order terms, a model map gives only a representative (not universal) picture.

Nonsymmetric flip–NS: a case study

Let us first motivate the selection of the perturbation terms. The rotational symmetry was introduced by the normal form computations and assuming a rigid rotation in ϕ. Now we will include the dynamics of the ψ-variable in our study. Moreover we consider the situation close to a relatively low resonance, where the bifurcation structure is most pronounced. The lowest-order resonance, which is permitted by the bifurcation analysis in Proposition 3.47 is the 1:7 resonance. This motivates the introduction of a third parameter μ_3 to control the detuning of the rotation and the addition of the term $\varepsilon_1 \bar{z}^{q-1}$ with $q = 7$ and ε_1 a fixed number to the second component of (3.85). The resulting map is still equivariant under the transformation $x \mapsto -x$. To distort this reflection symmetry, we also add the term $\varepsilon_2 x^6$ to the first component of the map. Now the symmetries are broken, but the planes $x = 0$ and $z = 0$ are still invariant. A final perturbation proportional to ε_3 is, therefore, introduced to eliminate this invariance as well. The coefficients are chosen small such that the invariant objects under study are relatively large. This natural approach was first used by Vitolo (2003) and Broer, Simó, and Vitolo (2008a,b). Our model map is now given by

$$
\begin{aligned}
F_1 : \begin{pmatrix} x \\ z \end{pmatrix} \mapsto & \begin{pmatrix} -x(1+\mu_1) \\ z e^{i\mu_3}(1+\mu_2) \end{pmatrix} \\
& + \begin{pmatrix} x\left(f_{300}x^2 + f_{111}|z|^2 + f_{500}x^4 + f_{311}x^2|z|^2 + f_{122}|z|^4\right) \\ z e^{i\mu_3}\left(\hat{g}_{210}x^2 + \hat{g}_{021}|z|^2 + \hat{g}_{410}x^4 + \hat{g}_{221}x^2|z|^2 + \hat{g}_{032}|z|^4\right) \end{pmatrix} \\
& + \begin{pmatrix} \varepsilon_2 x^6 \\ \varepsilon_1 \bar{z}^6 \end{pmatrix} + \varepsilon_3 \begin{pmatrix} \Re(z)^6 + \Im(z)^6 \\ x^6 + \Im(z)^6 + i(x^6 + \Re(z)^6) \end{pmatrix}.
\end{aligned}
\tag{3.93}
$$

One could take the seventh iterate of (3.93) in cylindrical coordinates (x, r, ψ) and study the tongue of the period-7 cycle. Then, for parameter values near the expected doubling of the single closed invariant curve, one may divide out the dynamics in the r-variable to restrict the map NF_1 to a local cylinder. For such a map the effects of the lowest-order perturbation terms are studied by Takens and Wagener (2000). The theoretical picture shows a richness of bifurcations involving several codim 2 bifurcations of fixed points and/or invariant curves leading to global bifurcation phenomena. We refer to Krauskopf and Oldeman (2004); Saleh and Wagener (2010) and Saleh (2005) for a similar setting for vector fields, where, moreover, some global bifurcation curves have been computed.

First we choose the coefficients (see Table 3.2). The cubic coefficients are such that the unfolding involves the $\widetilde{CD}$ and $\widetilde{CN}$ bifurcations. Then the fifth-

Table 3.2 Numerical values of the coefficients of map (3.93) such that the heteroclinic bifurcation occurs.

$f_{300} =$	-0.1	$f_{500} =$	-0.5	$\hat{g}_{410}=$	-0.00125
$f_{111} =$	0.3	$f_{311} =$	0	$\hat{g}_{221}=$	-0.00075
$\hat{g}_{210} =$	$-0.25 + 0.05i$	$f_{122} =$	0	$\hat{g}_{032}=$	$-0.0001125 - 0.025i$
$\hat{g}_{021} =$	$-0.1 - 0.015i$				

Table 3.3 Color coding for Lyapunov exponents.

Lyapunov exponents	Color	Dynamical object
$\lambda_1 > \lambda_2 \geq 0 > \lambda_3$	Red	Strange attractor
$\lambda_1 = \lambda_2 = 0 > \lambda_3$	Magenta	Invariant 2-torus
$\lambda_1 = 0 > \lambda_2 = \lambda_3$	Blue	Invariant circle of focus type
$\lambda_1 = 0 > \lambda_2 > \lambda_3$	Cyan	Invariant circle of node type
$0 > \lambda_1 \geq \lambda_2 \geq \lambda_3$	Green	Fixed point

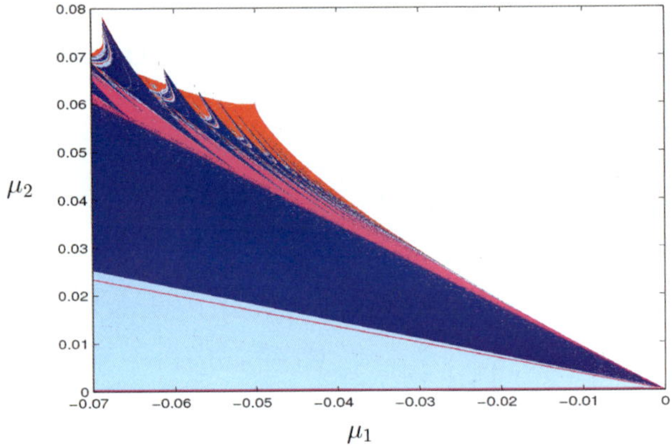

Figure 3.43 Bifurcation diagram in the (μ_1, μ_2)-plane near a heteroclinic connection with fixed $\mu_3 = 0.9411$.

order coefficients can be chosen such that the 2-torus is stable. The perturbation terms are arbitrarily fixed, $\varepsilon_1 = 0.5$ is larger than $\varepsilon_2 = \varepsilon_3 = 0.05$ since we want the phase locking to be visible.

We will start our numerical studies by constructing a bifurcation diagram in the two amplitude parameters (μ_1, μ_2). We fix arbitrarily $\mu_3 = 0.9411$ and compute the Lyapunov exponents, which are color-coded according to Table 3.3 (see Figure 3.43).

Take μ_1 negative and small, then for μ_2 small and positive we have a single stable closed invariant curve of node type (cyan). Increasing μ_2, we notice a

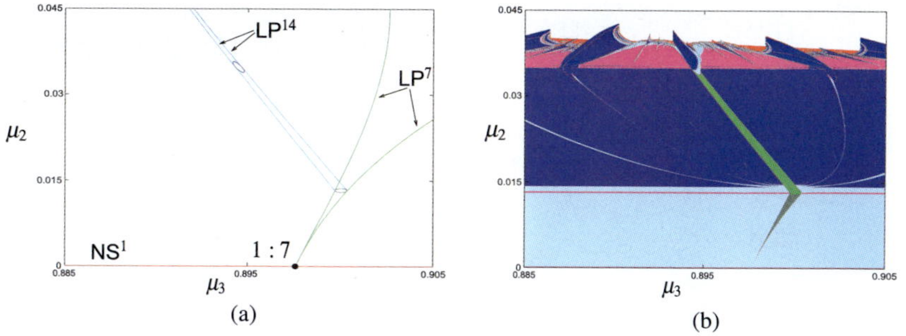

Figure 3.44 Bifurcation diagram for the flip–NS bifurcation. LP^k and NS^k denote fold and NS bifurcation of the period-k cycle, respectively. The pink circle is a period-doubling bifurcation of period-7 cycles. The dark blue circle is an NS bifurcation of period-14 cycles. See enlargements in Figures 3.45 and 3.46.

few isolated blue dots where two of its exponents become equal, however, this is not a bifurcation. Going further, we see a pink stripe where the $\widetilde{CD}$ bifurcation set is present. Locally, the dynamics occurs on a cylinder, which is why we code this by the same color as the 2-torus. The doubled invariant curve is initially of node type, but becomes of focus type quickly when we get to the blue region. This doubled invariant curve bifurcates into a 2-torus via the $\widetilde{CN}$ transition. In the upper part of Figure 3.43, stripes of blue are visible, corresponding to the resonant dynamics on the 2-torus. Increasing μ_2, we cross the heteroclinic wedge encountering attractors of high period and chaos (red), and then no attractor is found anymore. The diagram in Figure 3.43 should be compared with the sketch given in Figure 3.42.

Next we fix the two-parameter plane $\mu_1 = -0.04$ in the three-dimensional parameter space, which intersects the bifurcation set in such a way that we encounter all interesting dynamics. The (μ_2, μ_3)-plane cuts the bubbles of the fractal-like bifurcation sets $\widetilde{CD}$ and $\widetilde{CN}$. We present the diagrams with bifurcations of fixed points/cycles and the Lyapunov exponents next to each other, since they provide complementary information.

Let us start with Figure 3.44(a), since the local bifurcation analysis guides us through the figures with the Lyapunov exponents. Actually, it produces key subsets of the whole fractal bifurcation set. Starting at the NS^1-curve, we find a 1:7 resonance tongue. By continuation of the tongue boundaries LP^7 we arrive at the $\widetilde{CD}$-set, where we encounter two fold–flip points (see also Figure 3.45). They are of different type, where one of them involves a Neimark–Sacker, NS^{14}, and heteroclinic bifurcations, the latter not displayed. The period-doubling curve is composed of two circle segments. On both seg-

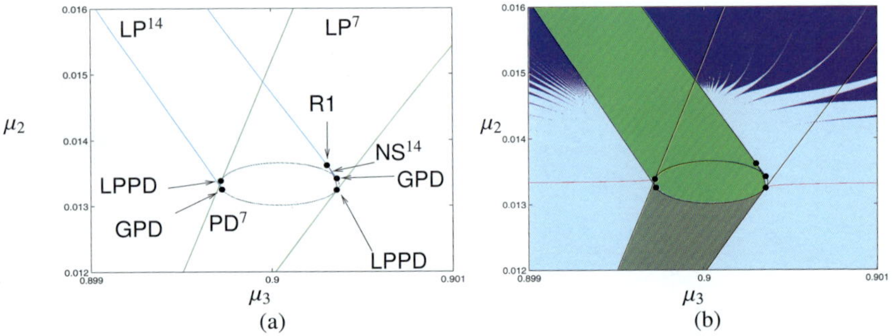

Figure 3.45 A subset of the bifurcation diagram near the $\widetilde{CD}$ boundary. NS^n denotes NS bifurcation of the period-n cycles; other codim 1 curves are labeled as in Figure 3.44. The codim 2 points of period-7 cycles are marked LPPD = fold–flip, GPD = generalized flip, R1 = 1:1 resonance.

ments there is a degenerate flip bifurcation, where a fold curve of period-14 cycles is rooted. On one of the fold curves there is a 1:1 resonance, where the NS bifurcation curve of period-14 cycles ends. For $\hat{g}_{210} = -0.25 - 0.06i$ we found that both degenerate flip points occurred on the upper-half of the period-doubling circle. Then on both fold-curves of the period-14 cycle a resonance 1:1 occurred.

In fact, the period-14 cycles correspond to a resonance on the doubled invariant curve. Following these fold curves while increasing μ_2, we then arrive at the $\widetilde{CN}$ boundary, resulting in two fold–NS points. Here the NS curve is an ellipse connecting these codim 2 points. On the NS curve of period-14 cycles there are two Chenciner points. Thus, the birth of the doubled torus is associated with the appearance of period-14 saddle and node cycles on the torus. The fold curves of period 14 terminate on the curve NS^2 for $\mu_2 \approx 0.1381$, which is not displayed. Somewhere in between heteroclinic bifurcations occur, which are also not visible here.

Now we turn to Figure 3.44(b), where the sequence of bifurcations of the invariant tori for increasing μ_2 is the same. On the invariant tori we also have resonant dynamics. We can compare this with Figure 3.44(a), where bifurcations of fixed points and cycles are displayed. In the green regions we have attracting cycles of period-7 and -14 on the invariant curves. The motivation for this parameter plane becomes apparent when we come to the region where the 2-torus exists. On the 2-torus, there is resonant dynamics resulting in circle attractors. These circle attractors undergo period-doubling bifurcations resulting in chaotic attractors near the heteroclinic bifurcations.

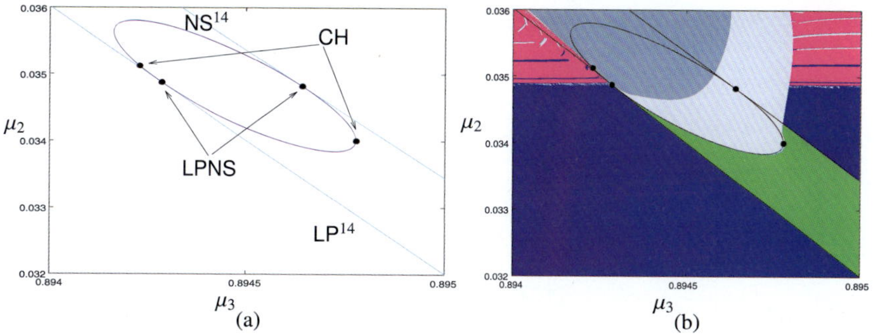

Figure 3.46 A subset of the bifurcation diagram near the $\widetilde{CN}$ bifurcation set. Here codim 1 curves are labeled as in Figure 3.44. The codim 2 points are marked by LPNS = fold–NS and CH = Chenciner bifurcation.

Let us now analyze the bubbles in more detail. We start with the case where the phase locking of period-7 interacts with the $\widetilde{CD}$ bifurcation set (see Figures 3.45(a) and 3.45(b)). At the bottom, we start with a single closed invariant curve of node type, either with or without phase locking. Without phase locking the doubling of the curve happens in the smooth quasi-periodic manner. The dynamics on a cylinder is visible as a pink stripe where two Lyapunov exponents are almost zero. Following the phase locking, however, the doubled curve exists only when we have crossed the circle of period-doubling of period-7 completely and are not near the R1-point. Further up, the transition of the doubled curve from node to focus type is clearly visible. The stripes where the doubled curve is of node type go up all the way (see Figure 3.44(b)). These give rise to resonances which complicate the diagram near the $\widetilde{CN}$ boundary. Now the resonance of period-14 on the doubled curve has appeared and we may follow it until it disappears, i.e., at the NS^2 bifurcation. Before that there is another bubble, which we now examine.

Again at the bottom of Figure 3.46 there is the doubled invariant circle of focus type, phase-locked or not. Going up along the borders the 2-torus is born. This happens near $\mu_2 \approx 0.0348$, but the transition is not sharply visible as not only all three Lyapunov exponents are close to zero, also the amplitudes for the frequency analysis are very small, causing numerical difficulties. The stripes in this region represent 2-tori with a resonant frequency vector. If we go up in the region of resonance in $\widetilde{CN}$, we cross the circle of NS bifurcations. There is a Chenciner point on the lower arc of the bifurcation circle, such that there is always a stable invariant curve of 14 pieces surrounding the doubled invariant curve. We conclude that the existence of the 2-torus is delimited by two bifurcation curves, where a stable and an unstable invariant curve of period-14

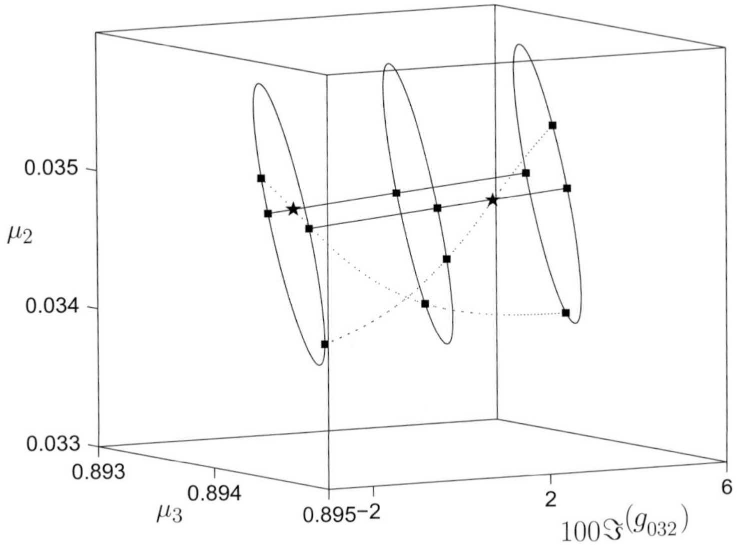

Figure 3.47 From one unfolding to another depending on the twist-term. The ellipses are NS bifurcations of period-14 cycles, the rather horizontal lines represent fold–NS bifurcations. The dotted lines are Chenciner bifurcation curves, which do not intersect each other. The stars indicate the position of the codim 3 bifurcations.

collide. This figure agrees well with Takens and Wagener (2000), and this is the situation we most frequently encountered. See also Vitolo (2003) and Broer, Simó, and Vitolo (2008a,b) for similar results of a bubble near a quasi-periodic Hopf bifurcation.

The bubbles in this two-parameter plane correspond to tubes in the three-dimensional (μ_1, μ_2, μ_3)-space. We now use the imaginary part of $\hat{g}_{032}$ as an extra parameter as it does not affect results for the amplitude map (3.90). We changed it from −0.025 to 0.055 and examined the bubble near the $\widetilde{CN}$ boundary. In $(\Im(\hat{g}_{032}), \mu_3, \mu_2)$-space, the circle of NS bifurcations is a cylinder with four bifurcation curves on it (see Figure 3.47). Two rather straight lines are curves of the fold–NS bifurcation, the dotted lines denote Chenciner bifurcations. Both straight curves intersect with one dotted curve, which corresponds to a codim 3 bifurcation, apparently a fold–Chenciner bifurcation. The role of the twist terms is to change from one two-parameter unfolding to another. We note that in a different setting a very similar picture is found by Saleh and Wagener (2010) and Saleh (2005).

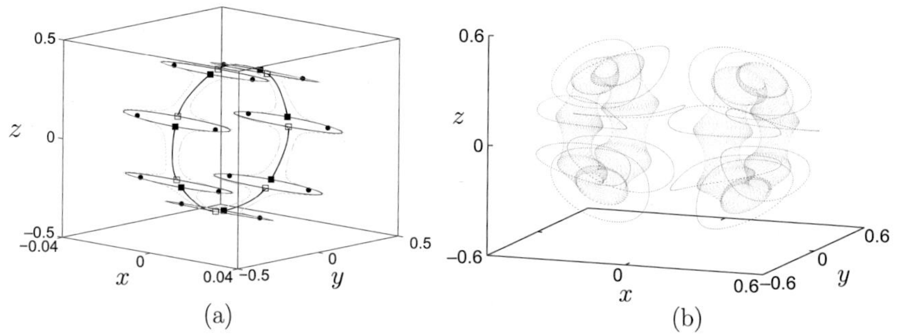

Figure 3.48 Phase portraits near bubbles: (a) $\mu_2 = 0.01335$, $\mu_3 = 0.9003383720148196$, ■, period-7 saddle cycle of which the one-dimensional unstable manifold (black curve) almost reconnects to the other period-7 saddle cycle denoted by □. Period-14 cycles are stable foci and denoted by • and the gray shows the doubled invariant curve. (b) $\mu_2 = 0.0352$, $\mu_3 = 0.89479$, 2-torus surrounded by a "period-14" invariant curve.

After the whole discussion of bifurcations of cycles and invariant curves we now give two phase portraits near the bubbles (see Figure 3.48). These show how coexisting attractors complicate the unfolding.

Nonsymmetric NS–NS: a case study

We use an approach for the double NS bifurcation similar to that for the flip–NS bifurcation. The natural way to perturb the normal form (3.86) is to introduce relatively low-order resonance terms. These will be $(\varepsilon_1 \bar{w}^{q_1-1}, 0)$ and $(0, \varepsilon_2 \bar{z}^{q_2-1})$. Now the choice $q_1 = 7, q_2 = 8$ gives the lowest orders compatible with Proposition 3.44. Then, as before, a final perturbation to break the invariance of the w- and z-planes is added. Thus, the model map takes the form

$$F_2\colon \begin{pmatrix} w \\ z \end{pmatrix} \mapsto \begin{pmatrix} we^{i\mu_3}(1 + \mu_1 + \hat{f}_{2100}|w|^2 + \hat{f}_{1011}|z|^2) \\ ze^{i\mu_4}(1 + \mu_2 + \hat{g}_{1110}|w|^2 + \hat{g}_{0021}|z|^2) \end{pmatrix}$$

$$+ \begin{pmatrix} we^{i\mu_3}\left(\hat{f}_{3200}|w|^4 + \hat{f}_{2111}|w|^2|z|^2 + \hat{f}_{1022}|z|^4\right) \\ ze^{i\mu_4}\left(\hat{g}_{2201}|w|^4 + \hat{g}_{1121}|w|^2|z|^2 + \hat{g}_{0032}|z|^4\right) \end{pmatrix}$$

$$+ \begin{pmatrix} \varepsilon_1 \bar{w}^6 + \varepsilon_3(i\Re(w)^6 + \Im(w)^6 + (1+i)\left(\Re(z)^6 + \Im(z)^6\right) \\ \varepsilon_2 \bar{z}^7 + \varepsilon_3((1+2i)\Re(w)^6 + (1+i)\Im(w)^6 + \Im(z)^6) \end{pmatrix}. \tag{3.94}$$

Next we choose the constants according to Table 3.4 and $\varepsilon_1 = 0.012i, \varepsilon_2 = 0.015, \varepsilon_3 = 0.0009$ such that also here we have the heteroclinic bifurcation.

The situation is much more complicated than in the previous case study, since controlling the remaining three parameters to study bifurcations of in-

Table 3.4 Numerical values of the coefficients of the map (3.94).

$\hat{f}_{2100} =$	$0.03 - 0.03i$	$\hat{f}_{3200} =$	$0.000729 - 0.00018i$
$\hat{f}_{1011} =$	$0.09 + 0.015i$	$\hat{f}_{2111} =$	$0.000576 + 0.0001125i$
$\hat{g}_{1110} =$	$-0.075 - 0.0147i$	$\hat{f}_{1022} =$	$0.000441 + 0.00027i$
$\hat{g}_{2210} =$	$0.000324 + 0.000576i$	$\hat{g}_{1121} =$	$0.000225 + 0.00036i$
$\hat{g}_{0032} =$	$-0.0045 - 0.00018i$	$\hat{g}_{0021} =$	$-0.03 + 0.006i$

variant objects leads to an overwhelming amount of information. Therefore, we will restrict ourselves to bifurcations of cycles. However, as before, the analysis of cycles dresses the skeleton provided by the study of the amplitude map. As before we fix the parameter $\mu_1 = -0.04$ and then increase μ_2 to control the bifurcation sequence, while we may use μ_3 and μ_4 to adjust the frequencies to the rationals.

We notice that this setup has been suggested by Baesens et al. (1991). There, there is a thorough exploration of the dynamics on the invariant 2-tori, sketches of curves of homoclinic tangency and portraits of resonance tongues. The resonances which we study on the 2-torus display a similar organization in parameter-space.

The main structure of the tongues is given in Figure 3.49. If we start with μ_2 small, then there is an invariant curve, possibly phase locked. In the lower part of the tongue there is a cycle of period-8 on the invariant curve. If we increase μ_2, this first resonance tongue will display a bubble near the $\widetilde{CN}$ bifurcation leading to a stable 2-torus. This bubble is similar to the one in the case study of the flip–NS bifurcation. There are fold–NS bifurcation points, where the ellipse touches the tongue, and Chenciner points from which two invariant curves originate. These collide along two complicated sets in three-dimensional parameter space.

There is no other way to track this phenomenon via continuation but to follow a fold bifurcation curve near this set. As we introduced two resonances, we were able to locate a cycle of period-56. From the NS bubble, two ellipses corresponding to fold bifurcations of the cycle of period-56 emerged. On both ellipses there are two 1:1 resonance points, which are pairwise connected by a NS bifurcation curve. We have not studied the orientation of these resonance points extensively; however, we note that near the $\widetilde{CN}$ bifurcation, also fold–NS bifurcation points were found, which disappeared while following the period-56 cycle for bigger μ_2. Apparently they merged with the 1:1 resonances in a triple-one bifurcation. We have not attempted to find global bifurcations, but they are certainly present.

The region where these resonances exist seems to be a torus. Indeed, with

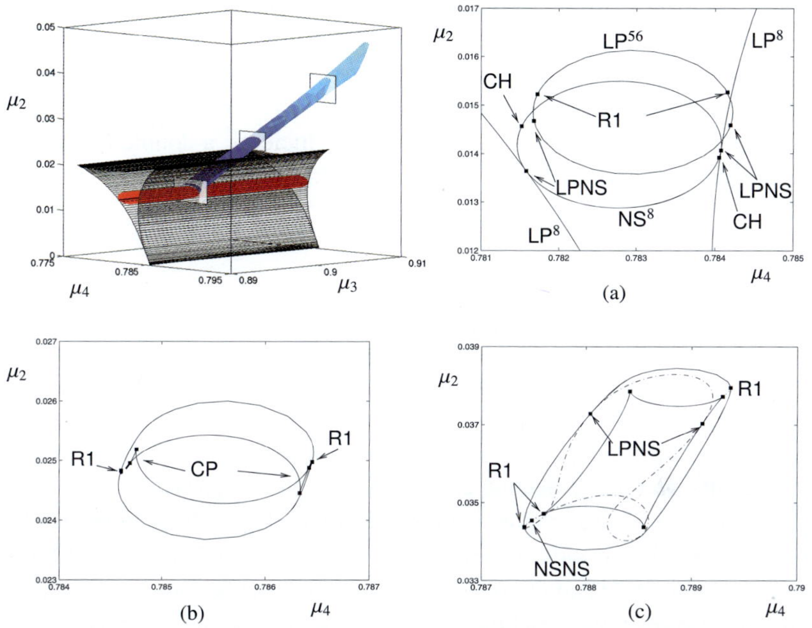

Figure 3.49 Bifurcation diagrams for $\mu_1 = -0.04$. The three-dimensional figures displays the period-8 tongue, the bubble inside and the emanating tube where period-56 cycles exist. The three slices correspond to the enlargements in figures (a), (b) and (c). The labels are as in previous figures, except for CP for cusp, and NSNS for double NS bifurcations. (a) $\mu_3 = 0.8972$; The period-8 tongue and the NS bubble with Chenciner points (CH) on it, similar to that in the flip–NS bifurcation. The upper ellipse are two curves of fold bifurcation of a period-56 cycle, which are indistinguishable in the figure. (b) $\mu_3 = 0.902$; the inner ellipse has developed two swallow-tails. (c) $\mu_3 = 0.9073$; an extra NS bifurcation curve is present, which intersects the one that connects two R1-points at NSNS.

increasing μ_2 we found two swallow-tail structures on the inner-ellipse, which then transformed into a flame, i.e., a region where there are coexisting attractors of period-56 (see (Peckham and Kevrekidis, 2002; Broer, Golubitsky, and Vegter, 2003)). The 1:1 resonances are still present and the connecting NS bifurcation curves as well.

Now let us take μ_2 near the $\widetilde{TT}$ boundary, i.e., where a motion on a 3-torus may be expected (see Figure 3.49(c)). On first sight we recognize the outer-ellipse and the inner-flame together with the 1:1 strong resonances. However, apart from the two NS bifurcation curves similar to those in Figures 3.49(a) and 3.49(b), a third curve (dashed) is present. It touches both fold bifurcation curves and makes two loops. On the left loop it intersects one of the NS bifurcation curves. Here we have a double NS bifurcation. We conjecture that it appears between two codim 3 bifurcations where we have two multipliers equal to one and a complex pair of modulus 1. We conclude this discussion with the remark that, for our choice of the coefficients, a double NS bifurcation of a cycle of period-56 appears in the unfolding of a double NS bifurcation of a fixed point.

3.6 Historical perspective

Here we give a brief overview of the literature concerning codim 1 and 2 bifurcations of discrete-time dynamical systems generated by iterated maps (first-order nonlinear difference equations), but for a mathematical formulation of these results we refer to Section 2.1 and this chapter, in particular Table 3.1.

There are three local codim 1 bifurcations by which a fixed point can lose stability: the *fold*, *period-doubling* (*flip*) and *NS* bifurcations. A proof for the theorems on the fold and flip bifurcation theorems can be found in Newhouse, Palis, and Takens (1983), but the results were known before that time.

The appearance of a closed invariant curve was first considered by Neimark (1959), later Melnikov (1963) and Sacker (1965) gave a complete exposition (see also (Iooss, 1979)). The theorems give non-degeneracy conditions and typical bifurcation diagrams. For generic families transversality with respect to parameters is always assumed. Descriptions of these bifurcations can be found in standard textbooks (Arnold, 1983; Arnold et al., 1994; Arrowsmith and Place, 1990; Kuznetsov, 2004). When non-degeneracy conditions do not hold or more instabilities appear, we have a codim 2 bifurcation.

The analysis of codim 2 bifurcations has a long tradition and in the course of roughly 30 years the local analysis has been more or less completed. While this chapter aims at providing a complete overview, one could also consult Rousseau (1990); Arrowsmith and Place (1990); Arnold et al. (1994), and Kuznetsov (2004) for the cases appearing in one and two dimensions.

The results for the *cusp* bifurcation are analogous to the case of vector fields and they already follow from Singularity Theory, but see Gheiner (2014). Results for the *generalized period-doubling* bifurcation were announced by Holmes and Whitley (1984a,b); Rousseau (1990); Arnold et al. (1994). Peckham and Kevrekidis (1991) used Lyapunov–Schmidt reduction for their analysis, while Gheiner (1998) investigated the topological equivalence for this case. For these one-dimensional cases, there is a weak conjugacy such that the truncated canonical unfoldings are actually topological normal forms (Arnold,

1983; Arnold et al., 1994; Katok and Hasselblatt, 1995; Balibrea, Oliveira, and Valverde, 2017). For all other cases the appearance of global bifurcations and resonances obstructs such an equivalence.

Then Chenciner (1985a,b, 1988) considered the case where the cubic non-degeneracy condition for the NS bifurcation does not hold. Due to a quasi-periodic bifurcation, the bifurcation set is dense in a certain region and very complicated. The book Arrowsmith and Place (1990) provides a very readable introductory exposition.

Another degeneracy for the NS bifurcation appears when the linear part of the map has a discrete rotational symmetry of order $q = 1, 2, 3, 4$. These codim 2 cases are known as the *strong resonances*. Both Neimark (1959), for the qth iterate of the map, and Takens (1974) (see also (Chow, Li, and Wang, 1994)), for the map composed with a symmetry transformation, showed that the normal forms can be approximated by time-1 flows of ODEs. Arnold (1977, 1983), on the other hand, came to the same ODEs by $2\pi q$-periodic normalization, while studying limit cycles with these resonances.

Concerning bifurcations of the approximating ODEs, Bogdanov (1975, 1976a,b) provides a complete analysis for the case $q = 1$. For $q = 2$ proofs are given by Takens (1974), Holmes and Rand (1977/78) and Horozov (1979), who also treats the case $q = 3$. For the cases $q = 1, 2, 3$ it has been shown that the results for these approximating vector fields are stable within symmetric perturbations.

For $q = 4$ the bifurcation diagram is rather complicated. Arnold (1977, 1983) obtained and predicted most bifurcation curves. Analytical results have been obtained by Neishtadt (1978), Wan (1978), Cheng (1990); Cheng and Sun (1992), and Zegeling (1993). Berezovskaia and Khibnik (1979, 1980) computed all boundaries corresponding to degenerate heteroclinic loops. Arnol'd conjectured that no other bifurcation sequences would appear, indeed, the (numerical) evidence by Krauskopf (1994, 1995, 1997, 2001) indicates that all sequences are now known.

Then, the results for the approximating ODE have to be carried over to the maps. In all cases, global bifurcations appear and therefore no topological equivalence can be obtained between the 1-flow and the original map. For $q = 1$ it has been shown by Broer, Roussarie, and Simó (1996a) that there is a narrow horn with all the complicated dynamics. In Paez Chavez (2010); Al-Hdaibat et al. (2018) the approximation of the homoclinic orbit in phase and parameter space is considered.

The analysis of the *fold–flip*, *fold–NS*, *flip–* and *double NS* bifurcations has only been taken up more recently. For the *fold–flip* bifurcation, Zholondek (1983) indicates some properties (see also (Arnold et al., 1994; Rousseau,

1990)). Gheiner (1994) presents a normal form and constructs invariant manifolds to investigate topological equivalence. In Kuznetsov, Meijer, and van Veen (2004) non-degeneracy of the unfolding was studied in full detail.

The analysis of the *fold–NS* bifurcation, requiring at least three-dimensional phase space, has been undertaken by Vitolo (2003); Broer, Simó, and Vitolo (2008b,a). After obtaining a bifurcation diagram, they specifically pay attention to resonance phenomena and strange attractors.

For the *flip–* and *double NS* bifurcations a large contribution came from Los (1988, 1989); Iooss and Los (1988). They are concerned with the primary quasi-periodic bifurcations occurring in these cases and proving that this occurs as in a normal form for a set of large measure. In Kuznetsov and Meijer (2006) we provided a complete description through a connection to the normal form results of the double Hopf bifurcation and studied how resonance phenomena complicate and unveil certain quasi-periodic bifurcations.

Appendices

3.A Proofs for Section 3.1

Proof of Theorem 3.2 The proof is essentially as in the ODE case (Kuznetsov, 2004, section 8.2.1, pp. 302–305).

Consider the restriction of (3.2) to a parameter-dependent one-dimensional center manifold near the cusp (CP) bifurcation

$$\xi \mapsto g(\xi, \alpha), \quad \xi \in \mathbb{R}, \ \alpha \in \mathbb{R}^2, \tag{3.95}$$

where $g\colon \mathbb{R} \times \mathbb{R}^2 \to \mathbb{R}$ is smooth near $(\xi, \alpha) = (0, 0)$. Suppose that $\xi = 0$ is a fixed point of (3.95) at $\alpha = 0$ and that

$$g(\xi, 0) = \xi + c_0\xi^3 + O(\xi^4), \quad c_0 \neq 0,$$

so that the conditions of Theorem 3.2 are satisfied. the Taylor expansion of g with respect to ξ at $x = 0$ can be written as

$$g(\xi, \alpha) = g_0(\alpha) + [1 + g_1(\alpha)]\xi + g_2(\alpha)\xi^2 + g_3(\alpha)\xi^3 + O(\xi^4),$$

where $g_0(0) = g_1(0) = g_2(0) = 0$ but $g_3(0) = c_0 \neq 0$ and the $O(\xi^4)$-terms in general smoothly depend on α.

Make now a shift of the coordinate $\xi = \eta + \delta$. The map (3.95) written in the η-coordinate takes the form

$$\begin{aligned}\eta \mapsto\ & [g_0(\alpha) + g_1(\alpha)\delta + \delta^2\varphi(\alpha, \delta)] + [1 + g_1(\alpha) + 2g_2(\alpha)\delta + \delta^2\phi(\alpha, \delta)]\eta \\ & + [g_2(\alpha) + 3g_3(\alpha)\delta + \delta^2\psi(\alpha, \delta)]\eta^2 + [g_3(\alpha) + \delta\theta(\alpha, \delta)]\eta^3 + O(\eta^4)\end{aligned}$$

Denote the coefficient in front of η^2 by $F(\alpha, \delta)$:

$$F(\alpha, \delta) := g_2(\alpha) + 3g_3(\alpha)\delta + \delta^2\psi(\alpha, \delta).$$

Since $F(0,0) = 0$ and $F_\delta(0,0) = 3g_3(0) = 3c_0 \neq 0$, the Implicit Function Theorem ensures the local existence and uniqueness of a smooth function $\delta\colon \mathbb{R}^2 \to \mathbb{R}$, such that $\delta(0) = 0$ and

$$F(\alpha, \delta(\alpha)) \equiv 0$$

for $\|\alpha\|$ small enough. The map for η with this $\delta = \delta(\alpha)$ will contain no η^2-term and can be written as

$$\eta \mapsto \beta_1(\alpha) + \eta + \beta_2(\alpha)\eta + c(\alpha)\eta^3 + O(\eta^4),$$

where

$$\begin{cases} \beta_1(\alpha) = g_0(\alpha) + g_1(\alpha)\delta(\alpha) + \delta^2(\alpha)\varphi(\alpha, \delta(\delta)), \\ \beta_2(\alpha) = g_1(\alpha) + 2g_2(\alpha)\delta(\alpha) + \delta^2(\alpha)\phi(\alpha, \delta(\alpha)) \end{cases}$$

and

$$c(\alpha) = g_3(\alpha) + \delta(\alpha)\theta(\alpha, \delta(\alpha)), \quad c(0) = g_3(0) = c_0 \neq 0.$$

Replacing η by ξ, we conclude that (3.95) is locally smoothly equivalent near the cusp bifurcation to (3.6) from Theorem 3.2. □

Proof of Lemma 3.4 Suppose that a fixed point of the map (3.2) exhibits a GPD bifurcation at $\alpha = 0$. Write the restriction of the critical map (3.2) to its one-dimensional center manifold near this fixed point as

$$\xi \mapsto f(\xi) = -\xi + a_2\xi^2 + a_3\xi^3 + a_4\xi^4 + a_5\xi^5 + O(\xi^6), \quad \xi \in \mathbb{R}. \tag{3.96}$$

Here ξ is a coordinate in the center manifold, and it is assumed the critical fixed point is located at $\xi = 0$. The locally invertible substitution

$$\xi = h(\eta) = \eta + h_2\eta^2 + h_4\eta^4 \tag{3.97}$$

transforms (3.96) near the origin into

$$\eta \mapsto g(\eta) = -\eta + b_2\eta^2 + b_3\eta^3 + b_4\eta^4 + b_5\eta^5 + O(\eta^6), \quad \eta \in \mathbb{R}. \tag{3.98}$$

To find b_k for $k = 2, 3, 4$, and 5, equate the coefficients of the η^k-terms in the expansion of the conjugacy identity

$$f(h(\eta)) = h(g(\eta)), \tag{3.99}$$

where the maps f, h and g are defined by (3.96), (3.97) and (3.98), respectively. The η^2-terms in (3.99) give

$$b_2 = -2h_2 + a_2.$$

The choice

$$h_2 = \frac{1}{2}a_2$$

ensures that $b_2 = 0$ in (3.98). With this value of h_2, collecting the η^3-terms in (3.99) implies

$$b_3 = a_2^2 + a_3.$$

Recall now that $b_3 = 0$ at the GPD bifurcation, i.e., $a_3 = -a_2^2$. Taking this relation into account, one obtains by collecting the η^4-terms in (3.99) that

$$b_4 = -2h_4 - \frac{5}{4}a_2^3 + a_4 \tag{3.100}$$

in (3.98). Thus, setting

$$h_4 = -\frac{5}{8}a_2^3 + \frac{1}{2}a_4,$$

ensures $b_4 = 0$ and finally gives

$$b_5 = -2a_2^4 + 3a_4a_2 + a_5. \tag{3.101}$$

Therefore, we have shown that (3.96) is locally smoothly equivalent to

$$\eta \mapsto -\eta + b_5\eta^5 + O(\eta^6),$$

where b_5 is specified in (3.101). Replacing η by ξ, we conclude that (3.96) is locally smoothly equivalent to (3.9) from Lemma 3.4 with $d_0 = b_5$. □

Proof of Theorem 3.5 See (Kuznetsov, 2004, lemma 9.3, p. 415).

Consider the restriction of (3.2) to a parameter-dependent one-dimensional center manifold near the GPD bifurcation

$$\xi \mapsto f(\xi, \alpha), \quad \xi \in \mathbb{R}, \ \alpha \in \mathbb{R}^2, \tag{3.102}$$

where $f\colon \mathbb{R}\times\mathbb{R}^2 \to \mathbb{R}$ is smooth near $(\xi, \alpha) = (0, 0)$. Since $\mu(0) = -1$, we have $f_\xi(0, 0) \neq 1$, so that there is a smooth continuation of the fixed point $\xi = 0$ for all sufficiently small $\|\alpha\|$. Shifting the origin to this fixed point, we can assume that $\xi = 0$ is a fixed point of (3.102) for all parameter values close to the critical ones. We can therefore consider

$$f(\xi, \alpha) = \mu(\alpha)\xi + a_2(\alpha)\xi^2 + a_3(\alpha)\xi^3 + a_4(\alpha)\xi^4 + a_5(\alpha)\xi^5 + O(\xi^6),$$

where $\mu(\alpha) = -(1 + \beta_1(\alpha))$ with $\beta_1(0) = 0$ and the $O(\xi^6)$-terms can smoothly depend on α.

Proceeding as in the proof of Lemma 3.4, we make a locally invertible substitution

$$\xi = h(\eta, \alpha) = \eta + h_2(\alpha)\eta^2 + h_4(\alpha)\eta^4, \quad \eta \in \mathbb{R}, \tag{3.103}$$

that transforms (3.96) near the origin into

$$\eta \mapsto g(\eta, \alpha) = \mu(\alpha)\eta + b_2(\alpha)\eta^2 + b_3(\alpha)\eta^3 + b_4(\alpha)\eta^4 + b_5(\alpha)\eta^5 + O(\eta^6). \quad (3.104)$$

For $\alpha = 0$, we recover the critical values a_k, b_k, and $h_{2,4}$ used in the proof of Lemma 3.4. Comparing the α-dependent coefficients of the η^k-terms in the expansion of

$$f(h(\eta, \alpha), \alpha) = h(g(\eta, \alpha), \alpha),$$

we get

$$b_2(\alpha) = -\mu^2(\alpha)h_2(\alpha) + \mu(\alpha)\, h_2(\alpha) + a_2(\alpha).$$

If we take

$$h_2(\alpha) = \frac{a_2(\alpha)}{\mu(\alpha)(\mu(\alpha) - 1)},$$

then $b_2(\alpha) \equiv 0$ near $\alpha = 0$. With this choice of $b_2(\alpha)$, we arrive at

$$b_3(\alpha) = a_3(\alpha) + \frac{2a_2^2(\alpha)}{\mu(\alpha)(\mu(\alpha) - 1)}.$$

At the GPD point, we have $a_3(0) = -a_2^2(0)$, i.e., $b_3(0) = 0$. Denote

$$\beta_2(\alpha) := b_3(\alpha), \quad \beta_2(0) = 0.$$

If we finally set

$$\begin{aligned} h_4(\alpha) = &\frac{1}{\mu(\alpha)(\mu^3(\alpha) - 1)}\, a_4(\alpha) \\ &+ \frac{3 - 2\mu(\alpha)}{\mu^2(\alpha)(\mu(\alpha) - 1)(\mu^3(\alpha) - 1)}\, a_2(\alpha)a_3(\alpha) \\ &+ \frac{1 - 4\mu(\alpha)}{\mu^3(\alpha)(\mu(\alpha) - 1)^2(\mu^3(\alpha) - 1)}\, a_2^3(\alpha)\,, \end{aligned}$$

the η^4-term in (3.104) vanishes and it takes the form

$$\eta \mapsto -(1 + \beta_1(\alpha))\eta + \beta_2(\alpha)\eta^3 + d(\alpha)\eta^5 + O(\eta^6),$$

where $\beta_{1,2}(\alpha)$ are defined above and such that $\beta_1(0) = \beta_2(0) = 0$, while

$$d(0) = c_2^0 = -2a_2^4(0) + 3a_4(0)a_2(0) + a_5(0) = d_0$$

at the GPD point, in accordance with (3.101). Replacing η by ξ, we conclude that (3.102) is locally smoothly equivalent to (3.10) from Theorem 3.5. □

3.B Proofs for Section 3.2

Proof of Lemma 3.7 As usual, write the restriction of (2.2) to the critical center manifold at α_0 in the complex form

$$w \mapsto e^{i\theta_0} w + \sum_{2\le j+k\le 5} \frac{1}{j!k!} g_{jk} w^j \bar{w}^k + O(|w|^6), \quad w \in \mathbb{C},$$

and introduce the map from $\mathbb{C}^2$ to $\mathbb{C}^2$ by the formula

$$\begin{pmatrix} w \\ \bar{w} \end{pmatrix} \mapsto \begin{pmatrix} \lambda_1 w + \displaystyle\sum_{2\le j+k\le 5} \frac{1}{j!k!} g_{jk} w^j \bar{w}^k \\ \lambda_2 \bar{w} + \displaystyle\sum_{2\le j+k\le 5} \frac{1}{j!k!} \bar{g}_{kj} w^j \bar{w}^k \end{pmatrix} + O(\|(w,\bar{w})\|^6), \tag{3.105}$$

where $\lambda_{1,2} = e^{\pm i\theta_0}$ and the second component is the complex conjugate of the first one, while w and $\bar{w}$ are treated as independent complex variables. We have $\lambda_1\lambda_2 = 1$, so that the only possible resonances of order less than 6 are

$$\lambda_1 = \lambda_1(\lambda_1\lambda_2), \quad \lambda_1 = \lambda_1(\lambda_1\lambda_2)^2,$$

and

$$\lambda_2 = \lambda_2(\lambda_1\lambda_2), \quad \lambda_2 = \lambda_2(\lambda_1\lambda_2)^2.$$

All other resonances are excluded by the imposed lemma conditions on θ_0. Then, Theorem 1.5 from Section 1.3 implies that the Poincaré normal form of the restriction to the critical center manifold is given by (3.14). Indeed, there is a smooth and close to identity transformation $(w, \bar{w}) \mapsto (z, \bar{z})$ that brings the first component of (3.105) to the form (3.14). □

Proof of Lemma 3.9 The proof consists of the following steps. We start by deriving the approximating vector field X in the z-plane for (3.17) composed with a rotation over an angle $\theta = -2\pi p/q$. Performing 13 Picard iterations, we get the vector field up to thirteenth order. Next we introduce polar coordinates and consider an annulus close to the bifurcating invariant curve. The resulting vector field can be brought into a perturbation of the pendulum equation by one more change of variables. We then list the lowest-order terms.

The composition of (3.17) with the reversed rotation over an angle $2\pi p/q$ results in a map close to identity. By a rotation in the z-plane we may assume that c_3 is real. For the Picard iterations we set the linear part of $X = (\beta_1 + i(\gamma\tau + a\beta_2))z$ with $\gamma = (2al_2 - d_1)$. The choice of γ simplifies some expressions. We will not show all steps in detail as the resulting expressions are lengthy. We only give the structure of the vector field with some general coefficients f_{ij}

with $i - j \mod 7 = 1$ for positive integers i, j as well as list the most important parts of the f_{ij}.

$$\dot{z} = f_{10}z + f_{21}z|z|^2 + f_{32}z|z|^4 + f_{06}\bar{z}^6 + f_{43}z|z|^6 + f_{80}z^8 + f_{17}z\bar{z}^7 + f_{54}z|z|^8 \\ + f_{28}|z|^4\bar{z}^6 + f_{91}z^8|z|^2 + f_{65}z|z|^{10} + f_{39}|z|^6\bar{z}^6 + f_{10,2}z^8|z|^4 + f_{76}z|z|^{12} + f_{0,13}\bar{z}^{13} \tag{3.106}$$

with

$$f_{10} = (\beta_1 + i(\tau + a\beta_2)), \quad f_{21} = (\beta_2 + ic_2 - 2ic_2\beta_1), \quad f_{32} = (l_2 + id_2 - 2c_2\beta_2),$$
$$f_{06} = c_3, \quad f_{80} = c_5 - \tfrac{i}{2}c_3d_1, \quad f_{17} = c_6 + 2ic_3d_1,$$
$$f_{0,13} = c_7 - \tfrac{1}{2}c_3c_6 - 3c_3c_5 - \tfrac{11i}{6}c_3^2d_1.$$

Next we convert to polar coordinates $z = re^{i\psi}$. Using (3.19) we introduce a new coordinate on an annulus $r = r^*\left(1 + \tau^1 2\left(\sigma + \tau l_2\mu_2\frac{a}{c_2}\right)\right)$ and new parameters $\beta_1 = \beta_1^*(1 + \tau^{1/2}\nu_1)$ and $\beta_2 = \beta_2^*(1 + \tau^{1/2}\nu_2)$. The intermediate approximating vector field on the annulus is given by

$$\begin{cases} \psi' = 2d_1\tau^{3/2}\sigma + \frac{1}{d_1}\tau^2\left(2al_2\nu_2 + d_1\sigma\right)^2 + \tau^{5/2}\Psi(\sigma, \psi, \tau, \nu), \\ \sigma' = \tau^3\left(c_3\cos(7\psi) + l_2(\nu_1 - 2\nu_2)\right) + \tau^7 2\Phi(\sigma, \psi, \tau, \nu), \end{cases} \tag{3.107}$$

where Ψ, Φ are $2\pi/q$-periodic functions in their second argument and of order $O(1)$ in τ. In order to remove the higher-order perturbations from the first equation we introduce $x = \sigma + \frac{1}{d_1}\tau^{1/2}\left(2al_2\nu_2 + d_1\sigma\right)^2 + \tau\Psi(\sigma, \psi, \tau, \nu)$. Differentiation of this relationship and substituting the approximate inverse $\sigma \approx x - \frac{\tau^{1/2}}{2}x^2$ leads to the following vector field in lowest order of τ

$$\begin{cases} \psi' = 2d_1\tau^{3/2}x, \\ x' = \tau^3\left(\delta_2 x\cos(7\psi) + \mu_2\right) + \tau^{7/2}\left(\gamma x^2 - h\cos(7\psi)x^2 + \mu_2 x\right) + \\ \qquad \tau^{7/2}\left(\delta_2 x\cos(7\psi) + \delta_3 x\sin(7\psi)\right) + \tau^5 c_3^2\sin(14\psi) + \Xi, \end{cases} \tag{3.108}$$

with coefficients and new parameters given by

$$\mu_1 = l_2(\nu_1 - 2\nu_2), \ \mu_2 = 2l_2(\nu_1 + 4(al_2 - d_1)\nu_2), \ \gamma = 4l_2, \ h = -35c_3, \ \delta_1 = c_3,$$
$$\delta_2 = -7c_3 + 7\tau^{1/2}\Re(c_5 - c_6), \ \delta_3 = -7\Im(c_5 + c_6), \ \delta_4 = \frac{35c_3^2}{2d_1} - 14\tau\Im(c_7).$$

Here we replaced $c_3 \to \tau^{1/2}c_3$ and $c_7 \to \tau^{-1}c_7$ to equalize the order in the small perturbation parameter τ. While some of these coefficients may still contain small terms, they are generically linearly independent for the original perturbation of the model map (3.17). The higher-order perturbation Ξ contains all terms depending quadratically on parameters and a possibly linear dependence on the parameters of the coefficients δ_i, γ is not indicated. It also contains any terms $O(x^3)$ as the periodic orbit along the annulus bifurcates for small x. Other higher-order harmonics are also considered a higher-order perturbation. Finally, rescaling time and redefining coefficients yields the vector field as stated in the lemma, except that we have set $h = 0$ as the corresponding term may be seen as the first harmonic perturbation of the x^2-term. □

Proof of Lemma 3.10 For equilibria we have $x = 0$ and $\delta_1 \cos(\psi) + \mu_1 = 0$. The Jacobian matrix is given by

$$A = \begin{pmatrix} 0 & 1 \\ -\delta_1 \sin(\psi) & \mu_2 + \delta_2 \cos(\psi) + \delta_3 \sin(\psi) + \delta_4 \sin(2\psi) \end{pmatrix}. \tag{3.109}$$

Requiring $\det(A) = 0$ for a fold bifurcation, we have $\sin\psi = 0$ so that $\mu_1 = \pm\delta_1$. The condition $\operatorname{tr}(A) = 0$ for a Hopf bifurcation leads to $\mu_2 = -\delta_2 \cos(\psi) - \delta_3 \sin(\psi) - \delta_4 \sin(2\psi)$ as asserted. It is not difficult to verify that the fold and Hopf curves coincide at the two points $(\mu_1, \mu_2) = \pm(\delta_1, \delta_2)$. Here we have Bogdanov–Takens bifurcations.

To verify the non-degeneracy conditions we compute the normal form coefficients. For the branch F_- with $\mu_1 = -\delta_1, x = 0, \psi = 0$, the right and left eigenvectors corresponding to the zero eigenvalue are given by $q = (1, 0)^T$ and $p = (\mu_2 + \delta_2, -1)$. The quadratic multilinear form is given by $B_-(u, v) = (0, -\delta_1 u_1 v_1 + (\delta_3 + 2\delta_4)(u_1 v_2 + u_2 v_1) + 2\gamma u_2 v_2)$. With this we find

$$a_{F_-} = \frac{\langle p, B_-(q, q)\rangle}{\langle p, q\rangle} = \frac{\delta_1}{\delta_2 + \mu_2}.$$

On the other branch F_+ we have $\mu_1 = \delta_1, x = 0, \psi = \pi$. We can then use $q = (1, 0)$ and $p = (\mu_2 - \delta_2, -1)$. The quadratic multilinear form is given by $B_+(u, v) = (0, \delta_1 u_1 v_1 + (-\delta_3 + 2\delta_4)(u_1 v_2 + u_2 v_1) + 2\gamma u_2 v_2)$. We can thus compute $a_{F_+} = \frac{\delta_1}{\delta_2 - \mu_2}$. We proceed to the normal form coefficients of the BT points where the eigenvectors are particularly convenient, i.e., $q_0 = (1, 0), q_1 = (0, 1)$ and $p_0 = (0, 1), p_1 = (1, 0)$. For the BT point on F_- we have

$$a_{BT,-} = \langle p_0, B_-(q_0, q_0)\rangle = -\delta_1, b_{BT,-} = \langle p_0, B_-(q_0, q_1)\rangle + \langle p_1, B(q_0, q_0)\rangle = \delta_3 + 2\delta_4,$$

while for the BT point on F_+ we find

$$a_{BT,+} = \delta_1, \quad b_{BT,+} = -\delta_3, +2\delta_4.$$

We require that the resonance tongue itself is non-degenerate, i.e., $\delta_1 \neq 0$. Therefore, a degenerate BT point can only occur if $b_{BT} = 0$.

Finally, for the Hopf bifurcation we use the eigenvectors $q = (1, \omega i)$ and $p = (-\omega i, 1)$ with $\omega = \sqrt{\delta_1 \sin(\psi)}$ when $\delta_1 \sin(\psi) > 0$. The multilinear forms are now given by

$$\begin{aligned} B(u, v) = (0, -\delta_1 \cos(\psi) u_1 v_1 &+ (-\delta_2 \sin(\psi) + \delta_3 \sin(\psi) \\ &+ 2\delta_4 \cos(2\psi))(u_1 v_2 + u_2 v_1) + 2\gamma u_2 v_2) \end{aligned}$$

and

$$C(u,v,w) = (0, \sin(\psi)u_1v_1w_1 + (-\delta_2\cos(\psi) - \delta_3\sin(\psi) - 4\delta_4\sin(2\psi)).$$

It is then straightforward to find the first Lyapunov coefficient l_1 using (1.31). Depending on the coefficients $\delta_1, \delta_2, \delta_3, \delta_4$, the first Lyapunov coefficient can vanish one or two times, leading to Bautin (generalized Hopf) points. □

3.C Proofs for Section 3.3

Proof of Lemma 3.13 Write the truncated normal form (3.25) as

$$N_\beta\colon \begin{pmatrix} \xi_1 \\ \xi_2 \end{pmatrix} \mapsto \begin{pmatrix} \xi_1 + \xi_2 \\ \xi_2 + \beta_1 + \beta_2\xi_2 + a\xi_1^2 + b\xi_1\xi_2 \end{pmatrix}, \tag{3.110}$$

where $a = a(\beta)$ and $b = b(\beta)$ but we will not indicate the parameter-dependence explicitly. The Jacobian matrix of this map with respect to (ξ_1, ξ_2) is

$$A(\xi,\beta) = \begin{pmatrix} 1 & 1 \\ 2\,a\xi_1 + b\xi_2 & b\xi_1 + \beta_2 + 1 \end{pmatrix}.$$

Fold bifurcation. A fixed point ξ of (3.110) exhibiting a fold bifurcation at some parameter values β satisfies

$$\begin{cases} N_\beta(\xi) - \xi & = & 0, \\ \det(A(\xi,\beta) - I_2) & = & 0. \end{cases}$$

Solving this system for (ξ_1, ξ_2, β_1) as functions of β_2 gives

$$\xi_1 = \xi_2 = 0, \;\; \beta_1 = 0,$$

implying (3.26).

Along the fold line, the Jacobian matrix

$$A = \begin{pmatrix} 1 & 1 \\ 0 & \beta_2 + 1 \end{pmatrix}$$

has eigenvector

$$q = \begin{pmatrix} 1 \\ 0 \end{pmatrix}$$

corresponding to eigenvalue $+1$, while the transposed matrix A^T has eigenvector

$$\tilde{p} = \begin{pmatrix} -\beta_2 \\ 1 \end{pmatrix}$$

corresponding to the same eigenvalue, so that $\langle \tilde{p}, q\rangle \neq 0$ for $\beta_2 \neq 0$. Then, we can normalize $\tilde{p}$ with respect to q and take

$$p = \begin{pmatrix} 1 \\ -\frac{1}{\beta_2} \end{pmatrix}$$

satisfying $\langle p, q\rangle = 1$. Then the fold normal form coefficient (2.7) is given by

$$\frac{1}{2}\langle p, B(q,q)\rangle = -\frac{a}{\beta_2}.$$

Since $a = a_0 + O(\|\beta\|)$ and $a_0 \neq 0$, this implies that the fold bifurcation is non-degenerate at $LP^{(1)}$.

Neimark–Sacker bifurcation. A fixed point ξ of (3.110) exhibiting an NS bifurcation at some parameter values β satisfies

$$\begin{cases} N(\xi,\alpha) - \xi = 0, \\ \det(A(\xi,\beta)) - 1 = 0. \end{cases}$$

Solving this system for (ξ_1, ξ_2, β_1) as functions of β_2 gives

$$\xi_1 = \frac{\beta_2}{2a-b},\ \xi_2 = 0,\ \beta_1 = \frac{a\beta_2^2}{(2a-b)^2},$$

which implies (3.27) after taking into account that $a = a_0 + O(|\beta|)$, $b = b_0 + O(|\beta|)$, and noticing that the eigenvalues of $A(\xi,\beta)$ are simple non-real only when $s\beta_2 > 0$ where $s = 2a_0(b_0 - 2a_0) \neq 0$.

Along the Neimark–Sacker curve $NS^{(1)}$, the Jacobian matrix

$$A = \begin{pmatrix} 1 & 1 \\ \dfrac{2a\beta_2}{2a-b} & 1 + \beta_2 + \dfrac{b\beta_2}{2a-b} \end{pmatrix}$$

has an eigenvalue

$$\mu = e^{i\theta_0} = 1 + \frac{a\beta_2}{2a-b} + \frac{i\sqrt{2\beta_2 a(-2a+b) - a^2\beta_2^2}}{2a-b}$$

with the eigenvector

$$q = \begin{pmatrix} 1 \\ \mu - 1 \end{pmatrix}.$$

The transposed matrix A^T has eigenvector

$$p = \begin{pmatrix} 1 \\ (\bar{\mu} - 1)\dfrac{2a-b}{2a\beta_2} \end{pmatrix}$$

corresponding to its eigenvalue $\bar{\mu}$. We normalize it relative to q, so that $\langle p, q\rangle = \bar{p}^T q = 1$, and compute

$$c_0 = \Re(e^{-i\theta_0} d_0) = \Re(\bar{\mu} d_0),$$

where d_0 is given by (2.18). Here, however, $C \equiv 0$, since no cubic term is present in (3.110), implying

$$d_0 = \frac{1}{2}\langle p, 2B(q,(I_n - A)^{-1}B(q,\bar{q})) + B(\bar{q},(\mu^2 I_n - A)^{-1}B(q,q))\rangle.$$

This leads to

$$\bar{\mu} d_0 = \frac{5i(b-2a)\sqrt{2a(b-2a)}}{6\beta_2^{3/2}} + \frac{(b-2a)^2}{2\beta_2} + O\left(\frac{1}{\sqrt{\beta_2}}\right).$$

The first term is purely imaginary. Therefore,

$$c_0 = \Re(\bar{\mu} d_0) = \frac{(b_0 - 2a_0)^2}{2\beta_2} + O(1)$$

and the NS bifurcation is non-degenerate along $NS^{(1)}$. □

Proof of Lemma 3.15 Denote by $F(u,\beta)$ the right-hand side of (3.29). The Jacobian matrix of F with respect to u is

$$A(u,\beta) = \begin{pmatrix} A_{11}(u,\beta) & A_{12}(u,\beta) \\ A_{21}(u,\beta) & A_{22}(u,\beta) \end{pmatrix},$$

where

$$A_{11}(u,\beta) = \left(\frac{b}{3} - \frac{a}{2}\right)\beta_1 - au_1 + \left(\frac{2a}{3} - \frac{b}{2}\right)u_2,$$

$$A_{12}(u,\beta) = 1 + \left(\frac{a}{5} - \frac{5b}{12}\right)\beta_1 - \frac{1}{2}\beta_2 + \left(\frac{2a}{3} - \frac{b}{2}\right)u_1 + \left(\frac{2b}{3} - \frac{a}{3}\right)u_2,$$

$$A_{21}(u,\beta) = \left(\frac{2a}{3} - \frac{b}{2}\right)\beta_1 + 2au_1 + (b-a)u_2,$$

$$A_{22}(u,\beta) = \left(\frac{b}{2} - \frac{a}{6}\right)\beta_1 + \beta_2 + (b-a)u_1 + \left(\frac{a}{3} - b\right)u_2.$$

(1) For $u = 0$ and $\beta_1 = 0$ matrix $A(u,\beta)$ reduces to

$$\begin{pmatrix} 0 & 1 - \frac{\beta_2}{2} \\ 0 & \beta_2 \end{pmatrix}$$

and has one simple eigenvalue $\lambda_1 = 0$ for $\beta_2 \neq 0$.

(2) Consider the defining system

$$\begin{cases} F(u,\beta) = 0, \\ \operatorname{tr} A(u,\beta) = 0. \end{cases}$$

Its solution corresponds to an equilibrium of (3.29) with eigenvalues satisfying $\lambda_1+\lambda_2 = 0$. Thus, this system implicitly defines Hopf bifurcation curve H. This curve in the β-plane has the following representation:

$$\beta_1 = -au_1^2 + O(u_1^3),$$
$$\beta_2 = (2\,a - b)\,u_1 + \left(-\frac{2\,a^2}{3} + \frac{5\,ab}{6}\right)u_1^2 + O(u_1^3),$$

implying

$$\beta_1 = -\frac{a_0}{(b_0 - 2a_0)^2}\beta_2^2 + O(\beta_2^3).$$

Moreover, if $2a_0(b_0 - 2a_0)\beta_2 > 0$ then $\det A(u,\beta) > 0$ for small β_2.

(3) Finally, for the homoclinic bifurcation curve we apply the singular rescaling

$$u_0 \to \frac{u_0}{a_0}\varepsilon^2, \quad u_1 \to \frac{u_1}{a_0}\varepsilon^3, \quad \beta_1 \to \frac{-4}{a_0}\varepsilon^4, \quad \beta_2 \to \frac{b_0\tau}{a_0}\varepsilon^2, \quad t \to \varepsilon s$$

to (3.29). This yields the following perturbation of a Hamiltonian vector field

$$\begin{pmatrix} u_0' \\ u_1' \end{pmatrix} = \frac{1}{a_0}\begin{pmatrix} u_1 \\ u_0^2 - 4 \end{pmatrix} + \frac{\varepsilon}{a_0}\begin{pmatrix} 2 - \frac{1}{2}u_0^2 \\ \frac{b_0 u_1(\tau + u_0)}{a_0} - u_0 u_1 \end{pmatrix} + O(\varepsilon^2),$$

which for $\varepsilon = 0$ has Hamiltonian $H := \frac{1}{a_0}\left(\frac{1}{2}u_1^2 + (4u_0 - \frac{1}{3}u_0^3\right)$. The level set $H = \frac{16}{3a_0}$ corresponds to a homoclinic orbit Γ of the saddle $u = (2,0)$. Now we can evaluate the Pontryagin–Melnikov integral (see Section 1.6)

$$\begin{aligned}
\Delta &= \int_\Gamma f_2 du_0 - f_1 du_1 = \int_\Gamma \left(f_2 + \left(\int \frac{\partial f_1}{\partial u_1} du_1\right)\right) du_0 \\
&= \int_\Gamma \frac{u_1}{a_0^2}\,(b_0(\tau + u_0) - 2a_0u_0)\,du_0 \\
&= \int_{-4}^{2} \frac{2}{a_0^2}(u_0 - 2)\,\sqrt{6(u_0 + 4)}\,(b_0(\tau + u_0) - 2a_0u_0)\,du_0 \\
&= \frac{-96}{35a_0^2}\,(7b_0\tau + 20a_0 - 10b_0)\,.
\end{aligned}$$

This shows that $\Delta = 0$ if $\tau = \frac{10(b_0-2a_0)}{7b_0}$. Hence, the homoclinic bifurcation curve can be approximated by

$$\begin{pmatrix} \beta_1 \\ \beta_2 \end{pmatrix} = \begin{pmatrix} -\dfrac{4}{a_0}\varepsilon^4 \\ \dfrac{10(b_0 - 2a_0)}{7a_0}\varepsilon^2 \end{pmatrix},$$

so that for $2a_0(b_0 - 2a_0)\beta_2 > 0$ we have

$$\beta_1 = -\frac{49}{25}\frac{a_0}{(b_0 - 2a_0)^2}\beta_2^2 + O(\beta_2^3).$$

This is the expansion for curve P in Lemma 3.15. This was shown by Al-Hdaibat et al. (2018) with a different method.

The non-degeneracy for small $\|\beta\|$ of all bifurcations above immediately follows from the non-degeneracy of the codim 2 BT bifurcation at $\beta = 0$, which is ensured by

$$s = 2a_0(b_0 - 2a_0) \neq 0$$

assumed in Lemma 3.15. □

Proof of Lemma 3.18 Write the map (3.34) as

$$N_\beta : \begin{pmatrix} \xi_1 \\ \xi_2 \end{pmatrix} \mapsto \begin{pmatrix} -\xi_1 + \xi_2 \\ -\xi_2 + \beta_1\xi_1 + \beta_2\xi_2 + c\xi_1^3 + d\xi_1^2\xi_2 \end{pmatrix}, \tag{3.111}$$

where $c = c(\beta)$ and $d = d(\beta)$. The Jacobian matrix of N_β with respect to the components of ξ is

$$A(\xi, \beta) = \begin{pmatrix} -1 & 1 \\ 3c\xi_1^2 + 2d\xi_1\xi_2 + \beta_1 & d\xi_1^2 + \beta_2 - 1 \end{pmatrix}.$$

Period-doubling bifurcation. If $\xi = 0$ and $\beta_1 = 0$, then

$$A = \begin{pmatrix} -1 & 1 \\ 0 & \beta_2 - 1 \end{pmatrix}$$

has one simple eigenvalue -1 for $\beta_2 \neq 0$. This implies the expression for $PD^{(1)}$ given in Lemma 3.18.

To verify the non-degeneracy of the PD bifurcation, notice that for $\beta_2 \neq 0$ the vectors

$$q = (1,\ 0) \quad \text{and} \quad p = \left(1,\ -\frac{1}{\beta_2}\right)$$

satisfy $Aq = -q$, $A^T p = -p$ and $\langle q, q\rangle = \langle p, q\rangle = 1$. The PD normal form coefficient is then given by formula (2.12). Since in our case $B(q, q) = 0$, the second term in (2.12) vanishes and the coefficient reduces to

$$\frac{1}{6}\langle p, C(q,q,q)\rangle = \frac{1}{6}\frac{d^3}{d\tau^3}\langle p, N_\beta(\tau q)\rangle\bigg|_{\tau=\beta_1=0} = -\frac{c}{\beta_2} = -\frac{c_0}{\beta_2} + O(1),$$

so that the period-doubling bifurcation on $PD^{(1)}$ is non-degenerate for sufficiently small $\beta_2 \neq 0$, since $c_0 \neq 0$.

Primary Neimark–Sacker bifurcation. Similarly, if $\xi = 0$ but $\beta_2 = -\beta_1$, then

$$A = \begin{pmatrix} -1 & 1 \\ \beta_1 & -1-\beta_1 \end{pmatrix}$$

has determinant equal to 1 with non-real eigenvalues $e^{\pm i\theta}$, provided that $\beta_1 < 0$. This gives the expression for $NS^{(1)}$ from Lemma 3.18.

Along the line $NS^{(1)}$, matrix A has an eigenvalue

$$\mu = e^{i\theta_0} = -1 - \frac{\beta_1}{2} + \frac{i}{2}\sqrt{-\beta_1^2 - 4\beta_1},$$

with the eigenvector

$$q = (1,\ \mu + 1).$$

The transposed matrix A^T has eigenvector

$$p = \left(1,\ \frac{\bar{\mu}+1}{\beta_1}\right)$$

corresponding to eigenvalue $\bar{\mu}$. We normalize p to have $\langle p, q\rangle = 1$. Compute now along the $NS^{(1)}$ line the quadratic and cubic coefficients in the Taylor expansion

$$\langle p, N_\beta(zq + \bar{z}\bar{q})\rangle = \mu z + \sum_{j+k\geq 2}\frac{1}{j!k!}g_{jk}z^j\bar{z}^k.$$

This gives $g_{20} = g_{11} = g_{02} = 0$, so that the normal form coefficient defined by (2.22) reduces to

$$\Re\left(\frac{\bar{\mu}g_{21}}{2}\right) = -\frac{1}{2}(d_0 + 3c_0) + O(\beta_1),$$

implying the nondegeneracy of the NS bifurcation of the fixed point of (3.34) along the line $NS^{(1)}$ for small $\beta_1 < 0$ due to the assumption

$$c_0(d_0 + 3c_0) \neq 0$$

made in Lemma 3.18. The NS bifurcation is supercritical if $d_0 + 3c_0 > 0$.

for $(\beta_1,\beta_2) = (-\varepsilon^2,\varepsilon^2)$ with $\varepsilon > 0$. Clearly, $g_{20} = g_{11} = 0$. Therefore, using (1.27) we obtain the first Lyapunov coefficient

$$l_1 = \frac{1}{2\omega}\Re(g_{21}) = -\frac{1}{2\varepsilon^3}(d_0 + 3c_0) + O\left(\frac{1}{\varepsilon}\right).$$

Thus, the Hopf bifurcation at $H^{(1)}$ is non-degenerate for small $\varepsilon > 0$, since $d_0 + 3c_0 \neq 0$. It is supercritical if $d_0 + 3c_0 > 0$ and subcritical if $d_0 + 3c_0 < 0$.

To study the Hopf bifurcation of a nontrivial equilibrium, first notice that the algebraic system

$$\begin{cases} f(\xi,\beta) = 0, \\ \operatorname{tr} f_\xi(\xi,\beta) = 0, \end{cases}$$

has a nontrivial solution,

$$\beta_2 = \frac{d_0 + 2c_0}{c_0}\beta_1, \quad \xi_1 = \sqrt{-\frac{\beta_1}{c_0}}, \quad \xi_2 = 0,$$

if $\beta_1 c_0 < 0$. The corresponding Jacobian matrix

$$A = \begin{pmatrix} \beta_1 & -1 - \frac{1}{3}\beta_1 \\ 2\beta_1 & -\beta_1 \end{pmatrix}$$

has purely imaginary eigenvalues if $\beta_1 > 0$. This verifies the expression for $H^{(2)}$ in part (c). Setting $\beta_1 = \varepsilon^2$ with $\varepsilon > 0$ results in $\lambda_{1,2} = \pm i\omega$, with

$$\omega = \varepsilon\sqrt{2 - \frac{\varepsilon^2}{3}} > 0.$$

Now, the vectors

$$q = \left(\frac{1}{2} + \frac{i\omega}{2\varepsilon^2},\ 1\right) \quad \text{and} \quad p = \left(1, -\frac{1}{2} - \frac{i\omega}{2\varepsilon^2}\right)$$

satisfy $Aq = i\omega q$ and $A^T p = -i\omega p$, respectively, and p can be normalized to achieve $\langle p, q\rangle = 1$. With such p, we compute the coefficients g_{20}, g_{11}, and g_{21} in the Taylor expansion

$$\langle p, f(\xi + zq + \bar{z}\bar{q}, \beta) = i\omega z + \sum_{j+k\le 2} \frac{1}{j!k!} g_{jk} z^j \bar{z}^k$$

along $H^{(2)}$, i.e., for

$$\beta = \left(\varepsilon^2,\ \frac{d_0 + 2c_0}{c_0}\varepsilon^2\right), \quad \xi = \left(\sqrt{-\frac{\varepsilon^2}{c_0}},\ 0\right)$$

with $\varepsilon > 0$ and $c_0 < 0$. The formula (1.27) gives

$$l_1 = \frac{1}{2\omega^2}\Re(ig_{20}g_{11} + \omega g_{21}) = \frac{\sqrt{2}(d_0 + 3c_0)}{4\varepsilon^3} + O\left(\frac{1}{\varepsilon}\right)$$

for the first Lyapunov coefficient. We conclude that the Hopf bifurcation of the nontrivial equilibrium at $H^{(2)}$ is non-degenerate for small $\varepsilon > 0$, since $d_0 + 3c_0 \neq 0$. It is supercritical if $d_0 + 3c_0 < 0$ and subcritical if $d_0 + 3c_0 > 0$.

(2) Starting with (3.37) we apply the singular rescaling

$$\xi_1 = \frac{\sqrt{-s\beta_1}}{\sqrt{sc_0}}x_1, \quad \xi_2 = \beta_1\sqrt{sc_0}x_2, \quad t = \frac{\sqrt{-s\beta_1}}{s}\tau.$$

This results in the perturbed Hamiltonian system

$$\begin{cases} \dot{x}_1 = x_2 + \delta g_1, \\ \dot{x}_2 = -sx_1 + sx_1^3 + \delta g_2, \end{cases}$$

with Hamiltonian for $\delta = 0$

$$H := \frac{sx_1^2}{2} + \frac{x_2^2}{2} - \frac{sx_1^4}{4} \tag{3.115}$$

and perturbation terms

$$\begin{aligned} g_1 :=& \frac{1}{2}\sqrt{-s\beta_1}x_1(1 - x_1^2) + \frac{\sqrt{-s\beta_1}\beta_1}{6sc_0}(3c_0 + 4d_0)x_1x_2^2 + \left(\frac{\beta_1}{3} + \frac{\beta_2}{2}\right)x_2 \\ &- \frac{\beta_1}{2c_0}(2c_0 + d_0)x_1^2x_2 + \frac{\beta_1^2}{30sc_0}(c_0 + 5d_0)x_2^3, \\ g_2 :=& -\frac{s(\beta_1 + 2\beta_2)}{2\sqrt{-s\beta_1}}x_2 + \frac{(3c_0 + 2d_0)}{2c_0\sqrt{-s\beta_1}}\beta_1x_1^2x_2 - \frac{\beta_1}{2c_0}(c_0 + 2d_0)x_1x_2^2 - \frac{\beta_1^2 d_0}{6c_0\sqrt{-s\beta_1}}x_2^3. \end{aligned}$$

Next we evaluate the Pontryagin–Melnikov integral (see Section 1.6):

$$\begin{aligned} \Delta =& \int_{\Gamma_0} g_2dx_1 - g_1dx_2 = \int_{\Gamma_0} g_2 + \left(\int \frac{\partial g_1}{x_1dx_2}\right)dx_1 \\ =& \int_{\Gamma_0} \frac{-x_2}{18c_0\sqrt{-s\beta_1}}\Big(18sc_0(\beta_1 + \beta_2) - 18(3c_0 + d_0)\beta_1sx_1^2 \\ &+ 27(c_0 + d_0)\beta_1\sqrt{-s\beta_1}x_1x_2 + (3c_0 + 7d_0)\beta_1^2x_2^2\Big)dx_1, \end{aligned}$$

where we use Green's Theorem, exploiting that the level set Γ_0 is closed and bounded.

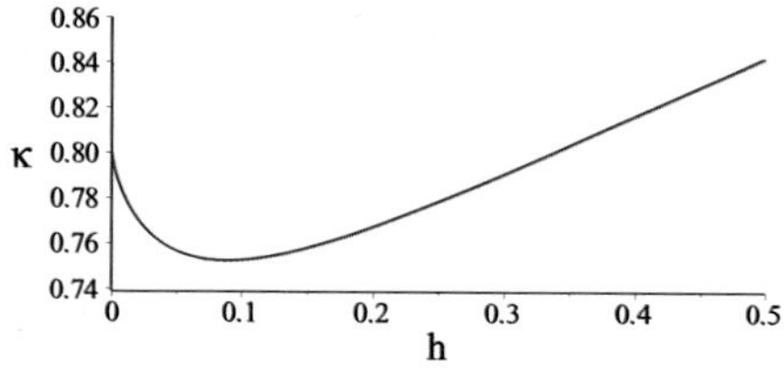

Figure 3.50 The function $\kappa(h)$ for the limit cycles for the R2 resonance has a minimum $\kappa_0 \approx 0.7522$ at $h_0 \approx 0.089$.

For $s = 1$, we have a heteroclinic connection $x_2 = \pm\frac{1}{2}\left(x_1^2 - 1\right)$ between the saddles $x_1 = \pm 1, x_2 = 0$. This leads to

$$\Delta_{het} = \frac{-2}{315c_0\sqrt{\beta_1}}\left(21(2c_0 - d_0)\beta_1 + 105c_0\beta_2 + (3c_0 + 7d_0)\beta_1^2\right).$$

So, a linear approximation to the heteroclinic curve C from part (a) is given by

$$\beta_2 = \frac{d_0 - 2c_0}{5c_0}\beta_1.$$

Now for $s = -1$ we use that on the homoclinic connection we have $x_2 = \pm x_1\sqrt{1 - \frac{1}{2}x_1^2}$ and $0 < x_1 \leq \sqrt{2}$. Accounting for the orientation of the curve the x_1x_2 term vanishes, we then find

$$\Delta_{hom} = \frac{-4}{315c_0\sqrt{\beta_1}}\left(21(7c_0 + 4d_0)\beta_1 - 105c_0\beta_2 + 2(3c_0 + 7d_0)\beta_1^2\right).$$

Solving $\Delta_{hom} = 0$ for β_2 results in the asserted linear approximation of the homoclinic connection curve P from part (b)

$$\beta_2 = \frac{7c_0 + 4d_0}{5c_0}\beta_1.$$

Finally, for the limit point of cycles, we note that the leading part for the parameters of Δ can be written as

$$\sqrt{\beta_1}(\beta_1 + \beta_2)\int_{C_h} x_2 dx_1 - \frac{3c_0 + d_0}{c_0}\int_{C_h} x_1^2 x_2 dx_1,$$

where C_h corresponds to a closed level curve with $H = h$. We define the integral $I_n(h) := \int_{C_h} x_1^n x_2 dx_1$ and the ratio $\kappa(h) := I_2(h)/I_0(h)$. These integrals cannot be determined analytically, so we show a numerical evaluation of $\kappa(h)$ in Figure 3.50. Then in a first approximation a limit cycle corresponds to a value of h and parameters β_1, β_2 such that

$$\beta_2 = \left(\kappa(h)\left(3 + \frac{d_0}{c_0}\right) - 1\right)\beta_1.$$

It is clear that for $\kappa_0 < \kappa < 0.8$ this relation is satisfied by two values of h leading to two coexisting limit cycles, one stable and one unstable. When $\kappa = \kappa_0$ these limit cycles collide, giving rise to the cyclic fold curve K from part (b). □

Proof of Lemma 3.23 When $\beta_1 = 0$, the truncated normal form (3.41) can be written as

$$z \mapsto e^{i\theta_0} z + b\bar{z}^2 + cz|z|^2,$$

where $\theta_0 = \theta_0(\beta_2) = (2\pi/3 + \beta_2)i$ and b, c are functions of β_2. For small $\beta_2 \neq 0$, the fixed point $z = 0$ of this map (in the real form) has a pair of eigenvalues $e^{\pm i\theta_0}$ exhibiting no strong resonance, i.e., we have a codim 1 NS bifurcation at the line $NS^{(1)}$ near $\beta = 0$.

The non-degeneracy of this bifurcation for small $\|\beta\|$ can be established using formula (2.22), where one should input $g_{20} = g_{11} = 0,\ g_{02} = 2b,\ g_{21} = 2c$. This gives

$$\Re\left(\frac{e^{-i\theta_0} g_{21}}{2}\right) - \frac{1}{4}|g_{02}|^2 = \Re(\bar{\mu}_0 c_0) - |b_0|^2 + O(\|\beta\|),$$

so that the NS bifurcation is non-degenerate for small $\|\beta\|$ by the lemma assumption.

□

Proof of Theorem 3.24 We have

$$R\,N_\beta(z) = e^{\beta_1 + i\beta_2} z + \bar{\mu}_0 b(\beta)\bar{z}^2 + \bar{\mu}_0 c(\beta) z|z|^2. \tag{3.116}$$

For small $\|\beta\|$, the map $z \mapsto R\,N_\beta(z)$ is close to identity and therefore can be approximated by the unit shift of a flow. The approximating ODE system has the same structure as (3.116), i.e.,

$$\dot{z} = (\beta_1 + i\beta_2)z + b_1(\beta)\bar{z}^2 + c_1(\beta)z|z|^2, \tag{3.117}$$

where b_1 and c_1 can be found by Picard iterations (1.20).

To obtain the expressions for $b_1(0)$ and $c_1(0)$, perform three Picard iterations for (3.117) at $\beta = 0$. This gives

$$\begin{aligned}
z^{(1)}(\tau) &= z,\\
z^{(2)}(\tau) &= z + b_1(0)\bar{z}^2\tau,\\
z^{(3)}(1) &= z + b_1(0)\bar{z}^2 + (|b_1(0)|^2 + c_1(0))z|z|^2 + O(|z|^4).
\end{aligned}$$

Comparing the coefficients in $z^{(3)}(1)$ with those in $R\,N_0$, we obtain

$$b_1(0) = \bar{\mu}_0 b_0, \quad c_1(0) = \bar{\mu}_0 c_0 - |b_0|^2$$

by taking into account that $\mu_0^3 = \bar{\mu}_0^3 = |\mu_0| = 1$. □

Proof of Theorem 3.25

(1) From the complex form (3.44), it is clear that for $\beta_2 \neq 0$ a Hopf bifurcation occurs at $\beta_1 = 0$. The first Lyapunov coefficient l_1 for this bifurcation can be computed directly using (1.27), i.e.,

$$l_1 = \frac{1}{2\omega_0^2} \Re(i g_{20} g_{11} + \omega_0 g_{21}),$$

where $\omega_0 = |\beta_2|$, $g_{20} = g_{11} = 0$ and $g_{21} = 2c_2$. This gives

$$l_1 = \frac{\Re(c_2)}{|\beta_2|} = \frac{c_1(0)}{|\beta_2|\,|b_1(0)|^2} + O(1) = \frac{\overline{\mu}_0 c_0 - |b_0|^2}{|\beta_2|\,|b_0|^2} + O(1),$$

which proves the non-degeneracy of the Hopf bifurcation for small $\beta_2 \neq 0$.

(2) For the approximation for the heteroclinic bifurcation curve C, we follow Chow, Li, and Wang (1994). First we apply a singular rescaling to (3.44)

$$\zeta = \varepsilon z, \quad \beta_1 = \gamma\varepsilon^2, \quad \beta_2\varepsilon, \quad t = \varepsilon\tau,$$

to arrive at the real form ($z = x + iy$)

$$\begin{pmatrix} x' \\ y' \end{pmatrix} = \begin{pmatrix} -y + x^2 - y^2 \\ x - 2xy \end{pmatrix} + \varepsilon \begin{pmatrix} x(\gamma + a_2(x^2 + y^2)) - b_2 y(x^2 + y^2) \\ y(\gamma + a_2(x^2 + y^2)) + b_2 x(x^2 + y^2) \end{pmatrix}. \quad (3.118)$$

For $\varepsilon = 0$ system (3.118) has a Hamiltonian $H = -\frac{1}{2}(x^2 + y^2) + x^2 y - \frac{1}{3}y^3$. The four equilibria are $(0,0)$, $(0,-1)$ and $(\pm\frac{1}{2}\sqrt{3}, \frac{1}{2})$. The nontrivial equilibria are saddles connected through three heteroclinic orbits. To find γ corresponding to the heteroclinic bifurcation we apply the Pontryagin–Melnikov method (see Section 1.6). We use the upper two saddles so that $x = -t$ with $-\frac{1}{2}\sqrt{3} \leq t \leq \frac{1}{2}\sqrt{3}$ and $y = \frac{1}{2}$. We have

$$\begin{aligned} \Delta &= \int_{\Gamma_0} f_1 dy - f_2 dx = \int_{-\frac{1}{2}\sqrt{3}}^{\frac{1}{2}\sqrt{3}} -\frac{1}{2}(\gamma + a_2(t^2 + \frac{1}{4})) + b_2 t(t^2 + \frac{1}{4}) dt \\ &= \frac{1}{2}\sqrt{3}\,(2\gamma + a_2)\,. \end{aligned}$$

Solving $\Delta = 0$ we have $\gamma = -\frac{a_2}{2}$ showing that $\beta_1 = \gamma\delta^2 = -\frac{a_2}{2}\beta_2^2$ is the first-order approximation of the heteroclinic bifurcation curve. The non-degeneracy of the heteroclinic bifurcation and the absence of other global bifurcations is shown by Horozov (1979). □

Proof of Theorem 3.28 We have

$$R\, N_\beta(z) = e^{\beta_1 + i\beta_2} z - ic(\beta) z|z|^2 - id(\beta)\bar{z}^3. \quad (3.119)$$

For small $\|\beta\|$, the map $z \mapsto R\, N_\beta(z)$ is close to identity and therefore can be approximated by the unit shift of a flow. The approximating ODE system has the same structure as (3.119), i.e.,

$$\dot{z} = (\beta_1 + i\beta_2)z + c_1(\beta)z|z|^2 + d_1(\beta)\bar{z}^3, \tag{3.120}$$

where c_1 and d_1 can be found by Picard iterations (1.20).

To obtain the expressions for $c_1(0)$ and $d_1(0)$, perform three Picard iterations for (3.120) at $\beta = 0$. This gives

$$z^{(1)}(\tau) = z^{(2)}(\tau) = z,$$
$$z^{(3)}(1) = z + c_1(0)z|z|^2 + d_1(0)\bar{z}^3 + O(|z|^4).$$

Comparing the coefficients in $z^{(3)}(1)$ with those in $R\, N_0$, we obtain

$$c_1(0) = -ic_0, \quad d_1(0) = -id_0.$$

□

Proof of Theorem 3.29 By considering the complex equation (3.53), it is obvious that the trivial equilibrium $\eta = 0$ undergoes a non-degenerate Hopf bifurcation at $\beta_1 = 0$ for small $\beta_2 \neq 0$.

To obtain linear approximations for the remaining bifurcation curves, it is sufficient to substitute (3.54) by the system

$$\begin{cases} \dot{\rho} = P(\rho, \varphi, \beta), \\ \dot{\varphi} = Q(\rho, \varphi, \beta), \end{cases} \tag{3.121}$$

with

$$P(\rho, \varphi, \beta) = \beta_1\rho + a_0\rho^3 + \rho^3\cos(4\varphi),$$
$$Q(\rho, \varphi, \beta) = \beta_2 + b_0\rho^2 - \rho^2\sin(4\varphi).$$

The Jacobian matrix of the right-hand side of (3.121) with respect to (ρ, φ) is

$$A(\rho, \varphi, \beta) = \begin{pmatrix} \beta_1 + 3a_0\rho^2 + 3\rho^2\cos(4\varphi) & -4\rho^3\sin(4\varphi) \\ 2b_0\rho - 2\rho\sin(4\varphi) & -4\rho^2\cos(4\varphi) \end{pmatrix}.$$

A fold bifurcation curve is then defined by the system

$$\begin{cases} P(\rho, \varphi, \beta) = 0, \\ Q(\rho, \varphi, \beta) = 0, \\ \det A(\rho, \varphi, \beta) = 0, \end{cases}$$

which has two exact solutions $(\varphi, \beta_1, \beta_2)$ parameterized by ρ on which

$$\beta_1 = -\frac{\rho^2}{a_0}\left(a_0^2 + \frac{b_0^2 \pm a_0 b_0 \sqrt{a_0^2 + b_0^2 - 1}}{a_0^2 + b_0^2} - 1\right)$$

$$\beta_2 = \rho^2\left(-b_0 + \frac{b_0 \pm a_0 \sqrt{a_0^2 + b_0^2 - 1}}{a_0^2 + b_0^2}\right)$$

implying

$$\frac{\beta_2}{\beta_1} = \frac{a_0 b_0 \pm \sqrt{\Delta}}{a_0^2 - 1},$$

where $\Delta = a_0^2 + b_0^2 - 1$ is assumed positive. This verifies the asymptotics of the fold bifurcation curves $T_{1,2}$.

Similarly, a Hopf bifurcation curve satisfies the system

$$\begin{cases} P(\rho, \varphi, \beta) = 0, \\ Q(\rho, \varphi, \beta) = 0, \\ \operatorname{tr} A(\rho, \varphi, \beta) = 0, \end{cases}$$

provided that $\det A(\rho, \varphi, \beta) > 0$ along the solution. This yields a solution branch $(\varphi, \beta_1, \beta_2)$ parameterized by ρ and such that

$$\beta_1 = -2a_0\rho^2,$$
$$\beta_2 = \rho^2\left(\operatorname{sign}(b_0)\sqrt{1 - a_0^2} - b_0\right)$$

implying

$$\frac{\beta_2}{\beta_1} = \frac{b_0 - \operatorname{sign}(b_0)\sqrt{1 - a_0^2}}{2a_0},$$

where $a_0^2 < 1$ is assumed. Moreover, along the solution branch we have

$$\det A(\rho, \varphi, \beta) = 8\rho^4\left(|b_0|\sqrt{1 - a_0^2} - (1 + a_0^2)\right) > 0$$

so that

$$|b_0| > \frac{1 + a_0^2}{\sqrt{1 - a_0^2}}$$

excludes neutral saddles. This verifies the asymptotic expression for the Hopf curve N'.

The non-degeneracy of the fold and Hopf bifurcations along the curves $T_{1,2}$ and N', respectively, can be verified by the standard methods. □

3.D Proofs for Section 3.4.1

Proof of Lemma 3.31

Step 1 (quadratic terms). Applying to (3.58) a polynomial coordinate transformation

$$\begin{cases} \xi_1 = x_1 + \frac{1}{2}G_{20}x_1^2 + G_{11}x_1x_2 + \frac{1}{2}G_{02}x_2^2, \\ \xi_2 = x_2 + \frac{1}{2}H_{20}x_1^2 + H_{11}x_1x_2 + \frac{1}{2}H_{02}x_2^2, \end{cases} \tag{3.122}$$

we obtain:

$$x_1 \mapsto x_1 + \frac{1}{2}g_{20}x_1^2 + (g_{11} + 2G_{11})x_1x_2 + \frac{1}{2}g_{02}x_2^2 + \cdots,$$

$$x_2 \mapsto -x_2 + \frac{1}{2}(h_{20} - 2H_{20})x_1^2 + h_{11}x_1x_2 + \frac{1}{2}(h_{02} - 2H_{02})x_2^2 + \cdots,$$

where dots stand for higher-order terms. By setting

$$G_{11} = -\frac{1}{2}g_{11}, \quad H_{20} = \frac{1}{2}h_{20}, \quad H_{02} = \frac{1}{2}h_{02}, \tag{3.123}$$

we eliminate as many quadratic terms as possible. The remaining quadratic terms are called *resonant*.

Step 2 (cubic terms). Assume now that step 1 is already done, so that (3.58) has only resonant quadratic and all cubic terms. Consider a polynomial transformation:

$$\begin{cases} \xi_1 = x_1 + \frac{1}{6}G_{30}x_1^3 + \frac{1}{2}G_{21}x_1^2x_2 + \frac{1}{2}G_{12}x_1x_2^2 + \frac{1}{6}G_{03}x_2^3, \\ \xi_2 = x_2 + \frac{1}{6}H_{30}x_1^3 + \frac{1}{2}H_{21}x_1^2x_2 + \frac{1}{2}H_{12}x_1x_2^2 + \frac{1}{6}H_{03}x_2^3. \end{cases} \tag{3.124}$$

Obviously, it does not change the quadratic terms. After this transformation, we get

$$\begin{aligned} x_1 \mapsto x_1 &+ \tfrac{1}{2}g_{20}x_1^2 + \tfrac{1}{2}g_{02}x_2^2 + \tfrac{1}{6}g_{30}x_1^3 + \tfrac{1}{2}(g_{21} + 2G_{21})x_1^2x_2 + \tfrac{1}{2}g_{12}x_1x_2^2 \\ &+ \tfrac{1}{6}(g_{03} + 2G_{03})x_2^3 + \cdots \end{aligned}$$

and

$$\begin{aligned} x_2 \mapsto &- x_2 + h_{11}x_1x_2 + \tfrac{1}{6}(h_{30} - 2H_{30})\,x_1^3 + \tfrac{1}{2}h_{21}x_1^2x_2 \\ &+ \tfrac{1}{2}(h_{12} - 2H_{12})\,x_1x_2^2 + \tfrac{1}{6}h_{03}x_2^3 + \cdots. \end{aligned}$$

By setting

$$G_{21} = -\frac{1}{2}g_{21}, \quad G_{03} = -\frac{1}{2}g_{03}, \quad H_{30} = \frac{1}{2}h_{30}, \quad H_{12} = \frac{1}{2}h_{12},$$

we eliminate four cubic terms. The remaining cubic terms are also called *resonant*. They are not altered by (3.124).

Step 3 (more cubic terms). The coefficients H_{11}, G_{20} and G_{02} of (3.122) do not affect quadratic terms of (3.58) but alter its cubic terms. Taking into account (3.123) while computing the cubic terms of the transformed map, we obtain

$$x_1 \mapsto x_1 + \tfrac{1}{2}g_{20}x_1^2 + \tfrac{1}{2}g_{02}x_2^2 + \tfrac{1}{6}\left(g_{30} + \tfrac{3}{2}g_{11}h_{20}\right)x_1^3$$
$$+\tfrac{1}{2}\left(2g_{02}H_{11} - g_{02}G_{20} + (g_{20} + 2h_{11})G_{02} + \tfrac{1}{2}g_{11}h_{02} + g_{12} - g_{11}^2\right)x_1x_2^2$$

and

$$x_2 \mapsto -x_2 + h_{11}x_1x_2 + \tfrac{1}{6}\left(3g_{02}H_{11} + 3G_{02}h_{11} + h_{03} + \tfrac{3}{2}h_{02}^2\right)x_2^3$$
$$+\tfrac{1}{2}\left(g_{20}H_{11} + h_{11}G_{20} - g_{11}h_{20} + \tfrac{1}{2}h_{02}h_{20} + h_{21}\right)x_1^2x_2,$$

where only the resonant quadratic and cubic terms are displayed . Thus, we can try to eliminate three altered terms by selecting H_{11}, G_{20} and G_{02}. This implies solving the following linear system:

$$\begin{pmatrix} 2g_{02} & -g_{02} & 2h_{11} + g_{20} \\ g_{20} & h_{11} & 0 \\ 3g_{02} & 0 & 3h_{11} \end{pmatrix}\begin{pmatrix} H_{11} \\ G_{20} \\ G_{02} \end{pmatrix} = \begin{pmatrix} -\dfrac{1}{2}g_{11}h_{02} - g_{12} + g_{11}^2 \\ g_{11}h_{20} - \dfrac{1}{2}h_{02}h_{20} - h_{21} \\ -h_{03} - \dfrac{3}{2}h_{02}^2 \end{pmatrix}. \tag{3.125}$$

The 3×3 matrix has zero determinant, so not all three terms can be removed. However, using the non-degeneracy condition $h_{11} \neq 0$, we can eliminate the resonant cubic terms in the second component of the normal form. Thus, we set

$$H_{11} = 0 \tag{3.126}$$

and obtain from the above linear system:

$$G_{20} = \frac{1}{h_{11}}\left(g_{11}h_{20} - h_{21} - \frac{1}{2}h_{02}h_{20}\right), \quad G_{02} = -\frac{1}{h_{11}}\left(\frac{1}{3}h_{03} + \frac{1}{2}h_{02}^2\right). \tag{3.127}$$

Step 4 (final transformation). Transform now the original map (3.58) using (3.122) with the coefficients (3.123) defined in step 1, and (3.126) and (3.127) defined in step 3. This results in a map with resonant quadratic terms, non-resonant cubic terms, and only two remaining resonant cubic terms in the first component. Transformation (3.124) from step 2 allows then to eliminate all non-resonant cubic terms, while keeping unchanged all remaining quadratic and cubic resonant terms. Finally, make the linear scaling

$$x_1 \mapsto \frac{x_1}{h_{11}}$$

to put the coefficient in front of x_1x_2 in the second component equal to 1. As a result, we obtain the expressions (3.60) and (3.61) for the critical normal form coefficients. □

Proof of Proposition 3.32 Expand F in ξ at $\xi = 0$ for any small $\|\alpha\|$:

$$\xi \mapsto F(\xi, \alpha) = \gamma(\alpha) + A(\alpha)\xi + R(\xi, \alpha),$$

where $\gamma(0) = 0$ and $R(\xi, \alpha) = O(\|\xi\|^2)$. The first assumption implies the existence of two eigenvectors, $q(\alpha)$ and $p(\alpha)$ in $\mathbb{R}^2$, such that

$$A(\alpha)q_1(\alpha) = \lambda_1(\alpha)q_1(\alpha), \quad A(\alpha)q_2(\alpha) = \lambda_2(\alpha)q_2(\alpha),$$

where $\lambda_1(0) = 1$ and $\lambda_2(0) = -1$. Note that, due to the simplicity of the eigenvalues ± 1 of $A(0)$, $\lambda_{1,2}$ depend smoothly on α, and q_1, q_2 can also be assumed to be smooth functions of α. Any $\xi \in \mathbb{R}^2$ can now be represented for all small $\|\alpha\|$ as

$$\xi = \eta_1 q_1(\alpha) + \eta_2 q_2(\alpha),$$

where $\eta = (\eta_1, \eta_2)^T \in \mathbb{R}^2$. One can compute the components of η explicitly:

$$\eta_1 = \langle p_1(\alpha), \xi \rangle, \quad \eta_2 = \langle p_2(\alpha), \xi \rangle,$$

where the adjoint eigenvectors satisfy

$$A^T(\alpha)p_1(\alpha) = \lambda_1(\alpha)p_1(\alpha), \quad A^T(\alpha)p_2(\alpha) = \lambda_2(\alpha)p_2(\alpha)$$

and $\langle p_1(\alpha), q_1(\alpha)\rangle = \langle p_2(\alpha), q_2(\alpha)\rangle = 1$. Since $\langle p_1(\alpha), q_2(\alpha)\rangle = \langle p_1(\alpha), q_1(\alpha)\rangle = 0$, the map F in the η-coordinates takes the form

$$\begin{pmatrix} \eta_1 \\ \eta_2 \end{pmatrix} \mapsto \begin{pmatrix} \sigma_1(\alpha) + \lambda_1(\alpha)\eta_1 + S_1(\eta, \alpha) \\ \sigma_2(\alpha) + \lambda_2(\alpha)\eta_2 + S_2(\eta, \alpha) \end{pmatrix}, \tag{3.128}$$

where

$$\begin{pmatrix} \sigma_1(\alpha) \\ \sigma_2(\alpha) \end{pmatrix} = \begin{pmatrix} \langle p_1(\alpha), \gamma(\alpha)\rangle \\ \langle p_2(\alpha), \gamma(\alpha)\rangle \end{pmatrix},$$

$$\begin{pmatrix} S_1(\eta, \alpha) \\ S_2(\eta, \alpha) \end{pmatrix} = \begin{pmatrix} \langle p_1(\alpha), R(\eta_1 q_1(\alpha) + \eta_2 q_2(\alpha), \alpha)\rangle \\ \langle p_2(\alpha), R(\eta_1 q_1(\alpha) + \eta_2 q_2(\alpha), \alpha)\rangle \end{pmatrix}.$$

Expanding $S_{1,2}(\eta, \alpha)$ further, we can write (3.128) as

$$\begin{pmatrix} \eta_1 \\ \eta_2 \end{pmatrix} \mapsto \begin{pmatrix} \sigma_1(\alpha) + \lambda_1(\alpha)\eta_1 + \displaystyle\sum_{i+j=2,3} \frac{1}{i!j!} g_{ij}(\alpha)\eta_1^i \eta_2^j \\ \sigma_2(\alpha) + \lambda_2(\alpha)\eta_2 + \displaystyle\sum_{i+j=2,3} \frac{1}{i!j!} h_{ij}(\alpha)\eta_1^i \eta_2^j \end{pmatrix} + O(\|\eta\|^4). \tag{3.129}$$

Now we want to put (3.129) to the form (3.63) up to and including cubic terms by an appropriate smooth coordinate transformation, also smoothly depending on parameters. Consider the following change of variables:

$$
\begin{cases}
\eta_1 = x_1 + \varepsilon_0(\alpha) + \varepsilon_1(\alpha)x_2 + \frac{1}{2}G_{20}(\alpha)x_1^2 + G_{11}(\alpha)x_1x_2 + \frac{1}{2}G_{02}(\alpha)x_2^2 \\
\qquad + \frac{1}{2}G_{21}(\alpha)x_1^2x_2 + \frac{1}{6}G_{03}(\alpha)x_2^3, \\
\eta_2 = x_2 + \delta_0(\alpha) + \delta_1(\alpha)x_1 + \frac{1}{2}H_{20}(\alpha)x_1^2 + \frac{1}{2}H_{02}(\alpha)x_2^2 \\
\qquad + \frac{1}{6}H_{30}(\alpha)x_2^3 + \frac{1}{2}H_{12}(\alpha)x_1x_2^2,
\end{cases}
\tag{3.130}
$$

where all coefficients are yet unknown smooth functions of α such that $\varepsilon_i(0) = \delta_i(0) = 0$ for $i = 0, 1$. Obviously, for $\alpha = 0$ (3.130) reduces to the transformation introduced in step 4 of the proof of Lemma 3.31 just before the final scaling.

Require now that the Taylor expansion of (3.129) in the x-coordinates up to and including cubic terms takes the form

$$
\begin{pmatrix} x_1 \\ x_2 \end{pmatrix} \mapsto \begin{pmatrix} \mu_1(\alpha) + (1 + \mu_2(\alpha))x_1 + A(\alpha)x_1^2 + B(\alpha)x_2^2 + C(\alpha)x_1^3 + D(\alpha)x_1x_2^2 \\ -x_2 + E(\alpha)x_1x_2 \end{pmatrix},
$$

where, $\mu_1(0) = \mu_2(0) = 0$. After all substitutions, this requirement translates into a system of algebraic equations:

$$
Q(\varepsilon_0, \varepsilon_1, \delta_0, \delta_1, \mu_1, \mu_2, G_{20}, G_{11}, G_{02}, G_{21}, G_{03}, H_{20}, H_{02}, H_{30}, H_{12}, A, B, C, D, E) = 0,
$$

where $Q\colon \mathbb{R}^{20} \to \mathbb{R}^{20}$ results from equating the corresponding Taylor coefficients. For the Jacobian matrix $J = DQ$ of this system evaluated at

$$
\varepsilon_0 = \varepsilon_1 = \delta_0 = \delta_1 = \mu_1 = \mu_2 = 0
$$

we have $\det(J) = -14\,7456h_{11}^3 \neq 0$. Therefore, the Implicit Function Theorem guarantees the local existence and smoothness of the coefficients of the transformation (3.130) as functions of α.

Moreover, one can show that

$$\begin{cases} \mu_1 = A_1\alpha_1 + A_2\alpha_2 + O(\|\alpha\|^2), \\ \mu_2 = \Big\{2\Big(g_{11}(0)h_{20}(0) - h_{21}(0) - \frac{1}{2}h_{02}(0)h_{20}(0)\Big)(A_1\alpha_1 + A_2\alpha_2) \\ \quad + [(g_{11}(0)B_1 + 2A_3)h_{11}(0) - (h_{02}(0)B_1 + 2B_3)g_{20}]\,\alpha_1 \\ \quad + [(g_{11}(0)B_2 + 2A_4)h_{11}(0) - (h_{02}(0)B_2 + 2B_4)g_{20}]\,\alpha_2\Big\}\Big/2h_{11}(0) \\ \quad + O(\|\alpha\|^2), \end{cases} \tag{3.131}$$

where A_i and B_i are defined by the following expansions of the coefficients of (3.129):

$$\begin{aligned} \sigma_1(\alpha) &= A_1\alpha_1 + A_2\alpha_2 + O(\|\alpha\|^2), & \lambda_1(\alpha) &= 1 + A_3\alpha_1 + A_4\alpha_2 + O(\|\alpha\|^2), \\ \sigma_2(\alpha) &= B_1\alpha_1 + B_2\alpha_2 + O(\|\alpha\|^2), & \lambda_2(\alpha) &= -1 + B_3\alpha_1 + B_4\alpha_2 + O(\|\alpha\|^2). \end{aligned} \tag{3.132}$$

The functions μ_1 and μ_2 can be taken as the unfolding parameters if the determinant of the Jacobian matrix

$$\Delta = \det\left(\frac{\partial\mu}{\partial(\alpha_1,\alpha_2)}\right)\Bigg|_{\alpha=0} \neq 0.$$

From (3.131) we have

$$\Delta = \frac{1}{2h_{11}}\Big[(A_1B_2 - A_2B_1)(g_{11}h_{11} - g_{20}h_{02}) + 2(A_1A_4 - A_2A_3)h_{11} + 2(A_2B_3 - A_1B_4)g_{20}\Big]_{\alpha=0}.$$

On the other hand, taking into account (3.132), we obtain by direct computations:

$$\det\left(\frac{\partial T}{\partial(\eta,\alpha)}\right)\Bigg|_{\eta=\alpha=0} = 4h_{11}\Delta,$$

where T is the map (3.62) written in the (η,α) coordinates, i.e., for the map (3.129). Thus, if $h_{11}(0) \neq 0$, the regularity of (3.62) at the origin is equivalent to $\Delta \neq 0$.

The scaling

$$x_1 \mapsto \frac{x_1}{E(\mu)}, \quad \mu_1 \mapsto \frac{\mu_1}{E(\mu)},$$

where $E(\mu) = h_{11}(0) + O(\|\alpha\|)$, gives finally (3.63). Obviously, the critical coefficients are the same as in Lemma 3.31. □

Proof of Proposition 3.35 The Jacobian matrix of (3.64) is

$$A(x,\mu) = N_x(x,\mu)$$
$$= \begin{pmatrix} 1 + \mu_2 + a(\mu)x_1 + 3c(\mu)x_1^2 + d(\mu)x_2^2 & b(\mu)x_2 + d(\mu)x_1x_2 \\ x_2 & -1 + x_1 \end{pmatrix}.$$

Fold bifurcation. The map (3.64) has a fixed point x with multiplier 1 if

$$\begin{cases} N(x,\mu) = x, \\ \det(A(x,\mu) - I_2) = 0. \end{cases}$$

Using the Implicit Function Theorem, we see that this algebraic system has a unique solution curve $LP^{(1)}$ near the origin that is given in statement 1. The critical (adjoint) eigenvectors are $q_1 = p_1 = (1,0)^T$, while

$$B(u,v) = \begin{pmatrix} a_0u_1v_1 + b_0u_2v_2 + O(\mu_2) \\ u_1v_2 + u_2v_1 \end{pmatrix}.$$

With (2.7) from Section 2.1, we obtain

$$a_{fold} = a_0 + O(\mu_2),$$

and if $a_0 \neq 0$ we have a non-degenerate (quadratic) fold when $\mu_2 \to 0$.

Flip bifurcation. Another look at the Jacobian matrix above shows that along the curve $PD^{(1)}$ defined in statement 2, the truncated normal form (3.64) has a fixed point with multiplier -1, i.e., this curve satisfies the algebraic system

$$\begin{cases} N(x,\mu) = x, \\ \det(A(x,\mu) + I_2) = 0. \end{cases}$$

Now we have $q_2 = p_2 = (0,1)^T$ as (adjoint) eigenvector. Clearly $C(q_2,q_2,q_2) = 0$ and we compute the flip coefficient (2.12) as

$$b_{flip} = \frac{1}{6}\langle p_2, 3B(q_2,(I_2 - A)^{-1}B(q_2,q_2))\rangle = -\frac{b_0}{\mu_2} + o(1),$$

where $o(1)$ is bounded as $\mu_2 \to 0$. Then, the flip bifurcation is non-degenerate if $b_0 \neq 0$.

Neimark–Sacker bifurcation. Considering the second iterate of (3.64) we solve for its fixed point with the determinant of the Jacobian matrix equal to 1, i.e.,

$$\begin{cases} N(N(x,\mu),\mu) = x, \\ \det A(N(x,\mu),\mu)\det A(x,\mu) - 1 = 0. \end{cases}$$

We find the following exact solution to this system:

$$x_1 = 0, \;\; b(\mu)x_2^2 + \mu_1 = 0, \;\; \mu_2 = \frac{d(\mu) + 2b(\mu)}{b(\mu)}\mu_1, \tag{3.133}$$

which implies the expansion for $NS^{(2)}$ in the statement 3. Evaluating the Jacobian matrix of the second iterate of (3.64) on (3.133), we find

$$A = (N(N(x,\mu),\mu))_x = \begin{pmatrix} 1 + 6\mu_1 + 4\mu_1^2 & 4b(\mu)\sqrt{-\frac{\mu_1}{b(\mu)}}(1+\mu_1) \\ -2\sqrt{-\frac{\mu_1}{b(\mu)}}(1+\mu_1) & 1 + 2\mu_1 \end{pmatrix}.$$

For small $\mu_1 < 0$, it has a complex eigenvalue

$$e^{i\theta_0} = 1 + 4\mu_1 + 2\mu_1^2 + \sqrt{\mu_1(2+\mu_1)(1+\mu_1)^2}.$$

Therefore, we find that in the case of $b_0 > 0, \mu_1 < 0$ there is a NS bifurcation.

We want to know the sign of the first Lyapunov coefficient c_{NS} along (3.133). We take

$$q = \left(b(\mu)\sqrt{-\frac{\mu_1}{b(\mu)}}\left(1 + \frac{\sqrt{\mu_1(2+\mu_1)}}{\mu_1}\right), 1 \right)^T$$

and

$$p = \left(-\frac{1}{2}\sqrt{\frac{\mu_1}{b(\mu)}}\left(-1 + \frac{\sqrt{\mu_1(2+\mu_1)}}{\mu_1}\right), 1 \right)^T,$$

such that $Aq = e^{i\theta_0}q$ and $A^T p = e^{-i\theta_0}p$. We should still scale p, since $\langle p, q\rangle \neq 1$. Next we compute the first Lyapunov coefficient c_{NS} on $NS^{(2)}$, using (2.19) from Section 2.1, where the multilinear forms B and C correspond to the *second iterate* of (3.64). This gives:

$$c_{NS} = 2(3b_0c_0 - 3a_0b_0 - 2b_0a_0^2 - d_0a_0) + o(\mu_1)$$

as $\mu_1 \to 0$. Therefore, the NS bifurcation of the period-2 cycle of (3.64) is non-degenerate near the origin, if (3.66) holds. □

Proof of Lemma 3.36 We construct φ^t as the first two-dimensional component of the flow

$$\xi \mapsto \phi^t(\xi) = \begin{pmatrix} \varphi^t(x,\mu) \\ \mu \end{pmatrix}, \;\; \xi = \begin{pmatrix} x \\ \mu \end{pmatrix} \in \mathbb{R}^4,$$

generated by a four-dimensional system with the parameters considered as constant variables:

$$\dot{\xi} = Y(\xi). \tag{3.134}$$

Here,

$$Y(\xi) = J\xi + Y_2(\xi) + Y_3(\xi) + \cdots, \quad J = \begin{pmatrix} 0 & 0 & 1 & 0 \\ 0 & 0 & 0 & 0 \\ 0 & 0 & 0 & 0 \\ 0 & 0 & 0 & 0 \end{pmatrix}, \quad Y_k(\xi) = \begin{pmatrix} X_k(\xi) \\ 0 \end{pmatrix},$$

where each X_k is an order-k homogeneous function from $\mathbb{R}^4$ to $\mathbb{R}^2$ with unknown coefficients. Define

$$M(\xi) = \begin{pmatrix} N(x,\mu) \\ \mu \end{pmatrix}$$

and introduce the 4×4 block-diagonal matrix

$$S = \begin{pmatrix} R & 0 \\ 0 & I_2 \end{pmatrix},$$

where R is given in (3.65). We look for a vector field Y such that $S\ M(\xi) = \phi^1(\xi) + O(\|\xi\|^4)$ (see (Takens, 1974)).

To find the vector field Y explicitly, perform three Picard iterations for (3.134) as described in Section 1.4. We start with setting $\phi_1^t(\xi) = \mathrm{e}^{Jt}\xi$. Then, clearly, the linear part of $S\ M(\xi)$ coincides with $\phi_1^1(\xi)$.

Since we know how the result ϕ_2^t of the second Picard iteration should look, we set some coefficients of Y_2 equal to zero immediately:

$$Y_2 = \begin{pmatrix} A_{10}\mu_1 x_1 + A_{01}\mu_2 x_1 + A_{20}x_1^2 + A_{02}x_2^2 \\ B_{11}x_1x_2 + B_{10}\mu_1 x_2 \\ 0 \\ 0 \end{pmatrix}.$$

Then

$$\begin{aligned} \phi_2^t(\xi) = \;& \mathrm{e}^{Jt}\xi + \int_0^t \mathrm{e}^{J(t-\tau)} Y_2(\phi_1^\tau(\xi))d\tau \\ = \;& \begin{pmatrix} x_1 + t\mu_1 + (A_{10}t + A_{20}t^2)\mu_1 x_1 + A_{01}\mu_2 x_1 t + A_{20}x_1^2 t + A_{20}x_2^2 t \\ x_2 + B_{11}x_1x_2 t + \left(\dfrac{1}{2}B_{11}t^2 + B_{10}t\right)\mu_1 x_2 \\ \mu_1 \\ \mu_2 \end{pmatrix} \\ & + O(\|\mu\|^2). \end{aligned}$$

Comparing quadratic terms in $S\ M(\xi)$ and $\phi_2^1(\xi)$, we find the coefficients of Y_2:

$$A_{10} = -a_0,\ A_{20} = a_0,\ A_{01} = 1,\ A_{02} = b_0,\ B_{10} = \frac{1}{2},\ B_{11} = -1.$$

Passing on to the cubic part we remark that we are only interested in certain cubic terms in x. Therefore, we put

$$Y_3 = \begin{pmatrix} A_{30}x_1^3 + A_{12}x_1x_2^2 \\ A_{21}x_1^2x_2 + B_{03}x_2^3 \\ 0 \\ 0 \end{pmatrix}$$

and get

$$\begin{aligned}
\phi_3^t(\xi) &= \mathrm{e}^{Jt}\xi + \int_0^t \mathrm{e}^{J(t-\tau)}\left[Y_2(\phi_2^\tau(\xi)) + Y_3(\phi_2^\tau(\xi))\right]d\tau \\
&= \begin{pmatrix} x_1 + t\mu_1 + \frac{1}{2}a_0(t^2-1)x_1\mu_1 + x_1\mu_2 + \frac{1}{2}a_0x_1^2 + \frac{1}{2}b_0x_2^2 \\ x_2 - x_1x_2 + \frac{1}{2}(1-t^2)\mu_1x_2 \\ \mu_1 \\ \mu_2 \end{pmatrix} \\
&\quad + \begin{pmatrix} \left(A_{30}t + A_{20}^2t^2\right)x_1^3 + \left(A_{12}t + t^2A_{02}(A_{20}+2B_{11})\right)x_1x_2^2 \\ \left(tB_{21} + \frac{1}{2}t^2B_{11}(B_{11}+A_{20})\right)x_1^2x_2 + \left(tB_{03} + \frac{1}{2}t^2A_{02}B_{11}\right)x_2^3 \\ 0 \\ 0 \end{pmatrix} \\
&\quad + O(\|\mu\|^2) + O(\|x\|^2\|\mu\|).
\end{aligned}$$

Comparing cubic terms in $S\ M(\xi)$ and $\phi_3^1(\xi)$, we find the coefficients of Y_3:

$$A_{30} = c_0 - a_0^2,\ A_{12} = d_0 + b_0(1-a_0),\ B_{21} = \frac{1}{2}(a_0-1),\ B_{03} = \frac{1}{2}b_0.$$

This gives the approximating ODE (3.68) from the proposition. □

Proof of Proposition 3.37 We first shift the x_1-coordinate in (3.68) with

$$x_1 \to x_1 + \left(\frac{\mu_1}{2} - \frac{\mu_2}{2a_0}\right).$$

Then we apply a singular rescaling

$$(x_1, x_2) \to \delta(x_1, x_2),\ \ (\mu_1, \mu_2) \to \delta^2(\mu_1, \mu_2),\ \ dt \to \frac{x_2^q}{\delta}dt \tag{3.135}$$

to obtain

$$\begin{cases} \dot{x}_1 = x_2^q\left(\beta_1 + \frac{1}{2}a_0x_1^2 + \frac{1}{2}b_0x_2^2 + \delta[d_1x_1^3 + d_2x_1x_2^2]\right), \\ \dot{x}_2 = x_2^q\left(-x_1x_2 + \delta[\beta_2x_2 + d_3x_1^2x_2 + d_4x_2^3]\right), \end{cases} \tag{3.136}$$

where

$$\beta_1 = \mu_1 + O(\|\mu\|^2), \ \ \beta_2 = \frac{\mu_2}{2a_0} + O(\|\mu\|^2).$$

The system (3.136) can be rewritten as

$$\dot{x} = f(x,\beta) + \delta g(x,\beta) \tag{3.137}$$

with

$$f(x,\beta) = x_2^q \begin{pmatrix} \beta_1 + a_0 x_1^2 + b_0 x_2^2 \\ -x_1 x_2 \end{pmatrix}, \ \ g(x,\beta) = x_2^q \begin{pmatrix} d_1 x_1^3 + d_2 x_1 x_2^2 \\ \beta_2 x_2 + d_3 x_1^2 x_2 + d_4 x_2^3 \end{pmatrix}.$$

For $\delta = 0$ and $q + 1 = 2a_0$, system (3.137) is Hamiltonian with

$$H(x) = x_2^{q+1} \left(\frac{\beta_1 + \frac{1}{2} a_0 x_1^2}{q+1} + \frac{b_0 x_2^2}{2(q+3)} \right). \tag{3.138}$$

We have $a_0, b_0 \neq 0$ as non-degeneracy conditions, therefore $q \neq -1$. Level curves of H for several values of a_0 are shown in Figure 3.51. Now we treat the term δg in (3.137) as a small perturbation of the Hamilton system. We should therefore evaluate the Pontryagin–Melnikov integral (see Section 1.6)

$$\Delta(h,\beta) = \oint_{\Gamma_h} f(x_h(\tau),\beta) \wedge g(x_h(\tau),\beta)\, d\tau,$$

where $x_h(\tau)$ is a periodic solution of the Hamiltonian system corresponding to a closed regular level set $\Gamma_h = \{x \colon H(x) = h\}$, while for $a_0 > 0$

$$\Delta(0,\beta) = \int_{-\infty}^{+\infty} f(x_0(\tau),\beta) \wedge g(x_0(\tau),\beta)\, d\tau,$$

where $x_0(\tau)$ is the nontrivial heteroclinic solution in the critical level set $H = 0$. Notice that $\lim_{h\to 0^-} \Delta(h,\beta) = \Delta(0,\beta)$, since the integral over the trivial heteroclinic connection equals zero. Then $\Delta(0,\beta) = 0$ defines a linear approximation to a curve on which the heteroclinic connection "survives" in (3.137) for small $\delta \neq 0$. Our computation is analogous to the one in Chow, Li, and Wang (1994).

Using Green's formula, we have

$$\begin{aligned} \Delta &= \oint_{\Gamma_h} x_2^q (d_1 x_1^3 + d_2 x_1 x_2^2) dx_2 - x_2^q (\beta_2 x_2 + d_3 x_1^2 x_2 + d_4 x_2^3) dx_1 \\ &= \oint_{\Gamma_h} x_2^q \left(d_1 x_1^3 + d_2 x_1 x_2^2 + (q+1) \left(\beta_2 x_1 + \frac{1}{3} d_3 x_1^3 \right) + (q+3) d_4 x_1 x_2^2 \right) dx_2. \end{aligned}$$

Note that along Γ_h we have $dH = x_2^q \left(\beta_1 + a_0 x_1^2 + b_0 x_2^2 \right) dx_2 + x_2^{q+1} x_1 dx_1 = 0$

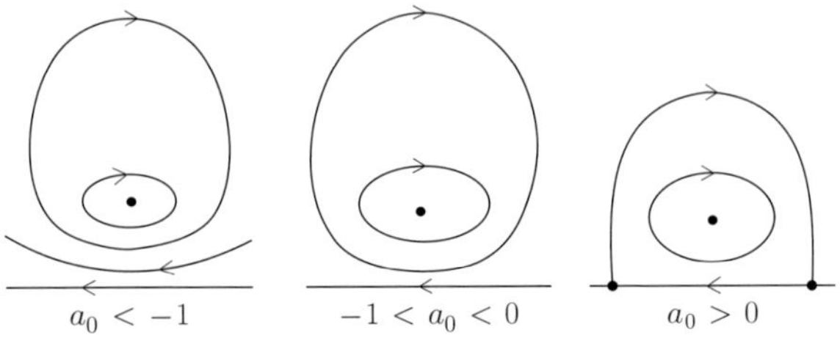

Figure 3.51 Level curves of H for several a_0.

and we continue

$$\begin{aligned}\Delta &= \oint_{\Gamma_h} x_2^q \left(a_0\beta_2 x_1 + \left(d_1 + \frac{1}{3}a_0 d_3\right)x_1^3 + (d_2 + (a_0+2)d_4)x_1x_2^2\right)dx_2 \\ &= \oint_{\Gamma_h} x_2^q \left(x_1\left(a_0\beta_2 - \frac{2}{b_0}\beta_1(d_2 + (a_0+2)d_4)\right)\right. \\ &\quad \left. + x_1^3\left(d_1 + \frac{1}{3}a_0 d_3 - \frac{a_0}{b_0}(d_2 + (a_0+2)d_4)\right)\right)dx_2 \\ &\quad - \oint_{\Gamma_h} \frac{2}{b_0}(d_2 + (a_0+2)d_4)x_1^2 x_2^{q+1}\,dx_1 \\ &= I_{1,h}\left(a_0\beta_2 - \frac{2}{b_0}\beta_1(d_2 + (a_0+2)d_4)\right) \\ &\quad + I_{3,h}\left(d_1 + \frac{a_0}{3}d_3 - \frac{a_0}{3b_0}(d_2 + (a_0+2)d_4)\right).\end{aligned}$$

Here we defined $I_{i,h} = \oint_{\Gamma_h} x_2^q x_1^i\,dx_2$ for $i = 1, 3$.

For $h = 0$ we can evaluate the Pontryagin–Melnikov integral as follows. We have $x_1^2 = -\frac{b_0 x_2^2}{1+a_0} - \beta_1$ and we get

$$I_{i,0} = 2\int_0^{\sqrt{\frac{-\mu_1(1+a_0)}{a_0 b_0}}} x_2^q \left(-\frac{b_0 x_2^2}{1+a_0} - \beta_1\right)^{i/2} dx_2.$$

The evaluation of $I_{i,0}$ yields a result with gamma functions. Computing the ratio $Q = I_{3,0}/I_{1,0}$, we find

$$Q = \frac{-3\beta_1}{a_0(3+2a_0)}.$$

Using the expressions for β_k and d_i, we find that the Melnikov integral $\Delta(0,\beta)$ has a zero if

$$\mu_2 = \frac{\mu_1}{a_0 b_0(2a_0+3)}\left(2a_0^2(b_0+d_0) + 2a_0 d_0 + 3b_0(a_0+c_0)\right) + o(\mu_1). \quad (3.139)$$

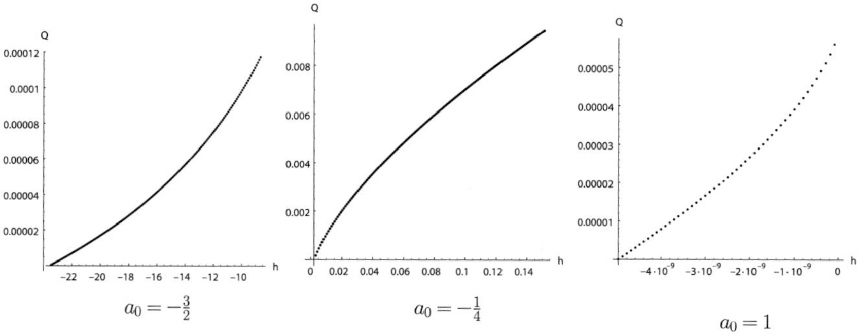

Figure 3.52 Ratio $Q(h) = I_{1,h}/I_{3,h}$.

This value of μ_2 asymptotically corresponds to the existence of a nontrivial heteroclinic orbit for the perturbed system (3.136). □

Remark 3.52 Moreover, for the vector field we have the uniqueness of the limit cycle. This may be verified as follows. We should evaluate the Pontryagin integral on a level curve of the Hamiltonian with $h \neq 0$. Now $Q(h) = I_{1,h}/I_{3,h}$ cannot be evaluated explicitly, but one can prove the monotonicity of Q as in Chow, Li, and Wang (1994). This implies the uniqueness of the limit cycle. We include some pictures, where we computed $Q(h)$ numerically, which illustrate the monotonicity (see Figure 3.52).

3.E Proofs for Section 3.4.2

Proof of Lemma 3.39 Suppose that a fixed point of map (3.2) exhibits an LPNS bifurcation at $\alpha = 0$. Write the restriction of (3.2) to its three-dimensional center manifold near this fixed point as

$$\begin{pmatrix} x \\ z \end{pmatrix} \mapsto \begin{pmatrix} x + \sum\limits_{2\le j+k+l\le 3} f_{jkl} x^j z^k \bar{z}^l \\ \mu z + \sum\limits_{2\le j+k+l\le 3} g_{jkl} x^j z^k \bar{z}^l \end{pmatrix} + O(\|(x,z)\|^4), \tag{3.140}$$

where $(x, z) \in \mathbb{R} \times \mathbb{C}$, $\mu = e^{i\theta_0}$ and $f_{jkl} = \bar{f}_{jlk}$. We assume further that $\mu^q \neq 1$ for $q = 1, 2, 3$ and 4.

To determine the general structure of the Poincaré normal form, we can consider (3.140) as a map on $\mathbb{R}\times\mathbb{C}^2$ with independent coordinates $(x, z, \bar{z})$, i.e.,

$$\begin{pmatrix} x \\ z \\ \bar{z} \end{pmatrix} \mapsto \begin{pmatrix} x + \sum\limits_{2\le j+k+l\le 3} f_{jkl} x^j z^k \bar{z}^l \\ \mu z + \sum\limits_{2\le j+k+l\le 3} g_{jkl} x^j z^k \bar{z}^l \\ \bar{\mu}\bar{z} + \sum\limits_{2\le j+k+l\le 3} \bar{g}_{jlk} x^j z^k \bar{z}^l \end{pmatrix} + O(\|(x, z, \bar{z})\|^4)$$

with the linearization matrix

$$A = \begin{pmatrix} 1 & 0 & 0 \\ 0 & \mu & 0 \\ 0 & 0 & \bar{\mu} \end{pmatrix}$$

that has simple eigenvalues $\lambda_1 = 1, \lambda_{2,3} = e^{\pm i\theta_0}$. It holds that $\lambda_2\lambda_3 = 1$. Applying the normalization technique described in Section 1.3, we see only the following low-order resonances:

$$\lambda_1 = \lambda_1^0(\lambda_2\lambda_3)^1 = \lambda_1^2(\lambda_2\lambda_3)^0 = \lambda_1^3(\lambda_2\lambda_3)^0 = \lambda_1^1(\lambda_2\lambda_3)^1$$

and

$$\lambda_2 = \lambda_1^1\lambda_2(\lambda_2\lambda_3)^0 = \lambda_1^2\lambda_2(\lambda_2\lambda_3)^0 = \lambda_1^0\lambda_2(\lambda_2\lambda_3)^1.$$

This immediately implies the normal form (3.71).

□

Proof of Theorem 3.40 When $\alpha = 0$, we write the map (3.71) as

$$\begin{pmatrix} x \\ z \end{pmatrix} \mapsto \begin{pmatrix} x + f_{200}x^2 + f_{011}z\bar{z} + f_{300}x^3 + f_{111}xz\bar{z} \\ e^{i\theta_0}z + g_{110}xz + g_{210}x^2z + g_{021}z^2\bar{z} \end{pmatrix} + O(\|(x, z)\|^4), \quad (3.141)$$

and make the transformation

$$\begin{cases} u = x + \varphi_{200}x^2, \\ w = z + \psi_{110}xz, \end{cases}$$

where $\varphi_{200} \in \mathbb{R}$ and $\psi_{110} \in \mathbb{C}$ are to be defined. In the new variables $(u, w) \in \mathbb{R}\times\mathbb{C}$, the map (3.141) can be written as

$$\begin{pmatrix} u \\ w \end{pmatrix} \mapsto \begin{pmatrix} u + F_{200}u^2 + F_{011}w\bar{w} + F_{300}u^3 + F_{111}uw\bar{w} \\ \mu w + G_{110}uw + G_{210}u^2w + G_{021}w^2\bar{w} \end{pmatrix} + O(\|(u, w)\|^4), \quad (3.142)$$

where $F_{200} = f_{200}$, $F_{011} = f_{011}$, $G_{110} = g_{110}$, $F_{300} = f_{300}$, but

$$\begin{aligned} F_{111} &= f_{111} + 2 f_{011}\phi_{200} - f_{011}\psi_{110} - f_{011}\bar{\psi}_{110}, \\ G_{021} &= g_{021} + \mu f_{011}\psi_{110}, \\ G_{210} &= g_{210} + \mu f_{200}\psi_{110} - g_{110}\phi_{200}. \end{aligned}$$

Since $f_{011} \neq 0$, we can set

$$\psi_{110} = -\frac{g_{021}}{\mu f_{011}} = -\frac{\bar{\mu} g_{021}}{f_{011}}$$

to ensure $G_{021} = 0$. With this setting,

$$F_{111} = f_{111} + 2 f_{011}\phi_{200} + 2 f_{011} \Re\left(\frac{\bar{\mu} g_{021}}{f_{011}}\right),$$

which vanishes if we properly select ϕ_{200}, namely

$$\phi_{200} = -\frac{f_{111}}{2 f_{011}} - \Re\left(\frac{\bar{\mu} g_{021}}{f_{011}}\right).$$

We finally get

$$G_{210} = g_{210} + \frac{f_{111} g_{110}}{2 f_{011}} + g_{110} \Re\left(\frac{\bar{\mu} g_{021}}{f_{011}}\right) - \frac{f_{200} g_{021}}{f_{011}}. \tag{3.143}$$

The map (3.141) becomes

$$\begin{pmatrix} u \\ w \end{pmatrix} \mapsto \begin{pmatrix} u + f_{200}u^2 + f_{011}|w|^2 + f_{300}u^3 \\ \mu w + g_{110}uw + G_{210}u^2 w \end{pmatrix} + O(\|(u, w)\|^4), \tag{3.144}$$

where G_{210} is given by (3.143). The linear scaling

$$u = \frac{x}{f_{200}}, \quad w = \frac{f_{200}\, z}{\sqrt{|f_{011}|}},$$

results in (3.73) at the critical parameter values.

To prove the theorem for small $\|\alpha\|$, one should start with a map that reduces to (3.140) when $\alpha = 0$, e.g.,

$$\begin{pmatrix} x \\ z \end{pmatrix} \mapsto \begin{pmatrix} \gamma(\alpha) + (1 + \nu(\alpha))x + \sum\limits_{2 \le j+k+l \le 3} f_{jkl}(\alpha) x^j z^k \bar{z}^l \\ \omega(\alpha) + \mu(\alpha) z + \sum\limits_{2 \le j+k+l \le 3} g_{jkl}(\alpha) x^j z^k \bar{z}^l \end{pmatrix} + O(\|(x, z)\|^4), \tag{3.145}$$

where $\gamma(0) = \nu(0) = 0, \omega(0) = 0, \mu(0) = e^{i\theta_0}$. Then, using the Implicit Function Theorem, it is possible to prove the existence of the parameter-dependent transformation

$$\begin{cases} x = u + \delta_0(\alpha) + \delta_1(\alpha)u + \delta_2(\alpha)w + \delta_3(\alpha)\bar{w} + \sum\limits_{2 \le j+k+l \le 3} \varphi_{ijk}(\alpha) u^j w^k \bar{w}^l, \\ z = w + \Delta_0(\alpha) + \delta_1(\alpha)u + \Delta_2(\alpha)w + \Delta_3(\alpha)\bar{w} + \sum\limits_{2 \le j+k+l \le 3} \psi_{ijk}(\alpha) u^j w^k \bar{w}^l, \end{cases}$$

where $\delta_j(0) = \Delta_j(0) = 0$ for $j = 0, 1, 2, 3$, that reduces (3.145) after a linear scaling to the parameter-dependent normal form (3.73), provided the specified non-degeneracy conditions hold. □

Proof of Lemma 3.41 If $\beta_1 = 0$, then the Jacobian matrix at the fixed point $(x,r) = (0,0)$ of

$$N_\beta\colon \begin{pmatrix} x \\ r \end{pmatrix} \mapsto \begin{pmatrix} x + \beta_1 + x^2 + sr^2 + c_0 x^3 \\ r + \beta_2 r + d_0 xr + d_1 x^2 r \end{pmatrix} \tag{3.146}$$

is

$$A = \begin{pmatrix} 1 & 0 \\ 0 & 1+\beta_2 \end{pmatrix}$$

with simple eigenvalue 1 provided that $\beta_2 \neq 0$. The one-dimensional critical center manifold W_0^c in (3.75) is tangent to the x-axis and the restriction of this map to W_0^c can be written as

$$x \mapsto x + x^2 + O(x^3).$$

This immediately implies that the fold bifurcation along line $LP^{(1)}$ is non-degenerate for $\beta_2 \neq 0$.

The system of algebraic equations

$$\begin{cases} N_\beta(x,r) = (x,r), \\ \det(A(x,r,\beta) - I_2) = 0, \end{cases}$$

where $A(x,r,\beta)$ is the Jacobian matrix of $N_\beta(x,r)$ with respect to (x,r), has a solution

$$x = -\frac{\beta_2}{d_0} + O(\beta_2^2), \quad r = 0, \quad \beta_1 = -\frac{\beta_2^2}{d_0^2} + O(\beta_2^3),$$

which is different from the fold solution. Since N_β is invariant under the transformation $r \mapsto -r$, this solution corresponds to a pitchfork bifurcation on $PF^{(1)}$.

To study the NS bifurcation when $s = 1$ and $d_0 < 0$, write $\beta_1 = -\epsilon^2$ with $\epsilon > 0$ and consider the system of equations

$$\begin{cases} N_\beta(x,r) = (x,r), \\ \det A(x,r,\beta) = 1. \end{cases}$$

This system is satisfied when there is a fixed point (x,r) with a pair of eigenvalues having product 1. The system admits a solution (cf. $NS^{(1)}$)

$$x = d_0\epsilon^2 + O(\epsilon^3), \quad r = \epsilon + O(\epsilon), \quad \beta_2 = -d_0^2\epsilon^2 + O(\epsilon^3), \tag{3.147}$$

at which the Jacobian matrix $A(x, r, \beta)$ can be written as

$$\begin{pmatrix} 1 + 2d_0\epsilon^2 + O(\epsilon^4) & 2\epsilon + O(\epsilon^2) \\ d_0\epsilon + O(\epsilon^3) & 1 + O(\epsilon^4) \end{pmatrix}.$$

The matrix without the O-terms, i.e.,

$$A = \begin{pmatrix} 1 + 2d_0\epsilon^2 & 2\epsilon \\ d_0\epsilon & 1 \end{pmatrix},$$

has $\det A \equiv 1$ and is sufficient for approximation of the critical eigenvalues and eigenvectors at the NS bifurcation. The eigenvalues are $\lambda_1 = \mu, \lambda_2 = \bar{\mu}$, where

$$\mu = 1 + d_0\epsilon^2 + i\epsilon\sqrt{-d_0(2 + \epsilon^2 d_0)}.$$

For $d_0 < 0$ and small $\epsilon > 0$, $\mu = \mathrm{e}^{i\theta_0}$ with $\theta_0 > 0$. The vectors

$$q = \left(\epsilon - i\sqrt{-\frac{2}{d_0} - \epsilon^2},\ 1\right),\ p = \frac{1}{2}\left(\epsilon d_0 - i\sqrt{-2d_0 - d_0^2\epsilon^2},\ 2\right)$$

satisfy $Aq = \mu q$ and $A^T p = \bar{\mu} p$ but should still be normalized such that $\langle p, q\rangle = \bar{p}^T q = 1$. After performing this normalization, we can define

$$g(z, \bar{z}) = \langle p, N_\beta(x + zq_1 + \bar{z}\bar{q}_1, r + zq_2 + \bar{z}\bar{q}_2) - (x, r)\rangle,$$

where $\beta_1 = -\epsilon^2$ and (x, r, β_2) are as in (3.147). Computing the Taylor expansion of g with respect to $(z, \bar{z})$, one can verify that the NS normal form coefficient (2.22) has the expansion

$$\frac{2(d_0 + d_1) - 3d_0(c_0 + d_0)}{d_0^2} + O(\epsilon) = -\frac{C_{NS}}{d_0^3} + O(\epsilon),$$

where C_{NS} is defined by (3.77). Thus, under the assumptions made in Lemma 3.41 and if we take into account that $d_0 < 0$, the NS bifurcation is supercritical along $NS^{(1)}$ near the LPNS point when $C_{NS} < 0$.

The cases with other sign combinations of d_0 and s are similar. □

Proof of Theorem 3.42 We construct φ_β^t as the first two-dimensional component of the flow

$$\xi \mapsto \phi^t(\xi) = \begin{pmatrix} \varphi_\beta^t(x, r) \\ \beta_1 \\ \beta_2 \end{pmatrix},\ \xi = \begin{pmatrix} x \\ r \\ \beta_1 \\ \beta_2 \end{pmatrix} \in \mathbb{R}^4,$$

generated by a four-dimensional system with the parameters considered as constant variables:

$$\dot{\xi} = Y(\xi). \tag{3.148}$$

Here,

$$Y(\xi) = J\xi + Y_2(\xi) + Y_3(\xi) + \cdots, \quad J = \begin{pmatrix} 0 & 0 & 1 & 0 \\ 0 & 0 & 0 & 0 \\ 0 & 0 & 0 & 0 \\ 0 & 0 & 0 & 0 \end{pmatrix}, \quad Y_k(\xi) = \begin{pmatrix} X_k(\xi) \\ 0 \end{pmatrix},$$

where each X_k is an order-k homogeneous function from $\mathbb{R}^4$ to $\mathbb{R}^2$ with unknown coefficients. Define

$$M(\xi) = \begin{pmatrix} N_\beta(x, r) \\ \beta \end{pmatrix}.$$

We look for a vector field Y such that $M(\xi) = \phi^1(\xi) + R(\xi)$, where $R(\xi) = O(\|\beta\|^2) + O(\|\beta\| \, \|(x, r)\|^) + O(\|(x, r)\|^4)$.

To find the vector field Y explicitly, perform three Picard iterations (1.20) for (3.148) as described in Section 1.4. We start with setting $\phi_1^t(\xi) = e^{Jt}\xi$. Then, clearly, the linear part of $M(\xi)$ coincides with $\phi_1^1(\xi)$.

Since we know how the result ϕ_2^t of the second Picard iteration should look, we set some coefficients of Y_2 equal to zero immediately:

$$Y_2 = \begin{pmatrix} A_{1010}x\beta_1 + A_{2000}x^2 + A_{0200}r^2 + A_{0020}\beta_1^2 \\ B_{0101}r\beta_2 + B_{0110}r\beta_1 + B_{1100}xr \\ 0 \\ 0 \end{pmatrix}.$$

Then

$$\phi_2^t(\xi) = e^{Jt}\xi + \int_0^t e^{J(t-\tau)} Y_2(\phi_1^\tau(\xi))d\tau$$

$$= \begin{pmatrix} x + t\beta_1 + x^2 A_{2000}t + t\,(A_{2000}t + A_{1010})\beta_1 x + A_{0200}r^2 t \\ r + B_{1100}rtx + \frac{1}{2}t\,(B_{1100}t + 2\,B_{0110})\,r\beta_1 + \beta_2 B_{0101}tr \\ \\ \beta_1 \\ \beta_2 \end{pmatrix}$$

$$+ \begin{pmatrix} +\frac{1}{6}t\left(2\,A_{2000}t^2 + 3\,A_{1010}t + 6\,A_{0020}\right)\beta_1^2 \\ 0 \\ 0 \\ 0 \end{pmatrix}.$$

Comparing quadratic terms in $M(\xi)$ and $\phi_2^1(\xi)$, we find the coefficients of Y_2

$$A_{0020} = \frac{1}{6}, A_{0200} = s, A_{1010} = -1, A_{2000} = 1, B_{0101} = 1, B_{0110} = -\frac{d_0}{2}, B_{1100} = d_0.$$

With this setting, we have

$$\phi_2^t(\xi) = \begin{pmatrix} x + t\beta_1 + x^2 t + t(t-1)\beta_1 x + sr^2 t \\ r + d_0 rtx + \frac{1}{2}d_0 t(t-1)r\beta_1 + \beta_2 tr \\ \\ \beta_1 \\ \beta_2 \end{pmatrix}.$$

To find the critical cubic coefficients in Y_3 we may perform the third Picard iteration with $\beta_1 = \beta_2 = 0$. Thus we look for Y_3 in the form

$$Y_3 = \begin{pmatrix} A_{3000}x^3 + A_{1200}xr^2 \\ B_{2100}x^2 r + B_{0300}r^3 \\ 0 \\ 0 \end{pmatrix}.$$

Then

$$\begin{aligned}\phi_3^1(\xi) &= \mathrm{e}^J \xi + \int_0^1 \mathrm{e}^{J(1-\tau)} \left[Y_2(\phi_2^\tau(\xi)) + Y_3(\phi_2^\tau(\xi)) \right] d\tau \\ &= \begin{pmatrix} x + \beta_1 + x^2 + sr^2 + (A_{3000}+1)x^3 + (A_{1200} + s + sd_0)xr^2 \\ r + d_0 rx + \beta_2 r + \left(B_{2100} + \frac{1}{2}d_0(d_0+1)\right)x^2 r + \left(B_{0300} + \frac{1}{2}sd_0\right)r^3 \\ 0 \\ 0 \end{pmatrix}.\end{aligned}$$

Comparing cubic terms in $M(\xi)$ and $\phi_3^1(\xi)$, we find the coefficients of Y_3:

$$A_{3000} = c_0 - 1,\ A_{1200} = -s(d_0+1),\ B_{2100} = d_1 - \frac{1}{2}d_0(d_0+1),\ B_{0300} = -\frac{1}{2}sd_0.$$

This gives the approximating ODE from Theorem 3.42. □

Proof of Lemma 3.43 The proof is analogous to the proof of Proposition 3.37. The Hamiltonian part is similar but the perturbation is different. We start with vector field (3.79) and apply the singular rescaling (3.135) to obtain

$$\begin{cases} \dot{x}_1 = x_2^q\left(\beta_1 + x_1^2 + sx_2^2 + \delta[-\beta_1 x + (c_0-1)x^3 - s(d_0+1)x_1 x_2^2]\right), \\ \dot{x}_2 = x_2^q\left(d_0 x_1 x_2 + \delta[\beta_2 x_2 - \frac{1}{2}d_0 x_2 \beta_1 + \left(d_1 - \frac{1}{2}d_0(d_0+1)\right)x_1^2 x_2 - \frac{1}{2}sd_0 x_2^3]\right). \end{cases} \tag{3.149}$$

We note that $\beta_1, d_0 < 0$ holds in this case. Setting $\delta = 0$ and $q = -\frac{d_0+2}{d_0}$ we have the Hamiltonian

$$H := -\frac{d_0}{2}x_2^{-\frac{2}{d_0}}\left(\beta_1 + x_1^2 + \frac{sx_2^2}{1-d_0}\right).$$

On the level curve $H = 0$, we have $x_1^2 = -\beta_1 - \frac{sx_2^2}{1-d_0}$, which defines a heteroclinic connection Γ_0 between two saddles. According to Theorem 1.16 in Section 1.6, we need to evaluate the Pontryagin–Melnikov integral and set to it zero to find the first-order approximation to the heteroclinic bifurcation curve. We define f as the unperturbed vector field and g as the perturbation, i.e., the $O(\delta)$-terms. We can then write

$$\Delta := \int_{\Gamma_0} g_1 dx_2 - g_2 dx_2 = \frac{1}{(1-d_0)} \int x_2^{-\frac{2}{d_0}} \Big(s d_0^2 x_2^2 + (1-d_0)\beta_2 + \\ \frac{1}{2}(d_0 - 1)\left(d_0^2 + 3d_0c_0 - 2d_0 - 2d_1\right) x_1^2 \Big) dx_2, \tag{3.150}$$

where we applied Green's Theorem to convert it to an integral over x_2 only. Substituting x_1^2 defined by $H = 0$, we then define the integral

$$I_n := \int_0^{\sqrt{s(d_0-1)\beta_1}/s} x_2^{n-\frac{2}{d_0}} dx_2 = \frac{d_0 \sqrt{s\beta_1(d_0-1)}}{s(d_0 n + d_0 - 2)} \left(\frac{\sqrt{s(d_0-1)\beta_1}}{s} \right)^{\frac{d_0 n - 2}{d_0}},$$

where $n = 0$ or $n = 2$. Now solving $\Delta = 0$ leads to

$$\beta_2 = \frac{-(3d_0c_0 - 2d_1)}{3d_0 - 2}\beta_1, \tag{3.151}$$

which completes the derivation of the required asymptotic. □

3.F Proofs for Section 3.5

Proof of Proposition 3.44 The proof proceeds in several steps. First, we consider the flip–NS bifurcation and then repeat the same steps for the double NS bifurcation, where only some details are different.

Case I: Flip–NS

Step 1. Consider a map $\xi \mapsto F(\xi, \alpha)$, $\xi = (x, z) \in \mathbb{R} \times \mathbb{C}$, whose linear part at $(\xi, \alpha) = (0, 0)$ is given by

$$A = \begin{pmatrix} -1 & 0 \\ 0 & e^{i\phi} \end{pmatrix}.$$

We can expand this map as

$$\begin{pmatrix} x \\ z \end{pmatrix} \mapsto \begin{pmatrix} -x + \sum\limits_{2\le j+k+l\le 5} F_{jkl} x^j z^k \bar{z}^l \\ \mu z + \sum\limits_{2\le j+k+l\le 5} G_{jkl} x^j z^k \bar{z}^l \end{pmatrix} + O(\|(x,z)\|^6), \tag{3.152}$$

where $(x, z) \in \mathbb{R} \times \mathbb{C}$, $\mu = e^{i\phi}$ and $F_{jkl} = \bar{F}_{jlk}$. We assume further that ϕ satisfies the non-resonant conditions specified in Proposition 3.44.

We may apply near-identity transformations to remove as many non-resonant monomials of degree 2 and higher and get the Poincaré normal form. To determine the general structure of this form, we can consider (3.152) as a map on $\mathbb{R} \times \mathbb{C}^2$ with independent coordinates $(x, z, \bar{z})$, i.e.,

$$\begin{pmatrix} x \\ z \\ \bar{z} \end{pmatrix} \mapsto \begin{pmatrix} -x + \sum\limits_{2 \le j+k+l \le 5} F_{jkl} x^j z^k \bar{z}^l \\ \mu z + \sum\limits_{2 \le j+k+l \le 5} G_{jkl} x^j z^k \bar{z}^l \\ \bar{\mu} \bar{z} + \sum\limits_{2 \le j+k+l \le 5} \bar{G}_{jlk} x^j z^k \bar{z}^l \end{pmatrix} + O(\|(x, z, \bar{z})\|^6)$$

with the linearization matrix

$$A = \begin{pmatrix} -1 & 0 & 0 \\ 0 & \mu & 0 \\ 0 & 0 & \bar{\mu} \end{pmatrix}$$

that has simple eigenvalues $\lambda_1 = -1, \lambda_{2,3} = e^{\pm i\phi}$. It holds that $\lambda_1^2 = 1$ and $\lambda_2 \lambda_3 = 1$. Applying the normalization technique described in Section 1.3, we encounter only the following low-order resonances:

$$\lambda_1 = \lambda_1 \lambda_1^2 = \lambda_1(\lambda_2 \lambda_3) = \lambda_1 \lambda_1^4 = \lambda_1 \lambda_1^2(\lambda_2 \lambda_3) = \lambda_1(\lambda_2 \lambda_3)^2$$

and

$$\lambda_2 = \lambda_1^2 \lambda_2 = \lambda_1^0 \lambda_2(\lambda_2 \lambda_3) = \lambda_1^4 \lambda_2 = \lambda_1^0 \lambda_2(\lambda_2 \lambda_3)^2.$$

Then we immediately find that (3.152) can be transformed to a map with the 5-jet:

$$\begin{pmatrix} x \\ z \end{pmatrix} \mapsto \begin{pmatrix} -x + f_{300} x^3 + f_{111} x|z|^2 + f_{500} x^5 + f_{311} x^3 |z|^2 + f_{122} x|z|^4 \\ e^{i\phi} z + g_{210} x^2 z + g_{021} z|z|^2 + g_{410} x^4 z + g_{221} x^2 |z|^2 + g_{032} z|z|^4 \end{pmatrix}, \tag{3.153}$$

where the coefficients f_{ijk} are real, while g_{ijk} are complex. The latter map coincides with NF_1 for $\alpha = 0$.

Step 2. Since 1 is not an eigenvalue of A, we may assume that the origin always is a fixed point. We write $F(\xi, \alpha) = A(\alpha)\xi + R(\xi, \alpha)$, where $R(\xi, \alpha) = O(\|\xi\|^2)$. Now we introduce the parameter-dependent eigenvectors of A and A^T,

$$A(\alpha) q_i(\alpha) = \lambda_i q_i(\alpha), \quad A^T(\alpha) p_i(\alpha) = \lambda_i p_i(\alpha), \tag{3.154}$$

where $\lambda_1(0) = -1, \lambda_2(0) = e^{i\phi}$ and $\lambda_3(0) = e^{-i\phi}$. The vectors $p_i(\alpha)$ can be scaled such that $\langle p_i(\alpha), p(\alpha)_j \rangle = \delta_{ij}$, the Kronecker delta for $i, j = 1, 2, 3$. Then the variable ξ may be written as $\xi = \eta_1 q_1(\alpha) + \eta_2 q_2(\alpha) + \bar{\eta}_2 \bar{q}_2(\alpha)$, where $\bar{\eta}_2$

is treated as an independent variable. The map F, written in the η-coordinates and truncated at fifth order, takes the form

$$F(\eta,\alpha) = \begin{pmatrix} \lambda_1\eta_1 + \sum_{2\le i+j+k\le 5} f_{ijk}(\alpha)\eta_1^i\eta_2^j\bar{\eta}_2^k \\ \lambda_2\eta_2 + \sum_{2\le i+j+k\le 5} g_{ijk}(\alpha)\eta_1^i\eta_2^j\bar{\eta}_2^k \end{pmatrix}. \tag{3.155}$$

We introduce a parameter-dependent coordinate transformation $\eta = G(\xi,\alpha)$ such that $Q := j^5(F(G(\xi,\alpha),\alpha) - G(NF_1(\xi,\alpha),\alpha)) = 0$, the 5-jet with respect to ξ. The general form of G is

$$G(\xi,\alpha) = \begin{pmatrix} \sum_{1\le i+j+k\le 5} F_{ijk}(\alpha)\xi_1^i\xi_2^j\bar{\xi}_2^k \\ \sum_{1\le i+j+k\le 5} G_{ijk}(\alpha)\xi_1^i\xi_2^j\bar{\xi}_2^k \end{pmatrix},$$

where $F_{100}(0) = G_{010}(0) = 1$ and $F_{010}(0) = F_{001}(0) = G_{100}(0) = G_{001}(0) = 0$. We collect all coefficients Q_{ijk} of the monomials $\xi_1^i\xi_2^j\bar{\xi}_2^k$. Such a Q_{ijk} is a function of F_{ijk}, G_{ijk}, μ_i and the critical normal form coefficients $\hat{F}_{ijk}$. Thus we interpret Q as a function from $\mathbb{R}^{55}\times\mathbb{C}^{55}$ to itself. The determinant DQ of the Jacobian matrix of Q with respect to F_{ijk}, G_{ijk}, μ_i and $\hat{F}_{ijk}$ evaluated at the critical point is given by

$$\begin{aligned} DQ = {} & -2^{51} i \sin(\phi) e^{\frac{i45\phi}{2}} \sin(\phi)\sin(3\phi)\sin(5\phi)\cos(\phi/2)(\cos(2\phi))^3(\cos(\phi))^{10} \\ & (\cos(6\phi)-1)^3(\cos(5\phi)+1)(\cos(4\phi)-1)^3(\cos(3\phi)+1)^2 \\ & (\cos(2\phi)-1)^{14}(\cos(\phi)+1)^3. \end{aligned} \tag{3.156}$$

Due to the non-resonance conditions imposed on ϕ, we see that $DQ \neq 0$.

Step 3. Finally we prove the regularity of the map $\tilde{G}\colon \alpha \mapsto \mu$. From (3.155) and (3.85) we write $\lambda_1 = -1 + \tilde{A}(\alpha)$ and $\lambda_2 = \cos(\phi) + \tilde{B}(\alpha) + i\sin(\phi) + i\tilde{C}(\alpha)$, or $|\lambda_2| = 1 + \cos(\phi)\tilde{B} + \sin(\phi)\tilde{C} + \text{h.o.t.}$ Thus we identify

$$\begin{pmatrix} \mu_1 \\ \mu_2 \end{pmatrix} = \begin{pmatrix} \tilde{A}(\alpha) \\ \cos(\phi)\tilde{B}(\alpha) + \sin(\phi)\tilde{C}(\alpha) \end{pmatrix} + h.o.t., \tag{3.157}$$

and we see that $\det\left(\frac{d\mu}{d\alpha}\right)_{\alpha=0} = \det\left(\frac{d(\lambda_1,|\lambda_2|)}{d\alpha}\right)_{\alpha=0} \neq 0$. Therefore we may use (3.85) as the unfolding for this bifurcation.

Case II: NS–NS

Step 1. Now consider a map $\xi \mapsto F(\xi,\alpha)$, $\xi = (w,z) \in \mathbb{C}^2$, which at $(\xi,\alpha) = (0,0)$ has the linear part given by

$$A = \begin{pmatrix} e^{i\phi_1} & 0 \\ 0 & e^{i\phi_2} \end{pmatrix}.$$

As earlier, we write

$$\begin{pmatrix} w \\ z \end{pmatrix} \mapsto \begin{pmatrix} \mu_1 w + \sum\limits_{2\le i+j+k+l\le 5} F_{ijkl} w^i \bar{w}^j z^k \bar{z}^l \\ \mu_2 z + \sum\limits_{2\le i+j+k+l\le 5} G_{ijkl} w^i \bar{w}^j z^k \bar{z}^l \end{pmatrix} + O(\|(w,z)\|^6), \tag{3.158}$$

where $(w, z) \in \mathbb{C}^2$, $\mu_{1,2} = e^{i\phi_{1,2}}$. We assume further that $\phi_{1,2}$ satisfy the non-resonant conditions assumed in Proposition 3.44.

We then introduce a map on $\mathbb{C}^4$ with independent coordinates $(w, \bar{w}, z, \bar{z})$, i.e.,

$$\begin{pmatrix} w \\ z \\ \bar{w} \\ \bar{z} \end{pmatrix} \mapsto \begin{pmatrix} \mu_1 w + \sum\limits_{2\le i+j+k+l\le 5} F_{ijkl} w^i \bar{w}^j z^k \bar{z}^l \\ \mu_2 z + \sum\limits_{2\le i+j+k+l\le 5} G_{ijkl} w^i \bar{w}^j z^k \bar{z}^l \\ \bar{\mu}_1 \bar{w} + \sum\limits_{2\le i+j+k+l\le 5} \bar{F}_{jilk} w^i \bar{w}^j z^k \bar{z}^l \\ \bar{\mu}_2 \bar{z} + \sum\limits_{2\le i+j+k+l\le 5} \bar{G}_{jilk} w^i \bar{w}^j z^k \bar{z}^l \end{pmatrix} + O(\|(w,z)\|^6),$$

with the linearization matrix

$$A = \begin{pmatrix} \mu_1 & 0 & 0 & 0 \\ 0 & \mu_2 & 0 & 0 \\ 0 & 0 & \bar{\mu}_1 & 0 \\ 0 & 0 & 0 & \bar{\mu}_2 \end{pmatrix}$$

that has simple eigenvalues $\lambda_{1,2} = e^{i\phi_{1,2}}, \lambda_{3,4} = e^{-i\phi_{1,2}}$. It holds that $\lambda_1\lambda_3 = 1$ and $\lambda_2\lambda_4 = 1$. Applying the normalization technique described in Section 1.3, we encounter only the following low-order resonances:

$$\lambda_1 = \lambda_1(\lambda_1\lambda_3) = \lambda_1(\lambda_2\lambda_4) = \lambda_1(\lambda_1\lambda_3)(\lambda_2\lambda_4) = \lambda_1(\lambda_1\lambda_3)^2 = \lambda_1(\lambda_2\lambda_4)^2$$

and

$$\lambda_2 = \lambda_2(\lambda_1\lambda_3) = \lambda_2(\lambda_2\lambda_4) = \lambda_2(\lambda_1\lambda_3)(\lambda_2\lambda_4) = \lambda_2(\lambda_1\lambda_3)^2 = \lambda_2(\lambda_2\lambda_4)^2.$$

Under the conditions stated, we find that the map (3.158) can be transformed into a map with the 5-jet

$$\begin{aligned} \begin{pmatrix} w \\ z \end{pmatrix} \mapsto & \begin{pmatrix} e^{i\phi_1} w + f_{2100} w|w|^2 + f_{1011} w|z|^2 \\ e^{i\phi_2} z + g_{1110} z|w|^2 + g_{0021} z|z|^2 \end{pmatrix} \\ & + \begin{pmatrix} f_{3200} w|w|^4 + f_{2111} w|w|^2|z|^2 + f_{1022} w|z|^4 \\ g_{2201} z|w|^4 + g_{1121} z|w|^2|z|^2 + g_{0032} z|z|^4 \end{pmatrix}. \end{aligned} \tag{3.159}$$

Note that all coefficients f_{ijkl} and g_{ijkl} are complex.

Step 2. For the double NS bifurcation a similar preparation but with two complex variables leads to a function $Q\colon \mathbb{C}^{250} \to \mathbb{C}^{250}$. The Jacobian DQ at the critical point is the product of expressions like $(\cos(3\phi_1 - \phi_2) - 1)^4$ and $(\cos(\phi_1 + 5\phi_2) - 1)$, but too lengthy to display here. From the non-resonance conditions required, we obtain that $DQ \neq 0$ and that these are the minimal set of such conditions.

Step 3. As before, we write $\lambda_1(\alpha) = \cos(\phi_1) + \tilde{A}(\alpha) + i\sin(\phi_1) + i\tilde{B}(\alpha)$ and $\lambda_2(\alpha) = \cos(\phi_2) + \tilde{C}(\alpha) + i\sin(\phi_2) + i\tilde{D}(\alpha)$ and at first order the map $\tilde{G}\colon \alpha \mapsto \mu$ is given as

$$\begin{pmatrix} \mu_1 \\ \mu_2 \end{pmatrix} = \begin{pmatrix} \cos(\phi_1)\tilde{A}(\alpha) + \sin(\phi_1)\tilde{B}(\alpha) \\ \cos(\phi_2)\tilde{C}(\alpha) + \sin(\phi_2)\tilde{D}(\alpha) \end{pmatrix} + h.o.t. \tag{3.160}$$

The map regularity of $\tilde{G}\colon \alpha \mapsto \mu$ at 0 is now ensured by $|D_\alpha(\mu)|_{\alpha=0} = |D_\alpha(|\lambda|)|_{\alpha=0} \neq 0$. □

Computations for Remark 3.45 It is sufficient to verify the statement for the normal forms, so DG_i should be nonzero. A straightforward calculation shows that

$$\det DG_1(0,0) = 16\sin^2(\phi)\left|\frac{d(\lambda_1, |\lambda_2|)}{d(\alpha_1, \alpha_2)}\right|.$$

For the regularity of G_2 we use DF_2 as in the beginning of Section 3.5.1 and introduce $B^T = C = \begin{pmatrix} 1 & 0 & 0 & 0 & 0 & 0 \\ 0 & 0 & 0 & 0 & 0 & 1 \end{pmatrix}$. Then indeed the 8×8 matrix in the remark is nonsingular. After some tedious algebra one finds

$$\frac{d(g_{11}, g_{22})}{d(\alpha_1, \alpha_2)} = \begin{pmatrix} \frac{\partial(\cos(\phi_1)\tilde{A}+\sin(\phi_1)\tilde{B})}{\partial\alpha_1} & \frac{\partial(\cos(\phi_1)\tilde{A}+\sin(\phi_1)\tilde{B})}{\partial\alpha_2} \\ \frac{\partial(\cos(\phi_2)\tilde{C}+\sin(\phi_2)\tilde{D})}{\partial\alpha_1} & \frac{\partial(\cos(\phi_2)\tilde{C}+\sin(\phi_2)\tilde{D})}{\partial\alpha_2} \end{pmatrix}.$$

In fact we have $\det DG_2(0,0) = 16(\cos(\phi_1) - 1)(\cos(\phi_2) - 1)\left|\frac{d(|\lambda_1|, |\lambda_2|)}{d(\alpha_1, \alpha_2)}\right|$. □

Proof of Proposition 3.46 We introduce a special change of coordinates M consisting of the resonant monomials in the 3-jet of (3.87). The transformation leads to a new map F with the same 3-jet but alters the 5-jet. Then we choose the $m_i, i = 1,2,3,4$ such that as many as possible of the $c_i, i = 1,2,\ldots,6$ are eliminated. We write

$$M(x,y) = \begin{pmatrix} x + m_1x^3 + m_2xy^2 \\ y + m_3x^2y + m_4y^3 \end{pmatrix},$$

$$F(x,y) = \begin{pmatrix} x + a_{11}x^3 + a_{12}xy^2 + c_1x^5 + c_2x^3y^2 + c_3xy^4 \\ y + a_{21}x^2y + a_{22}y^3 + c_4x^4y + c_5x^2y^3 + c_6y^5 \end{pmatrix}.$$

The transformation does not change x^5- and y^5-terms. The condition $j^5(H(M(x,y)) - M(F(x,y))) = 0$ leads to the following linear system:

$$2\begin{pmatrix} a_{12} & a_{21}-a_{11} & -a_{12} & 0 \\ 0 & a_{22} & 0 & -a_{12} \\ -a_{21} & 0 & a_{11} & 0 \\ 0 & -a_{21} & a_{12}-a_{22} & a_{21} \end{pmatrix}\begin{pmatrix} m_1 \\ m_2 \\ m_3 \\ m_4 \end{pmatrix} = \begin{pmatrix} h_{32}-c_2 \\ h_{14}-c_3 \\ h_{41}-c_4 \\ h_{23}-c_5 \end{pmatrix}. \tag{3.161}$$

The matrix of the system has determinant zero and its kernel is one-dimensional. We choose not to eliminate c_4 as this coincides with a natural non-degeneracy condition in the bifurcation analysis. Set $c_2 = c_3 = c_5 = m_3 = 0$ and solve (3.161) in m_1, m_2, m_4, c_4. We get the new coefficients

$$c_1 = h_{50}, \quad c_6 = h_{05}, \quad c_4 = h_{41} + a_{21}\left(\frac{h_{32}}{a_{12}} - \frac{h_{14}(a_{11}-a_{21})}{a_{12}(a_{12}-a_{22})} - \frac{h_{23}(a_{11}-a_{21})}{a_{21}(a_{12}-a_{22})}\right).$$

Then we apply a linear scaling $x \to x/\sqrt{|a_{11}|}, y \to y/\sqrt{|a_{22}|}$ and obtain the desired map. □

Proof of Proposition 3.47

1. The fixed point equation is given by $x = (1+\mu_1)x + s_1x^3 + c_1x^5$. For μ_1 small we have $x_+ = \sqrt{-\mu_1 s_1} + O(\mu_1)$. The extra fixed point exists for $\mu_1 s_1 < 0$. We evaluate the Jacobian matrix

$$\begin{aligned} DF_\mu(x_+,0) &= \begin{pmatrix} 1+\mu_1+3s_1x_+^2+5c_1x_+^4 & 0 \\ 0 & 1+\mu_2+s_1\delta x_+^2+c_4x_+^4 \end{pmatrix} \\ &= \begin{pmatrix} 1-2\mu_1+O(\mu_1^2) & 0 \\ 0 & 1+\mu_2-\delta\mu_1+O(\mu_1^2) \end{pmatrix}. \end{aligned}$$

   The multiplier in the x-direction is $\mu_{\hat{x}} = 1 - 2\mu_1 + O(\mu_1^2)$, so in this direction it is stable if $s_1 < 0$ and unstable if $s_1 > 0$. In the y-direction it is stable if $\mu_2 - \delta\mu_1 < 0$.
2. This is analogous to the preceding item.
3. For a nontrivial fixed point we search for nonzero (x, y) with

$$\begin{pmatrix} s_1 & s_2\theta \\ s_1\delta & s_2 \end{pmatrix}\begin{pmatrix} x^2 \\ y^2 \end{pmatrix} = -\begin{pmatrix} \mu_1 \\ \mu_2 \end{pmatrix} + \begin{pmatrix} c_1x^4 \\ c_4x^4 + c_6y^4 \end{pmatrix}.$$

   A solution is $(x^2, y^2) = (\rho_1^2, \rho_2^2) = \frac{1}{s_1s_2(\delta\theta-1)}(s_2(\mu_1 - \theta\mu_2), s_1(\mu_2 - \delta\mu_1)) + O(\|\mu\|^2)$. It exists if both components are positive. Now we turn to the stability of this point. From the Routh–Hurwitz criterion it follows that the roots of $\lambda^2 + a_1\lambda + a_0 = 0$ are strictly inside the unit circle if and only if

$$|a_1| < (1 + a_0), \tag{3.162a}$$

$$|a_0| < 1. \tag{3.162b}$$

We give the Taylor expansions in μ of the trace and determinant:

$$\begin{aligned} a_1 = \operatorname{tr}(DF_\mu(\rho_1,\rho_2)) &= 2 - \frac{2}{\delta\theta - 1}((\delta-1)\mu_1 + (\theta-1)\mu_2) \\ &\quad + z_{20}\mu_1^2 + z_{11}\mu_1\mu_2 + z_{02}\mu_2^2 + O(\|\mu\|^2), \\ a_0 = \det(DF_\mu(\rho_1,\rho_2)) &= 1 - \frac{2}{\delta\theta - 1}((\delta-1)\mu_1 + (\theta-1)\mu_2) \\ &\quad \hat{z}_{20}\mu_1^2 + \hat{z}_{11}\mu_1\mu_2 + \hat{z}_{02}\mu_2^2 + O(\|\mu\|^2), \end{aligned}$$

where

$$z_{20} - \hat{z}_{20} = \frac{-4\delta}{\delta\theta - 1}, z_{11} - \hat{z}_{11} = \frac{4(\delta\theta + 1)}{\delta\theta - 1}, z_{02} - \hat{z}_{02} = \frac{-4\theta}{\delta\theta - 1}.$$

A little algebra shows that conditions (3.162a) and (3.162b) are equivalent to

$$\frac{4}{\delta\theta - 1}(\delta\mu_1 - \mu_2)(\theta\mu_2 - \mu_1) = 4s_1s_2(\delta\theta - 1)\rho_1^2\rho_2^2 < 0 + O(\|\mu\|^2), \tag{3.163a}$$

$$-\frac{2}{\delta\theta - 1}((\theta\mu_2 - \mu_1) + (\delta\mu_1 - \mu_2)) = 2(s_1\rho_1^2 + s_2\rho_2^2) < 0 + O(\|\mu\|^2), \tag{3.163b}$$

if μ is sufficiently small.

4. Violation of (3.162b) corresponds to a NS bifurcation of the nontrivial fixed point. This can only occur if $s_1s_2 < 0$. Consider now μ_1 as a perturbation parameter, then a first-order approximation for the critical value is given by $\mu_{2,c} = -(\delta-1)\mu_1/(\theta-1)$. Now the first-order approximation of the multiplier is $\lambda = 1 + 2\mu_1\sqrt{(1-\delta\theta)}/(\theta-1)$, which is complex if $\delta\theta > 1$. For sufficiently small μ_1 we are not near strong resonances. Then we used one more order in the approximations of x, y, μ_2, λ and the (adjoint) eigenvectors p, q of the Jacobian matrix along the bifurcation curve. Then we verified transversality $\frac{d|\lambda|}{d\mu_2}|_{\mu_2=\mu_{2,c}} = \frac{1}{2}$, and for the non-degeneracy we computed the leading term of the first Lyapunov coefficient L_{NS} by (2.18):

$$L_{NS} = \mu_1 s_1\left(\frac{12(2\delta\theta-\delta-\theta)}{\theta(\theta-1)} + c_1\frac{8(2\delta\theta-\delta-1)}{(\theta-1)(\delta\theta-1)} - c_4\frac{8}{(\delta\theta-1)} + c_6\frac{8\delta(2\delta\theta-\theta-1)}{\theta(\theta-1)(\delta\theta-1)}\right) + O(\mu_1^2).$$

If $L_{NS} \neq 0$, a closed invariant curve appears, which is stable if L_{NS} is negative and unstable if L_{NS} is positive.

Proof of Proposition 3.49 We can make a singular rescaling of the vector field (3.92) and apply the Pontryagin–Melnikov method outlined in Section 1.6. The steps in our calculation closely follow the presentation in Chow, Li, and Wang (1994).

Step 1. We make a change of variables in (3.92) and perform a singular rescaling

$$(t, x^2, y^2, \mu_1, \mu_2) \mapsto \left(\tfrac{1}{2}\varepsilon^{-1}x^p y^q t, \varepsilon x, \varepsilon y, -s_1\varepsilon, \left(\tfrac{\delta-1}{\theta-1}\right) s_1\varepsilon + \varepsilon^2\nu\right),$$

which results in the following vector field

$$\begin{pmatrix} \dot{x} \\ \dot{y} \end{pmatrix} = s_1 x^{p-1} y^{q-1} \begin{pmatrix} x(-1 + x - \theta y), \\ y\left(\frac{\delta-1}{\theta-1} + \delta x - y\right) \end{pmatrix} + \varepsilon x^{p-1} y^{q-1} \begin{pmatrix} g_1(x, y) \\ g_2(x, y) \end{pmatrix} + O(\varepsilon^2), \tag{3.164}$$

where

$$g_1(x, y) = x\left(-\tfrac{1}{2} - y\theta\left(\tfrac{\delta+\theta-2}{\theta-1}\right) + \tilde{c}_1 x^2 + \tilde{c}_2 xy + \tilde{c}_3 y^2\right),$$
$$g_2(x, y) = y\left(\nu - \tfrac{1}{2}\left(\tfrac{\delta-1}{\theta-1}\right)^2 - x\delta\left(\tfrac{\delta+\theta-2}{\theta-1}\right) + \tilde{c}_4 x^2 + \tilde{c}_5 xy + \tilde{c}_6 y^2\right).$$

For $\varepsilon = 0$ this is a Hamiltonian system with

$$H(x, y) = s_1 x^p y^q (\delta\theta - 1)\left(\tfrac{1-x}{\theta-1} - \tfrac{y}{\delta-1}\right), \quad p = -\tfrac{(\delta-1)}{\delta\theta-1}, \; q = -\tfrac{(\theta-1)}{\delta\theta-1}.$$

Step 2. We have $p, q > 0$ due to the non-degeneracy conditions. Now we treat the term proportional to ε in (3.164) as a small perturbation of the Hamiltonian system. We should therefore evaluate the Pontryagin–Melnikov integral (see Section 1.6) for the critical level set $H = 0$, which consists of three heteroclinic orbits.

$$\Delta(h, \nu) = \oint_{\Gamma_h} dH(s(\tau), \nu) \wedge g(s(\tau), \nu)\, d\tau,$$

where $s(\tau)$ corresponds to a nontrivial heteroclinic solution in the positive quadrant of the Hamiltonian system on the level curve $H = h$. Then the equation $\Delta(0, \nu) = 0$ defines a quadratic approximation to a curve on which the heteroclinic connection "survives" in (3.164) for small $\varepsilon \neq 0$.

$$\begin{aligned}
\Delta(0, \nu) &= \oint_{\Gamma_0} g_1(x, y)dy - g_2(x, y)dx \\
&= \oint_{\Gamma_0} x^p y^{q-1}(-\tfrac{1}{2} - y\theta\left(\tfrac{\delta+\theta-2}{\theta-1}\right) + \tilde{c}_1 x^2 + \tilde{c}_2 xy + \tilde{c}_3 y^2)dy \\
&\quad - x^{p-1} y^q(\nu - \tfrac{1}{2}\left(\tfrac{\delta-1}{\theta-1}\right)^2 - x\delta\left(\tfrac{\delta+\theta-2}{\theta-1}\right) + \tilde{c}_4 x^2 + \tilde{c}_5 xy + \tilde{c}_6 y^2)dx \\
&= -I_{(p-1,q)}\left(\nu - \tfrac{p}{2q} - \tfrac{(\delta-1)^2}{2(\theta-1)^2}\right) + I_{(p,q)}\left(\tfrac{\delta(\theta+\delta-2)}{\theta-1}\right) \\
&\quad + I_{(p-1,q+1)}\left(\tfrac{\theta p(\theta+\delta-2)}{(q+1)(\theta-1)}\right) - I_{(p+1,q)}\left(\tilde{c}_4 + \tfrac{p+2}{q}\tilde{c}_1\right) \\
&\quad - I_{(p,q+1)}\left(\tilde{c}_5 + \tfrac{p+1}{q+1}\tilde{c}_2\right) - I_{(p-1,q+2)}\left(\tilde{c}_6 + \tfrac{p}{q+2}\tilde{c}_3\right),
\end{aligned} \tag{3.165}$$

where Green's formula is used and we defined $I_{(i,j)} = \int_{\Gamma_0} x^i y^j dx$. On the critical level curve we have $y = \frac{\delta-1}{\theta-1}(1 - x)$ and with this substitution we find

$$I_{i,j} = \left(\tfrac{\delta-1}{\theta-1}\right)^j \int_0^1 x^i (1 - x)^j dx = \left(\tfrac{\delta-1}{\theta-1}\right)^j \tfrac{\Gamma(1+i)\Gamma(1+j)}{\Gamma(2+i+j)}.$$

Using the definition of the Γ-function and substituting the $\tilde{c}_i$, we can now find ν from (3.165):

$$\nu = \frac{(\delta\theta-1)^2}{2(\theta-1)^3} + c_1\left(\frac{\delta(2\delta\theta-\delta-1)}{(2\delta\theta-\theta-\delta)(\theta-1)}\right) + c_4\left(\frac{\delta}{(2\delta\theta-\theta-\delta)}\right) - c_6\left(\frac{\theta(2\delta\theta-\theta-1)(\delta+1)^2}{(2\delta\theta-\theta-\delta)(\theta-1)^3}\right).$$

Since the time-1 flow approximates the map (3.90) up to order 2 in μ, this calculation shows that the curve

$$\mu_2 = -\frac{\delta-1}{\theta-1}\mu_1 + \nu\mu_1^2 + O(\mu_1^3)$$

is an approximation of the heteroclinic tangency in the map (3.90).

4
Center Manifold Reduction for Local Bifurcations

When a fixed point undergoes a bifurcation of low codimension, the scenarios are more or less known. The unfoldings of normal forms of minimal dimension for these cases have been studied and their principal structure is obtained (see Chapter 3). Although in many cases the bifurcation diagrams are in principle incomplete due to global phenomena, such as homoclinic or heteroclinic tangencies, many essential features of these diagrams are determined by the critical normal form coefficients.

However, a typical model from applications has a much larger dimension than this minimal one. Thus it is not straightforward to apply the theoretical results to an example. Fortunately, there is a center manifold to which the system can be restricted. This requires efficient algorithms for the computation of the critical normal form coefficients on the center manifold in terms of the original n-dimensional map (1.4). Even though neither the center manifold nor the critical normal form are unique, the qualitative conclusions do not depend on the choices that are made. Note that the genericity of a bifurcation in a given system also requires its non-degeneracy with respect to actual control parameters, which should unfold the singularity transversally. Once constructed, such algorithms provide useful initial guesses for where to search for new local and global phenomena in applications.

This chapter is devoted to the computation of the center manifold to have equipment for applications. First, we review a powerful normalization method originally developed for and applied to vector fields (see (Coullet and Spiegel, 1983; Elphick et al., 1987; Beyn et al., 2002; Kuznetsov, 1999; Ipsen, Hynne, and Sörensen, 1998)). It is, however, equally well applicable to maps. One can extend the description of the center manifold to include parameters as well, and we will use this to derive approximations to new local codim 1 curves that emanate from codim 2 bifurcation points. All formulas have been implemented in MatcontM, which is an adaptation of the continuation environment for ODEs MATCONT (Dhooge, Govaerts, and Kuznetsov, 2003b; Dhooge et al., 2003a, 2008). This is very useful as, when following codim 1 curves, one meets generically codim 2 points where non-degeneracy needs to be checked. Our formulas are suitable for checking non-degeneracy numerically and to switch to new branches. The new branches we predict are cycle bifurcations of two, three or four times the period of the bifurcating fixed point. For biological models, for example, these are the most interesting as these low-period scenarios are the most relevant and possible to observe. For such case studies we refer to Govaerts and Khoshsiar Ghaziani (2006); Davydova (2004).

4.1 The homological equation and its solutions

Consider a smooth map

$$x \mapsto f(x, \alpha), \quad f\colon \mathbb{R}^{n+2} \to \mathbb{R}^n, \tag{4.1}$$

and assume that it has a nonhyperbolic fixed point $x = x_0$ for a critical parameter value $\alpha = \alpha_0$. We use the expansion for (4.1) as in (3.3). The Center Manifold Theorem (Shoshitaishvili, 1975; Hirsch, Pugh, and Shub, 1977) guarantees the existence of stable, unstable and center invariant manifolds near the fixed point. The dynamics in the center manifold depends on both linear and nonlinear terms. Not all nonlinear terms are equally important, since some of them can be eliminated by an appropriate smooth coordinate transformation that puts the map restricted to the center manifold into a normal form, at least up to some finite order. Geometrically, by normalization we "rectify" our coordinates.

The center manifold can be proven to exist in several ways (Shoshitaishvili, 1975; Hirsch, Pugh, and Shub, 1977; Vanderbauwhede, 1988; Vanderbauwhede and van Gils, 1987), but the proofs do not give an actual approximation to the manifold. As 11 codim 2 cases need to be covered, it is important not to approach this question in an ad hoc manner, but to use a format that generates formulas for center manifold reduction sort of automatically. Indeed, in the late 1970s and early 1980s the formula for the direction of Hopf bifurcation received much attention (van Gils, 1984; Hassard, Kazarinoff, and Wan, 1981; Vanderbauwhede, 1982). At the time it was a big achievement as the derivation of such a formula could fill whole chapters of a book, or several pages for a single model in papers. Analyzing bifurcations can be approached in several ways, with for example Lyapunov–Schmidt or center manifold reduction. The first is a rather indirect method for verifying non-degeneracy and transversality. Let us elaborate a little on the latter.

1. When encountering a bifurcation, one can compute the center manifold (see, e.g., (Beyn and Kless, 1998)). After restricting the map to the center manifold and putting the linear part into Jordan normal form, one proceeds with normalizing the higher-order terms in the equations. Altogether this is a laborious approach.
2. One can also combine the reduction and normalization. While the center manifold was shown to exist in the 1970s, it took several years before it was noticed how to take advantage of the invariance of the center manifold. This idea can be found in the papers by Iooss and coworkers Coullet and Spiegel (1983); Iooss (1988); Elphick et al. (1987). A detailed description of this powerful normalization technique for ODEs can be found in Kuznetsov (1999); Ipsen, Hynne, and Sörensen (1998), while in Beyn et al. (2002) parameters are also accounted for.

When an event is encountered in the minimal possible dimension and the linear part is in Jordan normal form, there are formulas available for all cases (Kuznetsov, 2004; Kuznetsov, Meijer, and van Veen, 2004; Vitolo, 2003; Wen, Wang, and Chiu, 2006; Xie and Ding, 2005). However, the second approach does not require to transform the map f to Jordan normal form, but rather to evaluate the derivative tensor-vector products, and can also be useful in the minimal dimension. We have used this approach for critical normalization for maps (Kuznetsov and Meijer, 2005, 2006) and branch switching to cycle bifurcations for maps (Govaerts et al., 2007). This is a recursive approach in inner-product style, following the terminology of Murdock (2003), to which we also refer for a review of various aspects of normalization for ODEs.

A near-identity transformation of order 2 may be without effect at order 2, but useful to eliminate resonant terms of higher order. This is called hyper-normalization. As in these approaches nonlinear terms are removed one order at a time, we do not always use the possibility for further elimination. This is relevant for only a few cases, where bordered vectors play a role. See also a case study in Remark 4.3.

Suppose that the map obtained by restriction of map (4.1) to the center manifold can be transformed to some normal form

$$w \mapsto G(w,\beta), \quad G\colon \mathbb{R}^{n_c+2} \to \mathbb{R}^{n_c},$$

where n_c is the number of the critical eigenvalues (counting multiplicity), or, equivalently, the dimension of the center manifold. *A priori*, the type of the codim 2 singularity has been determined and the form of G is known, but not its coefficients. Locally the parameter-dependent center manifold can be parametrized by $w \in \mathbb{R}^{n_c}$ and $\beta \in \mathbb{R}^2$:

$$x = H(w,\beta), \quad H\colon \mathbb{R}^{n_c+2} \to \mathbb{R}^n.$$

Since the center manifold is invariant, we obtain the following *homological equation* for H:

$$f(H(w,\beta), V(\beta)) = H(G(w,\beta),\beta). \tag{4.2}$$

Now two key words come up:

Non-degeneracy: We want to find the coefficients $g_{\mu 0}$ of G for $\beta = 0$. First, we need to verify that the coefficients fulfill certain non-degeneracy conditions. If this is not the case, we are dealing with a degenerate situation, which can happen when symmetries are involved. Second, after this verification we can predict the bifurcation scenario, like sub- or supercriticality.

Transversality: For all generic local codim 2 bifurcations a representative unfolding of the critical normal form has been studied with parameters β. If the family of maps is transversal, then we may draw conclusions for the dynamics according to the bifurcation scenario determined from the non-degeneracy conditions when varying the parameters. If not, we need to be more careful.

Generally speaking, the coefficients $g_{\mu 0}$ may hint at the birth of invariant curves and/or homo- and heteroclinic orbits near codim 2 bifurcations. Also, in several cases, codim 1 curves of cycle bifurcations are rooted at these codim 2 points. Normal form analysis yields asymptotic expressions for such curves. The combination of this information with an approximation of the center manifold may actually provide a starting point for the continuation of these curves. However, for this we also need to step in the right direction in parameter space, so we also seek a relation $\alpha - \alpha_0 = V(\beta)$.

Now, to obtain an approximation of the solution of the homological equation, we write Taylor series for G, H and V:

$$G(w,\beta) = \sum_{|\mu|+|\nu|\geq 1} \frac{1}{\mu!\nu!} g_{\mu\nu} w^{\mu} \beta^{\nu}, \quad H(w,\beta) = \sum_{|\mu|+|\nu|\geq 1} \frac{1}{\mu!\nu!} h_{\mu\nu} w^{\mu} \beta^{\nu}, \tag{4.3}$$

$$V(\beta) = v_{10}\beta_1 + v_{01}\beta_2 + O(\|\beta\|^2), \tag{4.4}$$

where $g_{\mu\nu}$ are normal form coefficients and μ, ν are multi-indices. If we deal with just the critical coefficients, the index ν will sometimes be omitted. For a multi-index ν we have $\nu = (\nu_1, \nu_2, \ldots, \nu_n), \nu_i \in \mathbb{Z}_{\geq 0}, \ \nu! = \nu_1!\nu_2!\ldots\nu_n!, |\nu| = \nu_1 + \nu_2 + \cdots + \nu_n$ and $\tilde{\nu} \leq \nu$ if $\tilde{\nu}_i \leq \nu_i$ for all $i = 1, \ldots, n$. Inserting this ansatz into (4.2) we get a formal power series in w and β, the coefficients of which should be zero. This leads to an iterative procedure, where we obtain linear systems for each component h_ν of the center manifold

$$Lh_{\mu\nu} = R_{\mu\nu}, \tag{4.5}$$

where $L = A - \lambda I_n$ and λ a function (the sum or product) of the critical eigenvalues. It is easy to check that the right-hand side of (4.5) depends only on the quantities $h_{\tilde{\mu}\tilde{\nu}}, g_{\tilde{\mu}\tilde{\nu}}$ and $v_{\tilde{\nu}}$ where $\tilde{\mu} \leq \mu$ and $\tilde{\nu} \leq \nu$. Next, we have to solve (4.5) and the following important remark helps us to decide how to proceed.

Remark 4.1 **(Fredholm's Alternative)** If (4.5) has a solution, then either L is non singular or $R_{\mu\nu}$ is orthogonal to the adjoint eigenvectors of L corresponding to the zero eigenvalue of L.

Let us see how this helps. If L is non singular, then the monomial of order μ, ν is non-resonant and can be eliminated. The interesting case is when L is singular as we should then have

$$\langle p, R_{\mu\nu} \rangle = 0, \tag{4.6}$$

where p is any null-vector of the adjoint matrix $\bar{L}^T$ and $R_{\mu\nu}$ contains $g_{\mu\nu}$. Recall that we use the standard Hermitian inner-product $\langle a, b \rangle = \bar{a}^T b$ for vectors $a, b \in \mathbb{C}^n$. Given the linearization, we know for which multi-indices μ, ν the coefficients $g_{\mu\nu}$ appear, but the actual value comes only from this step. From the linear parts of (4.2), i.e., $|\mu| = 1$, we recover the classical eigenvalue and eigenvector problem. Note that if $w^{\nu_1\nu_2}$ is associated to some complex coordinate, then $h_{\nu_2\nu_1} = \overline{h_{\nu_1\nu_2}}$. The higher-order terms give both the critical coefficients and a better approximation of the center manifold. It is easy to see that if we want to compute an mth order coefficient, it suffices to have an $(m-1)$-th order approximation of the center manifold. Even better, we do not always need all the terms of the $(m-1)$-th order, thus reducing computing efforts and time.

Several codim 2 bifurcations are characterized by a vanishing normal form coefficient at some order. In such cases we usually have to normalize several more orders. But for that we will also need the vector x from equations $Mx = y$, where M is a singular matrix. We now discuss the *bordering technique* which allows us to find a solution to this problem.

Let M be a complex singular $n \times n$ matrix with a one-dimensional null-space spanned by q, an eigenvector corresponding to the simple eigenvalue 0 of M. Similarly, let p be an eigenvector corresponding to the eigenvalue 0 of the matrix $\overline{M}^T$. Then we construct the nonsingular $(n+1) \times (n+1)$ matrix

$$\begin{pmatrix} M & q \\ \bar{p}^T & 0 \end{pmatrix}. \tag{4.7}$$

Thus we are able to solve the system

$$\begin{pmatrix} M & q \\ \bar{p}^T & 0 \end{pmatrix}\begin{pmatrix} x \\ s \end{pmatrix} = \begin{pmatrix} y \\ 0 \end{pmatrix}, \tag{4.8}$$

where s is some complex auxiliary variable. We write $x = M^{INV}y$. First of all, we remark that in our applications the vector on the right-hand side is orthogonal to the critical eigenspace of the eigenvalue 0, i.e., $\langle p, y\rangle = 0$, due to (4.5). It then follows that $\langle p, x\rangle = s = 0$. This is sufficient and convenient for our studies as for codim 2 bifurcations this technique does not have to be applied for cases with non-semisimple linear part. One could add an appropriate second bordering vector to such matrices to find a solution.

Other choices for the vectors p and q can be made as long as the new matrix is nonsingular; however, we then no longer have $\langle p, x\rangle = s = 0$. Still, the solution vector x is not unique as any multiple of q may be added to x for the original equation $Mx = y$. One may wonder whether this leads to different results for the normal form computations. It does not; we will work out some cases in Remark 4.2. The property $\langle p, x\rangle = 0$ will be used to simplify some expressions in the non-degeneracy analysis. However it may be useful not to fix the vector but see what it can do at a higher order (see also Remark 4.3, where we perform a case study). In Section 4.3, where we consider branch-switching, we use this freedom in x to obtain simplified normal forms.

4.2 Critical normal form coefficients for local codim 2 bifurcations

One of the conditions to check for bifurcation analysis is non-degeneracy. Thus we need to have the critical normal form of the map. Most of these formulas have been presented earlier by Kuznetsov (2004). However, we briefly derive them here for completeness.

Cusp

In this case, there is only one critical eigenvalue $\lambda = +1$ and the coefficient b_0 given by (2.7) vanishes. Inserting the normal form (3.5), which we write as

$$w \mapsto w + c_0 w^3 + O(|w|^4),$$

into the homological equation (4.2) at $\beta = 0$ we obtain the following linear systems from the first three orders in the phase variable w:

$$(A - I_n)h_1 = 0, \tag{4.9}$$

$$(A - I_n)h_2 = -B(h_1, h_1), \tag{4.10}$$

$$(A - I_n)h_3 = 6c_0 q - C(h_1, h_1, h_1) - 3B(h_1, h_2). \tag{4.11}$$

The first equation (4.9) is solved with the eigenvector for the eigenvalue $\lambda = 1$. We will use q to denote right eigenvectors and we write $h_1 = q$. The second equation (4.10) has a singular matrix, thus we introduce the left (adjoint) eigenvector p satisfying $A^T p = p$. As λ is a simple eigenvalue of A, we can scale p such that $\langle p, q\rangle = 1$. Now we check the solvability condition (4.5)

$$\langle p, B(q, q)\rangle = 2b_0 = 0, \tag{4.12}$$

as we have a cusp bifurcation. Then the bordering technique supplies a solution for $h_2 = (I_n - A)^{INV} B(q, q)$. The coefficient c_0 of the critical normal form (3.5) is then found by applying the Fredholm Alternative to (4.11):

$$c_0 = \frac{1}{6}\langle p, C(q, q, q) + 3B(q, h_2)\rangle = \frac{1}{6}\langle p, C(q, q, q) + 3B(q, (I_n - A)^{INV} B(q, q))\rangle. \tag{4.13}$$

If $c_0 \neq 0$, then the cusp bifurcation (CP) is non-degenerate.

Generalized period-doubling

In this case, there is only one critical eigenvalue -1 and the coefficient c_0 given by (2.12) vanishes. We use the critical normal form (3.10)

$$w \mapsto -w + d_0 w^5 + O(|w|^6), \tag{4.14}$$

where $w \in \mathbb{R}$ is a properly selected local coordinate along the one-dimensional center manifold. As in the previous case we can find eigenvectors such that

$$Aq = -q, \qquad A^T p = -p, \qquad \langle p, q\rangle = 1. \tag{4.15}$$

Collecting the w^2-terms in (4.2) at $\beta = 0$, we obtain

$$(A - I_n)h_2 = -B(q, q), \tag{4.16}$$

which is a nonsingular linear system that has $h_2 = (I_n - A)^{-1} B(q, q)$ as a unique solution. We continue with the w^3-terms to get

$$(A + I_n)h_3 = -C(q, q, q) - 3B(q, h_2). \tag{4.17}$$

This singular system is solvable since

$$\langle p, C(q, q, q) + 3B(q, h_2)\rangle = 6c_0 = 0$$

according to (2.12). The fourth-order terms in (4.2) give

$$(A - I_n)h_4 = -\left(4B(q,h_3) + 3B(h_2,h_2) + 6C(q,q,h_2) + D(q,q,q,q)\right). \quad (4.18)$$

This is a nonsingular system and thus we can solve for h_4. Finally, the critical coefficient d_0 appears in the fifth-order terms:

$$\begin{aligned}(A + I_n)h_5 = {} & 120d_0q - (5B(q,h_4) + 10B(h_2,h_3) + 10C(q,q,h_3) \\ & + 15C(q,h_2,h_2) + 10D(q,q,q,h_2) + E(q,q,q,q,q)).\end{aligned}$$

The solvability of this singular linear system implies

$$\begin{aligned}d_0 = {} & \frac{1}{120}\langle p, 5B(q,h_4) + 10C(q,q,h_3) + 10B(h_2,h_3) + 15C(q,h_2,h_2) \\ & + 10D(q,q,q,h_2) + E(q,q,q,q,q)\rangle. \end{aligned} \quad (4.19)$$

If $d_0 \neq 0$, then the generalized period-doubling (GPD) is non-degenerate.

Remark 4.2 (**Choice of bordering vectors**) Suppose that we had chosen a different solution in (4.17) for h_3, denoted by $\tilde{h}_3$. Then from Section 4.1 it follows that $\tilde{h}_3 = h_3 + sq$ for some $s \in \mathbb{R}$. We also get a different vector h_4 and we have $\tilde{h}_4 = h_4 + 4(I_n - A)^{-}1B(q,sq) = h_4 + 4sh_2$. Evaluation of (4.19) with the new vectors $\tilde{h}_3$ and $\tilde{h}_4$ gives

$$\begin{aligned}\tilde{d}_0 &= \langle p, 5B(q,h_4 + 4sh_2) + 10C(q,q,h_3 + sq) + 10B(h_2,h_3 + sq) \\ &\quad + 15C(q,h_2,h_2) + 10D(q,q,q,h_2) + E(q,q,q,q,q)\rangle \\ &= d_0 + \langle p, 5B(q,4sh_2) + 10C(q,q,sq) + 10B(h_2,sq)\rangle \\ &= d_0 + 10s\langle p, C(q,q,q) + 3B(q,h_2)\rangle = d_0.\end{aligned}$$

Thus the final outcome of (4.19) is the same.

For the cusp bifurcation the argument is even simpler, as we get in a similar manner $\tilde{h}_2 = h_2 + sq$ from (4.10) for some $s \in \mathbb{R}$ and here q is an eigenvector from (4.9). Evaluation of (4.13) now gives

$$\begin{aligned}\tilde{c}_0 &= \langle p, C(q,q,q) + 3B(q,\tilde{h}_2)\rangle = \langle p, C(q,q,q) + 3B(q,h_2 + sq)\rangle \\ &= \langle p, C(q,q,q) + 3B(q,h_2)\rangle + s\langle p, B(q,q)\rangle = c_0.\end{aligned}$$

Thus, we see that the choice for a bordering vector is a matter of taste. Ours results in some vectors being orthogonal and simplifies some expressions.

Chenciner bifurcation

This bifurcation occurs when there is a pair of complex eigenvalues with modulus 1 and the first Lyapunov coefficient L_1 for the Neimark–Sacker bifurcation vanishes. It is also assumed that there are no other critical eigenvalues and no low-order resonances, i.e., $\lambda^k \neq 1$ for $k = 1, 2, 3, 4, 5, 6$. Under these conditions, the normal form of the map (3.14) restricted to the critical center manifold can be written as

$$w \mapsto e^{i\theta_0} w + c_1^0 w|w|^2 + c_2^0 w|w|^4 + O(|w|^6),$$

where $w \in \mathbb{C}$. We choose complex eigenvectors such that

$$\begin{aligned} Aq &= e^{i\theta_0} q, & A\bar{q} &= e^{-i\theta_0} \bar{q}, \\ A^T p &= e^{-i\theta_0} p, & A^T \bar{p} &= e^{i\theta_0} \bar{p}, \end{aligned}$$

where these are scaled such that $\langle p, q \rangle = \bar{p}^T q = 1$. Collecting the $w^j \bar{w}^k$-terms in (4.2) at $\beta = 0$, we find the following equations to be satisfied:

quadratic $(j + k = 2)$:

$$(A - e^{2i\theta_0} I_n) h_{20} = -B(q, q), \tag{4.20}$$

$$(A - I_n) h_{11} = -B(q, \bar{q}), \tag{4.21}$$

cubic $(j + k = 3)$:

$$(A - e^{3i\theta_0} I_n) h_{30} = -C(q, q, q) - 3B(q, h_{20}), \tag{4.22}$$

$$(A - e^{i\theta_0} I_n) h_{21} = 2c_1^0 q - C(q, q, \bar{q}) - B(\bar{q}, h_{20}) - 2B(q, h_{11}). \tag{4.23}$$

The equations from quadratic terms are easily dealt with. At cubic level we see that (4.23) is singular and thus the Fredholm Alternative gives as an intermediate result the familiar expression leading to (2.18) for the coefficient c_1^0:

$$c_1^0 = \frac{1}{2} \langle p, C(q, q, \bar{q}) + 2B(q, (I - A)^{-1} B(q, \bar{q})) + B(\bar{q}, (e^{2i\theta_0} I - A)^{-1} B(q, q)) \rangle.$$

The first Lyapunov coefficient

$$L_1 = \mathrm{Re}(e^{-i\theta_0} c_1^0) \tag{4.24}$$

vanishes at the codim 2 point, while its imaginary part may not vanish. The vector h_{21} is found with the bordering technique.

We proceed by listing the equations at fourth order and the one involving the resonant term leading to the normal form coefficient.

quartic ($j + k = 4$):

$$(A - e^{4i\theta_0} I_n)h_{40} = -\left[D(q,q,q,q) + 6C(q,q,h_{20}) + 3B(h_{20},h_{20}) + 4B(q,h_{30})\right], \tag{4.25}$$

$$\begin{aligned}(A - e^{2i\theta_0} I_n)h_{31} = & -\left[D(q,q,q,\bar{q}) + 3C(q,q,h_{11}) + 3C(q,\bar{q},h_{20}) + 3B(q,h_{21})\right. \\ & \left. +3B(h_{11},h_{20}) + B(\bar{q},h_{30})\right] + 3c_1^0 h_{20} e^{i\theta_0}, \end{aligned} \tag{4.26}$$

$$\begin{aligned}(A - I_n)h_{22} = & -\left[D(q,q,\bar{q},\bar{q}) + C(q,q,h_{02}) + C(\bar{q},\bar{q},h_{20}) + 4C(q,\bar{q},h_{11})\right. \\ & \left. + B(h_{20},h_{02}) + 2B(h_{11},h_{11}) + 2B(q,h_{12}) + 2B(\bar{q},h_{21})\right] \\ & + 2h_{11}(c_1^0 e^{-i\theta_0} + \bar{c}_1^0 e^{i\theta_0}), \end{aligned} \tag{4.27}$$

quintic ($j = 3, k = 2$):

$$\begin{aligned}(A - e^{i\theta_0} I_n)h_{32} = 12c_2^0 q - & \left[E(q,q,q,\bar{q},\bar{q}) + D(q,q,q,h_{02}) + B(h_{02},h_{30})\right. \\ & + 6B(h_{11},h_{21}) + 3B(h_{20},h_{12}) + 6C(q,h_{11},h_{11}) \\ & + 3C(q,q,h_{12}) + 6C(q,\bar{q},h_{21}) + 6C(\bar{q},h_{11},h_{20}) \\ & + C(\bar{q},\bar{q},h_{30}) + 3C(q,h_{20},h_{02}) + 2B(\bar{q},h_{31}) \\ & \left. + 3B(q,h_{22}) + 6D(q,q,\bar{q},h_{11}) + 3D(q,\bar{q},\bar{q},h_{20})\right] \\ & + 6h_{21}(c_1^0 + \tfrac{1}{2}\bar{c}_1^0 e^{2i\theta_0}). \end{aligned}$$

Note that the last line in (4.27) vanishes and that computational efforts can be minimized as $h_{ij} = \bar{h}_{ji}$ and h_{40}, h_{32} need not be computed. Finally, taking into account $\langle p, h_{21}\rangle = 0$, we obtain the expression

$$\begin{aligned}c_2^0 = & \frac{1}{12}\langle p, E(q,q,q,\bar{q},\bar{q}) + D(q,q,q,h_{02}) + 6D(q,q,\bar{q},h_{11}) + 3D(q,\bar{q},\bar{q},h_{20}) \\ & + 3C(q,h_{20},h_{02}) + 6C(q,h_{11},h_{11}) + 3C(q,q,h_{12}) + 6C(q,\bar{q},h_{21}) \\ & + 6C(\bar{q},h_{11},h_{20}) + C(\bar{q},\bar{q},h_{30}) + 3B(h_{20},h_{12}) + 6B(h_{11},h_{21}) \\ & + 3B(q,h_{22}) + B(h_{02},h_{30}) + 2B(\bar{q},h_{31})\rangle. \end{aligned} \tag{4.28}$$

The Chenciner bifurcation is non-degenerate if the second Lyapunov coefficient

$$L_2 = \mathrm{Re}(e^{-i\theta_0} c_2^0) + \frac{1}{2}\left[\mathrm{Im}(e^{-i\theta_0} c_1^0)\right]^2 \tag{4.29}$$

is nonzero.

Resonance 1:1

We have a 1:1 *resonance* if there is one double non-semisimple eigenvalue equal to 1 and no other critical eigenvalues exist. The critical normal form on the center manifold (3.23) can be written as

$$\begin{pmatrix} w_1 \\ w_2 \end{pmatrix} \mapsto \begin{pmatrix} w_1 + w_2 \\ w_2 + a_0 w_1^2 + b_0 w_1 w_2 \end{pmatrix} + O(\|w\|^3), \quad w \in \mathbb{R}^2.$$

We can find (generalized) eigenvectors of the Jacobian matrix A at $w = 0$ such that

$$Aq_0 = q_0, \qquad Aq_1 = q_1 + q_0 \tag{4.30}$$

and similarly for the transposed matrix A^T

$$A^T p_0 = p_0, \qquad A^T p_1 = p_1 + p_0. \tag{4.31}$$

Moreover, they can be chosen such that the following holds

$$\langle p_0, q_1 \rangle = \langle p_1, q_0 \rangle = 1, \quad \langle p_0, q_0 \rangle = \langle p_1, q_1 \rangle = 0. \tag{4.32}$$

Indeed following Kuznetsov (2005) we start with nonzero vectors satisfying (4.30) and (4.31). Computing $\mu = \sqrt{\langle q_0, q_0 \rangle}$ and defining new vectors

$$q_0 := \frac{1}{\mu} q_0, \quad q_1 := \frac{1}{\mu} q_1, \quad q_1 := q_1 - \langle q_0, q_1 \rangle q_0,$$

and analogously adjusting the adjoint vectors with $\nu = \langle p_1, q_0 \rangle$

$$p_0 := \frac{1}{\nu} p_0, \quad p_1 := p_1 - \langle p_1, q_1 \rangle p_0, \quad p_1 := \frac{1}{\nu} p_1,$$

makes them satisfy (4.32).

Collecting the w^2-terms in (4.2), we obtain the singular linear systems

$$(A - I_n)h_{20} = -B(q_0, q_0) + 2a_0 q_1, \tag{4.33}$$

$$(A - I_n)h_{11} = -B(q_0, q_1) + h_{20} + b_0 q_1, \tag{4.34}$$

$$(A - I_n)h_{02} = -B(q_1, q_1) + 2h_{11} + h_{20}. \tag{4.35}$$

The solvability of these systems requires that their right-hand sides should be orthogonal to p_0.

Thus, from (4.33) and (4.34) we obtain

$$a_0 = \frac{1}{2}\langle p_0, B(q_0, q_0) \rangle, \quad b_0 = \langle p_0, B(q_0, q_1) \rangle + \langle p_1, B(q_0, q_0) \rangle, \tag{4.36}$$

where we have used

$$\begin{aligned} \langle p_0, h_{20} \rangle &= \langle (A - I_n)^T p_1, h_{20} \rangle = \langle p_1, (A - I_n)h_{20} \rangle \\ &= \langle p_1, -B(q_0, q_0) + aq_1 \rangle = -\langle p_1, B(q_0, q_0) \rangle. \end{aligned}$$

For completeness, we show how to solve the third equation, although it is not necessary to do so in the codim 2 case. Solvability of (4.35) implies:

$$\begin{aligned}\langle p_0, (A - I_n)h_{02}\rangle &= \langle p_0, -B(q_1, q_1) + 2h_{11} + h_{20}\rangle \\ &= -\langle p_0, B(q_1, q_1)\rangle - \langle p_1, B(q_0, q_0)\rangle - 2\langle p_1, B(q_0, q_1)\rangle \\ &\quad + 2\langle p_1, h_{20}\rangle.\end{aligned}$$

When solving (4.33) we can shift $h_{20} \mapsto h_{20} + \delta_1 q_0$ by adding a multiple of q_0 to h_{20}. Now a proper selection of δ_1 takes care of the solvability of (4.35). For the critical normal form it is, however, not necessary to do so.

Resonance 1:2

Here we have one double non-semisimple eigenvalue -1 and no other critical eigenvalues. In this case, the normal form (3.33) on the center manifold at the critical parameter values reads

$$\begin{pmatrix} w_1 \\ w_2 \end{pmatrix} \mapsto \begin{pmatrix} -w_1 + w_2 \\ -w_2 + c_0 w_1^3 + d_0 w_1^2 w_2 \end{pmatrix} + O(\|w\|^4), \quad w \in \mathbb{R}^2.$$

As before, we introduce generalized eigenvectors of A and A^T

$$\begin{aligned} Aq_0 &= -q_0, & Aq_1 &= -q_1 + q_0, \\ A^T p_0 &= -p_0, & A^T p_1 &= -p_1 + p_0, \end{aligned}$$

satisfying the same normalization conditions (4.32) as for the 1:1 resonance. Collecting the quadratic w-terms in (4.2), we get

$$(A - I_n)h_{20} = -B(q_0, q_0), \tag{4.37}$$

$$(A - I_n)h_{11} = -B(q_0, q_1) - h_{20}, \tag{4.38}$$

$$(A - I_n)h_{02} = -B(q_1, q_1) - 2h_{11} + h_{20}. \tag{4.39}$$

Notice that the matrix $(A - I_n)$ is nonsingular, since A only has one double eigenvalue $\lambda = -1$ on the unit circle. Therefore, solutions for h_{20}, h_{11} and h_{02} are readily found.

From the cubic terms we obtain the equations

$$(A + I_n)h_{30} = 6c_0 q_1 - C(q_0, q_0, q_0) - 3B(q_0, h_{20}), \tag{4.40}$$

$$(A + I_n)h_{21} = 2d_0 q_1 + h_{30} - C(q_0, q_0, q_1) - 2B(q_0, h_{11}) - B(q_1, h_{20}), \tag{4.41}$$

$$(A + I_n)h_{12} = 2h_{21} - h_{30} - C(q_0, q_1, q_1) - 2B(q_1, h_{11}) - B(q_0, h_{02}), \tag{4.42}$$

$$(A + I_n)h_{03} = 3(h_{12} - h_{21}) + h_{30} - C(q_1, q_1, q_1) - 3B(q_1, h_{02}). \tag{4.43}$$

Now we can easily determine the critical coefficient

$$c_0 = \frac{1}{6}\langle p_0, C(q_0, q_0, q_0) + 3B(q_0, (I - A)^{-1}B(q_0, q_0))\rangle. \tag{4.44}$$

The equation for the critical coefficient d_0 involves the vector h_{30}. Taking the scalar product of (4.40) with p_1, we find similarly as for the 1:1 resonance that $\langle p_0, h_{30}\rangle = -\langle p_1, 3B(q_0, h_{20}) + C(q_0, q_0, q_0)\rangle$ from the equation at the w_1^3-term. Then the solvability of the equation for h_{21} implies

$$\begin{aligned} d_0 = {} & \frac{1}{2}\langle p_0, C(q_0, q_0, q_1) + 2B(q_0, h_{11}) + B(q_1, h_{20})\rangle \\ & + \frac{1}{2}\langle p_1, C(q_0, q_0, q_0) + 3B(q_0, h_{20})\rangle. \end{aligned} \tag{4.45}$$

In a similar manner as for the 1:1 resonance, it can be shown that by adding appropriate multiples of q_0 to h_{30} and h_{21}, (4.42) and (4.43) are solvable. However, as stated before, only the quadratic approximation of the center manifold is computed explicitly to obtain the critical normal form coefficients.

Resonance 1:3

Similar to the Chenciner bifurcation, the normal form for the resonance 1:3 has been studied in the complex form. However, one quadratic term in the normal form cannot be eliminated due to the appearance of two complex conjugated eigenvalues which are cubic roots of unity, i.e., the critical multipliers are $e^{i\theta_0}$ with $\pm\theta_0 = 2\pi/3$. The normal form (3.39) on the center manifold can be written at the critical parameter values as

$$w \mapsto e^{\pm i\theta_0} w + b_0 \bar{w}^2 + c_0 w|w|^2 + O(|w|^4), \qquad w \in \mathbb{C}.$$

We introduce complex eigenvectors such that

$$\begin{aligned} Aq &= e^{i\theta_0} q, & A\bar{q} &= e^{-i\theta_0}\bar{q}, \\ A^T p &= e^{-i\theta_0} p, & A^T \bar{p} &= e^{i\theta_0}\bar{p} \end{aligned}$$

and $\langle p, q\rangle = 1$. The quadratic part of the homological equation (4.2) gives

$$(A - e^{2i\theta_0} I_n)h_{20} = 2\bar{b}_0\bar{q} - B(q, q), \tag{4.46}$$

$$(A - I_n)h_{11} = -B(q, \bar{q}), \tag{4.47}$$

$$(A - e^{-2i\theta_0} I_n)h_{02} = 2b_0 q - B(\bar{q}, \bar{q}). \tag{4.48}$$

We notice that (4.46) and (4.48) are complex conjugated, so $\overline{h_{20}} = h_{02}$. Second, since $e^{i2\pi/3}$ is an eigenvalue of A, a singularity occurs. We apply the Fredholm condition to (4.48) to find

$$b_0 = \frac{1}{2}\langle p, B(\bar{q}, \bar{q})\rangle. \tag{4.49}$$

Then h_{02} can be computed using a bordered system. Collecting the cubic terms, we see that the w^3-terms involve the matrix $(A - I_n)$, which is nonsingular. Furthermore, if the equations for h_{30} and h_{21} are solvable, then so are those for h_{03} and h_{12} as they are complex conjugated. Therefore, we only present the $w^2\bar{w}$-terms in the homological equation, since this is all we need to find the critical coefficient c_0,

$$(A - e^{i\theta_0} I_n)h_{21} = 2c_0 q + e^{-i\theta_0}\bar{b}_0 h_{02} - 2B(q, h_{11}) - B(\bar{q}, h_{20}) - C(q, q, \bar{q}).$$

From (4.49) it follows that $\langle p, h_{02}\rangle = 0$, so that we obtain the expression for c_0, which is similar to that of the NS coefficient if $b_0 = 0$ (cf. (2.18)

$$\begin{aligned} c_0 = \frac{1}{2}\langle p, \big(C(q, q, \bar{q}) + 2B(q, (I - A)^{-1} B(q, \bar{q}))\\ -B(\bar{q}, \left(e^{2i\theta_0} I - A\right)^{INV} \left(2\bar{b}_0\bar{q} - B(q, q)\right)\big)\rangle. \end{aligned} \tag{4.50}$$

If $b_0 \neq 0$, the sign of $a_2(0) = \Re\left(e^{-i\theta_0} c_0/|b_0|^2 - 1\right)$ determines the stability of the bifurcating invariant closed curve.

Resonance 1:4

With $\theta_0 = \frac{1}{2}\pi$ we encounter one of the most difficult bifurcations to analyze, as not all non-degeneracy conditions can be expressed analytically. Still, it is straightforward to compute the coefficients of the critical normal form

$$w \mapsto iw + c_0 w|w|^2 + d_0\bar{w}^3 + O(|w|^4), \qquad w \in \mathbb{C},$$

which is (3.48) at the critical parameter values. The linear level of the homological equation asks to introduce complex eigenvectors satisfying

$$Aq = iq, \qquad A\bar{q} = -i\bar{q}, \qquad A^T p = -ip, \qquad A^T\bar{p} = i\bar{p}, \quad \langle p, q\rangle = 1.$$

The quadratic part of (4.2) gives

$$(A + I_n)h_{20} = -B(q, q), \tag{4.51}$$

$$(A - I_n)h_{11} = -B(q, \bar{q}). \tag{4.52}$$

Since ± 1 are not the eigenvalues of A, we can easily find h_{20}, h_{11}. Now, as previously, we only collect the equations from the resonant terms:

$$(A - i\, I_n)h_{21} = 2c_0 q - C(q, q, \bar{q}) - 2B(q, h_{11}) - B(\bar{q}, h_{20}), \tag{4.53}$$

$$(A - i\, I_n)h_{03} = 6d_0 q - C(\bar{q}, \bar{q}, \bar{q}) - 3B(\bar{q}, h_{02}). \tag{4.54}$$

The solvability conditions imply

$$c_0 = \frac{1}{2}\langle p, C(q, q, \bar{q}) + 2B(q, (I - A)^{-1} B(q, \bar{q})) - B(\bar{q}(I + A)^{-1} B(q, q))\rangle, \tag{4.55}$$

and

$$d_0 = \frac{1}{6}\langle p, C(\bar{q}, \bar{q}, \bar{q}) - 3B(\bar{q}, (I + A)^{-1} B(\bar{q}, \bar{q}))\rangle. \tag{4.56}$$

If $d_0 \neq 0$, then the quantity

$$A_0 = -\frac{ic_0}{|d_0|}$$

determines the bifurcation scenario near the 1:4 point. Some non-degeneracy conditions involving local bifurcations can be expressed analytically in terms of A_0 (see Section 3.3). As the global bifurcations in this unfolding have only been computed numerically, one should still be careful when classifying the bifurcation.

Fold–flip

This bifurcation is characterized by two simple multipliers on the unit circle, one $+1$ and one -1. As this may sound simple, it is an interesting case for two reasons. First, we need to apply the Fredholm alternative to all coefficients up to order 3 of the homological equation. Second, it is the first case in the list of Chapter 3 where hyper-normalization simplifies the bifurcation analysis.

Instead of the hyper-normal form (3.59) we use the 3-jet of the critical Poincaré normal form

$$\begin{pmatrix} w_1 \\ w_2 \end{pmatrix} \mapsto \begin{pmatrix} w_1 + a_1 w_1^2 + b_1 w_2^2 + c_1 w_1^3 + c_2 w_1 w_2^2 \\ -w_2 + e_1 w_1 w_2 + c_3 w_1^2 w_2 + c_4 w_2^3 \end{pmatrix}, \quad w \in \mathbb{R}^2, \tag{4.57}$$

in the homological equation.

We can introduce the vectors associated to the eigenvalues $+1$ and -1

$$Aq_1 = q_1, \; A^T p_1 = p_1, \; \langle p_1, q_1 \rangle = 1, \tag{4.58}$$

$$Aq_2 = -q_2, A^T p_2 = -p_2, \langle p_2, q_2 \rangle = 1. \tag{4.59}$$

Let us list first the equations obtained from (4.2) of order 2

$$(A - I_n)h_{20} = 2a_1 q_1 - B(q_1, q_1), \tag{4.60}$$

$$(A + I_n)h_{11} = e_1 q_2 - B(q_1, q_2), \tag{4.61}$$

$$(A - I_n)h_{02} = 2b_1 q_1 - B(q_2, q_2), \tag{4.62}$$

and those of order 3

$$(A - I_n)h_{30} = 6c_1 q_1 + 6a_1 h_{20} - C(q_1, q_1, q_1) - 3B(q_1, h_{20}), \tag{4.63}$$

$$\begin{aligned}(A - I_n)h_{12} = {} & 2c_2 q_1 + 2b_1 h_{20} - 2e_1 h_{02} \\ & - C(q_1, q_2, q_2) - 2B(q_2, h_{11}) - B(q_1, h_{02}),\end{aligned} \tag{4.64}$$

$$\begin{aligned}(A + I_n)h_{21} = {} & 2c_3 q_2 + 2\,(e_1 - a_1)\,h_{11} \\ & - C(q_1, q_1, q_2) - 2B(q_1, h_{11}) - B(q_2, h_{20}),\end{aligned} \tag{4.65}$$

$$(A + I_n)h_{03} = 6c_4 q_2 - 6b_1 h_{11} - C(q_2, q_2, q_2) - 3B(q_2, h_{02}). \tag{4.66}$$

Now we first present the standard scheme for a solution to (4.60)–(4.66). For the solvability requirement we apply the proper adjoint eigenvector to the left- and right-hand sides of each equation. Thus, one first computes

$$a_1 = \frac{1}{2}\langle p_1, B(q_1,q_1)\rangle, \quad e_1 = \langle p_2, B(q_1,q_2)\rangle, \quad b_1 = \frac{1}{2}\langle p_1, B(q_2,q_2)\rangle. \tag{4.67}$$

Then we have to compute the quadratic approximation of the center manifold. The bordering technique can be used to find

$$\begin{aligned} h_{20} &= (A - I_n)^{INV}(2a_1q_1 - B(q_1,q_1)), \\ h_{11} &= (A + I_n)^{INV}(e_1q_2 - B(q_1,q_2)), \\ h_{02} &= (A - I_n)^{INV}(2b_1q_1 - B(q_2,q_2)), \end{aligned}$$

where we now have

$$\langle p_1, h_{20}\rangle = \langle p_2, h_{11}\rangle = \langle p_1, h_{02}\rangle = 0. \tag{4.68}$$

The other critical normal form coefficients c_i for (4.57) are given by

$$\begin{aligned} c_1 &= \frac{1}{6}\langle q_1, C(q_1,q_1,q_1) + 3B(q_1,h_{20})\rangle, \\ c_2 &= \frac{1}{2}\langle q_1, C(q_1,q_2,q_2) + B(q_1,h_{02}) + 2B(q_2,h_{11})\rangle, \\ c_3 &= \frac{1}{2}\langle p_2, C(q_1,q_1,q_2) + B(q_2,h_{20}) + 2B(q_1,h_{11})\rangle, \\ c_4 &= \frac{1}{6}\langle p_2, C(q_2,q_2,q_2) + 3B(q_2,h_{02})\rangle. \end{aligned} \tag{4.69}$$

The final step is to express the non-degeneracy conditions in terms of the critical normal form coefficients. First, provided that $e_1 \neq 0$, we can now apply Lemma 3.31 to find the coefficients for (3.59)

$$a_0 = \frac{a_1}{e_1}, \; b_0 = b_1e_1, \; c_0 = \frac{c_1}{e_1^2}, \; d_0 = c_2 + \frac{2}{e_1}\left(b_1c_3 - (e_1 + a_1)c_4\right). \tag{4.70}$$

The non-degeneracy conditions are $a_0, b_0 \neq 0$. If $b_0 > 0$, then also $3b_0c_0 - 3a_0b_0 - 2b_0a_0^2 - d_0a_0 \neq 0$ is required.

Remark 4.3 The second scheme to compute the coefficients a_0, b_0, c_0, d_0 of the map (3.59) starts similarly as above. However, if we wish to insert the hyper-normal form (3.59) directly into the homological equation we should take care of the solvability of (4.63)–(4.66). To ensure this we use the freedom for the quadratic center manifold vectors. So we have

$$h_{20} = h_{20}^* + \delta_1q_1, \quad h_{11} = h_{11}^* + \delta_2q_2, \quad h_{02} = h_{02}^* + \delta_3q_1,$$

where $h_{20}^*, h_{11}^*, h_{02}^*$ satisfy (4.68). Denote c_i of (4.69) by c_i^*.

Then, (4.63)–(4.66) with $c_3 = c_4 = 0$ lead to

$$\begin{pmatrix} 6c_1 - 6c_1^* \\ 2c_2 + 2\delta_1 b_1 - 2(e_1 + a_1)\delta_3 - 4b_1\delta_2 - 2c_2^* \\ (2e_1 - 2a_1)\delta_2 - 2e_1\delta_2 - e_1\delta_1 - 2c_3^* \\ -6b_1\delta_2 - 3e_1\delta_3 - 6c_4^* \end{pmatrix} = 0.$$

The first observation is that c_1 cannot be altered, similar to what was shown for the cusp coefficient in Remark 4.2. If we represent this a little differently we make the connection with Lemma 3.31

$$\begin{pmatrix} 2b_1 & -b_1 & e_1 + a_1 \\ 2a_1 & e_1 & 0 \\ 2b_1 & 0 & e_1 \end{pmatrix} \begin{pmatrix} \delta_2 \\ \delta_1 \\ \delta_3 \end{pmatrix} = \begin{pmatrix} c_2 - c_2^* \\ -c_3^* \\ -c_4^* \end{pmatrix} \tag{4.71}$$

as the matrix is easily identified with the one in (3.125). Therefore we take $\delta_2 = 0$ and now the assumption $e_1 \neq 0$ allows one to obtain a solution for δ_1, δ_3 and c_2 from (4.71). The final result is the same as in the first computational scheme. Thus, this remark shows how hyper-normalization can be incorporated into center manifold reduction.

Fold–Neimark–Sacker

This bifurcation occurs for a fixed point having the algebraically simple multipliers $\{1, e^{i\theta_0}, e^{-i\theta_0}\}$ on the unit circle, where $(k\theta \bmod 2\pi) \neq 0$ for $k = 1, 2, 3, 4$. Then, there is a smooth local parametrization of the critical center manifold by $(w_1, w_2) \in \mathbb{R} \times \mathbb{C}$ such that the 3-jet of the map will assume the form

$$\begin{pmatrix} w_1 \\ w_2 \end{pmatrix} \mapsto \begin{pmatrix} w_1 + f_{011}w_2\bar{z} + f_{200}w_1^2 + f_{300}w_1^3 + f_{111}w_1w_2\bar{w}_2 \\ e^{i\theta_0}w_2 + g_{110}w_1w_2 + g_{210}w_1^2w_2 + g_{021}w_2^2\bar{w}_2 \end{pmatrix}. \tag{4.72}$$

We proceed by giving the formulas directly. We start with

$$\begin{aligned} Aq_1 &= q_1, & A^T p_1 &= p_1, & \langle p_1, q_1\rangle &= 1, \\ Aq_2 &= e^{i\theta_0}q_2, & A^T p_2 &= e^{-i\theta_0}p_2, & \langle p_2, q_2\rangle &= 1 \end{aligned}$$

and the quadratic coefficients are given by

$$f_{200} = \frac{1}{2}\langle p_1, B(q_1, q_1)\rangle, \quad f_{011} = \langle p_1, B(q_2, \bar{q}_2)\rangle, \quad g_{110} = \langle p_2, B(q_1, q_2)\rangle. \tag{4.73}$$

Then the following vectors should be computed

$$\begin{aligned} h_{200} &= (A - I_n)^{INV}(2f_{200}q_1 - B(q_1, q_1)), \\ h_{011} &= (A - I_n)^{INV}(f_{011}q_1 - B(q_2, \bar{q}_2)), \\ h_{110} &= (A - e^{i\theta_0}I_n)^{INV}(g_{110}q_2 - B(q_1, q_2)), \\ h_{002} &= (e^{2i\theta_0}I_n - A)^{-1}B(q_2, q_2), \end{aligned}$$

where the bordering vectors are of course chosen differently for $(A - I_n)^{INV}$ and $(A - e^{i\theta_0} I_n)^{INV}$. For the Poincaré normal form we then have

$$\begin{aligned} f_{300} &= \tfrac{1}{6}\langle p_1, C(q_1,q_1,q_1) + 3B(q_1,h_{200})\rangle, \\ f_{111} &= \langle p_1, C(q_1,q_2,\bar{q}_2) + B(q_1,h_{011}) + 2\Re(B(q_2,h_{101}))\rangle, \\ g_{210} &= \tfrac{1}{2}\langle p_2, C(q_1,q_1,q_2) + B(q_2,h_{200}) + 2B(q_1,h_{110})\rangle, \\ g_{021} &= \tfrac{1}{2}\langle p_2, C(q_2,q_2,\bar{q}_2) + B(q_2,h_{011}) + 2B(\bar{q}_2,h_{020})\rangle. \end{aligned} \tag{4.74}$$

Remark 4.4 If $f_{200} f_{011} \neq 0$, we may further transform (4.72) to the critical hyper-normal form with the 3-jet

$$\begin{pmatrix} x \\ z \end{pmatrix} \mapsto \begin{pmatrix} x + x^2 + s|z|^2 + c_0 x^3 \\ e^{i\theta} z + a_0 xz + b_0 z x^2 \end{pmatrix}, \tag{4.75}$$

where the coefficients are given by

$$\begin{aligned} b_0 &= \frac{1}{f_{011} f_{200}^2}\left(f_{011} g_{210} + g_{110}\left(\frac{1}{2} f_{111} + \Re\left(g_{021} e^{-i\theta}\right)\right) - f_{200} g_{021}\right), \\ a_0 &= \frac{g_{110}}{f_{200}}, \quad c_0 = \frac{f_{300}}{(f_{200})^2}, \quad s = \operatorname{sign}(f_{200} f_{011}). \end{aligned}$$

If we would not have fixed the vectors $h_{200}, h_{011}, h_{110}$, then hyper-normalization as in Remark 4.3 and scaling would have given the same result.

Flip–Neimark–Sacker

At this bifurcation the Jacobian matrix evaluated at the fixed point has three multipliers on the unit circle $\lambda \in \{-1, e^{i\theta_0}, e^{-i\theta_0}\}$, where $(k\theta_0 \bmod 2\pi) \neq 0$ for $k = 1,2,3,4,6$. This suffices for most cases; in the following we will specify what extra requirements might need to be satisfied. Write the 5-jet of the critical Poincaré normal form as

$$\begin{pmatrix} w_1 \\ w_2 \end{pmatrix} \mapsto \begin{pmatrix} w_1\left(-1 + f_{300} w_1^2 + f_{111}|z|^2 + f_{500} w_1^4 + f_{311} w_1^2 |w_2|^2 + f_{122}|w_2|^4\right) \\ w_2\left(e^{i\theta_0} + g_{210} w_1^2 + g_{021}|w_2|^2 + g_{410} w_1^4 + g_{221} w_1^2 |w_2|^2 + g_{032}|w_2|^4\right) \end{pmatrix}$$

with $(w_1, w_2) \in \mathbb{R} \times \mathbb{C}$.

As before, we now introduce the eigenvectors related to the multipliers $\lambda = -1$ and $\lambda = e^{i\theta_0}$

$$\begin{aligned} Aq_1 &= -q_1, & A^T p_1 &= -p_1, & \langle p_1, q_1\rangle &= 1, \\ Aq_2 &= e^{i\theta_0} q_2, & A^T p_2 &= e^{-i\theta_0} p_2, & \langle p_2, q_2\rangle &= 1. \end{aligned}$$

Collecting the quadratic w-terms in (4.2) at $\beta = 0$, we find

$$\begin{aligned} h_{200} &= (I_n - A)^{-1} B(q_1, q_1), \\ h_{011} &= (I_n - A)^{-1} B(q_2, \bar{q}_2), \\ h_{110} &= -(e^{i\theta_0} I_n + A)^{-1} B(q_1, q_2), \\ h_{020} &= (e^{2i\theta_0} I_n - A)^{-1} B(q_2, q_2), \end{aligned}$$

after which the cubic coefficients are computed as follows:

$$\begin{aligned} f_{300} &= \tfrac{1}{6}\langle p_1, C(q_1,q_1,q_1) + 3B(q_1,h_{200})\rangle, \\ f_{111} &= \langle p_1, C(q_1,q_2,\bar{q}_2) + B(q_1,h_{011}) + 2\Re(B(q_2,h_{101}))\rangle, \\ g_{210} &= \tfrac{1}{2}\langle p_2, C(q_1,q_1,q_2) + B(q_2,h_{200}) + 2B(q_1,h_{110})\rangle, \\ g_{021} &= \tfrac{1}{2}\langle p_2, C(q_2,q_2,\bar{q}_2) + B(q_2,h_{011}) + 2B(\bar{q}_2,h_{020})\rangle. \end{aligned} \tag{4.76}$$

Not surprisingly, we see a great similarity between these coefficients (4.76) and the third-order coefficients (4.74) for the fold-Neimark–Sacker bifurcation.

Remark 4.5 If, for this flip–Neimark–Sacker bifurcation, we find that

$$f_{300}\Re(e^{-i\theta_0}g_{021}) > 0 \quad \text{and} \quad f_{300}\Re(e^{-i\theta_0}g_{021}) - f_{111}\Re(e^{-i\theta_0}g_{111}) < 0$$

hold, then there is also an invariant 2-torus involved in the unfolding of the bifurcation. Its stability can only be determined if we proceed with quartic and quintic terms, which can be found in this chapter's Appendix.

Double Neimark–Sacker

When two pairs of complex conjugate multipliers $e^{\pm i\theta_{1,2}}$, $0 < \theta_2 < \theta_1 < \pi$ cross the unit circle, a double Neimark–Sacker bifurcation occurs. Similar to the previous case, we first require $(k\theta_i \bmod 2\pi) \neq 0$ for $k = 1,2,3,4$, $i = 1,2$ and $(\theta_1/\theta_2) \notin \{3, 2, \frac{3}{2}, 1\}$. We compute the (adjoint) eigenvectors for the multipliers and scale them properly

$$\begin{aligned} Aq_1 &= e^{i\theta_1}q_1, \quad A^T p_1 = e^{-i\theta_1}p_1, \quad \langle p_1, q_1\rangle = 1, \\ Aq_2 &= e^{i\theta_2}q_2, \quad A^T p_2 = e^{-i\theta_2}p_2, \quad \langle p_2, q_2\rangle = 1. \end{aligned}$$

Now we can compute the third-order coefficients of the critical normal form

$$\begin{pmatrix} w_1 \\ w_2 \end{pmatrix} \mapsto \begin{pmatrix} w_1\left(e^{i\theta_1} + f_{2100}|w_1|^2 + f_{1011}|w_2|^2\right) \\ w_2\left(e^{i\theta_2} + g_{1110}|w_1|^2 + g_{0021}|w_2|^2\right) \end{pmatrix} + \begin{pmatrix} w_1\left(f_{3200}|w_1|^4 + f_{2111}|w_1|^2|w_2|^2 + f_{1022}|w_2|^4\right) \\ w_2\left(g_{2201}|w_1|^4 + g_{1121}|w_1|^2|w_2|^2 + g_{0032}|w_2|^4\right) \end{pmatrix} + O(\|w\|^6),$$

where $w \in \mathbb{C}^2$, with the following formulas:

$$\begin{aligned} f_{2100} &= \tfrac{1}{2}\langle p_1, C(q_1,q_1,\bar{q}_1) + B(q_1,h_{1100}) + 2B(\bar{q}_1,h_{2000})\rangle, \\ f_{1011} &= \langle p_1, C(q_1,q_2,\bar{q}_2) + B(q_1,h_{0011}) + B(\bar{q}_2,h_{1010}) + B(q_2,h_{1001})\rangle, \\ g_{1110} &= \langle p_2, C(q_1,\bar{q}_1,q_2) + B(q_2,h_{1100}) + B(\bar{q}_1,h_{1010}) + B(q_1,h_{0110})\rangle, \\ g_{0021} &= \tfrac{1}{2}\langle p_2, C(q_2,q_2,\bar{q}_2) + B(q_2,h_{0011}) + 2B(\bar{q}_2,h_{0020})\rangle, \end{aligned} \tag{4.77}$$

where

$$\begin{aligned} h_{2000} &= (e^{2i\theta_1} I_n - A)^{-1} B(q_1, q_1), & h_{0020} &= (e^{2i\theta_2} I_n - A)^{-1} B(q_2, q_2), \\ h_{1100} &= (I_n - A)^{-1} B(q_1, \bar{q}_1), & h_{0011} &= (I_n - A)^{-1} B(q_2, \bar{q}_2), \\ h_{1010} &= (e^{i(\theta_1+\theta_2)} I_n - A)^{-1} B(q_1, q_2), & h_{1001} &= (e^{i(\theta_1-\theta_2)} I_n - A)^{-1} B(q_1, \bar{q}_2). \end{aligned} \tag{4.78}$$

Vectors h_{0110} and h_{0101} can be computed by complex conjugation. The fifth-order coefficients are only needed if one wants to investigate stability of a 3-torus, if it is present. For completeness they are listed in the Appendix.

4.3 Branch switching at local codim 2 bifurcations

When studying a dynamical system, the calculation of the critical normal form at the bifurcation gives qualitative information about what kind of new solutions may be expected in nearby families. The question "Nice to know, but where exactly?" is the one we will answer in this section, at least for local bifurcations.

As we mentioned in Chapter 3, it is known that from codim 2 points several new codim 1 branches emanate. In 6 out of 11 cases, branches of local bifurcations of cycles are involved. In these cases we can provide a useful prediction for such codim 1 branches, if the condition of transversality is satisfied. All other branches involve bifurcations of tori or homo- and/or heteroclinic orbits. Starting continuation of curves of such bifurcations is nontrivial, and we skip that here.

We first discuss the classical approach to branch-switching. It is also used in MatcontM to switch at a PD-point for the period-K cycle to the period-$2K$ cycle, since it corresponds to a branch point for $f^{(2K)}(x, \alpha) - x = 0$. Our formulas for the new branches coincide at lowest order with those found by linear branch switching. However, we often experienced that using only a linear approximation was not enough to start the continuation of the new branch. A higher-order predictor provided by center manifold reduction did work, but still the choice for the initial amplitude needs some experience.

4.3.1 Linear branch-switching

In this section we consider the approximation of a new cycle curve that emanates from a branch point BP for (3.2). The method is similar to that for branch points of equilibria, already used in the 1970s by Keller (1977); it is also used in content and Matcont.

A solution $X_0 = X(s_0)$ of

$$F(X) = g(x, \alpha) - x = 0 \tag{4.79}$$

is called a *simple singular point* if $F_X(X_0)$ has rank $n-1$. For system (4.79), we have $F_X^0 = [g_x(x_0, \alpha_0) - I_n, g_\alpha(x_0, \alpha_0)]$, and $X_0 = (x_0, \alpha_0)$ is a simple singular point if and only if, either

$$\dim N(g_x(x_0, \alpha_0) - I_n) = 1, \;\; g_\alpha(x_0, \alpha_0) \in R(g_x(x_0, \alpha_0) - I_n)$$

or

$$\dim N(g_x(x_0, \alpha_0) - I_n) = 2, \;\; g_\alpha(x_0, \alpha_0) \notin R(g_x(x_0, \alpha_0) - I_n).$$

The first case is a codim 2 situation, the second case has codimension 4, so for practical purposes we consider only the first case.

Suppose we have a solution branch $X(s)$ and let $X(s_0) = (x_0, \alpha_0)$ be a simple singular point. Then $N(F_X^0)$ is two-dimensional and can be written as span $\{\phi_1, \phi_2\}$ where $\phi_1, \phi_2 \in \mathbb{R}^{n+1}$ are linearly independent. Also, $N([F_X^0]^T)$ is one-dimensional and is spanned by a vector $\psi \in \mathbb{R}^n$. Let F_{YY}^0 be the bilinear form in the Taylor expansion of F about X_0. If $Y(s)$ is any solution branch of (4.79) with $Y(s_0) = X_0$, then $Y_s(s_0)$ can be written as $Y_s(s_0) = \alpha\phi_1 + \beta\phi_2$ for some $\alpha, \beta \in \mathbb{R}$. Differentiating the identity $F(Y(s)) = 0$ twice and computing the scalar product with ψ at s_0, we get

$$\langle \psi, F_{YY}^0(\alpha\phi_1 + \beta\phi_2)(\alpha\phi_1 + \beta\phi_2)\rangle = 0$$

or, equivalently,

$$c_{11}\alpha^2 + 2c_{12}\alpha\beta + c_{22}\beta^2 = 0, \tag{4.80}$$

where $c_{jk} = \langle \psi, F_{YY}^0 \phi_j \phi_k \rangle$ for $j, k = 1, 2$.

Equation (4.80) is called the *algebraic bifurcation equation* (ABE). The case $c_{12}^2 - c_{11}c_{22} < 0$ is impossible, since at least one branch goes through X_0. Thus, generically, $c_{12}^2 - c_{11}c_{22} > 0$, and (4.80) has two real nontrivial, independent solution pairs, (α_1, β_1) and (α_2, β_2), which are unique up to scaling. In this case we have a *simple branch point*, where two distinct branches pass through X_0.

The above procedure allows one to compute the normalized tangent vectors $Y_{1s}(s_0), Y_{2s}(s_0)$ of the two branches that pass through X_0. Now if

$$|\langle Y_{1s}(s_0), X_s(s_0)\rangle| < |\langle Y_{2s}(s_0), X_s(s_0)\rangle|$$

then we conclude that $Y_{1s}(s_0)$ is the tangent vector to the new branch; otherwise, $Y_{2s}(s_0)$ is the tangent vector.

Remark 4.6 Let us consider the normal form (3.33) for the 1:2 resonance. If the normal form coefficient c_0 is negative, there are two branches of codim 1

Neimark–Sacker bifurcation at this bifurcation; one of the original fixed point and one of a cycle of double period. If c_0 is positive, the branch corresponds to a neutral saddle. The defining system for this bifurcation is

$$F(x,\alpha,k) = (g^2(x,\alpha) - x, s_{11}(x,\alpha,k), s_{12}(x,\alpha,k)),$$

(see Section 5.2 for the definition of s_{11} and s_{12}). Note that $(0,0,0,0,1)$ is a zero of this equation. The Jacobian matrix is given by

$$DF(0,0,0,0,1) = \begin{pmatrix} 0 & -2 & 0 & 0 & 0 \\ 0 & 0 & 0 & 0 & 0 \\ 0 & 0 & 4 & 0 & -2 \\ 0 & 0 & -4 & 4 & 4 \end{pmatrix}.$$

In the notation as above, its null-space is spanned by $\phi_1 = (0,0,1,-1,2)$ and $\phi_2 = (1,0,0,0,0)$ and $\psi = (0,1,0,0)$. Obviously ϕ_1 corresponds to the original branch, ϕ_2 to the new one.

4.3.2 Parameter-dependent center manifold reduction

Our approach here is similar to that of Beyn et al. (2002), where switching at some codim 2 bifurcations of equilibria in ODEs was considered. In all cases ahead, the map $g(x,\alpha)\colon \mathbb{R}^n \times \mathbb{R}^2 \to \mathbb{R}^n$, where g is defined by (3.2), satisfies $g(x_0,\alpha_0) = x_0$, and its Jacobian matrix $A = g_x(x_0,\alpha_0)$ has at most three multipliers on the unit circle. We know a parameter-dependent smooth normal form $G(w,\beta)$ for the corresponding bifurcation (see Chapter 3). With this setup all necessary vectors for a higher-order predictor can be obtained from the *homological equation* (4.2).

It will be convenient to introduce some notation. If A has an eigenvalue -1, then denote by p the eigenvector of A^T corresponding to the eigenvalue -1. We will then write $\Gamma\colon \mathbb{R}^{n+2} \to \mathbb{R}^n$ for $\Gamma(q,v) = \langle p, A_1(q,v) + B(q,(I_n - A)^{-1}J_1 v)\rangle$ and $\gamma_i = \Gamma(q,e_i)$ for the evaluation on the standard basis vectors in $\mathbb{R}^2$. If $\gamma_i \neq 0$ for $i = 1,2$, then $s_1 = \frac{1}{(\gamma_1^2+\gamma_2^2)}(\gamma_1,\gamma_2)$ and $s_2 = (-\gamma_2,\gamma_1)$ compose a new orthogonal basis in $\mathbb{R}^2$. Occasionally, we interpret $\beta_2 = \bar{\beta}_1$ as one complex parameter; in such cases: $v_{01} = \overline{v_{10}} \in \mathbb{C}^2$.

Generalized period-doubling

We start with the linear part of $V(\beta)$. The homological equation (4.2) provides the following systems to be solved

$$(A - I_n)[h_{010},\ h_{001}] = -J_1[v_{10},\ v_{01}],$$

where $[a,\ b]$ is a $2{\times}n$ matrix with columns a and b. This can be solved formally with $[h_{010},\ h_{001}] = (I_n - A)^{-1}J_1[v_{10},\ v_{01}]$ and we proceed with

$$\begin{aligned}(A + I_n)[h_{110},\ h_{101}] &= -\,[q,\ 0] - A_1(q,[v_{10},\ v_{01}]) - B(q,[h_{010},\ h_{001}])\\ &= -\,[q,\ 0] - A_1(q,[v_{10},\ v_{01}]) - B(q,(I_n - A)^{-1}J_1[v_{10},\ v_{01}]).\end{aligned}$$

First we see that the systems are singular and the right-hand side must be orthogonal to the adjoint eigenvector p. Second, we see the operator $\Gamma(q,v)$ appearing naturally. The system is now rewritten as

$$(\gamma_1,\ \gamma_2)[v_{10},\ v_{01}] = [-1,\ 0].$$

The general solution is given by

$$v_{10} = -s_1 + \delta_1 s_2, \qquad v_{01} = \delta_2 s_2, \qquad \delta_1, \delta_2 \in \mathbb{R}.$$

The constants δ_1, δ_2 can only be fixed at a higher order, so we proceed

$$\begin{aligned}(A - I_n)h_{210} &= +2h_{200} - \big[C(q,q,h_{010}) + 2B(q,h_{110}) + B(h_{200},h_{010})\\ &\qquad + B_1(q,q,v_{10}) + A_1(h_{200},v_{10})\big],\\ (A - I_n)h_{201} &= -\big[C(q,q,h_{001}) + 2B(q,h_{101}) + B(h_{200},h_{001})\\ &\qquad + B_1(q,q,v_{01}) + A_1(h_{200},v_{01})\big],\\ (A + I_n)h_{310} &= -3h_{300} - \big[D(q,q,q,h_{010}) + 3C(q,q,h_{110}) + 3C(h_{200},q,h_{010}),\\ &\qquad + 3B(h_{110},h_{200}) + 3B(h_{210},q) + B(h_{300},h_{010})\\ &\qquad - +C_1(q,q,q,v_{10}) + 3B_1(h_{200},q,v_{10}) + A_1(h_{300},v_{10})\big],\\ (A + I_n)h_{301} &= 6q - \big[D(q,q,q,h_{001}) + 3C(q,q,h_{101}) + 3C(h_{200},q,h_{001})\\ &\qquad + 3B(h_{101},h_{200}) + 3B(h_{201},q) + B(h_{300},h_{001})\\ &\qquad + C_1(q,q,q,v_{01}) + 3B_1(h_{200},q,v_{01}) + A_1(h_{300},v_{01})\big].\end{aligned} \tag{4.81}$$

Since v_{10}, v_{01} appear linearly in these equations (via the multilinear functions), we have a linear system to be solved for δ_1, δ_2.

Remark 4.7 From the expansion (3.13) it is easy to see that the next order to be computed is $O(\varepsilon^4)$. We will not do this here for the following reason. Consider the following extension of the normal form for (3.10):

$$w \mapsto -(1+\beta_1+r_1\beta_2^2)w + (\beta_2+r_2\beta_2^2)w^3 + (c_2+r_3\beta_2)w^5 + c_3w^7 + O(|w|^8), \quad w \in \mathbb{R}.$$

Taking $w = \varepsilon$ we find the following higher-order approximation for the parameters for the fold curve of the second iterate:

$$(\beta_1,\beta_2) = \left(-c_2(1+4c_2r_1)\varepsilon^4, -2c_2\varepsilon^2 + (4c_2r_3 - 4r_2c_2^2 - 3c_3)\varepsilon^4\right) + O(\varepsilon^5).$$

Thus, in the higher-order approximation the seventh-order coefficient c_3 appears. Therefore, we do not give a higher-order approximation here.

1:2 Resonance

As before, the first four linear systems are given as

$$\begin{aligned}(A - I_n)[h_{0010},\ h_{0001}] &= -J_1[v_{10},\ v_{01}],\\ (A + I_n)[h_{1010},\ h_{1001}] &= [q_1,\ 0] - A_1(q_0, [v_{10},\ v_{01}]) - B(q_0, [h_{0010},\ h_{0001}]).\end{aligned}$$

As for the degenerate flip, we use the formal solution

$$[h_{0010},\ h_{0001}] = (I_n - A)^{-1} J_1[v_{10},\ v_{01}].$$

The solution for v_{10} and v_{01} is now

$$v_{10} = s_1 + \delta_1 s_2, \quad v_{01} = \delta_2 s_2, \quad \delta_1, \delta_2 \in \mathbb{R}.$$

The two remaining systems at linear order in phase variables are

$$\begin{aligned}(A + I_n)[h_{0110},\ h_{0101}] = [h_{1010},\ q_1 + h_{1001}] &- A_1(q_1, [v_{10},\ v_{01}])\\ &- B(q_1, [h_{0010},\ h_{0001}]).\end{aligned}$$

Now we insert s_i and write

$$\begin{aligned}Q_1 &= \langle p_1, A_1(q_0, s_1) + B(q_0, (A - I_n)^{-1} J_1 s_1)\rangle, Q_2 = \Gamma(q_1, s_1),\\ Q_3 &= \langle p_1, A_1(q_0, s_2) + B(q_0, (A - I_n)^{-1} J_1 s_2)\rangle, Q_4 = \Gamma(q_1, s_2).\end{aligned}$$

A little algebra shows that

$$\delta_1 = -\left(\frac{Q_1 + Q_2}{Q_3 + Q_4}\right), \quad \delta_2 = \frac{1}{Q_3 + Q_4}.$$

One can check that the transversality of the original family of maps to the bifurcation manifold coincides with the condition $\gamma_1\gamma_2(Q_3 + Q_4) \neq 0$.

1:3 Resonance

We follow a slightly different procedure here. We want to find $V(\beta) = v\beta + \bar{v}\bar{\beta}$, where $\beta = \beta_1 + i\beta_2$. Then we treat β and $\bar{\beta}$ as independent variables which makes it slightly easier to find the solutions. As the final $V(\beta)$ should be real, it follows that $v = v_{10} = \bar{v}_{01}$.

Let $\lambda = e^{2i\pi/3}$ and introduce $Aq = \lambda q$, $A^T p = \bar{\lambda} p$, $\langle p, q\rangle = 1$. As before, the first linear systems resulting from (4.2) are given by

$$\begin{aligned}(A - I_n)[h_{0010},\ h_{0001}] &= -J_1[v_{10},\ v_{01}],\\ (A - \lambda I_n)[h_{1010},\ h_{1001}] &= [q,\ 0] - A_1(q, [v_{10},\ v_{01}]) - B(q, [h_{0010},\ h_{0001}]),\end{aligned}$$

and two complex conjugated systems for h_{0101} and h_{0110}. With the same approach we will now find complex γ_i; rewriting the system for $v = v_{10} = \bar{v}_{01}$ we have $(\gamma_1, \gamma_2)v = 1, (\gamma_1, \gamma_2)\bar{v} = 0$, with $v = (\bar{\gamma}_2, -\bar{\gamma}_1)/(\gamma_1\bar{\gamma}_2 - \gamma_2\bar{\gamma}_1)$ as the solution. Finally $x = zq + \bar{z}\bar{q}$ relates the coordinates of the normal form and the original map.

1:4 Resonance

Replacing $\lambda = i$ we can repeat the procedure for the case of 1:3 resonance.

Fold–Flip

Let $Aq_{1,2} = \pm q_{1,2}$, $A^T p_{1,2} = \pm p_{1,2}$, $\langle p_1, q_1\rangle = \langle p_2, q_2\rangle = 1$. The necessary systems to solve from the homological equation (4.2) are

$$(A - I_n)[h_{0010},\ h_{0001}] = [q_1,\ 0] - J_1[v_{10},\ v_{01}], \tag{4.82}$$

$$(A - I_n)[h_{1010},\ h_{1001}] = [h_{2000},\ q_1] - A_1(q_1, [v_{10},\ v_{01}]) - B(q_1, [h_{0010},\ h_{0001}]), \tag{4.83}$$

$$(A + I_n)[h_{0110},\ h_{0101}] = [h_{1100},\ 0] - A_1(q_2, [v_{10},\ v_{01}]) - B(q_2, [h_{0010},\ h_{0001}]). \tag{4.84}$$

First notice that all matrices in the left-hand side are singular. If we take $(\gamma_1, \gamma_2) = p_1^T J_1$ and form the orthogonal vectors s_1 and s_2 as before, then $v_{10} = s_1 + \delta_1 s_2$ and $v_{01} = \delta_2 s_2$ solve system (4.82). Bordering the singular matrix $(A - I_n)$ one can solve for h_{0010} and h_{0001}. Any multiple of q_1 can be added to h_{0010} and h_{0001}, so we use $h_{0010} = (A - I_n)^{INV}(q_1 - J_1 v_{10}) + \delta_3 q_1$ and $h_{0010} = -(A - I_n)^{INV}(J_1 v_{01}) + \delta_4 q_1$. We will use this freedom to solve equations (4.83) and (4.84) simultaneously for all δs. Note that h_{2000} and h_{1100} are also found using bordered systems chosen, but such that $\langle p_1, h_{2000}\rangle = \langle p_2, h_{1100}\rangle = 0$.

Then we obtain the following four-dimensional system

$$\begin{pmatrix} L & 0_2 \\ 0_2 & L \end{pmatrix} \begin{pmatrix} \delta_1 \\ \delta_3 \\ \delta_2 \\ \delta_4 \end{pmatrix} = \begin{pmatrix} -\langle p_1, A_1(q_1, s_1) + B(q_1, (A - I_n)^{INV}(q_1 - J_1 s_1)) \\ -\langle p_2, A_1(q_2, s_1) + B(q_2, (A - I_n)^{INV}(q_1 - J_1 s_1)) \\ 1 \\ 0 \end{pmatrix}, \tag{4.85}$$

where 0_2 is the (2×2) zero matrix and L is defined by

$$L = \begin{pmatrix} \langle p_1, A_1(q_1, s_2) + B(q_1, (I_n - A)^{INV} J_1 s_2)\rangle & \langle p_1, B(q_1, q_1)\rangle \\ \langle p_2, A_1(q_2, s_2) + B(q_2, (I_n - A)^{INV} J_1 s_2)\rangle & \langle p_2, B(q_1, q_2)\rangle \end{pmatrix}. \tag{4.86}$$

Notice that $2a(0) = \langle p_1, B(q_1, q_1)\rangle$ and that q_1 can be chosen such that

$e(0) = \langle p_2, B(q_1, q_2)\rangle = 1$. The condition $\gamma_1\gamma_2 \det(L) \neq 0$ is equivalent to the transversality to the bifurcation manifold of the family $g(x,\alpha)$.

Flip–Neimark–Sacker

Introduce $Aq_1 = q_1$, $A^T p_1 = p_1$, $\langle p_1, q_1\rangle = 1$ and $Aq_2 = \mathrm{e}^{i\theta_0} q_2$, $A^T p_2 = \mathrm{e}^{-i\theta_0} p_2$, $\langle p_2, q_2\rangle = 1$. The linear systems obtained from the homological equation (4.2) are

$$
\begin{aligned}
(A - I_n)[h_{00010},\ h_{00001}] &= -\ J_1[v_{10},\ v_{01}],\\
(A + I_n)[h_{10010},\ h_{10001}] &= [-q_1,\ 0] - A_1(q_1, [v_{10},\ v_{01}])\\
&\quad - B(q_1, [h_{00010},\ h_{00001}]),\\
(A - \mathrm{e}^{i\theta_0} I_n)[h_{01010},\ h_{01001}] &= [0,\ q_2\mathrm{e}^{i\theta_0}] - A_1(q_2, [v_{10},\ v_{01}])\\
&\quad - B(q_2, [h_{00010},\ h_{00001}]).
\end{aligned}
$$

The same approach as for the degenerate flip and 1:2 resonance cases is to substitute the formal solution of the first equation into the second and we write

$$v_{10} = -s_1 + \delta_1 s_2, \quad v_{01} = \delta_2 s_2,$$

where the constants δ_i are to be found from the last equation. We compute

$$Q_i = \langle p_2, A_1(q_2, s_i) + B(q_2, (I_n - A)^{-1} J_1 s_i)\rangle$$

for $i = 1, 2$. We proceed similarly to Roose and Hlavacek (1985, Appendix) but adapt to the case of maps. We find

$$\delta_1 = \frac{\Re(\mathrm{e}^{-i\theta_0} Q_1)}{\Re(\mathrm{e}^{-i\theta_0} Q_2)}, \quad \delta_2 = -\frac{1}{\Re(\mathrm{e}^{-i\theta_0} Q_2)}. \tag{4.87}$$

Appendix 4: Fifth-order coefficients for flip–Neimark–Sacker and double Neimark–Sacker

Here we give the expressions for the fourth-order vectors and fifth-order coefficients. We list only those fourth-order vectors that are necessary to obtain the fifth-order coefficients. Also, vectors not listed below but present in the expressions can be found by complex conjugation. For the flip–Neimark–Sacker bifurcation it is assumed that $(k\theta_0 \bmod 2\pi) \neq 0$ for $k = 1,2,3,4,5,6,8,10$, while for the double Neimark–Sacker bifurcation $(k\theta_i \bmod 2\pi) \neq 0$ for $i = 1, 2$ and $k = 1,2,3,4,5,6$ and $(\theta_1/\theta_2) \notin \pm\{5,4,3,2,\frac{3}{2},1,\frac{2}{3},\frac{1}{2},\frac{1}{3},\frac{1}{4},\frac{1}{5}\}$.

Flip–Neimark–Sacker

Fourth-order vectors:

$$\begin{aligned}
h_{400} =& -(A-I_n)^{-1}\,[\,D(q_1,q_1,q_1,q_1)+6C(h_{200},q_1,q_1)+4B(h_{300},q_1)\\
&+3B(h_{200},h_{200})+24h_{200}f_{300}\,], \qquad (4.88)\\
h_{310} =& -(A+e^{i\theta_0}I_n)^{-1}\,[\,D(q_1,q_1,q_1,q_2)+3C(h_{110},q_1,q_1)+3C(h_{200},q_1,q_2)\\
&+3B(h_{210},q_1)+3B(h_{110},h_{200})+B(h_{300},q_2)+6h_{110}(f_{300}e^{i\theta_0}-g_{210})\Big], \qquad (4.89)\\
h_{220} =& -(A-e^{2i\theta_0}I_n)^{-1}\,[\,D(q_1,q_1,q_2,q_2)+4C(h_{110},q_1,q_2)+C(h_{200},q_2,q_2)\\
&+C(h_{020},q_1,q_1)+2B(h_{210},q_2)+2B(h_{120},q_1)+2B(h_{110},h_{110})\\
&+B(h_{020},h_{200})+4h_{020}e^{i\theta_0}g_{210}\Big], \qquad (4.90)\\
h_{211} =& -(A-I_n)^{-1}\,[\,D(q_1,q_1,q_2,\bar{q}_2)+2C(h_{101},q_1,q_2)+2C(h_{110},q_1,\bar{q}_2)\\
&+C(h_{200},q_2,\bar{q}_2)+C(h_{011},q_1,q_1)+2B(h_{111},q_1)+2B(h_{101},h_{110})\\
&+B(h_{011},h_{200})+B(h_{201},q_2)+B(h_{210},\bar{q}_2)+2h_{200}f_{111}\,], \qquad (4.91)\\
h_{130} =& -(A-e^{3i\theta_0}I_n)^{-1}\,[\,D(q_1,q_2,q_2,q_2)+3C(h_{020},q_1,q_2)+3C(h_{110},q_2,q_2)\\
&+3B(h_{120},q_2)+3B(h_{020},h_{110})+B(h_{030},q_1)], \qquad (4.92)\\
h_{121} =& -(A+e^{i\theta_0}I_n)^{-1}\,[\,+D(q_1,q_2,q_2,\bar{q}_2)+2C(h_{011},q_1,q_2)+2C(h_{110},q_2,\bar{q}_2)\\
&+C(h_{101},q_2,q_2)+C(h_{020},q_1,\bar{q}_2)+2B(h_{111},q_2)+2B(h_{011},h_{110})\\
&+B(h_{020},h_{101})+B(h_{120},\bar{q}_2)+B(h_{021},q_1)-2h_{110}(f_{111}e^{i\theta_0}-g_{021})\Big], \qquad (4.93)\\
h_{031} =& -(A-e^{2\theta_0}I_n)^{-1}\,[\,D(q_2,q_2,q_2,\bar{q}_2)+3C(h_{011},q_2,q_2)+3C(h_{020},q_2,\bar{q}_2)\\
&+3B(h_{020},h_{011})+3B(h_{021},q_2)+B(h_{030},\bar{q}_2)-6h_{020}e^{i\theta_0}g_{021}\Big], \qquad (4.94)\\
h_{022} =& -(A-I_n)^{-1}\,[\,D(q_2,q_2,\bar{q}_2,\bar{q}_2)+4C(h_{011},q_2,\bar{q}_2)+B(h_{020},h_{002})\\
&+2B(h_{011},h_{011})+2\Re(C(h_{020},\bar{q}_2,\bar{q}_2))+2\Re(2B(h_{021},\bar{q}_2))]. \qquad (4.95)
\end{aligned}$$

Fifth-order coefficients for (3.85):

$$\begin{aligned}
f_{500} =& \frac{1}{120}\langle p_1, E(q_1,q_1,q_1,q_1,q_1)+10D(h_{200},q_1,q_1,q_1)+15C(h_{200},h_{200},q_1)\\
&+10C(h_{300},q_1,q_1)+10B(h_{200},h_{300})+5B(h_{400},q_1)\rangle, \qquad (4.96)
\end{aligned}$$

$$
\begin{aligned}
f_{311} = \frac{1}{6}\langle p_1, & E(q_1, q_1, q_1, q_2, \bar{q}_2) + 3D(h_{200}, q_1, q_2, \bar{q}_2) + 3D(h_{110}, q_1, q_1, \bar{q}_2) \\
& + 3D(h_{101}, q_1, q_1, q_2) + 6C(h_{101}, h_{110}, q_1) + 3C(h_{101}, h_{200}, q_2) \\
& + 3C(h_{011}, h_{200}, q_1) + 3C(h_{110}, h_{200}, \bar{q}_2) + 3C(h_{210}, q_1, \bar{q}_2) \\
& + 3C(h_{201}, q_1, q_2) + 3C(h_{111}, q_1, q_1) + C(h_{300}, q_2, \bar{q}_2) + 3B(h_{111}, h_{200}) \\
& + 3B(h_{110}, h_{201}) + 3B(h_{211}, q_1) + 3B(h_{101}, h_{210}) + B(h_{310}, \bar{q}_2) \\
& + B(h_{011}, h_{300}) + B(h_{301}, q_2)\rangle,
\end{aligned} \tag{4.97}
$$

$$
\begin{aligned}
f_{122} = \frac{1}{6}\langle p_1, & E(q_1, q_2, q_2, \bar{q}_2, \bar{q}_2) + 4D(h_{011}, q_1, q_2, \bar{q}_2) + 2D(h_{110}, q_2, \bar{q}_2, \bar{q}_2) \\
& + 2D(h_{101}, q_2, q_2, \bar{q}_2) + D(h_{002}, q_1, q_2, q_2) + D(h_{020}, q_1, \bar{q}_2, \bar{q}_2) \\
& + 4C(h_{011}, h_{101}, q_2) + 4C(h_{011}, h_{110}, \bar{q}_2) + 4C(h_{111}, q_2, \bar{q}_2) \\
& + 2C(h_{021}, q_1, \bar{q}_2) + 2C(h_{002}, h_{110}, q_2) + 2C(h_{020}, h_{101}, \bar{q}_2) \\
& + 2C(h_{012}, q_1, q_2) + 2C(h_{011}, h_{011}, q_1) + C(h_{020}, h_{002}, q_1) \\
& + C(h_{120}, \bar{q}_2, \bar{q}_2) + C(h_{1,0,2}, q_2, q_2) + 4B(h_{011}, h_{111}) + 2B(h_{021}, h_{101}) \\
& + 2B(h_{121}, \bar{q}_2) + 2B(h_{112}, q_2) + 2B(h_{012}, h_{110}) + B(h_{020}, h_{102}) \\
& + B(h_{002}, h_{120}) + B(h_{022}, q_1)\rangle,
\end{aligned} \tag{4.98}
$$

$$
\begin{aligned}
g_{410} = \frac{1}{24}\langle p_2, & E(q_1, q_1, q_1, q_1, q_2) + 6D(h_{200}, q_1, q_1, q_2) + 4D(h_{110}, q_1, q_1, q_1) \\
& + 6C(h_{210}, q_1, q_1) + 12C(h_{110}, h_{200}, q_1) + 4C(h_{300}, q_1, q_2) + B(h_{400}, q_2) \\
& + 3C(h_{200}, h_{200}, q_2) + 6B(h_{200}, h_{210}) + 4B(h_{110}, h_{300}) + 4B(h_{310}, q_1)\rangle,
\end{aligned} \tag{4.99}
$$

$$
\begin{aligned}
g_{221} = \frac{1}{4}\langle p_2, & E(q_1, q_1, q_2, q_2, \bar{q}_2) + 4D(h_{110}, q_1, q_2, \bar{q}_2) + 2D(h_{011}, q_1, q_1, q_2) \\
& + 2D(h_{101}, q_1, q_2, q_2) + D(h_{200}, q_2, q_2, \bar{q}_2) + D(h_{020}, q_1, q_1, \bar{q}_2) \\
& + 4C(h_{011}, h_{110}, q_1) + 4C(h_{101}, h_{110}, q_2) + 4C(h_{111}, q_1, q_2) \\
& + 2C(h_{110}, h_{110}, \bar{q}_2) + 2C(h_{120}, q_1, \bar{q}_2) + 2C(h_{210}, q_2, \bar{q}_2) \\
& + 2C(h_{020}, h_{101}, q_1) + 2C(h_{011}, h_{200}, q_2) + C(h_{201}, q_2, q_2) \\
& + C(h_{020}, h_{200}, \bar{q}_2) + C(h_{021}, q_1, q_1) + 4B(h_{110}, h_{111}) + 2B(h_{120}, h_{101}) \\
& + 2B(h_{121}, q_1) + 2B(h_{011}, h_{210}) + 2B(h_{211}, q_2) + B(h_{220}, \bar{q}_2) \\
& + B(h_{020}, h_{201}) + B(h_{021}, h_{200})\rangle,
\end{aligned} \tag{4.100}
$$

$$\begin{aligned} g_{032} = \frac{1}{12}\langle p_2, & E(q_2, q_2, q_2, \bar{q}_2, \bar{q}_2) + 6D(h_{011}, q_2, q_2, \bar{q}_2) + 3D(h_{020}, q_2, \bar{q}_2, \bar{q}_2) \\ & + D(h_{002}, q_2, q_2, q_2) + 6C(h_{020}, h_{011}, \bar{q}_2) + 6C(h_{021}, q_2, \bar{q}_2) \\ & + 6C(h_{011}, h_{011}, q_2) + 3C(h_{020}, h_{002}, q_2) + 3C(h_{012}, q_2, q_2) \\ & + C(h_{030}, \bar{q}_2, \bar{q}_2) + 6B(h_{011}, h_{021}) + 3B(h_{020}, h_{012}) + 3B(h_{022}, q_2) \\ & + 2B(h_{031}, \bar{q}_2) + B(h_{030}, h_{002})\rangle. \end{aligned} \tag{4.101}$$

Double Neimark–Sacker

Fourth-order vectors:

$$\begin{aligned} h_{3100} = & -(A - e^{2i\theta_1} I_n)^{-1} \left[D(q_1, q_1, q_1, \bar{q}_1) + 3C(h_{2000}, q_1, \bar{q}_1) + 3C(q_1, q_1, h_{1100}) \right. \\ & \left. + 3B(h_{2000}, h_{1100}) + 3B(q_1, h_{2100}) + B(h_{3000}, \bar{q}_1) - 6h_{2000} e^{i\theta_1} f_{2100} \right], \end{aligned} \tag{4.102}$$

$$\begin{aligned} h_{2200} = & -(A - I_n)^{-1} \left[D(q_1, q_1, \bar{q}_1, \bar{q}_1) + 4C(q_1, h_{1100}, \bar{q}_1) + C(h_{2000}, \bar{q}_1, \bar{q}_1) \right. \\ & + C(h_{0200}, q_1, q_1) + 2B(h_{1100}, h_{1100}) + 2B(q_1, h_{1200}) + 2B(h_{2100}, \bar{q}_1) \\ & \left. + B(h_{2000}, h_{0200}) \right], \end{aligned} \tag{4.103}$$

$$\begin{aligned} h_{2110} = & -(A - e^{i(\theta_1+\theta_2)} I_n)^{-1} \left[D(q_1, q_1, \bar{q}_1, q_2) + 2C(q_1, h_{1010}, \bar{q}_1) \right. \\ & + 2C(q_1, q_2, h_{1100}) + C(h_{2000}, q_2, \bar{q}_1) + C(q_1, q_1, h_{0110}) + 2B(q_1, h_{1110}) \\ & + 2B(h_{1010}, h_{1100}) + B(q_2, h_{2100}) + B(h_{2000}, h_{0110}) + B(h_{2010}, \bar{q}_1) \\ & \left. - 2h_{1010}(f_{2100} e^{i\theta_2} - e^{i\theta_1} g_{1110}) \right], \end{aligned} \tag{4.104}$$

$$\begin{aligned} h_{2101} = & -(A - e^{i(\theta_1-\theta_2)} I_n)^{-1} \left[D(q_1, q_1, \bar{q}_1, \bar{q}_2) + 2C(q_1, \bar{q}_2, h_{1100}) \right. \\ & + 2C(q_1, h_{1001}, \bar{q}_1) + C(h_{2000}, \bar{q}_2, \bar{q}_1) + C(h_{0101}, q_1, q_1) + B(\bar{q}_2, h_{2100}) \\ & + 2B(h_{1001}, h_{1100}) + 2B(q_1, h_{1101}) + B(h_{2001}, \bar{q}_1) + B(h_{0101}, h_{2000}) \\ & \left. - 2h_{1001}(e^{i\theta_1} g_{1101} - f_{2100} e^{-i\theta_2}) \right], \end{aligned} \tag{4.105}$$

$$\begin{aligned} h_{1120} = & -(A - e^{2i\theta_2} I_n)^{-1} \left[D(q_1, \bar{q}_1, q_2, q_2) + 2C(q_2, h_{1010}, \bar{q}_1) + 2C(q_1, q_2, h_{0110}) \right. \\ & + C(q_1, h_{0020}, \bar{q}_1) + C(q_2, q_2, h_{1100}) + 2B(q_2, h_{1110}) + 2B(h_{0110}, h_{1010}) \\ & \left. + B(h_{0020}, h_{1100}) + B(h_{1020}, \bar{q}_1) + B(q_1, h_{0120}) - 2h_{0020} e^{i\theta_2} g_{1110} \right], \end{aligned} \tag{4.106}$$

$$\begin{aligned} h_{1111} = & -(A - I_n)^{-1} \left[D(q_1, \bar{q}_1, q_2, \bar{q}_2) + C(h_{0101}, q_1, q_2) + C(q_1, \bar{q}_2, h_{0110}) \right. \\ & + C(q_2, \bar{q}_2, h_{1100}) + C(q_1, h_{0011}, \bar{q}_1) + C(q_2, h_{1001}, \bar{q}_1) + C(\bar{q}_2, h_{1010}, \bar{q}_1) \\ & + B(q_2, h_{1101}) + B(\bar{q}_2, h_{1110}) + B(q_1, h_{0111}) + B(h_{1001}, h_{0110}) \\ & \left. + B(h_{1011}, \bar{q}_1) + B(h_{0101}, h_{1010}) + B(h_{0011}, h_{1100}) \right], \end{aligned} \tag{4.107}$$

$$\begin{aligned} h_{1021} =& - (A - \mathrm{e}^{i(\theta_1+\theta_2)})^{-1} \big[D(q_1, q_2, q_2, \bar{q}_2) + 2C(q_2, \bar{q}_2, h_{1010}) \\ &+ 2C(q_1, q_2, h_{0011}) + C(q_1, \bar{q}_2, h_{0020}) + C(q_2, q_2, h_{1001}) + 2B(q_2, h_{1011}) \\ &+ 2B(h_{0011}, h_{1010}) + B(\bar{q}_2, h_{1020}) + B(q_1, h_{0021}) + B(h_{0020}, h_{1001}) \\ &- 2h_{1010}(f_{1011}\mathrm{e}^{i\theta_2}\mathrm{e}^{i\theta_1} g_{0021})\big], \end{aligned} \tag{4.108}$$

$$\begin{aligned} h_{1012} =& - (A - \mathrm{e}^{i(\theta_1-\theta_2)})^{-1} \big[D(q_1, q_2, \bar{q}_2, \bar{q}_2) + 2C(q_1, \bar{q}_2, h_{0011}) \\ &+ 2C(q_2, \bar{q}_2, h_{1001}) + C(q_1, q_2, h_{0002}) + C(\bar{q}_2, \bar{q}_2, h_{1010}) + 2B(\bar{q}_2, h_{1011}) \\ &+ 2B(h_{0011}, h_{1001}) + B(q_2, h_{1002}) + B(h_{0002}, h_{1010}) + B(q_1, h_{0012}) \\ &- 2h_{1001}(\mathrm{e}^{i\theta_1} g_{0012} - f_{1011}\mathrm{e}^{-i\theta_2})\big], \end{aligned} \tag{4.109}$$

$$\begin{aligned} h_{0031} =& - (A - \mathrm{e}^{2i\theta_2})^{-1} s\big[D(q_2, q_2, q_2, \bar{q}_2) + 3C(q_2, q_2, h_{0011}) + 3C(q_2, \bar{q}_2, h_{0020}) \\ &+ 3B(q_2, h_{0021}) + 3B(h_{0011}, h_{0020}) + B(\bar{q}_2, h_{0030}) - 6h_{0020}\mathrm{e}^{i\theta_2} g_{0021}\big], \end{aligned} \tag{4.110}$$

$$\begin{aligned} h_{0022} =& - (A - I_n)^{-1} \big[D(q_2, q_2, \bar{q}_2, \bar{q}_2) + 4C(q_2, \bar{q}_2, h_{0011}) + C(\bar{q}_2, \bar{q}_2, h_{0020}) \\ &+ C(q_2, q_2, h_{0002}) + 2B(\bar{q}_2, h_{0021}) + 2B(h_{0011}, h_{0011}) + 2B(q_2, h_{0012}) \\ &+ B(h_{0002}, h_{0020})\big]. \end{aligned} \tag{4.111}$$

Fifth-order coefficients for (3.86):

$$\begin{aligned} f_{3200} =& \frac{1}{12}\langle p_1, E(q_1, q_1, q_1, \bar{q}_1, \bar{q}_1) + 6D(q_1, q_1, h_{1100}, \bar{q}_1) + 3D(h_{2000}, q_1, \bar{q}_1, \bar{q}_1) \\ &+ D(h_{0200}, q_1, q_1, q_1) + 6C(q_1, h_{1100}, h_{1100}) + 6C(h_{2000}, h_{1100}, \bar{q}_1) \\ &+ 6C(q_1, h_{2100}, \bar{q}_1) + 3C(q_1, q_1, h_{1200}) + 3C(h_{2000}, h_{0200}, q_1) \\ &+ C(h_{3000}, \bar{q}_1, \bar{q}_1) + 6B(h_{2100}, h_{1100}) + 3B(h_{2000}, h_{1200}) \\ &+ 3B(q_1, h_{2200}) + 2B(h_{3100}, \bar{q}_1) + B(h_{0200}, h_{3000})\rangle, \end{aligned} \tag{4.112}$$

$$\begin{aligned} f_{1022} =& \frac{1}{4}\langle p_1, E(q_1, q_2, q_2, \bar{q}_2, \bar{q}_2) + 4D(q_1, q_2, \bar{q}_2, h_{0011}) + 2D(q_2, \bar{q}_2, \bar{q}_2, h_{1010}) \\ &+ 2D(q_2, q_2, \bar{q}_2, h_{1001}) + D(q_1, \bar{q}_2, \bar{q}_2, h_{0020}) + D(q_1, q_2, q_2, h_{0002}) \\ &+ 4C(\bar{q}_2, h_{0011}, h_{1010}) + 4C(q_2, \bar{q}_2, h_{1011}) + 4C(q_2, h_{0011}, h_{1001}) \\ &+ 2C(q_1, q_2, h_{0012}) + 2C(\bar{q}_2, h_{0020}, h_{1001}) + 2C(q_2, h_{0002}, h_{1010}) \\ &+ 2C(q_1, h_{0011}, h_{0011}) + 2C(q_1, \bar{q}_2, h_{0021}) + C(q_1, h_{0002}, h_{0020}) \\ &+ C(q_2, q_2, h_{1002}) + C(\bar{q}_2, \bar{q}_2, h_{1020}) + 4B(h_{1011}, h_{0011}) \\ &+ 2B(\bar{q}_2, h_{1021}) + 2B(h_{0021}, h_{1001}) + 2B(q_2, h_{1012}) + 2B(h_{0012}, h_{1010}) \\ &+ B(h_{0020}, h_{1002}) + B(h_{0002}, h_{1020}) + B(q_1, h_{0022})\rangle, \end{aligned} \tag{4.113}$$

$$f_{2111} = \frac{1}{2}\langle p_1, E(q_1, q_1, \bar{q}_1, q_2, \bar{q}_2) + 2D(q_1, q_2, h_{1001}, \bar{q}_1) + 2D(q_1, \bar{q}_2, h_{1010}, \bar{q}_1)$$
$$+ 2D(q_1, q_2, \bar{q}_2, h_{1100}) + D(h_{2000}, q_2, \bar{q}_2, \bar{q}_1) + D(q_1, q_1, \bar{q}_2, h_{0110})$$
$$+ D(h_{0101}, q_1, q_1, q_2) + D(q_1, q_1, h_{0011}, \bar{q}_1) + 2C(q_1, h_{0011}, h_{1100})$$
$$+ 2C(\bar{q}_2, h_{1010}, h_{1100}) + 2C(q_1, q_2, h_{1101}) + 2C(q_1, h_{1001}, h_{0110})$$
$$+ 2C(q_2, h_{1001}, h_{1100}) + 2C(q_1, h_{1011}, \bar{q}_1) + 2C(q_1, \bar{q}_2, h_{1110})$$
$$+ 2C(h_{1001}, h_{1010}, \bar{q}_1) + 2C(h_{0101}, q_1, h_{1010}) + C(q_2, \bar{q}_2, h_{2100})$$
$$+ C(h_{0101}, h_{2000}, q_2) + C(q_1, q_1, h_{0111}) + C(h_{2001}, q_2, \bar{q}_1)$$
$$+ C(h_{2000}, \bar{q}_2, h_{0110}) + C(\bar{q}_2, h_{2010}, \bar{q}_1) + C(h_{2000}, h_{0011}, \bar{q}_1)$$
$$+ 2B(h_{1010}, h_{1101}) + 2B(h_{1110}, h_{1001}) + 2B(h_{1011}, h_{1100})$$
$$+ 2B(q_1, h_{1111}) + B(\bar{q}_2, h_{2110}) + B(h_{0101}, h_{2010}) + B(h_{2001}, h_{0110})$$
$$+ B(h_{2000}, h_{0111}) + B(q_2, h_{2101}) + B(h_{0011}, h_{2100}) + B(h_{2011}, \bar{q}_1)\rangle, \tag{4.114}$$

$$g_{1121} = \frac{1}{2}\langle p_2, E(q_1, \bar{q}_1, q_2, q_2, \bar{q}_2) + 2D(q_2, \bar{q}_2, h_{1010}, \bar{q}_1) + 2D(q_1, q_2, \bar{q}_2, h_{0110})$$
$$+ 2D(q_1, q_2, h_{0011}, \bar{q}_1) + D(q_1, \bar{q}_2, h_{0020}, \bar{q}_1) + D(q_2, q_2, h_{1001}, \bar{q}_1)$$
$$+ D(q_2, q_2, \bar{q}_2, h_{1100}) + D(h_{0101}, q_1, q_2, q_2) + 2C(q_2, h_{1001}, h_{0110})$$
$$+ 2C(q_2, h_{0011}, h_{1100}) + 2C(q_1, q_2, h_{0111}) + 2C(q_2, \bar{q}_2, h_{1110})$$
$$+ 2C(h_{0011}, h_{1010}, \bar{q}_1) + 2C(q_2, h_{1011}, \bar{q}_1) + 2C(\bar{q}_2, h_{0110}, h_{1010})$$
$$+ 2C(q_1, h_{0011}, h_{0110}) + 2C(h_{0101}, q_2, h_{1010}) + C(\bar{q}_2, h_{1020}, \bar{q}_1)$$
$$+ C(q_1, h_{0021}, \bar{q}_1) + C(q_2, q_2, h_{1101}) + C(q_1, \bar{q}_2, h_{0120})$$
$$+ C(h_{0101}, q_1, h_{0020}) + C(h_{0020}, h_{1001}, \bar{q}_1) + C(\bar{q}_2, h_{0020}, h_{1100})$$
$$+ 2B(h_{1110}, h_{0011}) + 2B(h_{1011}, h_{0110}) + 2B(h_{0111}, h_{1010})$$
$$+ 2B(q_2, h_{1111}) + B(h_{0021}, h_{1100}) + B(h_{1001}, h_{0120}) + B(h_{0101}, h_{1020})$$
$$+ B(h_{0020}, h_{1101}) + B(q_1, h_{0121}) + B(h_{1021}, \bar{q}_1) + B(\bar{q}_2, h_{1120})\rangle, \tag{4.115}$$

$$g_{0032} = \frac{1}{12}\langle p_2, E(q_2, q_2, q_2, \bar{q}_2, \bar{q}_2) + 6D(q_2, q_2, \bar{q}_2, h_{0011}) + 3D(q_2, \bar{q}_2, \bar{q}_2, h_{0020})$$
$$+ D(q_2, q_2, q_2, h_{0002}) + 6C(q_2, \bar{q}_2, h_{0021}) + 6C(q_2, h_{0011}, h_{0011})$$
$$+ 6C(\bar{q}_2, h_{0011}, h_{0020}) + 3C(q_2, q_2, h_{0012}) + 3C(q_2, h_{0002}, h_{0020})$$
$$+ C(\bar{q}_2, \bar{q}_2, h_{0030}) + 6B(h_{0021}, h_{0011}) + 3B(h_{0012}, h_{0020})$$
$$+ 3B(q_2, h_{0022}) + 2B(\bar{q}_2, h_{0031}) + B(h_{0030}, h_{0002})\rangle, \tag{4.116}$$

$$
\begin{aligned}
g_{2210} = \frac{1}{4}\langle p_2, & E(q_1, q_1, \bar{q}_1, \bar{q}_1, q_2) + 4D(q_1, q_2, h_{1100}, \bar{q}_1) + 2D(q_1, q_1, h_{0110}, \bar{q}_1) \\
& + 2D(q_1, h_{1010}, \bar{q}_1, \bar{q}_1) + D(h_{2000}, q_2, \bar{q}_1, \bar{q}_1) + D(h_{0200}, q_1, q_1, q_2) \\
& + 4C(h_{1010}, h_{1100}, \bar{q}_1) + 4C(q_1, h_{0110}, h_{1100}) + 4C(q_1, h_{1110}, \bar{q}_1) \\
& + 2C(q_2, h_{2100}, \bar{q}_1) + 2C(h_{2000}, h_{0110}, \bar{q}_1) + 2C(q_2, h_{1100}, h_{1100}) \\
& + 2C(h_{0200}, q_1, h_{1010}) + 2C(q_1, q_2, h_{1200}) + C(h_{2000}, h_{0200}, q_2) \\
& + C(q_1, q_1, h_{0210}) + C(h_{2010}, \bar{q}_1, \bar{q}_1) + 4B(h_{1110}, h_{1100}) \\
& + 2B(h_{2100}, h_{0110}) + 2B(h_{1010}, h_{1200}) + 2B(h_{2110}, \bar{q}_1) + 2B(q_1, h_{1210}) \\
& + B(h_{0200}, h_{2010}) + B(h_{2000}, h_{0210}) + B(q_2, h_{2200})\rangle. \qquad (4.117)
\end{aligned}
$$

Part Two

Software

5
Numerical Methods and Algorithms

In this chapter we describe several numerical algorithms for continuation of bifurcations, growing of manifolds and Lyapunov exponents. Continuation itself and the implementation in MatcontM is discussed in the next chapter.

5.1 Continuation of cycles

The iteration of (1.4) gives rise to a sequence of points $\{x_1, x_2, x_3, \ldots, x_{K+1}\}$ in which $x_{j+1} = f(x_j, \alpha)$. Each point x of a cycle of period-K then satisfies the fixed point equation for the Kth iterate

$$f^{(K)}(x, \alpha) - x = 0,$$

which we rewrite using (3.2) as

$$g(x, \alpha) - x = 0. \tag{5.1}$$

In MatcontM branches of period-K cycles are computed by a variant of the Gauss–Newton continuation algorithm from Allgower and Georg (1990) applied to (5.1) (see Section 6.3).

Table 5.1 Detection of codim 1 bifurcations of cycles.

Bifurcation	Test function(s)
LP	$v_{n+1} = 0, \det\begin{pmatrix} F_X \\ v^T \end{pmatrix} \neq 0$
PD	$\det(A + I_n) = 0$
NS	$\det(A \odot A - I_m) = 0$
BP	$\det\begin{pmatrix} F_X \\ v^T \end{pmatrix} = 0$

To detect the bifurcations introduced in Section 2.1, as well as branch points of (5.1), we use the standard test functions listed in Table 5.1, where v is the tangent vector to the curve (5.1) in the continuation X-space for $X = (x, \alpha)$, $F(X) = g(x, \alpha) - x$, $A = g_x$ is the Jacobian matrix of $g = f^{(K)}$ and $\odot$ is the *bi-alternate matrix product*. For our purposes, it is sufficient to remark that $A \odot A$ is an $m \times m$ matrix with $m = \frac{1}{2}n(n-1)$, where elements are labeled by *index pairs* $(p, q), (r, s)$ and can be computed by

$$(A \odot A)_{(p,q),(r,s)} := a_{pr}a_{qs} - a_{qr}a_{ps},$$

where $n \geq p > q \geq 1$ and $n \geq r > s \geq 1$. If A has eigenvalues $\lambda_1, \lambda_2, \ldots, \lambda_n$, then $A \odot A$ has eigenvalues $\lambda_p\lambda_q$, where $n \geq p > q \geq 1$ (see e.g., (Govaerts, 2000, section 4.4.4)). The test function for a saddle node LP relies on observing changes in the parameter direction along the curve, while for a branch point BP one tries to detect a rank deficiency. We note that these two cases are distinguished by the Implicit Function Theorem. The test functions for PD and NS points use the Jacobian matrix A to detect changes in multipliers. We notice that $\det(A \odot A - I_m)$ also vanishes if there is a pair of real eigenvalues with product 1. Such points are called *neutral saddles*. We have to exclude them when processing the NS-points.

5.2 Continuation of codimension 1 bifurcation curves

The usual way to setup defining systems for the LP, PD and NS bifurcation curves for period-K cycles is to add the corresponding eigenvalue problem to the fixed point (cycle) equation. The resulting system is called the *maximally extended defining system*. In this approach, the *limit point curve* and *period-doubling curve* are computed using the following systems

$$\begin{cases} g(x, \alpha) - x = 0, \\ (g_x(x, \alpha) \mp I_n)q = 0, \\ c^T q - 1 = 0, \end{cases} \tag{5.2}$$

where the continuation variable is $X = (x, q, \alpha) \in \mathbb{R}^{2n+2}$, so we need two active (free) system parameters. The function g is given by (3.2), while $c \in \mathbb{R}^n$ is a reference vector such that $c^T q \neq 0$. One should take the "−" sign in (5.2) for the LP-curve and the "+" sign for the PD-curve. Notice that the matrices $(A \mp I_n)$, where $A = g_x(x, \alpha)$, are singular at solutions of (5.2).

A natural maximally extended defining system for the *Neimark–Sacker curve* is the system

$$\begin{cases} g(x,\alpha) - x = 0, \\ (g_x(x,\alpha) - e^{i\theta} I_n)q = 0, \\ c^T q - 1 = 0, \end{cases} \tag{5.3}$$

where the continuation variable is $X = (x, q, \theta, \alpha) \in \mathbb{R}^n \times \mathbb{C}^n \times \mathbb{R}^3$ and includes the argument θ. A reference vector $c \in \mathbb{C}^n$ is used to normalize the eigenvector $q \in \mathbb{C}^n$ and should satisfy $\Re(c^T q)\Im(c^T q) \neq 0$. Writing $q = q_1 + iq_2$ and $c = c_1 + ic_2$ with $q_{1,2}, c_{1,2} \in \mathbb{R}^n$, we obtain an equivalent to (5.3) real defining system

$$\begin{cases} g(x,\alpha) - x = 0, \\ g_x(x,\alpha)q_1 - (\cos\theta)q_1 + (\sin\theta)q_2 = 0, \\ g_x(x,\alpha)q_2 - (\sin\theta)q_1 - (\cos\theta)q_2 = 0, \\ c_1^T q_1 + c_2^T q_2 - 1 = 0, \\ c_2^T q_1 - c_1^T q_2 = 0, \end{cases} \tag{5.4}$$

where the continuation variable is now $X = (x, q_1, q_2, \theta, \alpha) \in \mathbb{R}^{3n+3}$. Notice that the matrix $(A^2 - 2(\cos\theta)A + I_n)$, where $A = g_x(x,\alpha)$, has rank defect 2 at solutions of (5.4). This follows from the fact that both q_1 and q_2 are linearly independent null vectors of this matrix. Thus, the real dimension of the maximally extended defining systems for LP and PD bifurcations is $2n + 2$, while that for NS is $3n+3$, where n is the dimension of the phase space, and this may become computationally expensive.

In MatcontM, the LP, PD and NS curves for period-K cycles are computed using *minimally extended defining systems* (see also (Govaerts, 2000)). That is, the algebraic problem for the continuation contains a minimal number of additional equations to define the problem, typically one or two. A *standard minimally extended defining system* would include vanishing of a determinant of some matrix as an additional equation. Indeed, the LP and PD curves can be computed using

$$\begin{cases} g(x,\alpha) - x = 0, \\ \det(g_x(x,\alpha) \mp I_n) = 0, \end{cases} \tag{5.5}$$

where the continuation variable is merely $X = (x, \alpha) \in \mathbb{R}^{n+2}$. However, as determinants scale badly, this approach is not suitable for problems with large dimensions. Moreover, the continuation algorithm requires partial derivatives of the defining system with respect to components of (x, α), for which no simple expressions exist if determinants are involved. Such derivatives have to be computed numerically, reducing the accuracy. Therefore, in MatcontM we have adopted the use of *minimally extended systems based on bordering methods*, which is free from such problems and which we describe below.

The LP and PD curves are both defined in MatcontM by the following system

$$\begin{cases} g(x,\alpha) - x = 0, \\ s(x,\alpha) = 0, \end{cases} \tag{5.6}$$

where $(x,\alpha) \in \mathbb{R}^{n+2}$. The function g is given by (3.2), while s is obtained by solving the algebraic system

$$\begin{pmatrix} g_x(x,\alpha) \mp I_n & w_{bor} \\ v_{bor}^T & 0 \end{pmatrix} \begin{pmatrix} v \\ s \end{pmatrix} = \begin{pmatrix} 0_n \\ 1 \end{pmatrix}, \tag{5.7}$$

where the bordering vectors $w_{bor}, v_{bor} \in \mathbb{R}^n$ are chosen such that the matrix

$$M(x,\alpha) := \begin{pmatrix} g_x(x,\alpha) \mp I_n & w_{bor} \\ v_{bor}^T & 0 \end{pmatrix}$$

is nonsingular. As above, one should take the "−" sign in (5.7) for the LP curve and the "+" sign for the PD curve. In fact, $s(x,\alpha)$ is proportional to $\det(g_x(x,\alpha) \mp I_n)$. Indeed, by Cramer's rule

$$s(x,\alpha) = \frac{\det(g_x(x,\alpha) \mp I_n)}{\det M(x,\alpha)}.$$

The derivatives of s can also be obtained easily from the derivatives of $g_x(x,\alpha)$:

$$s_z = -w^T (g_x)_z v, \tag{5.8}$$

where z is a state variable or an active parameter and w is defined as the solution of the linear system

$$\begin{pmatrix} g_x^T(x,\alpha) \mp I_n & v_{bor} \\ w_{bor}^T & 0 \end{pmatrix} \begin{pmatrix} w \\ s \end{pmatrix} = \begin{pmatrix} 0_n \\ 1 \end{pmatrix}. \tag{5.9}$$

We note that the quantities called s in (5.7) and (5.9) are the same since they are both equal to the bottom-right element of the inverse of the square matrix $M(x,\alpha)$.

The NS (both *Neimark–Sacker* and *neutral saddle*) curves are defined in MatcontM by the following system

$$\begin{cases} g(x,\alpha) - x = 0, \\ s_{i_1 j_1}(x,\alpha,\kappa) = 0, \\ s_{i_2 j_2}(x,\alpha,\kappa) = 0, \end{cases} \tag{5.10}$$

Table 5.2 Detection of codim 2 bifurcations of cycles.

	LP	PD	NS
CP	$b_0 = 0$		
GPD		$c_0 = 0$	
CH			$L_1 = 0$
R1	$\langle w, v\rangle = 0$		$\det(A - I_n) = \kappa - 1 = 0$
R2		$\langle w, v\rangle = 0$	$\det(A + I_n) = \kappa + 1 = 0$
R3			$\kappa + \frac{1}{2} = 0$
R4			$\kappa = 0$
LPPD	$\det(A + I_n) = 0$	$\det(A - I_n) = 0$	
LPNS	$\det(A \odot A - I_m) = 0$		$\det(A - I_n) = 0, \kappa - 1 \neq 0$
PDNS		$\det(A \odot A - I_m) = 0$	$\det(A + I_n) = 0, \kappa + 1 \neq 0$
NSNS			$\det(A_1 \odot A_1 - I_{m_1}) = 0$

i.e., by $n + 2$ equations for the $(n + 3)$ unknowns $x \in \mathbb{R}^n$, $\alpha \in \mathbb{R}^2$, $\kappa \in \mathbb{R}$. Here, $(i_1, j_1, i_2, j_2) \in \{1, 2\}$ and $s_{i,j}$ are the components of S:

$$S = \begin{pmatrix} s_{11} & s_{12} \\ s_{21} & s_{22} \end{pmatrix},$$

which is obtained by solving

$$\begin{pmatrix} (g_x)^2(x,\alpha) - 2\kappa g_x + I_n & W_{bor} \\ V_{bor}^T & O \end{pmatrix} \begin{pmatrix} V \\ S \end{pmatrix} = \begin{pmatrix} 0_{n,2} \\ I_2 \end{pmatrix}, \tag{5.11}$$

where $V_{bor}, W_{bor} \in \mathbb{R}^{n\times 2}$ are chosen (and can be adapted) so that the matrix in (5.11) is nonsingular. Along the Neimark–Sacker curve, $\kappa = \cos\theta$ is the real part of the critical multipliers $e^{\pm i\theta}$. The derivatives of s_{ij} can be obtained easily from the derivatives of $g_x(x, \alpha)$, as before.

Table 5.2 specifies test functions used in MatcontM to detect and locate all local codim 2 singularities along the codim 1 bifurcation curves. Here, a_0, b_0 and c_0 are the critical normal form coefficients given by (2.7), (2.12) and (2.18). The matrix A_1 is defined as $A_1 = A|_{N^C}$, where $A = g_x$ and N^C is the orthogonal complement in $\mathbb{R}^n$ of the two-dimensional eigenspace associated with the pair of multipliers with unit product of A^T; $m_1 = \frac{1}{2}(n-2)(n-3)$.

It is possible and sometimes necessary to adapt the defining system while continuing a bifurcation curve, i.e., to update the auxiliary variables used in the defining system of the computed branch. The bordering vectors v_{bor} and w_{bor} may require updating since they must ensure that the matrices in (5.7) and (5.9) are nonsingular. Updating is done in MatcontM by replacing v_{bor} and w_{bor} with the normalized vectors v, w computed in (5.7) and (5.9), respectively. Updating of V and W in (5.11) is done similarly, while (i_1, j_1, i_2, j_2) are updated in such a way that the linearized system of (5.10) is as well-conditioned as possible.

5.3 Computation of normal form coefficients

When a codim 2 bifurcation is located, the normal form is computed as in Section 4.2. The tensor-vector products are computed with symbolic derivatives if available, otherwise they are computed with finite differences. After the computation, the non-degeneracy conditions are reported to the user. We have also written routines that provide predictors to new codim 1 branches, as in Section 4.3.2. These predictors also compute the normal form coefficients. The user has to specify an initial amplitude when calling the routines as this might need some (experienced) tuning.

5.3.1 Symbolic derivatives with respect to phase variables

The iteration of (3.1) gives rise to a sequence of points

$$\{x^1, x^2, x^3, \ldots, x^{K+1}\},$$

where $x^{J+1} = f^{(J)}(x^1, \alpha)$ for $J = 1, 2, \ldots, K$. Suppose that symbolic derivatives of f up to order 5 can be computed at each point. We write

$$A(x^J)_{i,j} = \frac{\partial f_i}{\partial x_j}(x^J),\ B(x^J)_{i,j,k} = \frac{\partial^2 f_i}{\partial x_j \partial x_k}(x^J),\ C(x^J)_{i,j,k,l} = \frac{\partial^3 f_i}{\partial x_j \partial x_k \partial x_l}(x^J),$$

and similarly for $D(x^J)$ and $E(x^J)$.

We want to find recursive formulas for the derivatives of the composition (3.2), i.e., the coefficients of the multilinear functions in (3.3) that we now denote by $A^{(J)}$, $B^{(J)}$ and $C^{(J)}$ to indicate the iterate explicitly:

$$(A^{(J)})_{i,j} = \frac{\partial (f^{(J)}(x^1))_i}{\partial x_j},\ (B^{(J)})_{i,j,k} = \frac{\partial^2 (f^{(J)}(x^1))_i}{\partial x_j \partial x_k},$$

and $C^{(J)}$, $D^{(J)}$ and $E^{(J)}$ are analogously defined. What follows is a straightforward application of the Chain Rule.

For $J = 1$ we have $A^{(1)} = A(x^1)$, $B^{(1)} = B(x^1)$ and $C^{(1)} = C(x^1)$ and these are known. Now, as

$$A^{(J)}_{i,j} = \sum_k \frac{\partial f_i}{\partial x_k}(f^{(J-1)}(x^1)) \frac{\partial (f^{(J-1)}(x^1))_k}{\partial x_j} = \sum_l A(x^J)_{i,k} A^{(J-1)}_{k,j}, \tag{5.12}$$

we see that

$$(F(x, \alpha))_x = A(x^K) A(x^{K-1}) \cdots A(x^1) - I_n, \tag{5.13}$$

where $F(x, \alpha) = f^{(K)}(x, \alpha) - x$.

For the second-order derivatives we first write $B^{(J)}$ once in coordinates

$$\begin{aligned} B^{(J)}_{i,j,k} &= \frac{\partial}{\partial x_j}\frac{\partial}{\partial x_k} f_i(f^{(J-1)}(x)) \\ &= \sum_{l,m} \frac{\partial^2 f_i}{\partial x_l \partial x_m}(x^J)\frac{\partial (f^{(J-1)})_m}{\partial x_j}\frac{\partial (f^{(J-1)})_l}{\partial x_k} + \sum_l \frac{\partial f_i}{\partial x_l}(x^J)\frac{\partial^2 (f^{(J-1)})_l}{\partial x_j \partial x_k}. \end{aligned}$$

For any two vectors q_1 and q_2, we can multiply the previous expression by $(q_1)_j(q_2)_k$ and sum over (k,l) to obtain

$$B^{(J)}(q_1,q_2) = B(x^J)(A^{(J-1)}q_1, A^{(J-1)}q_2) + A(x^J)B^{(J-1)}(q_1,q_2). \tag{5.14}$$

As $A(x^J)$ and $B(x^J)$ are known, (5.14) allows us to compute the multilinear form $B^{(K)}(q_1,q_2)$ recursively.

Let $q_i, i = 1,2,3,4,5$, be given vectors. Multilinear forms with higher-order derivatives can be computed with

$$\begin{aligned} C^{(J)}(q_1,q_2,q_3) &= C(x^J)(A^{(J-1)}q_1, A^{(J-1)}q_2, A^{(J-1)}q_3) \\ &\quad + B(x^J)(B^{(J-1)}(q_1,q_2), A^{(J-1)}q_3)^* \\ &\quad + A(x^J)(C^{(J-1)}(q_1,q_2,q_3)), \end{aligned} \tag{5.15}$$

where * means that all combinatorially different terms have to be included, i.e.,

$$\begin{aligned} B(x^J)(B^{(J-1)}(q_1,q_2), A^{(J-1)}q_3)^* &= B(x^J)(B^{(J-1)}(q_1,q_2), A^{(J-1)}q_3) \\ &\quad + B(x^J)(B^{(J-1)}(q_1,q_3), A^{(J-1)}q_2) \\ &\quad + B(x^J)(B^{(J-1)}(q_2,q_3), A^{(J-1)}q_1). \end{aligned}$$

For $D^{(J)}$, we get

$$\begin{aligned} D^{(J)}(q_1,q_2,q_3,q_4) &= D(x^J)(A^{(J-1)}q_1, A^{(J-1)}q_2, A^{(J-1)}q_3, A^{(J-1)}q_4) \\ &\quad + C(x^J)(B^{(J-1)}(q_1,q_2), A^{(J-1)}q_3, A^{(J-1)}q_4)^* \\ &\quad + B(x^J)(B^{(J-1)}(q_1,q_2), B^{(J-1)}(q_3,q_4))^* \\ &\quad + B(x^J)(C^{(J-1)}(q_1,q_2,q_3)), A^{(J-1)}q_4)^* \\ &\quad + A(x^J)D^{(J-1)}(q_1,q_2,q_3,q_4). \end{aligned} \tag{5.16}$$

Finally, for $E^{(J)}$, it holds

$$\begin{aligned} E^{(J)}(q_1,q_2,q_3,q_4,q_5) &= E(x^J)(A^{(J-1)}q_1, A^{(J-1)}q_2, A^{(J-1)}q_3, A^{(J-1)}q_4, A^{(J-1)}q_5) \\ &\quad + D(x^J)(B^{(J-1)}(q_1,q_2), A^{(J-1)}q_3, A^{(J-1)}q_4, A^{(J-1)}q_5)^* \\ &\quad + C(x^J)(B^{(J-1)}(q_1,q_2), B^{(J-1)}(q_3,q_4), A^{(J-1)}q_5)^* \\ &\quad + C(x^J)(C^{(J-1)}(q_1,q_2,q_3), A^{(J-1)}q_4, A^{(J-1)}q_5)^* \\ &\quad + B(x^J)(C^{(J-1)}(q_1,q_2,q_3), B^{(J-1)}(q_4,q_5))^* \\ &\quad + B(x^J)(D^{(J-1)}(q_1,q_2,q_3,q_4))(A^{(J-1)}q_5)^* \\ &\quad + A(x^J)(E^{(J-1)}(q_1,q_2,q_3,q_4,q_5)). \end{aligned}$$

The multilinear forms $A^{(K)}(q_1)$, $B^{(K)}(q_1,q_2)$, $C^{(K)}(q_1,q_2,q_3)$, $D^{(K)}(q_1,q_2,q_3,q_4)$ and $E^{(K)}(q_1,q_2,q_3,q_4,q_5)$ are then used in the computations of the normal form coefficients for codim 1 and codim 2 bifurcations of period-K cycles and also in the algorithms for branch switching.

5.3.2 Symbolic derivatives with respect to parameters

If enough symbolic derivatives of f are available, then MatcontM computes the expressions involving J_1 and A_1 in (3.3) symbolically. The idea is as follows. Taking the derivative of (3.2) with respect to α_k gives

$$\frac{\partial(f^{(J)}(x^1,\alpha))}{\partial\alpha_k} = \frac{\partial f}{\partial\alpha_k}(x^J,\alpha) + \frac{\partial f}{\partial x}(x^J,\alpha)\frac{\partial(f^{(J-1)}(x^1,\alpha))}{\partial\alpha_k}, \tag{5.17}$$

which is recursively computable. Also, mixed derivatives, which are necessary for continuation and branch switching, can be found recursively:

$$\frac{\partial^2(f^{(J)}(x^1,\alpha))}{\partial\alpha_k\partial x} = \frac{\partial^2 f}{\partial\alpha_k\partial x}(x^J,\alpha) + \frac{\partial^2 f}{\partial x^2}(x^J,\alpha)\frac{\partial(f^{(J-1)}(x^1,\alpha))}{\partial\alpha_k}. \tag{5.18}$$

In fact, the recursion is not applied to (5.18) itself, but to its product with a fixed vector.

This is sufficient for all continuations of fixed points and their codim 1 bifurcations. It is also sufficient for all cases of branch switching from codim 2 points, except for the case of degenerate flip. For this case, we fall back to a finite difference approximation. Since it is only used in the prediction step for which high accuracy is not needed, this seems acceptable.

5.3.3 Recursive formulas for derivatives of the defining systems for continuation

For the continuation of fixed points and cycles, we need the derivatives of (3.2), which can be computed from (5.13) and (5.17).

Now, we consider the derivatives of s (as defined in (5.6)) with respect to z, a state variable or parameter. The flip and Neimark–Sacker cases can be handled in a similar way. Let M be the matrix in (5.7). By taking derivatives of (5.7) with respect to z, we obtain

$$M\begin{bmatrix} v_z \\ s_z \end{bmatrix} + \begin{bmatrix} A_z^{(K)} & 0 \\ 0 & 0 \end{bmatrix}\begin{bmatrix} v \\ s \end{bmatrix} = 0. \tag{5.19}$$

Using (5.9) we obtain

$$s_z = -w^T(A^{(K)})_z v. \tag{5.20}$$

If z represents one of the state variables, then $s_{x_i} = -\langle w, B^{(K)}(e_i, v)\rangle$ as computed before. When z is a parameter α_k we can write

$$s_{\alpha_k} = \sum_{J=1}^{K} C_J, \tag{5.21}$$

where

$$C_J = -w^T f_x(x^K) \cdots (f_x(x^J))_{\alpha_k} f_x(x^{J-1}) \cdots f_x(x^1)v \tag{5.22}$$

where $J = 1, \ldots, K$. In this expression

$$(f_x(x^J))_{\alpha_k} = [f_x(f^{(J)}(x^1, \alpha))]_{\alpha_k} = f_{x\alpha}(x^J, \alpha) + B(x^J)T_J, \tag{5.23}$$

where T_J is a vector that can be recursively defined by

$$T_J = f_{\alpha_k}(x^{J-1}, \alpha) + A(x^{J-1})T_{J-1}, \quad T_1 = 0. \tag{5.24}$$

Summarizing, for the computation of s_α we need to compute $f_x, f_{\alpha_k}, f_{xx}, f_{x\alpha_k}$ in all iteration points $x^1, \ldots, x^K$, and given these compute T_J for $J = 1, \ldots K$. Then

$$C_J = -w^T A(x^K) \cdots (f_{x\alpha_k}(x^J) + B(x^J)T_J)A(x^{J-1}) \cdots A(x^1)v \tag{5.25}$$

and s_{α_k} is computed via (5.21).

5.3.4 Algorithmic differentiation for directional derivatives

Another method to compute derivatives is algorithmic differentiation (AD), also known as automatic differentiation. Every function is built from elementary functions. For every elementary function, including arithmetic operations, trigonometric functions and exponentials, the derivative can be written using the Chain Rule. The basic idea of AD is to use the chain rule to build up a finite Taylor expansion of the function while evaluating the function. As we study maps and its iterates, this is a natural approach to construct the higher-order derivatives. We discuss the *forward mode* of AD as implemented by Pryce et al. (2010) for a single variable only. For more details on AD we refer to Griewank, Juedes, and Utke (1996) and Griewank and Walther (2008).

A special class of variables `adtayl` has been set up in MatcontM. An `adtayl` object `x` has one field `tc` consisting of an $m \times n \times p + 1$ array of Taylor coefficients. For a scalar variable $x = x(t)$, the array `tc` consist of the Taylor coefficients $x_0, x_1, \ldots, x_p$. In the scalar case $m = n = 1$, while if $m = 1$ or $n = 1$ this corresponds to a row or column vector, respectively. Singleton dimensions are always squeezed. The call `t=adtayl(a,p)` creates an object representing the series $a_0 + a_1 t + \cdots + a_p t^p$ with $a_0 = a$ and $a_1 = 1$. New variables based on t will automatically involve the correct series. The following example creates an `adtayl` object and produces a new series:

```
s=adtayl(1,2);pol=(2+s)*(3+s*s)
```

The result mentions $a_0 = 12, a_1 = 10, a_2 = 5$, which is as expected.

Suppose we want to compute the directional derivative $Av = \frac{d}{dt} f^{(J)}(x0 + tv)\big|_{t=0}$. We need a function `func` to evaluate f and to do this J times. We also have a base point `x0`, a direction vector `v` and parameters `par`. The following code can now be used to compute `Av`:

```
s=adtayl(0,p)            % Create adtayl-object of order p
y1=x0+s*v;               % Base point + perturbation
for i=1:J                % Evaluate the map J times
  y1 = func(0,y1,par{:});
end
ytayl=tcs(y1);          % Extract Taylor coefficients
```

The desired result can be extracted as `Av=ytayl(:,2)` as the jth column corresponds to the $(j-1)$-th derivative. This can be generalized to higher-order derivatives using polarization identities (see Section 5.3.5). For complex vectors v, one should compute the Taylor coefficients separately on the real and imaginary part and compose the final result.

Numerical experiments (Pryce et al., 2010) have shown that AD yields the same normal form coefficients within machine precision as symbolic derivatives. There is, however, a trade-off in computational time between the setup for the `adtayl` class and the iteration number. If both symbolic derivatives are available, this is much faster for low iteration numbers. For higher-order derivatives the computational complexity of AD scales better with iteration numbers. In practice, it is therefore a good idea to use symbolic derivatives for low iteration numbers and switch to AD for iteration numbers above a given threshold.

5.3.5 Numerical computation of the directional derivatives

If symbolic derivatives of the original map are not available, then also finite differences can be used. This is an option of last resort which is not accurate for high-order derivatives and very high iterates. In those cases, numerical errors easily propagate and results must be interpreted with care. To determine critical normal form coefficients we do not need the full tensors, but the multilinear forms evaluated on vectors, which can be computed with directional derivatives and central finite differences. Here we use so-called polarization identities to compute the directional derivatives. For instance, for second-order derivatives one has $B(u+v,u+v) = B(u,u) + 2B(u,v) + B(v,v)$ and $B(u-v,u-v) = B(u,u) - 2B(u,v) + B(v,v)$. Then the mixed directional derivative can be computed as

$$B(u,v) = \frac{1}{4}\left(B(u+v,u+v) - B(u-v,u-v)\right).$$

Similarly, the expansion

$$\begin{aligned} C(u+v+w,u+v+w,u+v+w) &= C(u,u,u) + C(v,v,v) + C(w,w,w) \\ &\quad + 3C(u,u,v) + 3C(u,u,w) + 3C(u,v,v) \\ &\quad + 3C(u,w,w) + 3C(v,v,w) + 3C(v,w,w) \end{aligned}$$

can be used to derive the compact forms

$$C(u,u,v) = \frac{1}{6}\left(C(u+v,u+v,u+v) + C(u-v,u-v,u-v) - 2C(v,v,v)\right)$$

and

$$\begin{aligned} C(u,v,w) = \frac{1}{24} &\left(C(u+v+w,u+v+w,u+v+w)\right. \\ &+ C(u+v-w,u+v-w,u+v-w) \\ &- C(u-v+w,u-v+w,u-v+w) \\ &+ \left. C(u-v-w,u-v-w,u-v-w)\right). \end{aligned}$$

Similar formulas have been derived and implemented for higher-order directional derivatives. For a general discussion of directional derivatives we refer to Kuznetsov (2004). Similar to Hassard, Kazarinoff, and Wan (1981), an analysis based on the assumption that higher-order derivatives are of the same order of magnitude as the function itself leads to $h_{min} \approx \epsilon_m^{1/(k+2)}$, where ϵ_m denotes machine-precision. Using double-precision we have $\epsilon_m \approx 10^{-15}$ and thus $h \equiv h_1 = 10^{-5}$, which is the default value in MatcontM. In MatcontM, the **Increment** ($= h_1$) can be adjusted by the user. The increments of the higher-order derivatives are then adapted according to the above formulas.

5.4 Computation of one-dimensional invariant manifolds of saddle fixed points

5.4.1 Computing an unstable manifold

We use the algorithm for computing the global one-dimensional unstable manifold of a saddle point of a map proposed by Krauskopf and Osinga (1998a,b). We suppose f is invertible and orientation-preserving, otherwise we consider

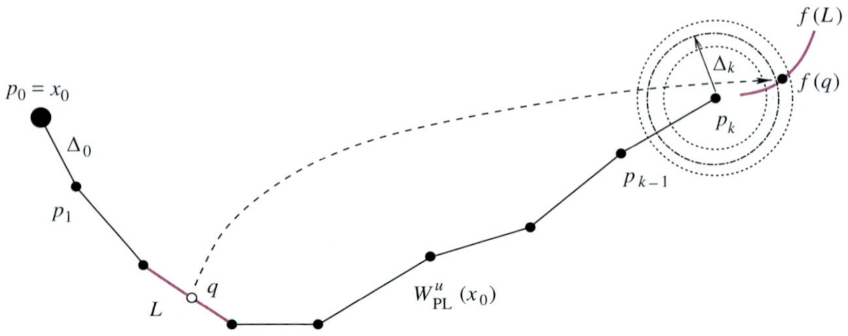

Figure 5.1 Growing the unstable manifold.

its second iterate. Let x_0 be a saddle point of f. The unstable manifold of x_0 can now be defined as

$$W^u(x_0) = \left\{x \in \mathbb{R}^2 : f^{-j}(x) \to x_0 \text{ as } j \to \infty\right\}. \tag{5.26}$$

The Unstable Manifold Theorem 1.1 guarantees the existence of the local unstable manifold

$$W^u_{loc}(x_0) = \left\{x \in W^u(x_0) : f^{-j}(x) \in U \text{ for all } j \in \mathbb{N}\right\} \tag{5.27}$$

in a suitable neighborhood U of x_0. Furthermore, it states that $W^u_{loc}(x_0)$ is tangent to the unstable eigenspace $E^u(x_0)$ corresponding to $\lambda > 0$.

The idea is to grow the global unstable manifold $W^u(x_0)$ independently of the dynamics in steps as a list of ordered points. At each step a new point is added at a prescribed distance Δ_k from the last point. New points are found as f-images of suitable points from the part already computed. The algorithm starts with a linear approximation of the local manifold and grows the manifold up to a pre-specified arclength l with a speed depending on the local curvature of the manifold.

We now briefly describe a single step of the algorithm and suppose that the sequence of points $M = \{p_0, p_1, p_2, \ldots, p_k\}$ approximating the unstable manifold is already computed, where the point p_1 is taken at a distance Δ_0 from $p_0 = x_0$. The point p_1 can be obtained by iterations, starting from a point in the unstable eigenspace (line) $E^u(x_0)$ at a small distance δ_{init} to x_0.

The next point p_{k+1} should have the property that the line segment $[p_k, p_{k+1}]$ accurately approximates $W^u(x_0)$, thus providing the next segment of its *piecewise-linear* approximation $W^u_{\text{PL}}(x_0)$. In order to achieve a good approximation, the distance Δ_k between p_k and p_{k+1} must be adjusted from step to step according to the curvature of the manifold.

We want to find p_{k+1} near a sphere centered at p_k with radius Δ_k. To this end, we search in $W^u_{\mathrm{PL}}(x_0)$ a line segment L that is mapped by f to a curve which intersects the sphere. We start with the line segment in $W^u_{\mathrm{PL}}(x_0)$ that contains the pre-image of p_k and move linearly through $W^u_{\mathrm{PL}}(x_0)$. Once L is found, we use bisection to find a point $q \in L$ such that

$$(1 - \epsilon)\Delta_k < \|f(q) - p_k\| < (1 + \epsilon)\Delta_k.$$

The point $f(q)$ is a candidate for the next point p_{k+1} in M (see Figure 5.1). If it is acceptable according to the curvature criterion, then $p_{k+1} := f(q)$ is added to M, the segment $[p_k, p_{k+1}]$ is added to $W^u_{\mathrm{LP}}(x_0)$, and the step is completed. The next Δ_{k+1} could be increased. However, if $f(q)$ fails the curvature test, then we reject this point, decrease the distance Δ_k and repeat the procedure. The curvature criterion means that

$$\alpha_{\min} \le \alpha_k \le \alpha_{\max} \quad \text{and} \quad (\Delta\alpha)_{\min} \le \Delta_k \alpha_k \le (\Delta\alpha)_{\max},$$

where α_k is the angle between $(p_k - p_{k-1})$ and $(p_{k+1} - p_k)$. It is also ensured that

$$\Delta_{\min} \le \Delta_k \le \Delta_{\max}.$$

The constants $\alpha_{\min,\max}$, $(\Delta\alpha)_{\min,\max}$ and $\Delta_{\min,\max}$, as well as δ_{init} and ϵ, are defined by the user. This algorithm is presented in pseudo-code by Krauskopf and Osinga (1998b).

5.4.2 Computing a stable manifold

For the stable manifold we discuss a generalization of the method from England, Krauskopf, and Osinga (2004). Their algorithm grows a one-dimensional stable manifold without using the explicit inverse at all for planar maps. In Bruschi (2010) a higher-dimensional version was proposed which uses the inverse in the form of Newton iterations. This implicitly assumes that the manifold is locally unique. As for the unstable manifold, we assume that the eigenvalue corresponding to the stable direction is positive, which can be ensured by considering the second iterate.

The initial part of the algorithm produces a point p_1 at a distance Δ_0 from $p_0 = x_0$. Here, iterations with the linear inverse, starting from a point in the stable eigenspace (line) $E^s(x_0)$ at a small distance δ_{init} to $p_0 = x_0$, can be used.

The main part of the algorithm is a two-step procedure with prediction and correction. Let p_k be the last computed point of the sequence

$$M = \{p_0, p_1, p_2, \ldots, p_k\}$$

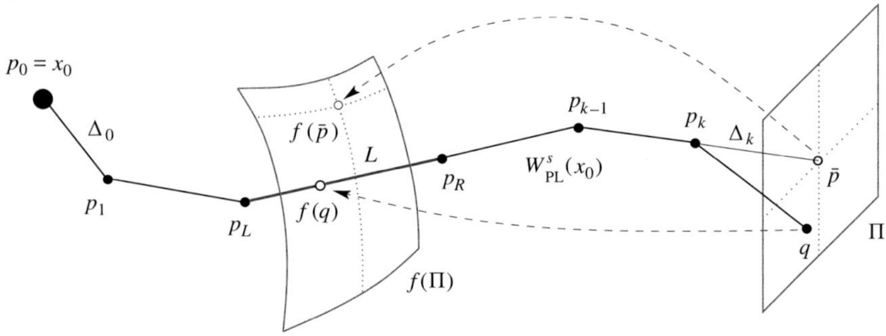

Figure 5.2 Growing the stable manifold.

defining the piecewise-linear approximation $W^s_{\rm PL}(x_0)$ of the global stable manifold $W^s(x_0)$. First, we predict a new point

$$\bar{p} = p_k + \Delta_k \frac{p_k - p_{k-1}}{\|p_k - p_{k-1}\|}$$

(see Figure 5.2). We want to find a new point q in a hyperplane Π orthogonal to $(p_k - p_{k-1})$ and such that $f(q)$ belongs to a previously computed segment $L = [p_L, p_R]$ in $W^s_{\rm PL}(x_0)$, i.e.,

$$\begin{cases} f(q) - \tau(p_R - p_L) = p_R, \\ \langle q - \bar{p}, p_k - \bar{p}\rangle = 0, \end{cases} \tag{5.28}$$

where $\tau \in [0, 1]$. In practice, the Jacobian matrix of (5.28) is ill-conditioned. We first use $f(\bar{p})$ to estimate τ, and then apply Newton corrections starting at $\bar{p}$ to find q. The resulting point q is a candidate for the next point in M. If it is acceptable according to the same curvature criterion as in the computation of the unstable manifold, then $p_{k+1} := q$ is added to M. The adaptation of Δ_k is also similar to the unstable manifold case.

5.5 Continuation of connecting orbits

We adapt our presentation from Khoshsiar Ghaziani et al. (2009).

Assume that the eigenvalues of $A(\alpha) := (f^J(x_1, \alpha))_x$ of the cycle $\{x_1, x_2, ..., x_{J-1}\}$ are ordered as follows:

$$|\lambda_1| \leq \cdots \leq |\lambda_m| < 1 < |\lambda_{m+1}| \leq \cdots \leq |\lambda_n|.$$

The algorithm requires the evaluation of two projections associated with the eigenspaces of $A(\alpha)$. These projections are constructed using the real Schur factorizations

$$A(\alpha) = Q^{(1)}R^{(1)}[Q^{(1)}]^T, \quad A(\alpha) = Q^{(N)}R^{(N)}[Q^{(N)}]^T,$$

where $Q^{(1)}, R^{(1)}, Q^{(N)}$ and $R^{(N)}$ are $n \times n$ matrices. The first factorization has been chosen so that the first m columns $q_1^S, \ldots, q_m^S$ of $Q^{(1)}$ form an orthonormal basis of the right invariant subspace S_1 of $A(\alpha)$, corresponding to $\lambda_1, \ldots, \lambda_m$ and the remaining $n - m$ columns $q_{m+1}^U, \ldots, q_n^U$ of $Q^{(1)}$ form an orthonormal basis of $S_1^\perp$. Similarly, the first $l = n - m$ columns $q_1^U, \ldots, q_l^U$ of $Q^{(N)}$ form an orthonormal basis of the right invariant subspace U_1 of $A(\alpha)$, corresponding to $\lambda_{m+1}, \ldots, \lambda_n$ and the remaining $n - l - m$ columns $q_{l+1}^U, \ldots, q_n^U$ of $Q^{(1)}$ form an orthonormal basis of $S_N^\perp$.

When dealing with homoclinic connections, we want to find a sequence of points $(x_k)_{k=1,\ldots,N}$ satisfying

- the cycle condition

$$f^J(x_1, \alpha) - x_1 = 0; \tag{5.29}$$

- the iteration conditions

$$f^J(x_k, \alpha) - x_{k+1} = 0, \quad k = 2, 3, \ldots, N-2; \tag{5.30}$$

- the left projection boundary conditions

$$(x_2 - x_1)^T q_{n_U+i}^U(\alpha) = 0, \quad i = 1, \ldots, n - n_U; \tag{5.31}$$

- the right projection boundary conditions

$$(x_{N-1} - x_1)^T q_{n_S+i}^S(\alpha) = 0, \quad i = 1, \ldots, n - n_S. \tag{5.32}$$

A regular zero of a system of equations (5.29), (5.30), (5.31) and (5.32) corresponds to a transversal homoclinic orbit to a hyperbolic fixed point. Thus, a zero for this system can be continued in one parameter. A limit point of this system of equations corresponds to a homoclinic tangency (Beyn and Kleinkauf, 1997).

We note that for continuation the Jacobian matrix is required. As it is sparse, it is advantageous to compute this efficiently. These details can be found in Khoshsiar Ghaziani et al. (2009).

5.5.1 Continuation of invariant subspaces

Equations (5.31) and (5.32) imply that we should have the stable and unstable eigenspaces of the map at the fixed points x_1 available at each step of the continuation. In principle, the Schur decomposition can be computed at every step. This is, however, computationally expensive and does not guarantee smoothness of the bases. Instead, we add the computation of the subspaces to the continuation. This was originally introduced by Demmel, Dieci, and Friedman (2000). In a continuation context it is more natural to reparameterize the Jacobian matrix $A(\alpha)$ by the so-called pseudo-arclength variable s. So as the parameter $\alpha(s)$ varies along a solution branch we can see this as a matrix-valued function $A: s \in \mathbb{R} \to \mathbb{R}^{n\times n}$, and we write $A(s) := A(\alpha(s))$.

We first consider the unstable eigenspace of $A(s)$. Suppose that initially we have the (real) block Schur factorization

$$A(0) = Q(0)R(0)Q^T(0), \quad Q(0) = [Q_1(0) \;\; Q_2(0)], \tag{5.33}$$

where $A(0), R(0)$ and $Q(0)$ are $n \times n$ matrices, $Q(0)$ is orthogonal, $Q_1(0)$ has dimensions $n \times n_U$ and $R(0)$ is block upper triangular

$$R(0) = \begin{bmatrix} R_{11}(0) & R_{12}(0) \\ 0 & R_{22}(0) \end{bmatrix}, \tag{5.34}$$

where $R_{11}(0)$ and $R_{22}(0)$ are $n_U \times n_U$ and $(n - n_U) \times (n - n_U)$ matrices, respectively; $R_{ii}(0), i = 1, 2$, are not required to be triangular. The columns of $Q_1(0)$ span the unstable invariant subspace $S(0)$ of $A(0)$, and the columns of $Q_2(0)$ span the orthogonal complement $S^{\perp}(0)$. We want to obtain a block Schur factorization for the matrix $A(s)$, close to $A(0)$.

Suppose that the matrix $A(s)$ has two groups of eigenvalues, $\Lambda_1(s)$ (with modulus > 1) and $\Lambda_2(s)$ (with modulus < 1), which stay disjoint for all s around 0. Then, in a neighborhood of $s = 0$, we need a smooth factorization

$$A(s) = Q(s)R(s)Q^T(s), \quad Q(s) = [Q_1(s) \;\; Q_2(s)], \tag{5.35}$$

where $R(s)$ is in block Schur form

$$R(s) = \begin{bmatrix} R_{11}(s) & R_{12}(s) \\ 0 & R_{22}(s) \end{bmatrix}. \tag{5.36}$$

Here, R_{11} has eigenvalues $\Lambda_1(s)$ and R_{22} has eigenvalues $\Lambda_2(s)$. As shown by Dieci and Eirola (1999), it is always possible to obtain a smooth path of block Schur factorizations that satisfies (5.35) and (5.36). However, this smooth path is usually not unique.

Thus we can write

$$Q(s) = Q(0)U(s), \quad \text{with} \quad U(0) = I, \tag{5.37}$$

so that we only need to compute the $n \times n$ matrix $U(s)$. We partition $U(s)$ in blocks of the same size as $R(0)$ in (5.34):

$$U(s) = [U_1(s) \;\; U_2(s)] = \begin{bmatrix} U_{11}(s) & U_{12}(s) \\ U_{21}(s) & U_{22}(s) \end{bmatrix}, \tag{5.38}$$

so that $U_{11}(s)$ and $U_{22}(s)$ are $n_U \times n_U$ and $(n-n_U)\times(n-n_U)$ matrices, respectively. By redefining $Q(s)$ and $R(s)$ if necessary, one can assume $U_{11}(s)$ and $U_{22}(s)$ are symmetric positive-definite matrices.

Since $U(0) = I$, there is an open interval about 0, call it I_0, where we can require that U_1 has the structure

$$U_1(s) = \begin{bmatrix} I \\ U_{21}(s)U_{11}^{-1}(s) \end{bmatrix} U_{11}(s). \tag{5.39}$$

Next, for all $s \in I_0$, we define

$$Y_U(s) = U_{21}(s)U_{11}^{-1}(s). \tag{5.40}$$

One can show that $I + Y_U^T Y_U = U_{11}^{-T}(s)U_{11}^{-1}(s)$, hence U_{11}^{-1} is the unique square root of $I + Y_U^T Y_U$. This implies that we can rewrite (5.39) in terms of Y_U and choose U_{11} symmetric, to obtain

$$U_1 = \begin{bmatrix} I \\ Y_U \end{bmatrix} (I + Y_U^T Y_U)^{-\frac{1}{2}}. \tag{5.41}$$

Similarly, using $U_2^T U_2 = I$ and $U_1^T U_2 = 0$ for U_2, so that eventually we obtain for every $s \in I_0$

$$U(s) = \left[\begin{pmatrix} I \\ Y_U \end{pmatrix} (I + Y_U^T Y_U)^{-\frac{1}{2}} \quad \begin{pmatrix} -Y_U^T \\ I \end{pmatrix} (I + Y_U Y_U^T)^{-\frac{1}{2}} \right]. \tag{5.42}$$

Hence, the columns of

$$Q_U(s) = Q_U(0) \begin{bmatrix} I \\ Y_U \end{bmatrix} \tag{5.43}$$

form a base for the unstable eigenspace at x_1 and the columns of

$$Q_U^{\perp}(s) = Q_U(0) \begin{bmatrix} -Y_U^T \\ I \end{bmatrix} \tag{5.44}$$

form a base for the orthogonal complement of the unstable eigenspace. We note that these bases are in general not orthogonal. Thus, we need to find the matrix $Y_U \in \mathbb{R}^{(n-n_U)\times n_U}$ in (5.42). For any given $s \in I_0$, define $\hat{R}_{11}, \hat{R}_{12}, E_{21}$ and $\hat{R}_{22}$ by

$$Q^T(0)A(s)Q(0) = \begin{bmatrix} \hat{R}_{11} & \hat{R}_{12} \\ E_{21} & \hat{R}_{22} \end{bmatrix}, \tag{5.45}$$

where $\hat{R}_{11}$ is of size $n_U \times n_U$ and $\hat{R}_{22}$ is an $(n-n_U)\times(n-n_U)$ matrix. Substituting (5.45) and $Q(s)$ from (5.37) and (5.42) into the invariant subspace relation

$Q_2^T(s)A(s)Q_1(s) = 0$, we obtain the following *algebraic Riccati equation* for Y_U:

$$F(Y_U) = 0, \quad F(Y_U) := \hat{R}_{22}Y_U - Y_U\hat{R}_{11} + E_{21} - Y_U\hat{R}_{12}Y_U. \tag{5.46}$$

In the same way we can compute a right invariant (stable) n_S-dimensional subspace $S(s)$ of $A(s)$. First, we consider $Q(s) = [Q_1(s)\ Q_2(s)] \in \mathbb{R}^{n\times n}$, $Q_1(s) \in R^{n\times n_S}$, $Q_2(s) \in R^{n\times(n-n_S)}$ so that $Q_1(s)$ spans $S(\alpha)$ and $Q_2(s)$ spans the orthogonal complement $S^{\perp}(\alpha)$.

Using the same procedure as in the computation of the unstable subspace for x_1, we can obtain the relations

$$Q(s) = Q(0)S(s), \quad \text{with } S(0) = I, \tag{5.47}$$

and

$$S(s) = \left[\begin{pmatrix} I \\ Y_S \end{pmatrix}(I + Y_S^T Y_S)^{-\frac{1}{2}} \quad \begin{pmatrix} -Y_S^T \\ I \end{pmatrix}(I + Y_S Y_S^T)^{-\frac{1}{2}}\right], \tag{5.48}$$

and eventually the algebraic Riccati equation for Y_S

$$F(Y_S) = 0, \quad F(Y_S) := \hat{R}_{22}Y_S - Y_S\hat{R}_{11} + E_{21} - Y_S\hat{R}_{12}Y_S \tag{5.49}$$

to compute the stable invariant subspace and its orthogonal complement for x_1. Solving (5.49) for Y_S of size $(n - n_S) \times n_S$, enables us to compute the span of the stable invariant subspace of x_N and its orthogonal complement. If $Q_S(0)$ is the orthogonal matrix from the starting homoclinic orbit, related to the stable invariant subspace, then a basis for the stable eigenspace in the new step at x_N is given by the columns of

$$Q_S(s) = Q_S(0)\begin{bmatrix} I \\ Y_S \end{bmatrix}. \tag{5.50}$$

A basis for the orthogonal complement of the subspace in the new step $Q_S^{\perp}$ is given by the columns of

$$Q_S^{\perp}(s) = Q_S(0)\begin{bmatrix} -Y_S^T \\ I \end{bmatrix}. \tag{5.51}$$

These bases are in general not orthogonal.

5.5.2 The defining system

To continue a heteroclinic connection of the cycle x_1 we need the following:

- **Continuation variables.** There are $K = Nn + (n - n_U)n_U + (n - n_S)n_S + 1$ continuation variables consisting of:
  - an n-vector with the coordinates of the initial fixed point;

– $(N-2)$ n-vectors with the coordinates of the mesh points $x_2, \ldots, x_{N-1}$;
– an n-vector x_N with the coordinates of the final fixed point;
– the vector Y_U^v, i.e., column-wise vectorized Y_U;
– the vector Y_S^v, i.e., column-wise vectorized Y_S;
– an active parameter ap, i.e., α_a.

- **Defining system.** The defining system consists of $K-1 = Nn+(n-n_U)n_U+(n-n_S)n_S$ equations:
  – the fixed point constraint $f^J(x_1, \alpha) - x_1 = 0$;
  – the constraints $f^J(x_{j-1}, \alpha) - x_j = 0, \quad j = 3, \ldots, N-1$;
  – the row-wise vectorized Riccati equation (5.46) for Y_U;
  – the row-wise vectorized Riccati equation (5.49) for Y_S;
  – the initial boundary conditions (5.31);
  – the final boundary conditions (5.32).

During continuation, the basis Q_U and $Q_U^\perp$ will no longer be orthogonal. To restore this, one can periodically compute the singular value decomposition $[U, S, V] = svd(Q_U)$, where U and V are unitary matrices of sizes $n \times n$ and $n_U \times n_U$, respectively, and S is a diagonal matrix of size $n \times n_U$. To adapt the continuation problem, one can set $Q_U = U$ and $Y_U = 0$.

5.5.3 Finding initial data for connecting orbits

The continuation of connecting orbits requires special starting data as in addition to the fixed point also an orbit piece is needed. Finding accurate values can be a tedious task, but the computation of invariant manifolds helps to determine a suitable approximation to a connecting orbit. We describe two algorithms to find intersection points between two manifolds.

The first algorithm is for planar maps. In this case we assume we have a stable and an unstable manifolds which are both one-dimensional curves. As each manifold consists of line segments, one can test each possible combination of segments for intersections, as was done by Bruschi (2010). A more efficient approach is to employ a modified line sweep algorithm as implemented for MatcontM (Vanhulle, 2017). First we create a list with the manifolds sorted according to their x-coordinate. Given a starting point of a segment on the unstable manifold, we look for all segments of the stable manifold that strictly overlap with this segment, i.e., not just with one point but more. For these overlapping points we also check whether they overlap in the y-direction. For such points we then check whether they have an intersection point. This method has complexity $O((2n+k)\log(2n))$ with n the number of segments and k the number of intersection points. Once the intersection points have been determined, they have to be assembled into orbit pieces. We start with one point as part of an orbit. This point is iterated by the map and then we check if it is sufficiently close to some other computed intersection point. If this is the case, that intersection point is added to the current orbit. Otherwise, the current orbit is finished and new orbit pieces are assembled with the remaining intersection points. The result consists typically of two longer orbit pieces which are primary connecting orbits, while shorter pieces come from secondary tangencies.

This method only works for planar maps and requires that an intersection exists. Sometimes the connecting orbit exists for a very small region in parameter space. Then it will be difficult to find the intersection points. Also for maps of dimension $n > 2$ one needs a different strategy. Typically, one of the manifolds is one-dimensional. This leads to the following second method, as discussed by Neirynck et al. (2018). Suppose that we have a cycle with a one-dimensional unstable manifold. Also suppose that the manifold returns close to the cycle. We can compute this manifold using methods from Section 5.4. Next we approximate the other manifold by an $(n-1)$-dimensional hyperplane. It is straightforward to find intersections, if any, of the curve and the hyperplane. Now we can use the intersection point closest to the cycle together with its pre-images as the initial data. The latter is possible as during the computation of the invariant manifolds one can add information about how points from one segment are mapped to other segments. This method is easily generalized to the case of a stable manifold or a heteroclinic orbit. The final initial data for the continuation may be less accurate. Numerical experiments have shown that a few Newton corrections including parameter corrections still lead to an actual connecting orbit if that is sufficiently close (Neirynck et al., 2018).

As a final improvement of the orbits we mention that one can extends the orbit pieces by adding points along the linear leading eigenspaces so that the start and end point of the orbit is close to the fixed point, or cycle.

5.6 Bifurcations of homoclinic orbits

We consider bifurcations of homoclinic orbits of a fixed point or cycle x_0 with n_s multipliers with modulus smaller than 1 and n_u multipliers with modulus larger than 1. We assume that there are no critical multipliers with modulus 1, as then the dimension of the problem would change, so $n_u + n_s = n$ equals the dimension of the phase space. We order the multipliers by their modulus $\mu^s_{n_s} < \cdots < \mu^s_2 < \mu^s_1 < 1 < \mu^u_1 < \mu^u_2 \ldots < \mu^u_{n_u}$. The multipliers μ^s_1 and μ^u_1 are the leading multipliers. In addition, we denote by $(x_1, x_2, \ldots, x_N)$ the points on the computed part of the homoclinic orbit.

The primary codim 1 bifurcation of homoclinic orbits is the homoclinic tangency LP_HO. This is a limit point of a curve of homoclinic orbits (Beyn and Kleinkauf, 1997). The test function is $\phi = v_K$, i.e., the parameter component of the tangent vector to the curve of homoclinic orbits. There are two more cases based on the leading multipliers. Depending on whether the bifurcation occurs for the stable or unstable eigenspace, we have either x=S or x=U below.

- **Neutral saddle homoclinic orbit** HO_NS. This bifurcation occurs when the product of the leading eigenvalues equals 1, i.e., the test function $\phi = |\mu_1^s \mu_1^u| - 1$ vanishes.
- **Belyakov homoclinic orbit** HO_Bx. This bifurcation occurs when the leading eigenvalues change from real to complex, i.e., $\mu_1^x = \mu_2^x$ with $x = s$ or $x = u$.

The above cases may also occur when following a curve of homoclinic tangencies, leading to HT_NS and HT_Bx. When computing curves of homoclinic tangencies, there are two more cases that may occur due to properties of the stable and unstable manifolds. The importance of these cases is that a return map close to the codim 2 point is given by the generalized Hénon map with implications for the dynamics of cycles of any sufficiently high iteration number. These cases were studied by Gonchenko, Gonchenko, and Tatjer (2002c). As this bifurcation may occur with respect to either the stable or the unstable manifold, we get four additional cases. We describe the cases for the stable manifold assuming the unstable manifold is one-dimensional. The orbit structure and the local manifolds are shown in Figure 5.3.

- **Stable foliation homoclinic tangency** HT_SF. This generalized tangency defined as a tangency is the unstable manifold to the strong stable foliation of the stable manifold.

  To come to a test function for this bifurcation we need that the tangent vector to the unstable manifold is orthogonal to the leading stable left eigenvector v_s^L. The continuation only computes the points of the connecting orbit, i.e., $x_1, \ldots, x_N$, but not the tangent vector to the unstable manifold. To compute the tangent vector we start with the unstable eigenvector v_u at the saddle point. Now as tangent vectors are mapped by the linearization, i.e., $v_u^{m+1} \approx f_x(x_m) v_u^m$, then the tangent vector at x_m is given by

  $$v_u^m = f_x(x_{m-1}) f_x(x_{m-2}) \cdots f_x(x_1) v_u.$$

  So we can choose $\phi_{SF} = (v_s^L)^T v_u^N$ as a test function.
- **Extended stable homoclinic tangency** HT_ES. For this bifurcation we consider an extended unstable manifold W^{ue} and assume that the leading multi-

pliers are real. Its tangent plane at every point x_m of the connecting orbit is spanned by the tangent vector to the unstable manifold v_u^m and the mapped leading stable eigenvector v_s^m. Then a bifurcation occurs if W^{ue} is tangent to

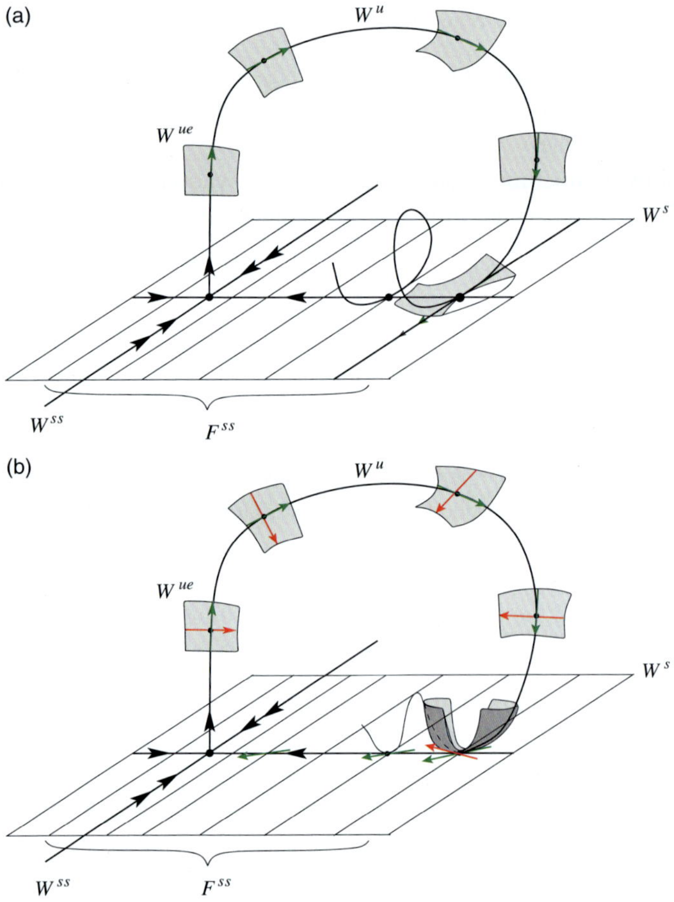

Figure 5.3 Geometry of orbits and local manifolds at generalized homoclinic tangencies. W^s, W^{ss}, W^u, W^{ue} denote the stable, strong stable, unstable and unstable-extended manifolds, respectively. The green vectors are tangent vectors along the homoclinic orbit. (a) The HT_SF bifurcation. Starting from the unstable eigenvector, the tangent vector is eventually tangential to the strong stable foliation F^{ss} (thick gray arrow). (b) The HT_ES bifurcation. The red vector is the stable eigenvector mapped along the connecting orbit. Eventually the red vector is tangent to W^s, leading to a double tangency. The green vector converges to a vector in the leading stable direction. Figures adapted from Gonchenko, Gonchenko, and Tatjer (2002c).

Table 5.3 Test functions for detection of codim 2 bifurcations of homoclinic orbits along the LP_HO curve. We have either x=S or x=U.

Bifurcation	Test function(s)
HT_NS	$\lvert\mu_1^s\mu_1^u\rvert - 1,$
HT_Bx	$\begin{cases} (\mu_1^x - \mu_2^x)^2, & \text{if } \mu_{1,2}^x \text{ are real,} \\ (-\Im(\mu_1^x - \mu_2^x))^2 & \text{if } \mu_{1,2}^x \text{ are complex.} \end{cases}$
HT_xF	$\phi_{SF} = (v_s^L)^T v_u^N, \quad \phi_{UF} = (v_u^L)^T v_s^1$
HT_Ex	$\phi_{ES} = (v_u^L)^T v_s^N, \quad \phi_{EU} = (v_s^L)^T v_u^1$

the stable manifold W^s. As we consider a homoclinic tangency, this means that not just v_u^m, but also v_s^m is tangential to W^s. Defining v_s^N similarly as v_u^N above, we arrive at the test function $\phi_{ES} = (v_u^L)^T v_s^N$.

Similar test functions for the unstable case can be formulated but in that case we must consider mapping a vector backward along the connecting orbit from the last to the first point on the orbit

$$v_s^m = f_x(x_{m+1})^{-1} f_x(x_{m+2})^{-1} \cdots f_x(x_N)^{-1} v_s,$$

where we assume each linear mapping to be invertible. A homoclinic tangency as well as a neutral saddle tangency may occur in two-dimensional maps. The Belyakov case and the two generalized homoclinic tangencies can only occur in maps of three dimensions and higher. We summarize the test functions for the homoclinic codim 2 bifurcations in Table 5.3.

5.7 Computation of Lyapunov exponents

Lyapunov exponents measure the rate of separation of orbits near a reference orbit $\{x_k\}$. We consider $f(x(k) + \delta v) \approx x(k+1) + \delta Df(x_k)v$ for some small constant δ and unit vector $v \in \mathbb{R}^n$. The growth rate r_k at the kth step is $\|Df(x_k)v\|$. Next this is averaged over many iterates $\lambda(v) = \lim_{N\to} \frac{1}{N}\sum_{k=1}^N r_k$, provided this limit exists. This is an approximation only and a finite N must be chosen with care.

To compute all Lyapunov exponents we use n unit vectors spanning $\mathbb{R}^n$. In practice, these vectors will align after some iterations to the vector corresponding to the largest Lyapunov exponent. Therefore, one must orthogonalize the vectors after some iterations. This is done efficiently using QR-factorization as it also provides the rates. This leads to the following algorithm.

0a. Transients: define the initial point x_0 and iterate it a number of times until the orbit is close to the attractor.

0b. Initialization: set $V_0 = I_n$ the identity matrix and create a vector L with n zeros.
1. Iterate the map N-times; at every step compute $V_{k+1} = Df(x(k))V_k$.
2. Compute the QR-factorization of V_N. Update $L_i \leftarrow L_i + \log(|R_{ii}|)$ and set $V_0 = Q$.
3. Monitor the convergence of the estimate of the exponents $\hat{\lambda} = L/M$ with M the current number of steps. If sufficiently accurate, or the maximum number of steps is reached, the output is $\lambda = L/M$. Otherwise repeat steps 1 and 2.

Table 5.4 Interpretation of Lyapunov exponents.

Configuration of λ	Interpretation
$\lambda_1 > 0 > \lambda_2 \geq \lambda_3$	Chaos
$\lambda_1 > \lambda_2 > 0 > \lambda_3$	Hyperchaos
$\lambda_1 = 0 > \lambda_2 > \lambda_3$	Invariant curve of node type
$\lambda_1 = 0 > \lambda_2 = \lambda_3$	Invariant curve of focus type
$\lambda_1 = \lambda_2 = 0 > \lambda_3$	2-torus
$\lambda_1, \lambda_2, \lambda_3 < 0$	Fixed point, cycle

If multiple Lyapunov exponents are close to zero, one must set M to a high value. Otherwise it may not be possible to determine whether an exponent is zero or not as secondary zero Lyapunov exponents converge more slowly.

Next we order the Lyapunov exponents as $\lambda_1 \geq \lambda_2 \geq \cdots \geq \lambda_n$ and classify the dynamics based on them. Positive exponents correspond to chaos, while negative exponents indicate stability. Zero exponents enlarge the dimension of the attractor and correspond typically to tori. In two dimensions only a single exponent can be positive, indicating chaos, or zero, indicating an invariant curve. For a three-dimensional phase space one can classify the dynamics according to Table 5.4. Lyapunov exponents are only an indication of this dynamics and not a proof. Yet, they provide information that is otherwise hard to obtain.

6

Features and Functionality of MatcontM

In this chapter we describe the software MatcontM, a continuation environment in matlab (2016). It continues fixed points of an iterate of a map and handles the bifurcations mentioned in Chapter 3. It also deals with global bifurcations and invariant curves. Its versatile functionality equips the user with all current methods of numerical bifurcation analysis of maps.

There are several other software packages supporting bifurcation analysis of iterated maps. Orbits of maps and one-dimensional invariant manifolds of saddle fixed points can be computed and visualized using dynamics (Nusse and Yorke, 1998) and DsTool (Back et al., 1992). Location and continuation of fixed point bifurcations is implemented in auto (Doedel et al., 1997-2000) and the LBFP-version of LocBif (Khibnik et al., 1993). The latter program computes the critical normal form coefficient at LP points and locates some codim 2 bifurcations along branches of codim 1 fixed points and cycles. In auto the continuation uses maximally extended systems (see Chapter 5). Minimally extended systems based on bordering were first implemented, together with the standard extended defining systems, in content (Govaerts, Kuznetsov, and Sijnave, 1999). content (Kuznetsov and Levitin, 1995–1997) was the first software that computed the critical normal form coefficients for all three codim 1 bifurcations of fixed points and cycles and allowed continuing these bifurcations in two parameters and to detect all 11 codim 2 singularities along them (see Table 5.2).

Branch switching at PD and BP points is also implemented in auto, LocBif and content. However, only trivial branch switching is possible at codim 2 points and only for two (cusp and 1:1 resonance) of 11 codim 2 bifurcations are the critical normal form coefficients computed by content. No other software supports switching at codim 2 points to the continuation of the double-, triple- and quadruple-period bifurcation curves.

Continuation of connecting orbits and their bifurcations is also supported by

HomMap (Yagasaki, 1998). The algorithm was known to experts much earlier (Kawakami, 1981) and used (Mira, 1987). MatcontM's support for global bifurcations is more extensive, and allows a more user-friendly way to generate initial data.

6.1 General description of MatcontM

MatcontM is a toolbox for numerical bifurcation analysis of iterated maps in matlab. It enables the user to explore the dynamics of a model and its bifurcations. The typical result is a two-dimensional bifurcation diagram with various codim 1 bifurcation curves and codim 2 bifurcation points. Such analysis builds on the results of the previous Chapter. The user can verify the non-degeneracy of a bifurcation, as well as switch branches to other bifurcation curves, based on numerical values for the critical normal form coefficients.

The computational core of the toolbox is the Continuer, which executes the numerical continuation. The continuation requires a mapfile, (discussed in the following), a curve definition, as well as continuation settings. For every numerical continuation, the user has to choose a curve type, as well as initial data, i.e., values for state variables, parameters and the iteration number. There are eight different curve definitions as shown in Figures 6.1–6.3 (see codim 0 and 1). Their implementation is described in Chapter 5 and is based on Govaerts et al. (2007); Khoshsiar Ghaziani et al. (2009). The numerical settings

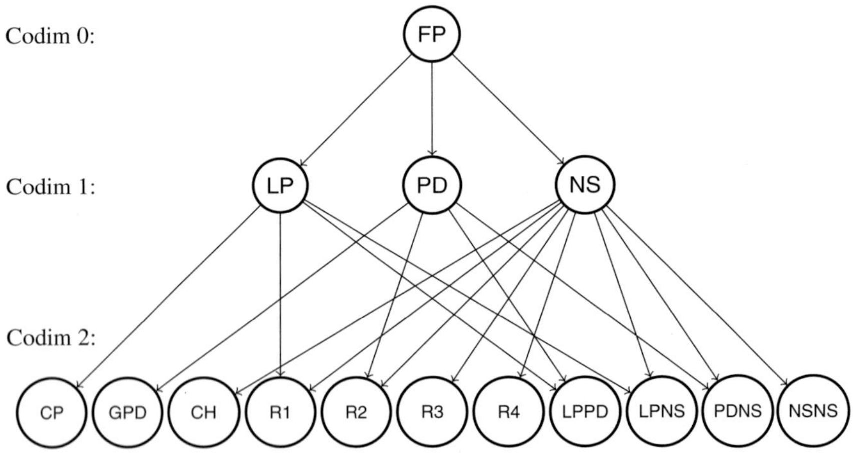

Figure 6.1 The hierarchy of local bifurcations. The arrows indicate which bifurcation points can be detected from a certain bifurcation curve.

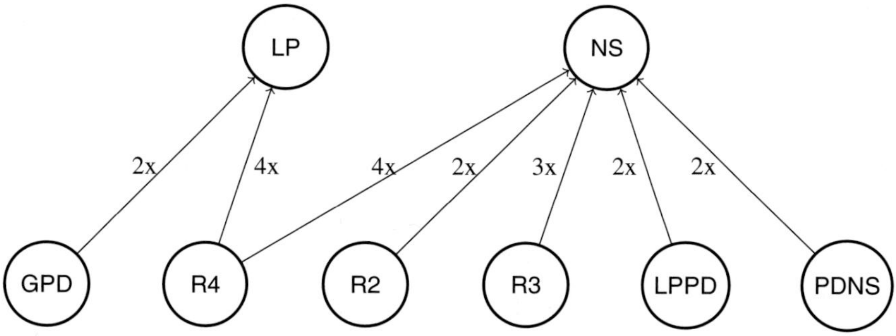

Figure 6.2 Possibilities for branch switching, i.e., starting continuation of a codim 1 bifurcation curve from a codim 2 point. Numbers indicate how the iteration number is multiplied.

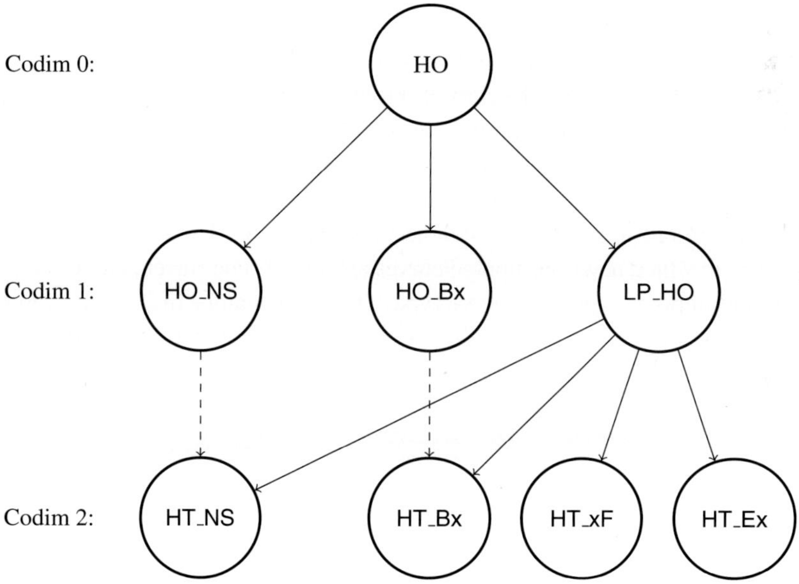

Figure 6.3 Detection graph of the various homoclinic bifurcations. The arrows indicate which bifurcation points can be detected from a certain bifurcation curve; x stands either for S (stable) or U (unstable). Dashed curves indicate that such a bifurcation exists, but its detection is not supported. For heteroclinic orbits, only continuation of the orbit and its tangency is supported, i.e., HE and LP_HE.

influence the speed and accuracy of the continuation process. These are best understood from the description of numerical continuation in Section 6.3. The user may specify values for continuation options, or use default values. In the following we will be more specific.

The toolbox can be used in two ways: from the command-line (CL) or via the graphical user interface (GUI). CL-use allows full control and scripting via m-files. The main advantages of the GUI are as follows:

- The user can focus on exploring the model and its bifurcations. The continuation process as well as the numerical analysis for bifurcations is largely hidden from the user. The process itself is interactive.
- A new model can be set up in a simple way.
- Initialization of continuation and switching to different branches is straightforward.
- Handling and storage of the data resulting from continuation is automated.
- Visualization and inspection of the data of bifurcation curves is supported.

Two aspects of the GUI stand out. First, it has been developed as a layer on top of the CL version. That is, the computational core is always the same. So the GUI allows a flexible access to various computations and also checks the input to call the Continuer with the correct settings. Second, for the GUI itself, there is a model that views and controls data. Via the GUI, a user can generate and manage the data. The data consist of a current system and a current curve. The current system specifies the map currently analyzed and special user functions to be monitored in addition of standard test functions. There is also a current diagram to which newly computed curves are added. The current curve consists of an initial point type and a curve type. These data can be inspected using the Data Browser (see Figure 6.4). We have the following natural hierarchy in the data:

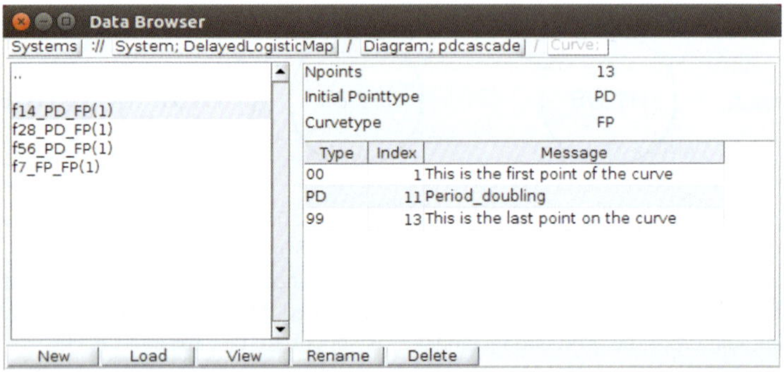

Figure 6.4 A screenshot of the Data Browser window. At the top, the hierarchy in the data is visible. At the bottom, one can manage the data at the current level. At the top level of a system, one can specify new systems or edit existing ones. In between, one selects diagrams or curves, while at the bottom level of a curve, one may inspect the continuation data in depth.

System. This is the map that we currently analyze. A new system can be added or an existing one deleted. For each map there is a *mapfile* that contains the definition of the map as well as its derivatives with respect to state variables and parameters up to some order.

Diagram. Computed curves are stored in an active diagram, i.e., directory. The user can select one active diagram among multiple diagrams. Data can be transferred between diagrams as well.

Curves. A diagram consists of computed curves. The automatic naming scheme adopted is as follows: dn_PT_CT(x). Here, d stands for the direction along the curve (forward or backward), n is the iteration number, PT is the initial point type, CT is the curve type and x is the number of the curve among all curves of its particular type. A curve can be loaded to start the continuation of a similar curve. We can also rename or delete curves.

(Special) points. A curve consists of a set of (ordered) points. We can browse these by loading them into the MATLAB array editor or selecting special points. These special points correspond to bifurcations of higher codimension along the curve. The first and last point are also always given. We can check the settings of the Starter and Continuer by inspecting the Starterdata and Continuerdata, respectively.

Typically, the data come from numerical continuation. The data may also come from simulations, computation of (un)stable manifolds or Lyapunov exponents. The flowchart in Figure 6.5 gives an impression of the various processes that can be executed and controlled for numerical continuation. Managing the data via the GUI is done through several *functions*.

Starter. This function controls the initial point, i.e., the state variables and parameters, and the iteration number. Here we select active parameters and adjust settings for derivatives.

Continuer. This function controls the general parameters of the continuation, such as minimal and maximal step sizes, the number of points to be computed, when to decide that the computation of a bifurcation point was successful and so on.

Branch Manager. This function monitors the selection of the initializer currently chosen. There are certain codim 2 bifurcations from which branches with higher iteration numbers can be started (see Figure 6.2). Therefore, the correct iteration number is managed by this function as well.

Curve Manager. This function decides how a computed curve is named and where it is saved. It further decides (using Options) whether old curves are deleted. The user can move curves to different diagrams, and rename or delete them.

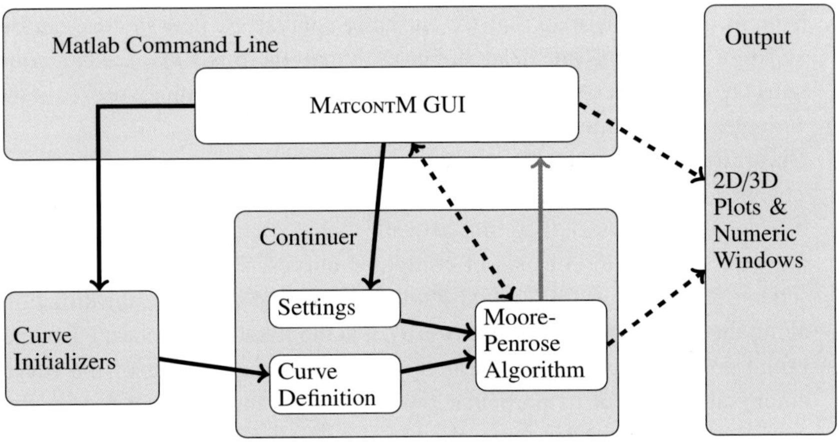

Figure 6.5 Workflow of MatcontM: a continuation is initialized and options for the Continuer are set. Next the Continuation computes a curve. This data are returned to either the Workspace for CL-usage (gray arrow) or stored automatically through the GUI. All steps may be executed via the CL or GUI (solid arrows), except that the GUI allows interactive control of the Continuer and Visualization (dashed arrows).

Plot Manager. Computed points are always shown in a dialog window. In addition, the user may want to have numerical or graphical output during the continuation. This function enables the interaction between the session data and the output windows.

In addition, there is an internal Windowmanager that keeps track of the various windows. Together, these functions allow the user to set up the computations. From a user perspective, there are three windows that matter; the main window and the windows of the Starter and Continuer. The latter are described above. The main menu allows opening the Data Browser for selecting or editing the map, selecting the type of initial point and the curve type. Some global settings can be changed via the main menu, which can also be used to fix various various continuation settings and start the continuation.

6.2 The mapfile

Most functionality of MatcontM is generic, only the mapfile is specific to the system. This mapfile implements the actual function (1.4). As may be clear from the numerical methods, we also need the derivatives of the map with respect to coordinates and parameters. A mapfile is an m-file that can be generated using the GUI automatically. There is also a standalone version to generate a mapfile. The coordinates are organized into a single vector, named `kmrgd` if generated by MatcontM to avoid confusion with other variable names. The parameters are passed separately at every evaluation. The name of the parameter has the prefix `par`. As an example of a mapfile, we consider the Delayed Logistic Map from Tutorial 1 (see Section 7.1). The following code would be a valid mapfile with derivatives implemented up to second order.

```
function out = DelayedLogisticMap
out{1} = [];        %Only used in ODE setting
out{2} = @fun_eval;
out{3} = @jacobian;
out{4} = @jacobianp;
out{5} = @hessians;
out{6} = @hessiansp;
out{7} = [];       %der3 would be 3rd order wrt state
out{8} = [];       %der4 would be 4th order wrt state
out{9} = [];       %der5 would be 5th order wrt state
% ------------------------------------------------
function dydt = fun_eval(t,kmrgd,par_r,par_eps)
  dydt=[par_r*kmrgd(1)*(1-kmrgd(2))+par_eps;
        kmrgd(1);];
% ------------------------------------------------
function jac = jacobian(t,kmrgd,par_r,par_eps)
  jac=[ -par_r*(kmrgd(2) - 1) , -kmrgd(1)*par_r ; 1 , 0 ];
% ------------------------------------------------
function jacp = jacobianp(t,kmrgd,par_r,par_eps)
  jacp=[ -kmrgd(1)*(kmrgd(2) - 1) , 1 ; 0 , 0 ];
% ------------------------------------------------
function hess = hessians(t,kmrgd,par_r,par_eps)
  hess(:,:,1)=[ 0 , -par_r ; 0 , 0 ];
  hess(:,:,2)=[ -par_r , 0 ; 0 , 0 ];
  %------------------------------------------------------
function tens3  = der3(t,kmrgd,par_r,par_eps)
%------------------------------------------------------
function tens4  = der4(t,kmrgd,par_r,par_eps)
%------------------------------------------------------
function tens5  = der5(t,kmrgd,par_r,par_eps)
```

6.3 Numerical continuation

The general idea is to formulate the computation of fixed points, cycles and their bifurcations as a suitable *algebraic problem* (AP) of the form

$$F(u, p) = 0, \tag{6.1}$$

where $u \in \mathbb{R}^N$ is composed of state variables and possibly other variables characterizing the system. Numerical continuation allows producing one-dimensional solution branches of (6.1) under the variation of a component of p, called *free* parameters. This requires N equations for the $N + 1$ unknowns. During continuation it is better not to distinguish between state variables and parameters. Therefore, write $x = (u, p)$ for the continuation variables in the continuation space $\mathbb{R}^{N+1}$

$$F(x) = 0, \tag{6.2}$$

with $F\colon \mathbb{R}^{N+1} \to \mathbb{R}^N$. The system (6.2) is called the *defining system*. Given an initial point $x^{(0)}$ (approximately) satisfying (6.2) we want to generate a sequence of points $\{x^{(k)}\}_{k\geq 0}$ satisfying (6.2) with a certain tolerance and then join them, i.e., by line segments, thus obtaining an approximation of the one-dimensional solution branch (curve). Starting with a regular point $x^{(0)}$ on the curve[1] and a tangent vector $v^{(0)}$, we predict a new point $X^0 = x^{(i)} + hv^{(i)}$ with step size h from the previous point $x^{(i)}$.

To find the next point $x^{(i+1)}$ on the curve nearest to X^0, one can try to solve the following optimization problem

$$\min_x\{\|x - X^0\| \mid F(x) = 0\}.$$

For X^0 close to $x^{(i)}$, this problem is equivalent to solving the system

$$\begin{cases} F(x) = 0, \\ v^T(x - X^0) = 0, \end{cases}$$

where $F_x(x)v = 0$ and $\|v\| = 1$. The linearization of this system about X^0 gives

$$\begin{cases} F(X^0) + F_x(X^0)(X - X^0) = 0, \\ [V^0]^T(X - X^0) = 0, \end{cases} \tag{6.3}$$

implying

$$\begin{pmatrix} F_x(X^0) \\ [V^0]^T \end{pmatrix}(X - X^0) = -\begin{pmatrix} F(X^0) \\ 0 \end{pmatrix}.$$

[1] A point $x \in \mathbb{R}^{N+1}$ satisfying $F(x) = 0$ is called *regular* if rank $F_x(x) = N$.

Therefore, define

$$X^1 = X^0 - \begin{pmatrix} F_x(X^0) \\ [V^0]^T \end{pmatrix}^{-1} \begin{pmatrix} F(X^0) \\ 0 \end{pmatrix}.$$

Then compute V^1 satisfying

$$F_x(X^1)V^1 = 0, \quad \|V^1\| = 1,$$

and set

$$X^2 = X^1 - \begin{pmatrix} F_x(X^1) \\ [V^1]^T \end{pmatrix}^{-1} \begin{pmatrix} F(X^1) \\ 0 \end{pmatrix},$$

etc.

In general, the *Moore–Penrose corrections* are defined by

$$X^{k+1} = X^k - \begin{pmatrix} F_x(X^k) \\ [V^k]^T \end{pmatrix}^{-1} \begin{pmatrix} F(X^k) \\ 0 \end{pmatrix}, \tag{6.4}$$

where

$$F_x(X^k)V^k = 0, \quad \|V^k\| = 1. \tag{6.5}$$

Each correction occurs within the plane orthogonal to the null-space of $F_x(X^k)$ at X^k (see Figure 6.6(a)). If both the norm of the correction $\|\Delta X\| = \|X^k - X^{k-1}\|$ and the residue $\|F(X^k)\|$ are sufficiently small, we can take $x^{(i+1)} = X^k$ as the new point on the curve and V^k as the new tangent vector. This process is repeated for the next correction. If not, the step size is halved, and a new attempt to compute $x^{(i+1)}$ is made.

A disadvantage of the described Moore–Penrose correction algorithm is that one needs to compute the null-vector V^k by setting up and solving (6.5) at each X^k. One can avoid this by looking for X^{k+1} within the plane through X^k that is orthogonal to the *previous* kernel, i.e., V^{k-1} (see Figure 6.6(b)).

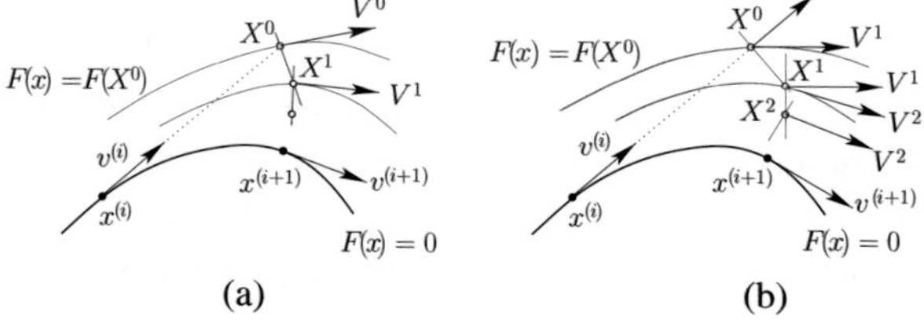

Figure 6.6 Moore–Penrose corrections: (a) exact; (b) approximate.

Let $V^0 = v^{(i)}$ with $||V^0|| = 1$. As in the exact Moore–Penrose algorithm, set

$$X^1 = X^0 - \begin{pmatrix} F_x(X^0) \\ [V^0]^T \end{pmatrix}^{-1} \begin{pmatrix} F(X^0) \\ 0 \end{pmatrix}.$$

To find V^1 satisfying $F_x(X^0)V^1 = 0$, compute first

$$W = \begin{pmatrix} F_x(X^0) \\ [V^0]^T \end{pmatrix}^{-1} \begin{pmatrix} 0 \\ 1 \end{pmatrix},$$

which amounts to solving a linear system

$$\begin{cases} F_x(X^0)W = 0, \\ \langle V^0, W \rangle = 1, \end{cases}$$

with exactly the same matrix as used to compute X^1. The vector W spans the kernel of $F_x(X^0)$. Now we can set

$$V^1 = \frac{W}{||W||}$$

and repeat the procedure.

This leads to the following *approximate Moore–Penrose corrections*:

$$\begin{cases} X^{k+1} = X^k - \begin{pmatrix} F_x(X^k) \\ [V^k]^T \end{pmatrix}^{-1} \begin{pmatrix} F(X^k) \\ 0 \end{pmatrix}, \\ W^{k+1} = \begin{pmatrix} F_x(X^k) \\ [V^k]^T \end{pmatrix}^{-1} \begin{pmatrix} 0 \\ 1 \end{pmatrix}, \quad V^{k+1} = \dfrac{W^{k+1}}{||W^{k+1}||}. \end{cases}$$

As in the exact Moore–Penrose case, the vectors V^k converge to the next tangent vector $v^{(i+1)}$. Notice that

$$W^{k+1} = V^k - \begin{pmatrix} F_x(X^k) \\ [V^k]^T \end{pmatrix}^{-1} \begin{pmatrix} F_x(X^k)V^k \\ 0 \end{pmatrix}.$$

Both exact and approximate Moore–Penrose corrections converge quadratically to a point $x^{(i+1)}$ near $x^{(i)}$ in the curve (6.2), provided the curve is regular and the step size h is sufficiently small.

For Moore–Penrose continuation the tangent vector is updated at every correction step. Another popular method is *pseudo-arclength continuation*. Although the derivation is different, the correction step is quite similar except that the next point is searched in the plane orthogonal to the original tangent vector. Both methods can pass along folds of solution curves. Close to an actual zero of (6.2) the derivative F_x will not differ too much for the next Newton iteration. In that case one may apply a Newton-chord update to speed up the computations. Here, one evaluates the function F at the corrected point, but uses the Jacobian matrix that is already available. In MatcontM, one Newton-chord update is used before recomputing the derivative. For the continuation an initial tangent vector is needed. If this vector is unknown, we try to correct the initial point using the standard base vectors, cycling through them until the corrections are successful. The continuation variable is ordered such that the state variables come first. If along the curve the change is largest in another variable such as a parameter, it may be better to start with the final vector and cycle backwards. This mainly matters in the case that many state variables are involved, as with global bifurcations or invariant curves.

At the CL level, there is a global variable/structure `cds` that contains all settings for the Continuer. There is also another internal global variable that sets some variables for the specific curve type. In order to use a different setting one may create a custom options structure by calling `options=contset`. This creates a variable `options` with default variables. A particular field can then be modified by calling `options=contset(options,variable,value)`. Next, these options are passed to the Continuer (see the following). We finish with a list of the most important fields for the Continuer.

InitStepsize The initial step size h (default 0.01).
MinStepsize The minimum step size to compute the next point on the curve. Continuation stops if it fails with this step size (default 10^{-5}).
MaxStepsize The maximum step size. Upon a successful correction step, step size is increased by 1.3 (default 0.1).
MaxCorrIters The maximum number of correction iterations (default 10).
MaxNewtonIters The maximum number of Newton iterations before switching to Newton-chord in the corrector (default 10).
MaxTestIters Maximum number of iterations to locate a zero of a test function (default 10).
Increment The increment to compute numerical derivatives (default 10^{-5}).
FunTolerance Tolerance of function values $\|F(X)\|$ (default 10^{-6}).
VarTolerance Tolerance of correction $\|\Delta X\|$ (default 10^{-6}).
TestTolerance Tolerance of test functions (default 10^{-5}).
MaxNumPoints The number of points to compute (default 300).
Multipliers Boolean indicating whether the eigenvalues of the linearization must be computed (default 0).
Singularities Boolean indicating whether singularities should be monitored (default 0).
IgnoreSingularity Vector of indices corresponding to singularities to be ignored (default empty).

Backward Boolean indicating the direction along the curve (default 0, if 1 reverse tangent vector).

MoorePenrose Boolean indicating the use of Moore–Penrose continuation (default 1, if 0 then pseudo-arclength continuation is used).

TSearchOrder Boolean indicating whether we cycle in increasing index through the basis vectors for the first point, or in decreasing order (default 1, increasing).

Adapt Integer indicating to update auxiliary variables during continuation (default 3, no adaptation if 0).

AutDerivative Boolean indicating the use of algorithmic differentiation (AD) for normal form coefficients (default 1).

AutDerivativeIte Integer indicating the use of AD if the iteration number of the cycle equals or exceeds this number (default 24).

6.4 Calling the Continuer

The syntax of the Continuer is

```
[x,v,s,h,f]=cont(@curve,x0,v0,options)
```

The Continuer takes four arguments. The first argument is the curve definition file. `curve` can be any of the following: `fixedpointmap`, `limitpointmap`, `perioddoublingmap` and `neimarksackermap` for local bifurcations and we have `heteroclinic`, `heteroclinicT`, `homoclinic` and `homoclinicT` for global bifurcations. The second argument `x0` is the initial continuation point consisting of state variables, parameters and auxiliary variables. These are assembled through a special initialization function (see Section 6.4.1). The tangent vector `v0` can be specified, e.g., from a previous continuation or a branch-switching algorithm. If this variable is empty, then the Continuer will try to use standard base vectors to start the continuation. The final argument is the options variable for the Continuer, as discussed in the previous section.

The output of the Continuer is as follows

- `x` and `v` are the points and their tangent vectors along the curve. Each column corresponds to a point.
- `s` is a structure with additional information about special points. It has the

following fields

`s.index`	index of the singularity in `x`
`s.label`	label indicating the type of singularity
`s.data`	additional information, e.g., normal form coefficients
`s.msg`	message string for this singularity

- `h` is an array containing output of the Continuer for each point with the following components step size, half the number of corrections, user function values and test function values. The types of available test functions differs between curve types. `h` will thus depend on the curve type and the selected singularities to monitor as well as user functions.
- `f` is an array whose columns contain the multipliers of the fixed point or cycle, if these are computed.

We should also mention that it is possible to extend a given curve with a slightly different syntax:

```
[x,v,s,h,f]=cont(x,v,s,h,f,cds)
```

The output variables `x,v,s,h` and `f` contain the input variables with the new points appended.

6.4.1 Initializing the Continuer

Before any curve can be continued, the algebraic problem (6.1) has to be specified. To accomplish this there is a vast number of initializer routines. For each of the arrows in Figures 6.1–6.3 there is an initializer file called `init_PT_CT`, where PT indicates the type of initial point (codimension 0, 1, 2) and CT specifies the curve type (any of FPm, LPm, PDm, NSm, Hom, HomT, Het, HetT). The syntax for the initializer is

```
[x0,v0]=init_PT_CT(@mapfile,x,p,ap,n)
```

The mapfile corresponds to the m-file defining the system that we currently explore. The input `x` is a column vector specifying coordinates of the fixed point or cycle for local bifurcations, while for connecting orbits it is an orbit piece. In the latter case, the first (and final for heteroclinic orbits) entries correspond to the saddle cycle. The input `p` is a vector of parameter values. The argument `ap` is an index specifying which parameters are active, i.e., the free parameters. The final argument `n`, which is specific for maps, is a positive integer indicating the iteration number of the cycle. It equals 1 for fixed points.

The resulting output is an array `x0` that contains the fixed point, or connecting orbit, as well as some auxiliary variables. The latter are computed by the

Table 6.1 The output reported by MatcontM for each local codim 1 and 2 bifurcation, indicating which formulas are used. Also a reference to the theorem and a figure of the unfolding is included.

Label	Reported Coefficients	Formulas	Theorem	Figure
LP	b_0	(2.7)	2.2	2.1
PD	c_0	(2.12)	2.5	2.2
NS	L_1	(2.18), (4.24)	2.9	2.3
CP	c_0	(4.13)	3.3	3.1
GPD	d_0	(4.19)	3.6	3.2
CH	L_2	(4.28), (4.29)	3.8	3.3
R1	$s = 2a_0(b_0 - 2a_0)$	(4.36)	3.13	3.6
R2	c_0, d_0	(4.44), (4.45)	3.18	3.9, 3.10
R3	$a_2(0)$	(4.49), (4.50), (3.46)	3.25	3.14
R4	A_0	(4.55), (4.56), (3.55)	3.29	3.16, 3.17
LPPD	$a_0, b_0, L_1^{(2)}$	(4.70)	3.35	3.24 – 3.27
LPNS	s, a_0, b_0, c_0	(4.75)	3.41	3.31 – 3.34
PDNS	$a_{11}, a_{12}, a_{21}, a_{22}$	(4.76) and Section 3.5.1	3.46	3.37 – 3.40
NSNS	$a_{11}, a_{12}, a_{21}, a_{22}$	(4.77) and Section 3.5.1	3.46	3.37 – 3.40

initializer and appended to the input `x`. So the size of `x` and `x0` may differ. For most initializers, the tangent vector is set to be empty. For those that involve asymptotics from codim 2 bifurcations the asymptotics are differentiated with respect to the expansion parameter yielding a tangent vector to the branch. Meanwhile the initializer sets up two global variables: `cds` for the Continuer options and an internal one for the curve definition. For instance, the order of symbolic derivatives depends on the mapfile, and this order is set at this stage.

6.4.2 Detection and processing of bifurcations

Let us consider a smooth parameterized solution curve of the continuation $X = X(s)$ with $s = s_0$ corresponding to a bifurcation. To detect this bifurcation we use a *test function* $\psi: \mathbb{R}^{N+1} \to \mathbb{R}$ such that $\psi(X(s_0)) = 0$. In this setting, we also refer to bifurcations as *singularities*. The test function should have a regular zero, i.e., $\frac{d\psi}{ds} \neq 0$, such that the corresponding bifurcation can be detected as a sign-change of ψ. So for two successive points X_0, X_1 a bifurcation corresponds to $\psi(X_0)\psi(X_1) < 0$. Upon detection of a sign-change, the Continuer usually switches to a secant method to accurately locate a point $\psi(X) = 0$ on the curve. However, there is one exception: To accurately locate a *branch point* (BP), a special regular defining system (Govaerts, 2000, Section 7.8.4) is solved by Newton's method, since the convergence tube for each branch shrinks when BP is approached.

Every bifurcation requires at least one test function, while some bifurcations require a specific test function not to vanish to distinguish cases. The complete set of test functions used in MatcontM is given in Tables 5.1–5.3. Suppose that along a bifurcation curve we may expect n_s different types of singularities, which are characterized through n_t test functions. To detect all singularities we use a singularity matrix S of size $n_s \times n_t$ with entries

$$S_{ij} = \begin{cases} 0 & \text{means for singularity } i \text{ test function } j \text{ must vanish,} \\ 1 & \text{means for singularity } i \text{ test function } j \text{ must not vanish,} \\ 8 & \text{means ignore test function } j \text{ for singularity } i. \end{cases} \tag{6.6}$$

When a singularity has been found, there is some additional processing of the singularity. Multipliers and eigenvectors are computed when necessary. Coefficients of the algebraic bifurcation equation and vectors tangent to branches intersecting at the singular point, as well as the critical normal form coefficients, are computed for branch points and local codim 1 and 2 bifurcations following the methods of Chapter 4. Specifically, we report only the non-degeneracy conditions as specified in Table 6.1 such that the theorem for that specific bifurcation can be consulted for the unfolding. Such information is stored in the `s.data` field.

As an example we show a fixed point continuation for the Delayed Logistic Map (7.1). We set the initial condition $(x, y) = (0, 0)$, active parameter r starting at $r = 0.4$ and fix $eps = 0$.

```
init; opt=contset;  %Add folders to path, create options
opt=contset('opt','Singularities',1);      %Modify an option
x=[.15;.15];n=1; % initial point and iteration number
p0=[.4,.1];ap=1; % initial parameter and active parameter
[x0,v0]=init_FPm_FPm(@DelayedLogisticMap,x,p0,ap,n);
[x,v,s,h,f]=cont(@fixedpointmap,x0,v0,opt);
```

These commands lead to the following output in the Command window. They show a successful run with a bifurcation detected and processed. It shows that a supercritical Neimark–Sacker bifurcation has been detected.

```
first point found
tangent vector to first point found
label = NS  , x = ( 0.550000 0.550000 1.818182 )
 normal form coefficient of NS = -6.993007e-01
elapsed time  = 0.5 secs
npoints curve = 300
```

7
MatcontM Tutorials

7.1 Tutorial I: iteration of maps and continuation of fixed points and cycles

This tutorial shows how to use MatcontM to compute orbits of maps

$$u \mapsto f(u, \alpha), \quad u \in \mathbb{R}^n, \alpha \in \mathbb{R}^m,$$

at fixed parameter values, and to continue fixed points and periodic orbits (cycles) in one parameter.

We study the following recurrent relation (Maynard Smith, 1968)

$$x_{k+1} = r x_k (1 - x_{k-1}) + \epsilon,$$

where x_k is the density of a population at year k, r is the growth rate and ϵ is the migration rate. It is assumed that the growth is determined not only by the current population density but also by its density one year before. Introduce $y_k = x_{k-1}$. The population dynamics can then be studied by iterating the planar map

$$\begin{pmatrix} x \\ y \end{pmatrix} \mapsto \begin{pmatrix} rx(1-y) + \epsilon \\ x \end{pmatrix}. \tag{7.1}$$

This map is called the *Delayed Logistic Map*.

7.1.1 Installation and system specification

Obtain the latest file `matcontmxxx`.zip.[1] Extract `matcontmxxx` and change the working directory of MATLAB to the root of the extracted folder. You will see the file `matcontm.m`. You start MatcontM by executing in MATLAB:

[1] For example, at http://sourceforge.net/projects/matcont/files/matcontm/. This tutorial has been tested on MATLAB R2017b with `matcontm5p4`.

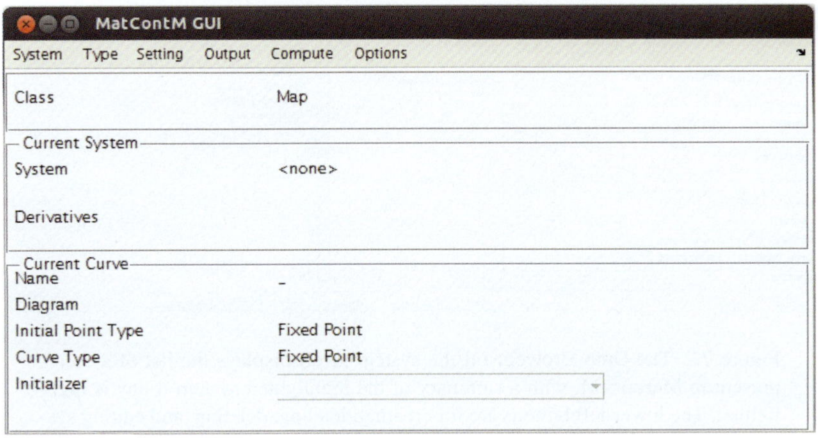

Figure 7.1 Main window of MatcontM (empty).

```
>> matcontm
```

This opens the main MatcontM window (Figure 7.1).

Select the menu item **System** in the upper left corner and then select **System Browser** (Figure 7.2).

In what follows, we will use the notation **System|System Browser**.

A Data Browser window appears, showing the list of available systems (maps). The right panel contains a summary of the highlighted system in the list, provided that any system is highlighted.

Underneath the list is a series of buttons that perform system-related actions. From left to right they are named **New, Load, Delete, Edit, Add/Edit Userfunctions**. Press the **New** button to add a new system. An empty System window appears.

Specify the Delayed Logistic Map by entering the data shown in Figure 7.3. Fill in the name, coordinates and parameters of the system. In our case, we have requested that the derivatives up to third order are calculated symbolically in advance by the MATLAB symbolic toolbox.

Press **OK**. Adding a system can take a few seconds. This map is now configured and selected for further use. After configuration, the default Starter and Continuer windows appear. (Note: If an existing system is edited then it is necessary to also load it for use.)

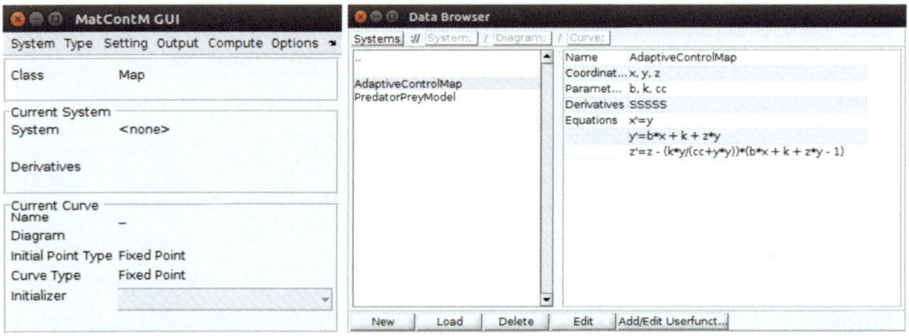

Figure 7.2 The Data Browser (at the system level) displays the list of systems present in MatcontM, with a summary of the highlighted system if any is highlighted. The lower-left buttons are for creating, loading, deleting, and editing systems.

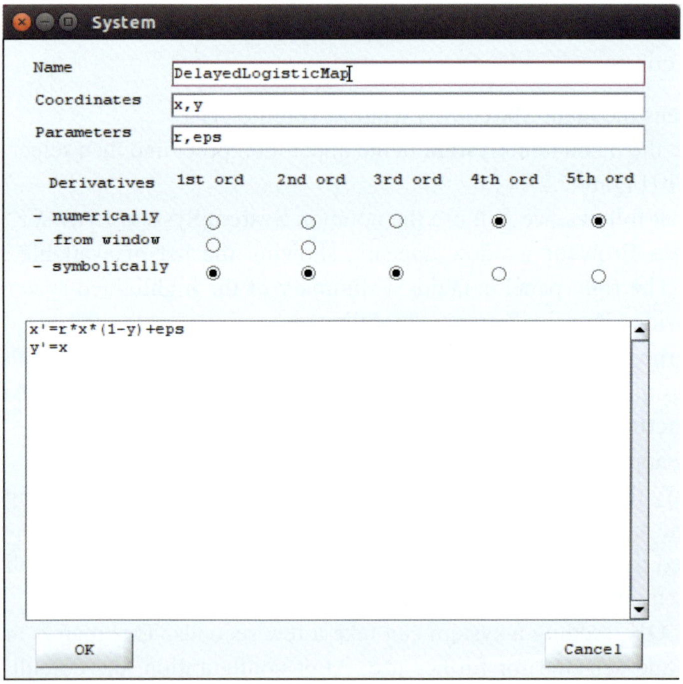

Figure 7.3 The System input window.

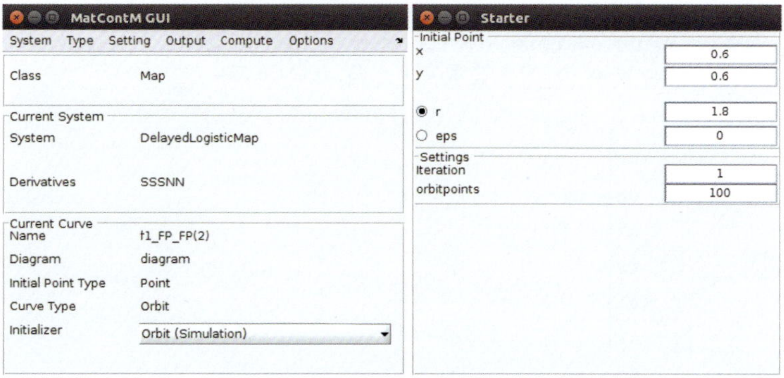

Figure 7.4 Main window and a configured Starter window.

7.1.2 Orbit iteration

Select **Type|Initial Point|Point** in the main window (Figure 7.1). Select the Starter window[2] and set initial data for iterations (Figure 7.4):

`x`	0.6
`y`	0.6
`r`	1.8
`eps`	0

Select **Output|Graphic|2D plot** in the main window to open a graphic window (see Figure 7.5). You can change attributes of the open Plot2D window by selecting the **MatContM|Layout** menu in the upper-right corner. No changes to the default layout are needed in this example.[3]

Select **Compute|Forward** in the main window.[4] The Output window will appear (this can be seen in Figure 7.8(a)). This window allows you to pause, stop, and resume computation. This window also displays information during computation. For now, you only need to close this window after computation is finished.

The computations produce an orbit converging to a fixed point (Figure 7.6(a)). Clear the figure by selecting **MatContM|Clear**. Increase the number of computed points by setting the value of `orbitpoints` to 300 in the

[2] You can reopen this window by selecting **Setting|Starter** in the main window.

[3] The menu **MatContM|Orbit Options** in the Plot2D window allows you to add point numbers and connecting lines.

[4] Backward iteration is not supported.

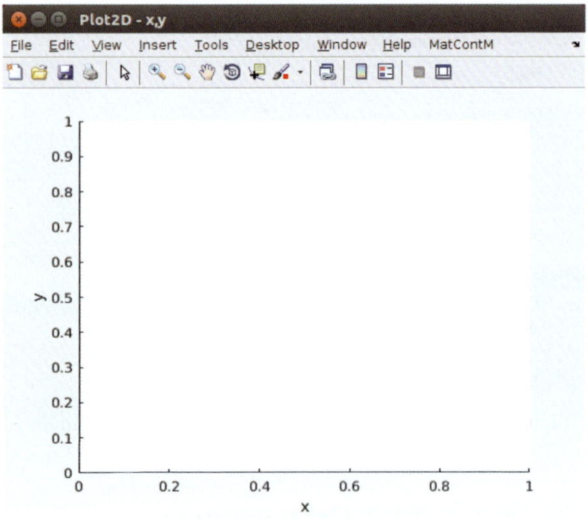

Figure 7.5 The default Plot2D window.

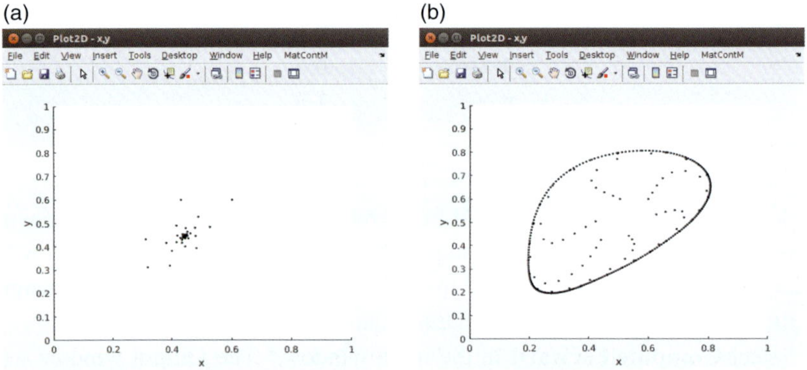

Figure 7.6 Convergence to a fixed point (a) and to a closed invariant curve (b).

Starter window. Increase now the value of r to 2.1 and iterate the map with **Compute|Forward**. The orbit now approaches a *closed invariant curve* (Figure 7.6(b)).

7.1.3 Fixed point continuation

Change the value of r back to 1.8 and recompute the orbit by selecting **Compute|Forward**. Select the **View Curve** button in the Output window to browse

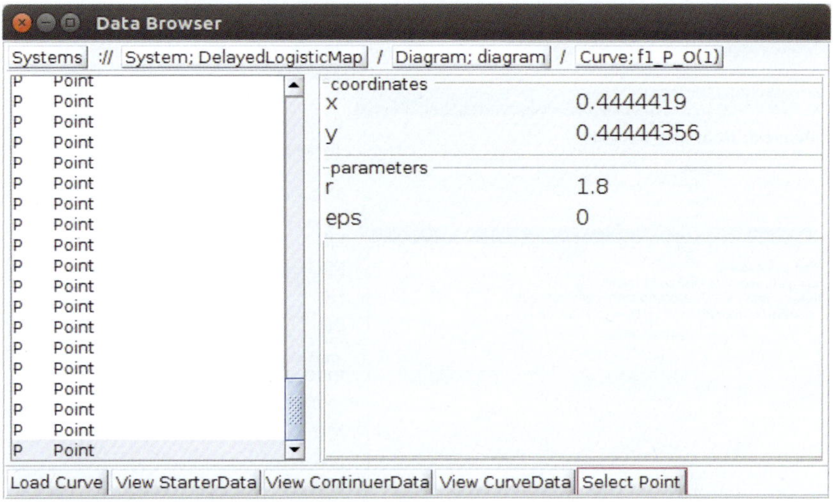

Figure 7.7 The last point is selected.

the latest computed curve. You can also obtain this window by right-clicking in the `Current Curve` section of the main window. Scroll down and highlight the last computed point (see Figure 7.7). Double-click or press the lower-right button **Select Point**. The values for x and y in the Starter window should now be close to 0.44444... .

Select **Type|Initial Point|Fixed Point**. A fixed point continuation is now selected. Activate the parameter r in the Starter window by clicking the radio button. Activate the `Calculate multipliers` option to monitor the eigenvalues (multipliers). Open the Numeric window by selecting **Output|Numeric** in the main window. In the same window, select **Options|Pause** and opt to suspend computation `At Each Point`, and press **OK**. Clear the Plot2D window.

Select **Compute|Forward**. Press the **Resume** button in the Output window after the computation of each point. Check that the fixed point becomes unstable as two multipliers cross the unit circle at $r = 2.0$ (Figure 7.8). This is a supercritical Neimark–Sacker (NS) bifurcation (normal form coefficient is -1.0) that gives rise to the stable closed invariant curve that we observed already. Notice that the moduli of the multipliers are close to 1, while their arguments $\theta_{1,2} \approx \pm 60^\circ$. Press the **Stop** button after detecting the Neimark–Sacker bifurcation and close the Output window. We permanently save the curve by giving it a unique name by right-clicking on the `Current Curve` panel in the main window. Select **Rename Curve** and enter, e.g., `Fixed Point(+)`.

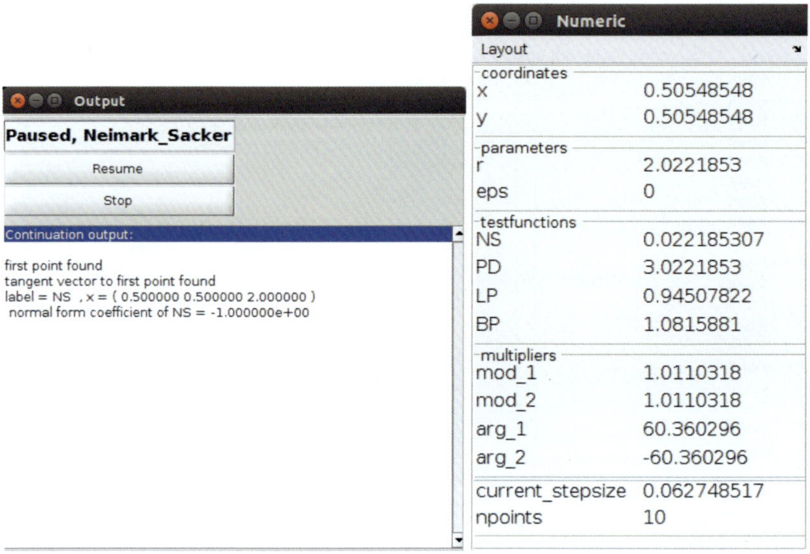

Figure 7.8 The Output window and the Numeric window during the computation of a fixed point continuation. A supercritical NS bifurcation was just detected.

7.1.4 Phase locking

Return now to simulations by selecting **Type|Initial Point|Point** but inputing $r = 2.177$ in the Starter window. Make sure `Orbit (Simulation)` is selected as `Initializer` in the main window (by selecting **Type|Curve|Orbit (Simulation)**). Clear the Plot2D window.

The iterations with **Compute|Forward** (which you have to resume at each point) will result in a stable 7-cycle This phenomenon is called phase locking: A quasi-periodic motion on the invariant curve is replaced by purely periodic motion.

Select a point on this cycle with maximal x-value ($x \approx 0.90$) using the Data Browser. For this, you can open the browser by pressing **View Curve** button. Then click on a point near the end of the list and see the corresponding numerical values in the right panel of the Data Browser window. When you have found a suitable point, press the **Select Point** button and close the Output window.

We will continue this period-7 cycle as a fixed point of the seventh iterate of the map with respect to the parameter r by selecting **Type|Initial Point|Fixed Point**. In the Starter window, activate the parameter `r` and set the value of `Iteration` to 7 (Figure 7.9(a)).

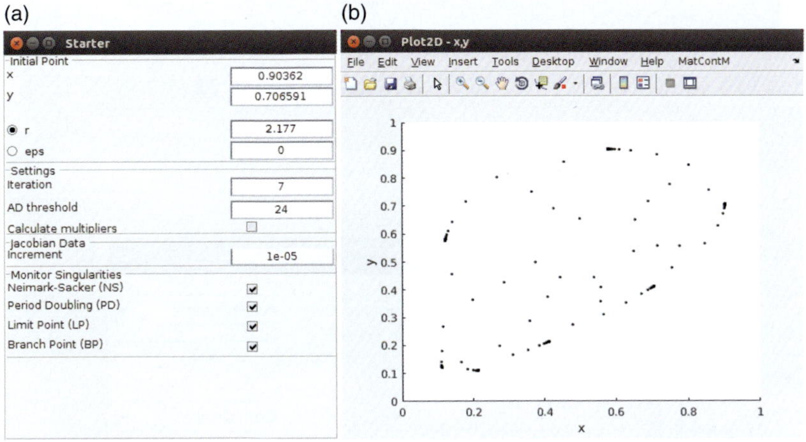

Figure 7.9 Starter window (a) for the continuation of the fixed point of the seventh iterate (b).

Select the Continuer window[5] and set `MaxStepsize` to 0.01. It is also useful to change the attributes of the 2D graphic window. Select **MatContM|Layout** and select parameter `r` and coordinate `x` as `Abscissa` and `Ordinate`, respectively. It is also advised to change the visibility range for `r` as seen in Figure 7.10. Make sure to clear the window using **MatContM|Clear**.

Before we start computations, it is advised to set **Options|Pause** to Suspend Computation `At special points`. Select **Compute|Forward**. You need to press **Resume** several times at detected *limit points* (labeled with LP) to get the closed curve presented in Figure 7.11. Note: The upper-left LP might not be detected during the computation of the closed curve. This bifurcation is too close to the starting point.

The seventh iterate of the map has seven stable fixed points and seven saddle fixed points that collide and disappear simultaneously at two limit point (fold) bifurcations, labeled with LP. Thus, the phase locking happens in a narrow parameter window between $r_1 = 2.174...$ and $r_2 = 2.201...$. Store the computed fixed point curve by renaming it (right-click in the `Current Curve` in the main window) to, e.g., `7-FixedPoint(+)`.

7.1.5 PD cascade

One can compute a parameter value r where the period-7 cycle exhibits a period-doubling bifurcation. For this, right-click on the `Current Curve` panel

[5] You can reopen this window by selecting **Setting|Continuer** in the main window.

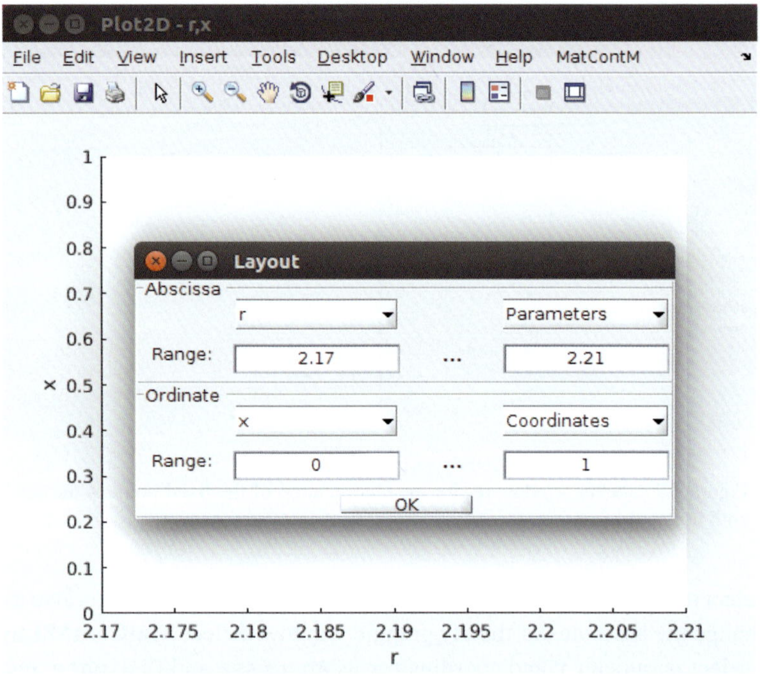

Figure 7.10 Layout window with the new attributes of the Plot2D window.

in the main window to display a context menu (see Figure 7.12), where you select **View Diagram**.

Then select the `7-Fixed point(+)` curve via the Diagram Browser. Highlight the `7-FixedPoint(+)` curve and press the **Load** button below. This makes this curve current, as is visible in the main window. Go to the Starter window and deactivate the parameter `r` but instead activate the parameter `eps`. Clear the Plot2D window (or create a new one) and change the visibility attributes as suggested in Figure 7.13.

We will create a new diagram by selecting `New diagram` in the menu shown in Figure 7.12. Enter the name `pdcascade`. The current diagram is now set to `pdcascade` instead of the default `diagram`. Curves will now be stored in this new diagram.

Select **Compute|Forward** until a period-doubling (flip) bifurcation (labeled PD) is detected (Figure 7.14).

Once we detect a PD bifurcation, we stop the computation by pressing **Stop**. The normal form coefficient of PD is positive, so we have detected a supercrit-

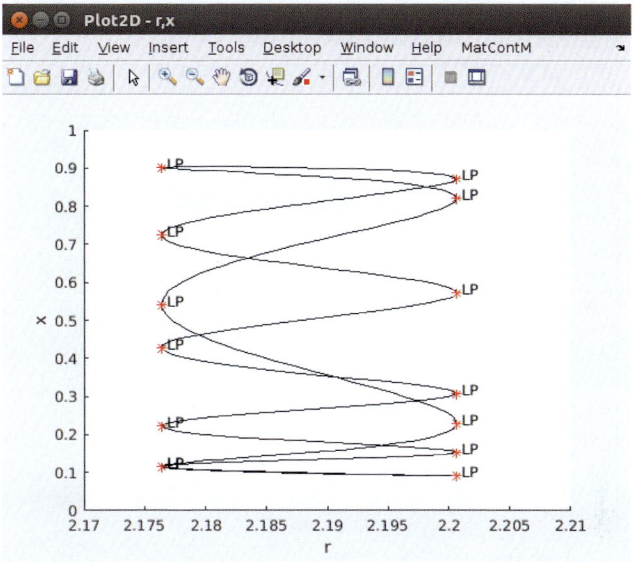

Figure 7.11 The continuation of the period-7 cycle as the fixed point of the seventh iterate of the map.

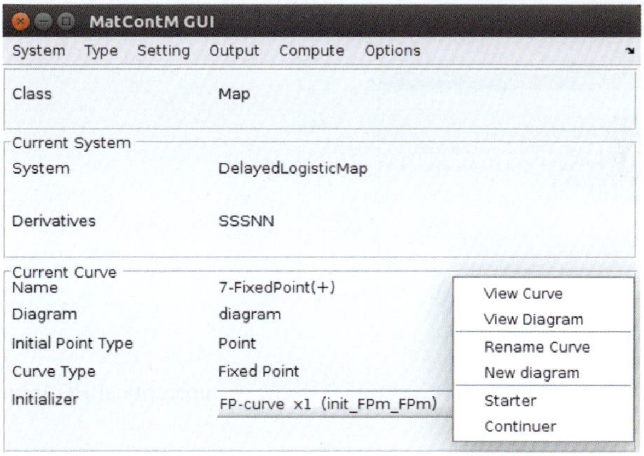

Figure 7.12 Context menu.

ical period-doubling bifurcation. Select this bifurcation as the initial point using the **Data Browser**. We will compute the period-doubled cycle (actually, the 14-cycle for the Delayed Logistic Map). We have to change the `Initializer`,

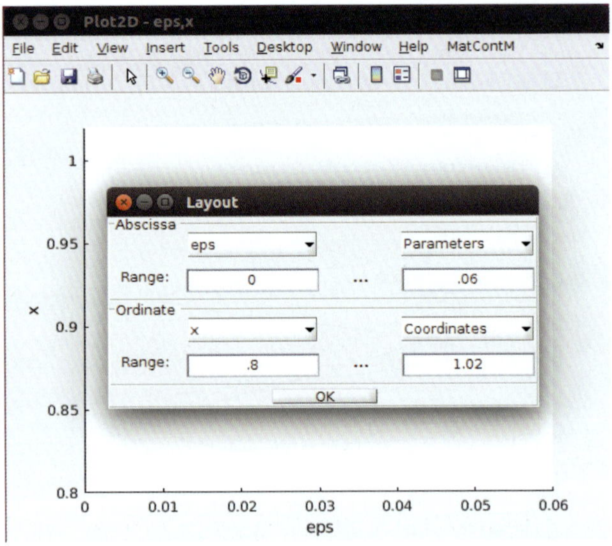

Figure 7.13 New visibility attributes.

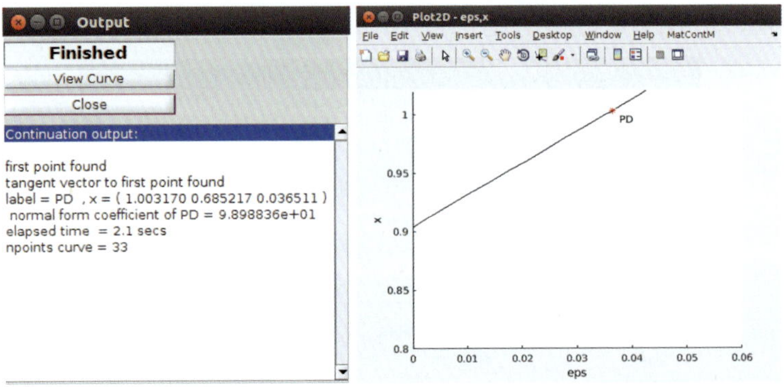

Figure 7.14 Continuation of a period-7 cycle: A supercritical PD bifurcation is detected.

since a two-parameter continuation of the PD bifurcation curve is selected by default. Change the `Initializer` to `FP-curve x2` (see Figure 7.15). This allows us to start the continuation of the bifurcating period-doubled cycle in parameter `eps`. Select **Compute|Forward**; you should obtain Figure 7.16. Stop after detecting the PD point.

We repeat the process. Select the last period-doubling bifurcation. You can

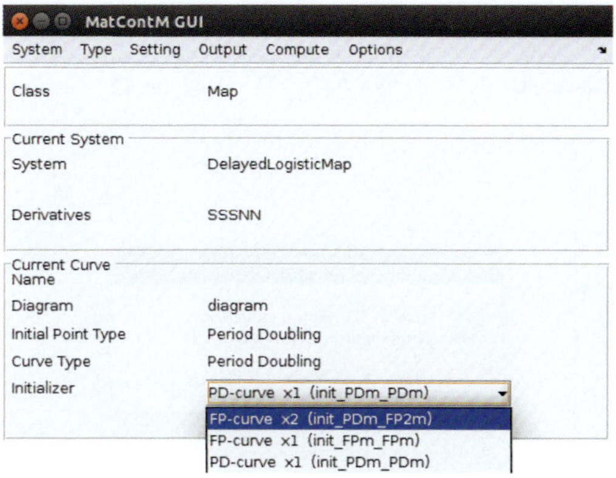

Figure 7.15 Selection of the initializer from `PD-curve` to `FP-curve x2`.

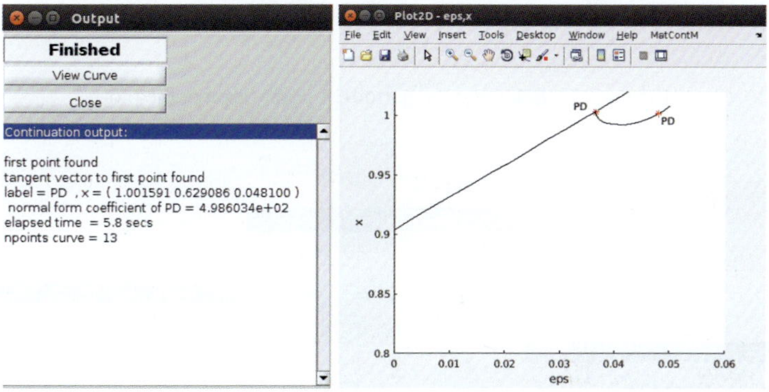

Figure 7.16 A supercritical period-doubling bifurcation on the doubled-curve (14-cycle) is detected.

also select bifurcations directly from plot windows by double-clicking on the labels (see Figure 7.17). A small window will appear which allows you to confirm your selection. Make sure no other MATLAB plot tools are enabled (Zoom, Pan, etc.) and all computations have finished or are terminated.

Ensure that the `Initializer` is set to `FP-curve x2` and select **Compute|Forward**. This computation will also detect some LP bifurcations, you may stop after detecting the first PD bifurcation.

You can repeat this process once more. However, because the curves are getting too close to each other, we need to sharpen some parameters. Set `Amplitude` to 0.0001 in the Starter window and set `InitStepSize` and `Max-`

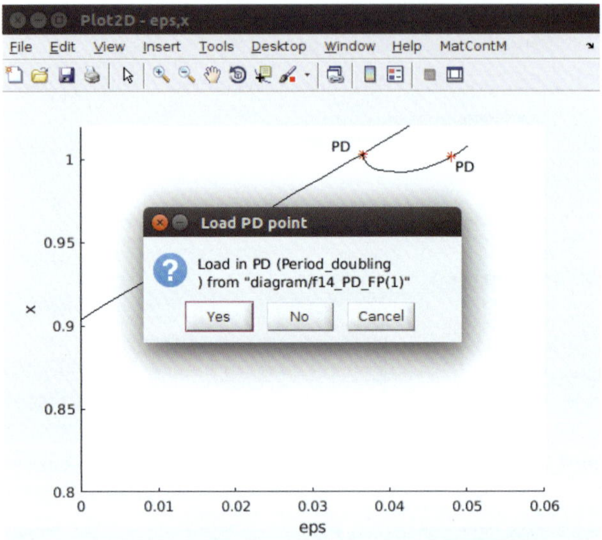

Figure 7.17 Selecting the period-14 point through the plot.

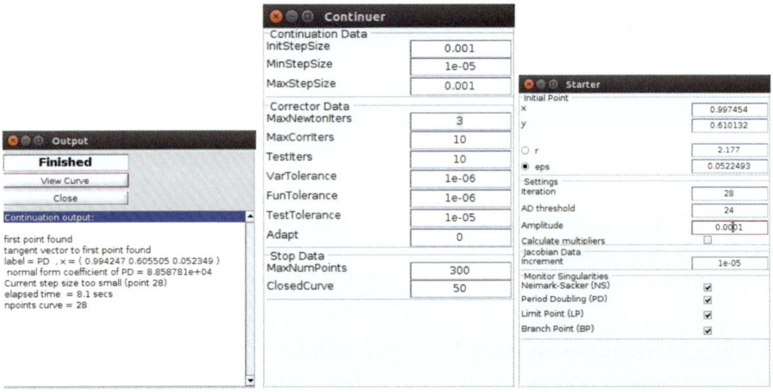

Figure 7.18 Computing the PD cascade gets increasingly hard because the curves lie too close to each other. We can find the fourth PD bifurcation (see Output window) on a 28-cycle using this setting.

`StepSize` to 0.001 in the Continuer window (Figure 7.18). The resulting cascade is shown in Figure 7.19.

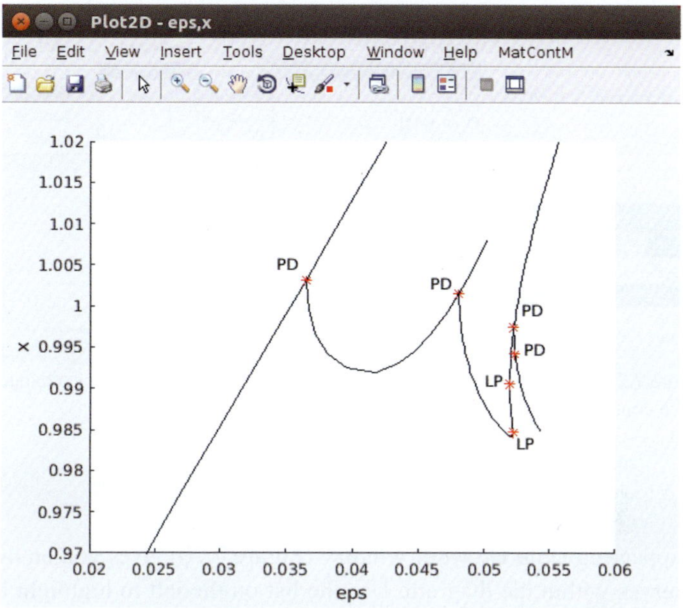

Figure 7.19 A zoom in the period-doubling cascade. The left curve is the original 7-cycle. The period doubles when a curve emanates from a PD bifurcation.

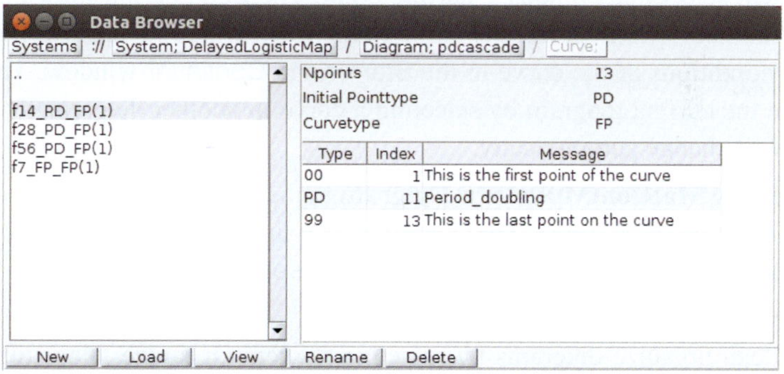

Figure 7.20 Diagram Data Browser.

7.1.6 Viewing and managing data

Right-clicking on the `Current Curve` panel opens the menu shown in Figure 7.12. Select **View diagram**.

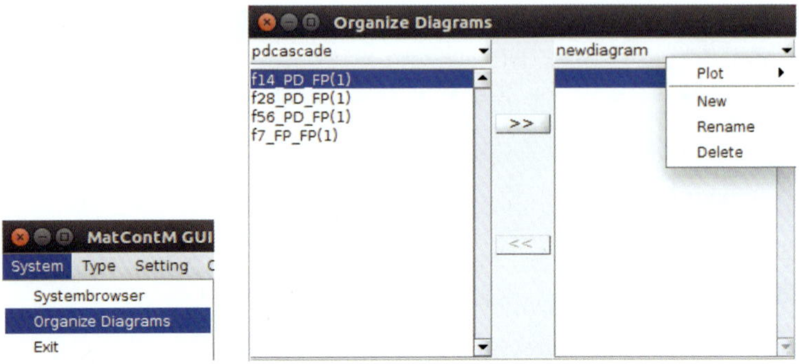

Figure 7.21 Diagrams `pdcascade` and `newdiagram` are selected for exchange. A curve context menu is visible.

The appearing Data Browser window (Figure 7.20) gives you an overview of the curves within the diagram. Use the list on the left to highlight a curve. Selecting **View** will open a window showing the raw data of the curve. By double-clicking on **..** you will go one level up and see an overview of the available diagrams. By double-clicking on a curve name you will go one level down and see the content of the curve, as seen in Figure 7.20. Double-clicking on a point will select that point as a starting point. The **Load** button shown in Figures 7.20 and 7.7 will select the curve in the main windows and will load the initial conditions of the curve in the Starter and Continuer window. You can change the current diagram by selecting a curve in another diagram. The diagram will change automatically.

Selecting **MatContM|Redraw Diagram** in the plot window will redraw the entire diagram. Sometimes the diagram can be littered with redundant curves. One can delete such curves in the Data Browser. It is also possible to move curves between diagrams.

You can organize diagrams by selecting **System|Organize Diagram** (Figure 7.21). This tool allows you to create new diagrams, move curves between diagrams and remove curves. You can right-click on the panels to obtain a list of actions. Curves can be removed by right-clicking and selecting the remove action. You can move curves between two diagrams by first selecting two diagrams using the lists at the top, left and right. You can move curves around by selecting them in one list and moving them using the arrow buttons in the middle. At each moment you can redraw the current diagram via **MatContM|Redraw Diagram** in the plot window.

The raw data can easily be exported to the MATLAB Command window. You need to create an interface and retrieve the data.

```
>> cli = CommandLineInterface        %create interface
>> data = cli.getCurve               %retrieve current curve
                                     %data as struct
data =
    x: [3x13 double]
    v: [3x13 double]
    s: [3x1 struct]
    h: [6x13 double]
    f: [NaN NaN NaN NaN NaN NaN NaN NaN NaN NaN NaN NaN NaN]

>> data.x
ans =
  Columns 1 through 11

    1.0028    0.9996    0.9966    0.9942    0.9925    0.9920    ...
    0.6843    0.6748    0.6653    0.6556    0.6459    0.6361    ...
    0.0365    0.0367    0.0372    0.0381    0.0396    0.0418    ...

  Columns 12 through 13

    1.0033    1.0081
    0.6297    0.6319
    0.0487    0.0502
```

The `getCurve` command will load the data of the curve selected in the main window. The 14-cycle was selected.

7.1.7 Additional problems

A. **Conservative dynamics**

Simulate the following *area-preserving* 2D-map (Hénon 1969)

$$\begin{pmatrix} x \\ y \end{pmatrix} \mapsto \begin{pmatrix} x\cos\alpha - (y - x^2)\sin\alpha \\ x\sin\alpha + (y - x^2)\cos\alpha \end{pmatrix}$$

for $\cos\alpha = 0.8, 0.4, 0.24, 0.02, 0$ in the domain $(x, y) \in [-1, 1] \times [-1, 1]$. Pay special attention to fixed points and cycles.

B. **Feigenbaum cascade in Ricker map**
Consider the following simple population model

$$x_{k+1} = \alpha x_k \mathrm{e}^{-x_k},$$

where $x_k \geq 0$ is the population density in year k, and $\alpha \geq 0$ is the growth rate. Study the fixed points and cycles of the corresponding scalar map

$$x \mapsto \alpha x \mathrm{e}^{-x}.$$

1. Starting from $x = 0$ at $\alpha = 0$, compute with MatcontM a branch of trivial fixed points of this map in the (x, α)-space. Detect a branch point, and switch to the nontrivial branch of the fixed points using an appropriate `Initializer` option.
2. Detect a PD bifurcation of this fixed point and switch to the continuation of the bifurcating period-2 cycle.
3. Find several further PD bifurcations and observe numerically the accumulation of the period-doubling bifurcations.
4. Study analytically the first period-doubling bifurcation, i.e., find the critical parameter value and compute the cubic normal form coefficient. Do they agree with your numerical observations?

7.2 Tutorial 2: two-parameter local bifurcation analysis

We perform a two-parameter bifurcation analysis on a discrete-time prey–predator model using MatcontM.[6] The map

$$\begin{pmatrix} x \\ y \end{pmatrix} \mapsto \begin{pmatrix} ax(1-x) - \dfrac{bxy}{1+\epsilon x} \\ \dfrac{dxy}{1+\epsilon x} \end{pmatrix} \tag{7.2}$$

models the year-to-year dynamics of a prey–predator ecosystem. It describes how predators respond to changes in prey availability. The coordinates x and y represent the population densities of the prey and the predator, respectively, while a, b, d and ϵ are the parameters of the system. Parameters a and d determine the growth rate of the prey and predator; b determines the rate at which prey are consumed; ϵ acts as a limitation parameter on the growth of the predator population for increasing prey population density.

The fixed points of system (7.2) can be determined algebraically by solving the fixed point equations. One trivial solution is the origin $F_1 = (0, 0)$, which

[6] This tutorial has been tested on MATLAB R2017b with `matcontm5p4`.

means having neither prey nor predators. For $y = 0$, we get the solution $F_2 = (\frac{a-1}{a}, 0)$, the situation where no predators exist, only prey. For $x \neq 0$ and $y \neq 0$, we get the solution

$$F_3 = \left(\frac{1}{d-\epsilon}, \frac{d}{d-\epsilon}\left(\frac{a}{b}\left(1 - \frac{1}{d-\epsilon}\right) - \frac{1}{b}\right)\right). \tag{7.3}$$

F_1, F_2 and F_3 are the only fixed points of (7.2).

7.2.1 System specification

This map (7.2) is already implemented as `PredatorPreyModel` in MatcontM. Select it using **System|System Browser** and highlighting `PredatorPreyModel`. Press **Edit** to check that the system is correct and note that all derivatives up to and including fifth order are generated symbolically. Press **OK**. Now you can actually load the system by pressing the **Load** button (see Figure 7.22).

Alternatively, you can implement the system yourself. Select **System|System Browser**, press **New** and enter the data shown in Figure 7.23. Here we use the name `PPModel`. In Figure 7.23 all derivatives up to and including fifth order are generated symbolically.

We recall that if derivatives are not provided symbolically, they will be generated by automatic differentiation with the same accuracy.

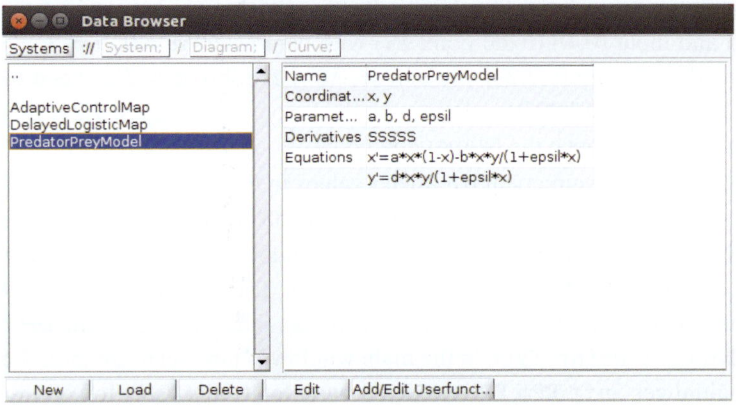

Figure 7.22 Press **Load** to select the highlighted system.

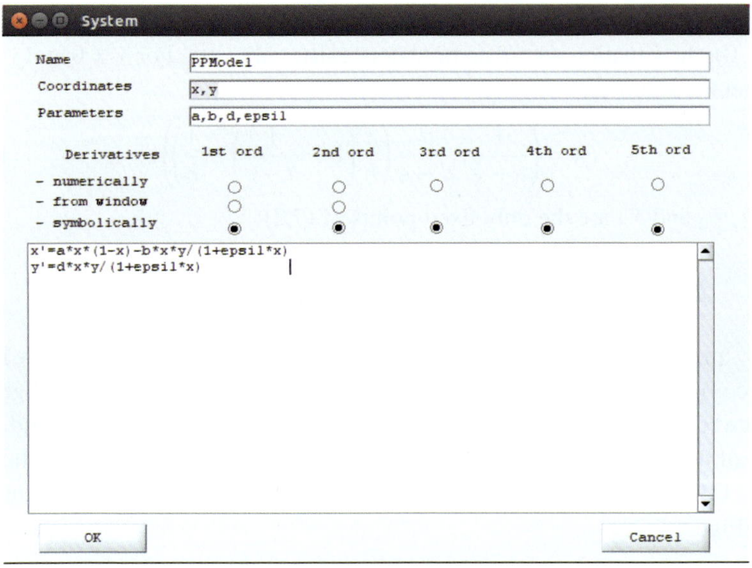

Figure 7.23 Entry for a new system. Press **OK** to add the system.

7.2.2 Continuation of fixed points

We will now compute a series of fixed-point curves.

Verify that the `Initial Point Type` is set to `Fixed Point` and the `Initializer` is set to `init_FPm_FPm`. Select the Starter window, activate parameter a and input $(0, 0)$ (fixed point F_1) with parameters $a = 0$, $b = 3$, $d = 3.5$ and epsil $= 1$ (Figure 7.24, you can set `AD threshold` $= 7$ or keep the default 24). Open a Plot2D window with the layout as in Figure 7.25. It is also recommended to open the Numeric window.

Ensure that the numerical parameter values in the Continuer window are as follows: `InitStepSize` $= 0.01$ while `MaxStepSize` $= 0.1$.

Select **Compute|Forward** in the main window and stop the computation after detecting the first `Branch Point` (BP). This occurs at $a = 1$. A BP occurs generically when two fixed point curves intersect. Select the BP point. The `Initial Point Type` in the main window is now set to `Branch Point`. The initializer `init_BPm_FPm` (selected by default) allows us to continue the intersecting curve of fixed points. We note that when a point is selected on a curve then the Starter and Continuer data of new curves started from that point are by default those of the curve on which the point was selected. Of course, the user can change these.

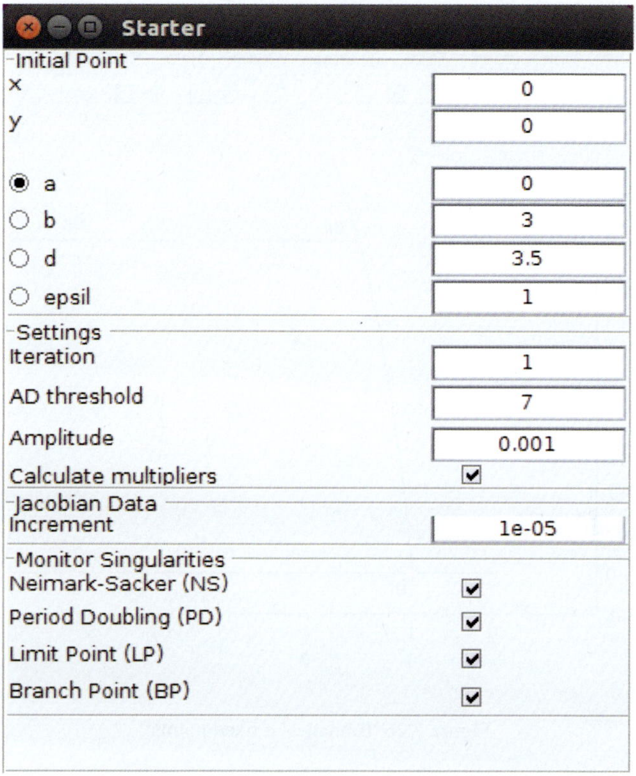

Figure 7.24 Starter for fixed point branch.

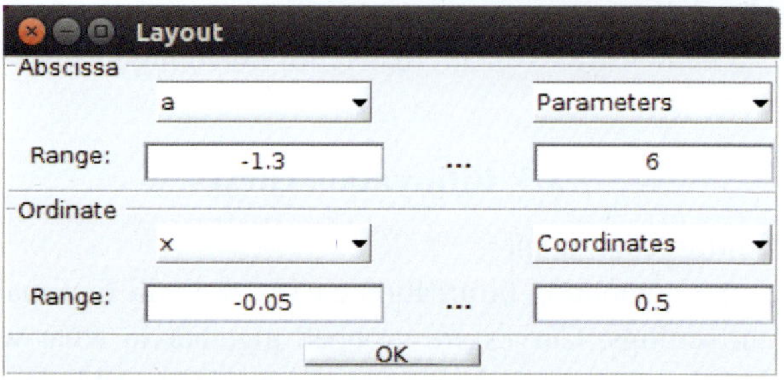

Figure 7.25 New plot layout options.

Select **Compute|Forward** and keep an eye on the Numeric window. Stop the computation after detecting the first BP point (at $a = \frac{5}{3}$). The fixed points of this new branch are F_2 fixed points. Select the latest BP point and select

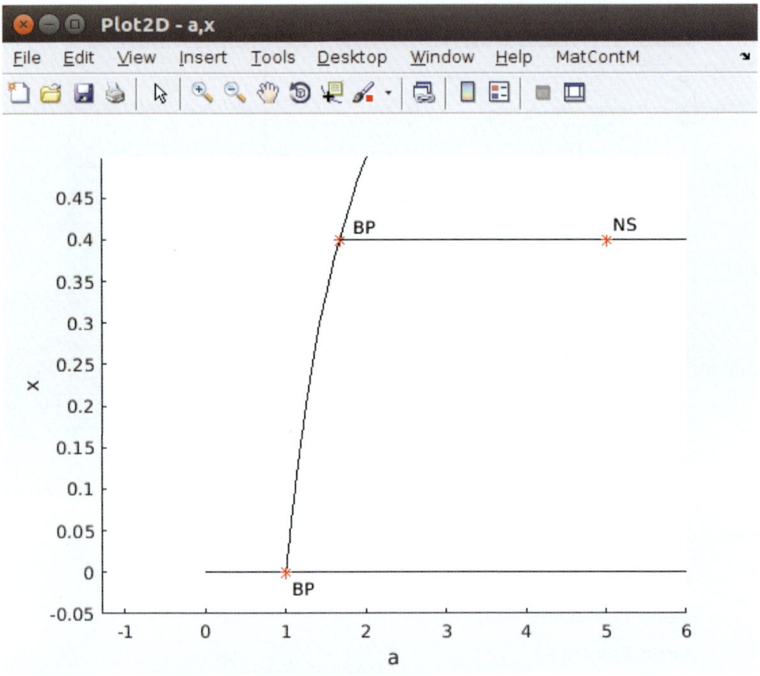

Figure 7.26 Branches of fixed points.

Compute|Forward. This branch consists of F_3 fixed points. Stop the computation after detecting an NS bifurcation at $a = 5$. The resulting diagram is shown in Figure 7.26.

Section 7.2.5 demonstrates an alternative for obtaining this NS bifurcation.

7.2.3 Bifurcation curves

Select the NS bifurcation point.

Before we start computing bifurcation curves, we will first make a change in our diagram settings. Curves are grouped together in what we call "diagrams." The current diagram in which the curves are stored is the default diagram named `diagram`. We can create a new diagram using the context menu (right-click) in the `Current Curve` part of the main window, select **New Diagram** and enter the name `bifcurves`. All new curves will now be stored in that diagram.

The NS bifurcation is now selected for continuation (see main window), an NS bifurcation curve has been selected as curve (`init_NSm_NSm`). Note that

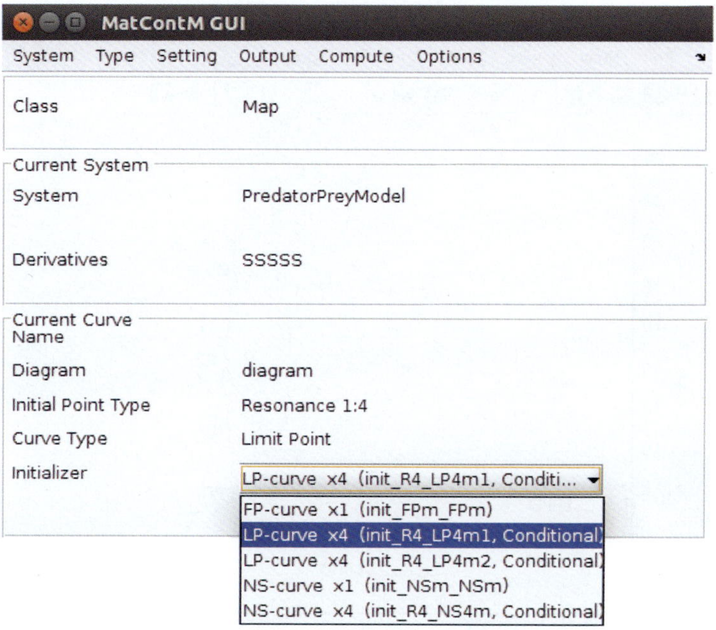

Figure 7.32 Selecting new branches from a R4 point.

We note that a change in any of the fields in the Starter or Continuer windows will be ignored in the computation of an extension.

If you extend several times you will find another LPPD bifurcation and a Resonance 1:1 (R1) bifurcation. Stop the computation.

You can rescale your plot window with the MATLAB tools and the Layout window. When you change the scale of the graph, you might want to redraw all curves because the labels seem misplaced. By using the MatcontM menu in the Plot2D window, you can select **Redraw Diagram** to redraw all the curves in the current diagram. **Redraw Curve** only redraws the curve selected in the main window.

You can use these two draw buttons to visualize curves after they have been computed and stored. You just need to load the desired curve as the current curve, open and configure a plot window and press **Redraw Curve/Diagram**.

The result is shown in Figure 7.33. You see two green LP curves emanating from the R4 point on the NS curve.

We now select the first LPPD bifurcation, located on the first LP^4 curve. The easiest way is to double-click on the LPPD label. Make sure all tools are disabled. This bifurcation lies on an LP curve of the fourth iterate of the

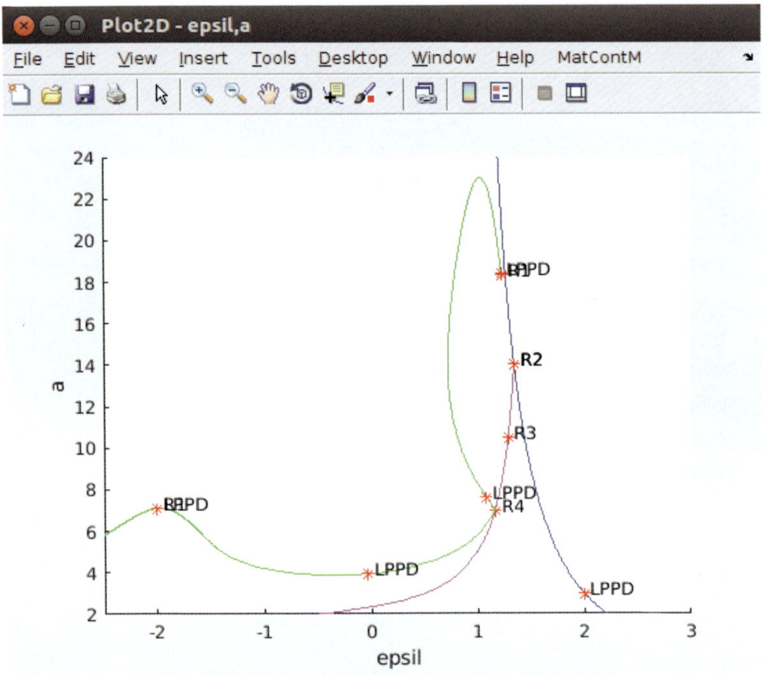

Figure 7.33 Computing the two LP^4 curves from R4.

map. This bifurcation indicates the presence of a PD curve passing through the LPPD point. Select the `PD-curve x1` initializer and perform a forward continuation, followed by a backwards continuation.

The loop emanating from R3 is computed by double-clicking on R3 and computing `NS-curve x3`. Notice that the NS curve here corresponds to a neutral saddle cycle of period 3, not to a Neimark–Sacker bifurcation.

The full picture of the `bifcurves` diagram is shown in Figure 7.34. The blue curve in this Figure is the PD^4 curve. The closed magenta curve is the NS^3 (neutral saddle) curve. Figure 7.35 is a zoom near the R3 and R4 points; Figure 7.36 is a more detailed zoom near the R4 point.

To create a three-dimensional plot, select **Output|Graphic|3D Plot**. Next select **MatContM|Layout** in the Plot3D window and use the layout settings as in Figure 7.37. Select **MatContM|Redraw Diagram** to draw all computed bifurcation curves. These are the curves located in the diagram `bifcurves`.

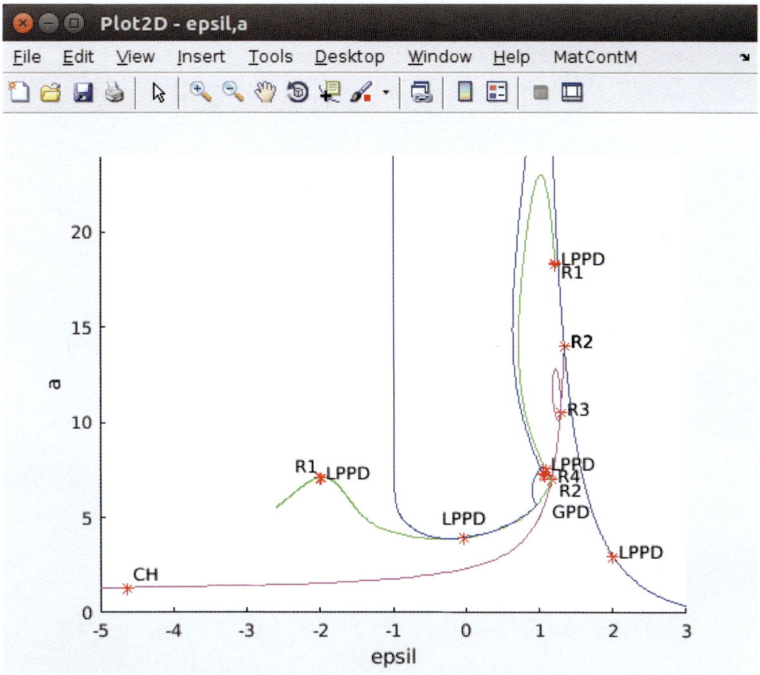

Figure 7.34 All the bifurcation curves in the diagram `bifcurves`.

7.2.4 User functions

User functions are custom test functions added by the user to a system. The continuation algorithm evaluates these functions at each point on the computed curve and checks for sign changes in order to detect zeros. When a zero is detected, a bisection-like algorithm is used to determine the exact point where the zero occurs. These zero points are reported as special points. A user function can use the coordinates and parameters and returns a real value. Such functions can be used to detect situations specific to the system or to add flags for certain values.

We can add, edit, and delete user functions by going to the System Browser by selecting **System|System Browser**. Select the system `PPModel` and click on the button **Add/Edit Userfunctions** below. The window for user functions appears (Figure 7.39). We can now add, edit, and delete user functions for the selected system. We note that no check is done on the input. So if the input is

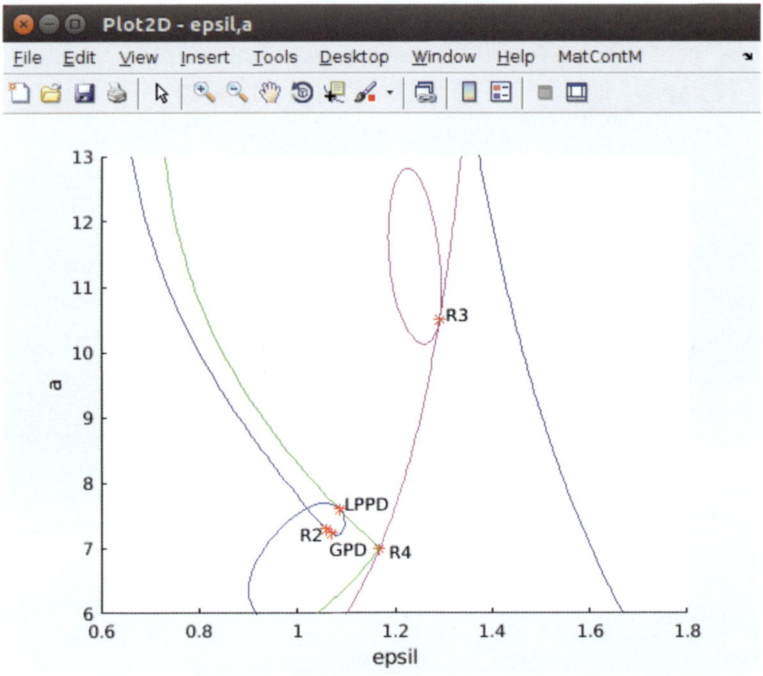

Figure 7.35 Zoom containing R3 and R4.

not grammatically correct in MATLAB then no error or warning will be generated but the session will probably crash later on.

For example, we add three user functions. We wish to know when a coordinate (population density) reaches the value 0.5. We start by filling in the `Name` field, we will name our function `halfX`. We fill in the label by entering E1. The formula for our user function has to be entered in the form: `res=...` (no spaces after `res`). We can use the coordinates x and y and the parameters a, b, d and `epsil` in our formulas.

We enter `res= x - 0.5`.

Click on the **Add** button to add the user function to the list in the upper-right corner.

We name the second function `halfY`, the label is E2 and the equation is `res=y - 0.5`. Press **Add**.

Our third function allows us to monitor when the parameter a has an integer value during computation.

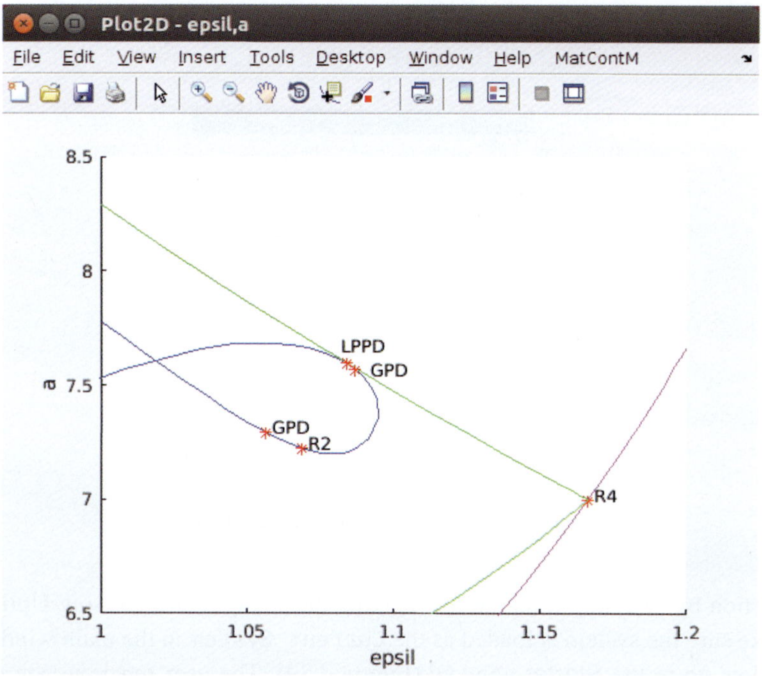

Figure 7.36 Zoom near R4.

We must design a function where $f(a) = 0$ if a is an integer. A sign change must occur in the zero point for it to be detected by the continuation algorithm. We can use the MATLAB built-in functions. We name the function `IntegerA`, give it label IA and use the equation:

```
res=abs(a-round(a))*(mod(floor(a),2)-not(mod(floor(a),2)))
```

The function values for the interval $[-2, 5]$ are shown in the following plot:

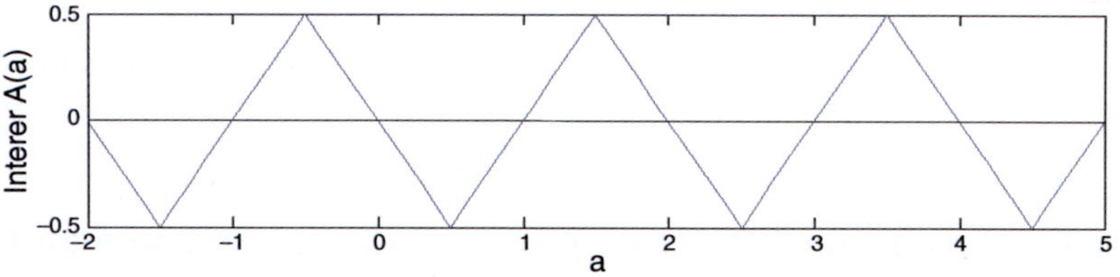

Press **Add**. Press **OK** to exit the window, the user functions are now specified.

You can always return to add, edit, and delete user functions. You can edit a

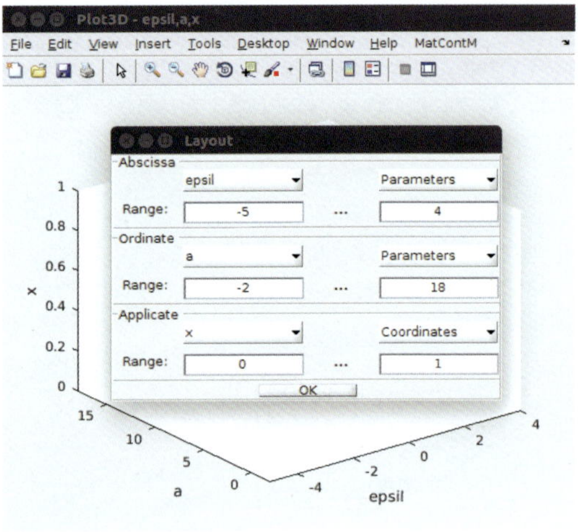

Figure 7.37 Layout options for the 3D plot.

function by selecting it on the list, making the changes and pressing **Update**. Make sure the system is loaded as the `Current System` in the main window.

Now go to the Starter window (Figure 7.39). The user functions are now added to the subpanel named `Monitor Userfunctions`, you can enable them or disable them. They are disabled by default so enable them all.

Enabled user functions become available in the plot Layout window and in the Numeric window (see Figure 7.40).

Using the settings from Figure 7.39, we compute a fixed-point curve by varying the parameter epsil.

User function zeros are treated in the same way as the bifurcations: They are displayed on plots and they can be selected in the Data Browser window when viewing a curve. When a user function zero is selected as the initial point for continuation, no initializers (branches) are available. This can be solved by declaring the type of the point, by selecting in the menu: **Type|Initial Point**. The results of the computation starting from the setting in Figure 7.39 are shown in Figure 7.41. Another computation is shown and Figure 7.42 shows what happens if two bifurcation points coincide.

7.2.5 Configuring the Starter window from the MATLAB Command window

Working with a graphical user interface (GUI) has many advantages, but also some disadvantages. One of the difficulties is entering many numbers in the

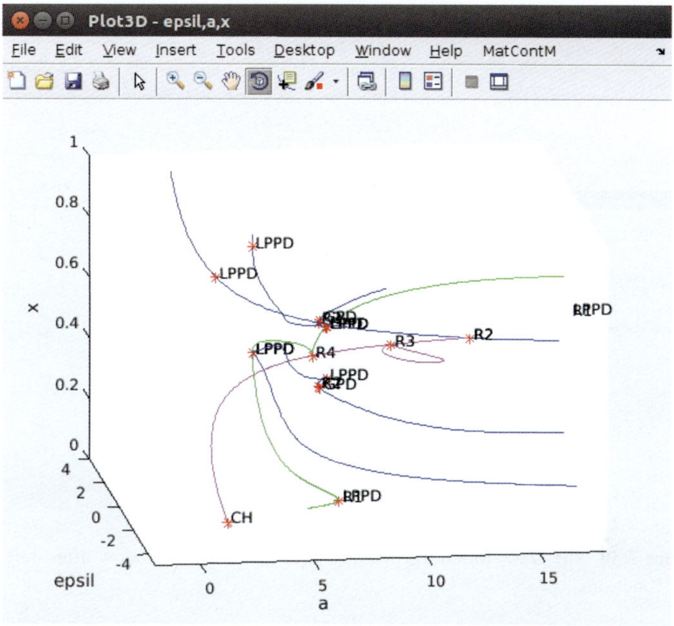

Figure 7.38 The MATLAB *rotation tool* has been selected.

Starter window. Once the continuation has started and bifurcations are being detected, typing becomes minimal.

While the GUI of MatcontM is running, the MATLAB command line (CL) stays available. The `CommandLineInterface` bridges the gap between GUI and CL; it allows for the transfer of data between the GUI and the CL workspace. We will show this by example for system (7.2).

First, select in the main window menu: **Type|Initial Point|Fixed Point**, to prepare for a fixed point continuation. The map `PPModel` (`PredatorPreyModel`) has three types of fixed points. F_1 and F_2 are simple to work with. F_3 is rather complicated. We will use the MATLAB *anonymous functions* to make computing the coordinates easy for different sets of parameters.

```
>> F3_point = @(a,b,d,epsil) [1/(d-epsil) ,
                        (d/(d-epsil))*((a/b)*(1-(1/(d-epsil)))-(1/b))];
>> F3 = @(Param) F3_point(Param(1) , Param(2) , Param(3) , Param(4));
```

We can now use the `F3_point` function to compute the coordinates for any parameters or use the `F3` function when the parameters are stored in an array.

```
>> param = [4.1,  3,  3.5,  1];
```

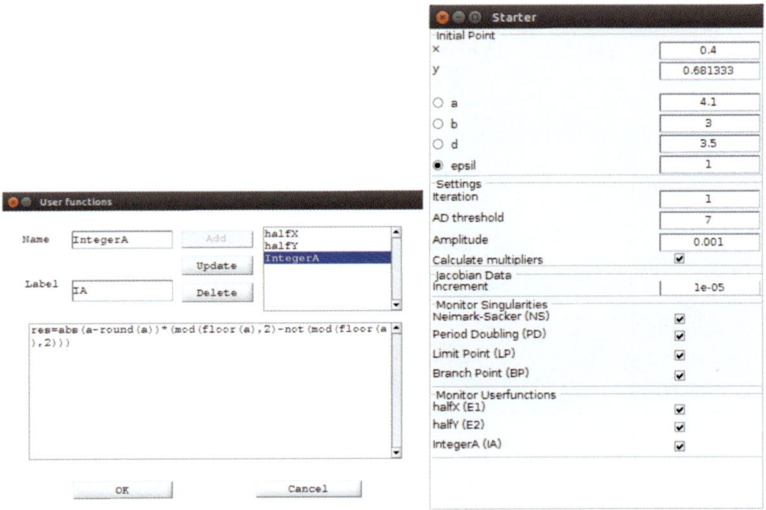

Figure 7.39 The User functions window and the Starter window after entering user functions.

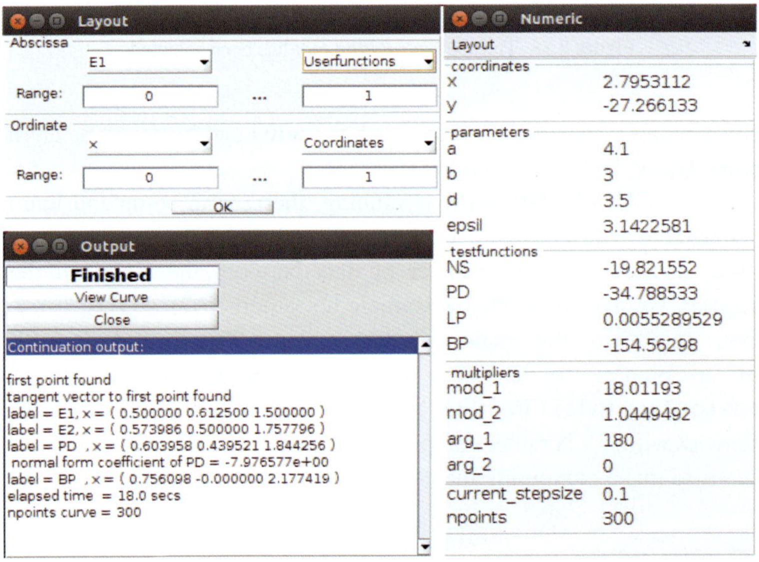

Figure 7.40 Numeric and Layout windows with user functions. The Output window displays the detection of user function zeros.

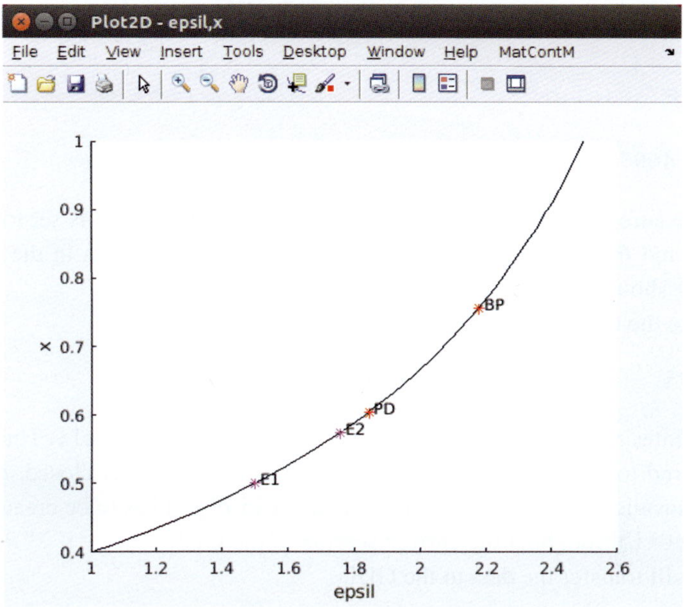

Figure 7.41 Fixed point continuation showing detection of zeros of user functions.

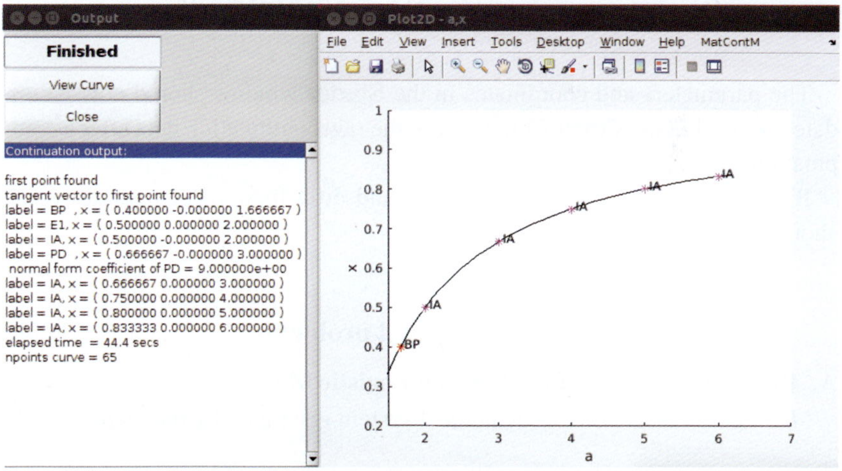

Figure 7.42 The E1 and first IA point as well as the PD and second IA point coincide.

```
>> coord = F3(param)

coord =

    0.4000    0.6813
```

Make sure system (7.2) is loaded and the initial point type is set to `Fixed Point` and the initializer is set to `FP-Curve x1`. `Iteration` in the Starter window should be set to 1.

While the GUI is running, we execute in the MATLAB prompt:

```
>> cli = CommandLineInterface;
```

This creates a CommandLineInterface object that is stored in `cli`. This object is tethered to the current GUI session. If the GUI session is closed, `cli` becomes invalid. If the GUI is restarted, a new `cli` object has to be created. You can use `cli` throughout the current session of the GUI.

We will transfer the data to the GUI:

```
>> cli.setPa(param);          %or:  cli.setParameters(param);
>> cli.setCo(coord);          %or:  cli.setCoordinates(coord);
>> %cli.setIP(coord, param) %alternative command,
                              %IP = InitialPoint
```

The parameters and coordinates in the Starter window should now be updated. Use `cli.getCurve()` to retrieve the raw continuation data after a computation.

If you select a as the free parameter and do a forward computation, you should find an NS bifurcation at $a = 5$.

7.2.6 Additional problems

A. **Bifurcation curves of the Delayed Logistic Map**

Revisit the discrete-time dynamical system generated by the map

$$\begin{pmatrix} x \\ y \end{pmatrix} \mapsto \begin{pmatrix} rx(1-y)+\epsilon \\ x \end{pmatrix}$$

studied in Tutorial 1. Use the located bifurcation points on the stored curves as initial data for the two-parameter continuation of these bifurcations to obtain Figure 7.43.

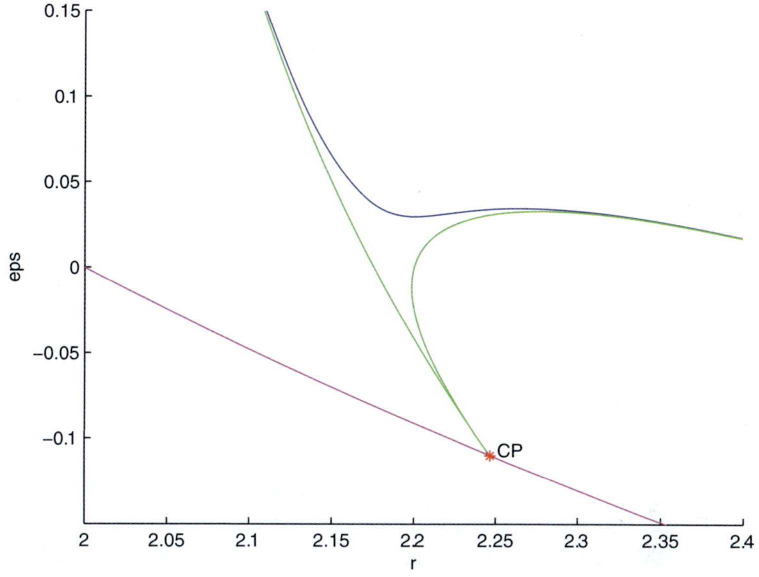

Figure 7.43 Bifurcation curves of Delayed Logistic Map: NS (magenta), LPC[7] for the period-7 cycles (green), and PD[7] for the period-7 cycles (blue). The point labeled as CP is actually close to the resonance 1:7 point on the NS curve.

B. **Discrete-time Lotka–Volterra model**

Consider the following simple prey–predator population model[8]

$$\begin{cases} x_{k+1} = ax_k(1 - x_k) - bx_k y_k, \\ y_{k+1} = dx_k y_k, \end{cases}$$

where (x_k, y_k) are the prey and predator densities in year k, and (a, b, d) are positive parameters. Study fixed points and cycles of the corresponding planar map

$$\begin{pmatrix} x \\ y \end{pmatrix} \mapsto \begin{pmatrix} ax(1 - x) - bxy \\ dxy \end{pmatrix}.$$

1. Study the map for $b = 0.2$ and $(a, d) \in [0, 12] \times [0, 4]$ using MatcontM:

   a. Find by simulations a positive fixed point of the map at $a = 3.5$ and $d = 2.67$. Continue this fixed point with respect to parameter d and detect two codim 1 bifurcations of this point, namely NS and PD.

[8] The map (7.2) reduces to this model for $\epsilon = 0$.

b. Continue the corresponding codim 1 bifurcation curves NS^1 and PD^1 in the (d, a)-plane and locate several codim 2 points, including strong resonances R2, R3 and R4 on the NS^1 curve.
c. Starting from the detected R4 point, compute two LP^4 bifurcation curves of 4-cycles rooted there.
d. Compute the PD^4 and NS^4 bifurcation curves of 4-cycles located between the LP^4 curves.

   Hint: To obtain starting points, first continue a stable period-4 cycle from one of the LP points with respect to parameter a and stop the computation somewhere between the LP^4 curves. Then continue the last stable 4-cycles with respect to parameter d to detect its NS and PD bifurcations.

2. Study the map analytically for positive (a, b, d):

   a. Find all nonnegative fixed points and derive explicit formulas for the NS^1 and PD^1 curves, as well as bifurcation curves of fixed points with at least one zero coordinate. What is the stability domain in the (d, a)-plane of the positive fixed point?
   b. Compute analytically the normal form coefficients for the PD and NS bifurcations along the curves NS^1 and PD^1. Predict what happens when crossing these curves at generic points.

Are your numerical and analytical results in agreement with each other ?

7.3 Tutorial 2: invariant manifolds and connecting orbits

This tutorial is devoted to the numerical construction of one-dimensional stable and unstable invariant manifolds of saddle fixed points of (orientation-preserving) maps

$$u \mapsto f(u, \alpha), \quad u \in \mathbb{R}^n, \quad \alpha \in \mathbb{R}^m,$$

at fixed parameter values, and to the continuation of their intersections (i.e., orbits asymptotically connecting saddles) with respect to one or two parameters.[9]

Notice that in this case a "point" on a continuation curve represents an entire connecting orbit. We must first obtain an initial connecting orbit. Such an orbit lies in the intersection of the unstable manifold of a fixed

[9] MatcontM is available at http://sourceforge.net/projects/matcont/files/matcontm/. This tutorial has been tested on MATLAB R2017b with `matcontm5p4`.

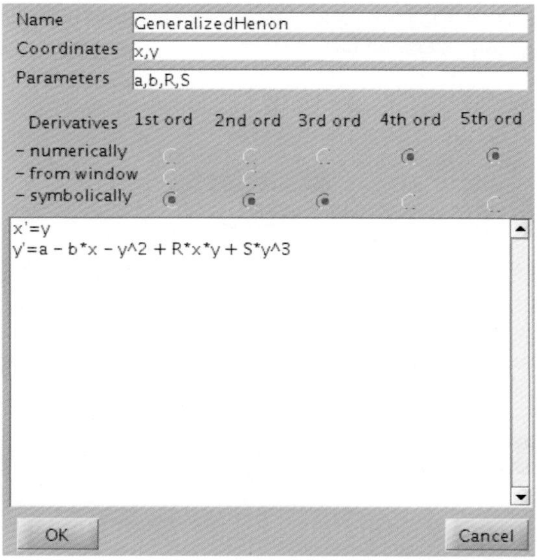

Figure 7.44 Using the System window, add the Generalized Hénon Map.

point u_1 and the stable manifold of a fixed point u_2. We note that there can be multiple connecting orbits between the same saddle points. In the case of a homoclinic orbit, u_1 and u_2 are the same point u_0. MatcontM can compute one-dimensional stable and unstable manifolds and in the case of two-dimensional maps find intersection points between these manifolds. The intersection points then lie on connecting orbits.

The user can also provide MatcontM with an orbit that has been obtained through other means, if the orbit is stored as a matrix in the matlab workspace.

As an example, we will use a map called the *Generalized Hénon Map* (GHM)

$$\begin{pmatrix} x \\ y \end{pmatrix} \mapsto \begin{pmatrix} y \\ \alpha - \beta x - y^2 + Rxy + S y^3 \end{pmatrix}. \tag{7.4}$$

This system may be already implemented under the name `GeneralizedHenon`. If not, you can re-enter it in MatcontM as in Figure 7.44.

7.3.1 Computation of one-dimensional invariant manifolds of a saddle

Once the system is loaded we can start by computing a connecting orbit for continuation. We do this by changing the type of the initial point. We select

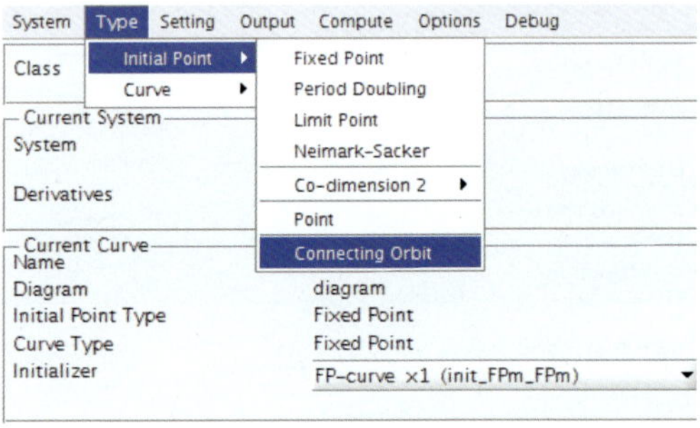

Figure 7.45 Set the type of the initial point to **Connecting Orbit**.

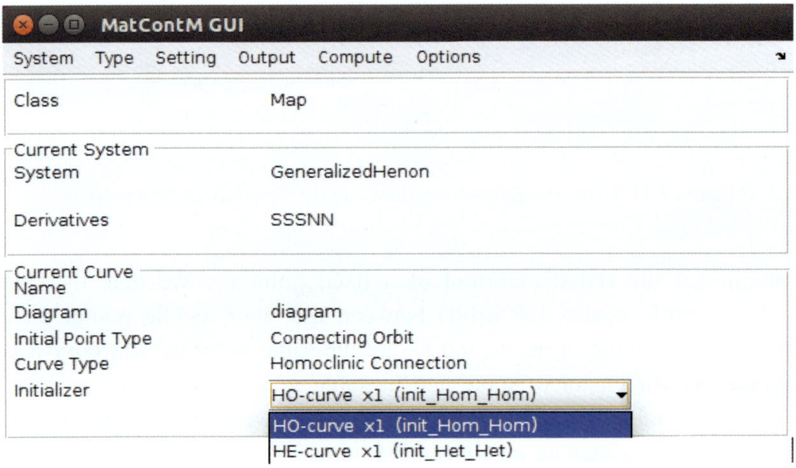

Figure 7.46 Select the homoclinic connection continuation (`HO-curve`).

in the main menu: **Type|Initial Point|Connecting Orbit** (Figure 7.45). Once the type of the initial point has been set to an orbit, you will be able to select or compute a connecting orbit. You can choose between a homoclinic orbit (`HO-curve`) and a heteroclinic orbit (`HE-curve`) (see Figure 7.46). Make sure the homoclinic orbit option is selected.

We select the Starter window. The `Initial Point` subpanel of the Starter window contains either zeros or a point that was selected on a curve of fixed points. This information is only used as a dummy fixed point for computing manifolds and plays no further role in the continuation of connecting orbits. Notice that a special panel has appeared on the Starter window when changing

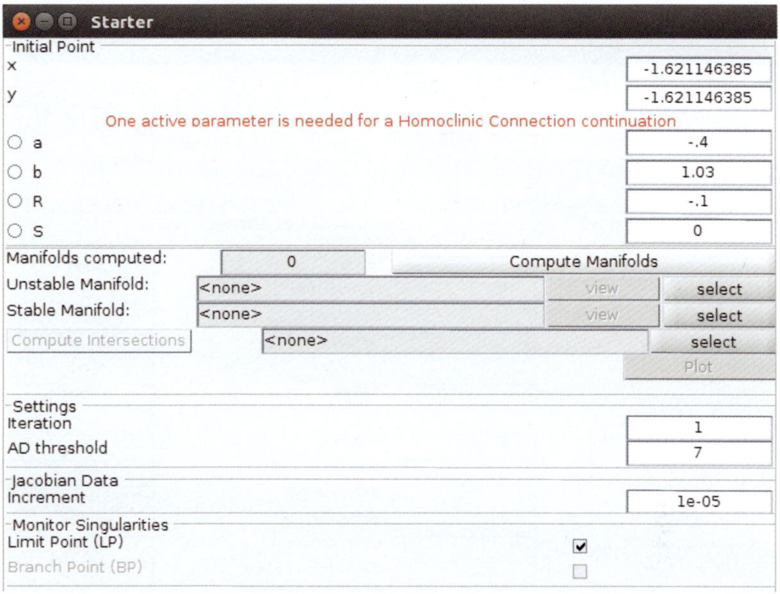

Figure 7.47 The Starter window after entering the parameter values.

the type of the initial point to connecting orbits. This panel will allow us to compute manifolds, find intersections and select connecting orbits.

Before we start computing manifolds, we first have to enter the values of the parameters in the Starter window. Set $a = -0.4$, $b = 1.03$, $R = -0.1$ and $S = 0$ (Figure 7.47).

Start computation of a stable and an unstable manifold by pressing the **Compute Manifolds** button in the Starter window. A Compute Manifolds window for computing manifolds will appear (Figure 7.48).

First we need to enter a fixed point. Enter the point

$$(x, y) = (-1.621146385, -1.621146385).$$

Proceed by pressing the **Select fixed point and proceed to configure** button. We can now configure the algorithm used to compute the manifold. The most important setting defines whether we want to compute a stable or unstable manifold. We will first compute the unstable manifold: next to `function`, we select `Unstable Manifold`. We will adjust some other settings as well:

- set `nmax` to `4000` (maximum number of points in the manifold);
- set `distanceInit` to `-1e-08` (initial displacement from the fixed point);
- set `deltaMax` to `0.01`.

(a)

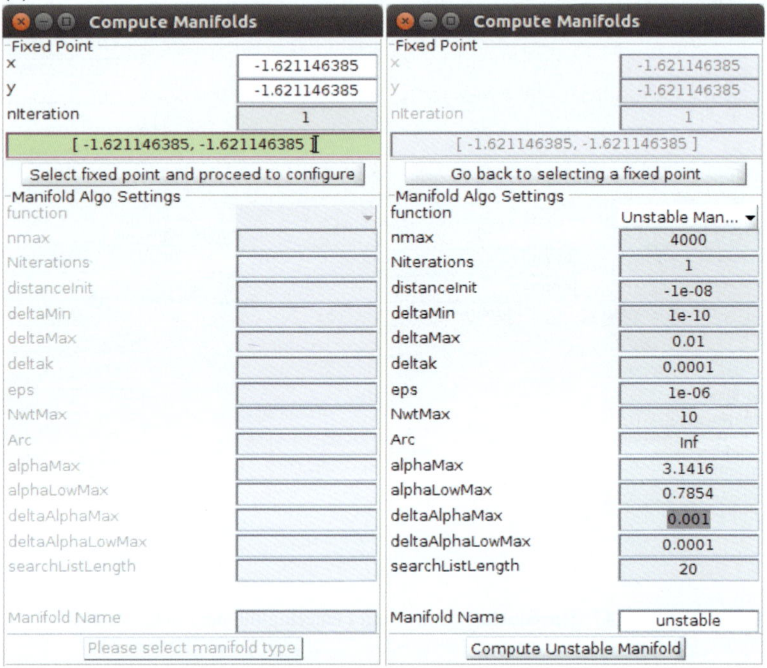

Figure 7.48 (a) Compute Manifolds window after entering a fixed point. (b) Compute Manifolds window configured for the computation of an unstable manifold.

It is important to set `distanceInit` *negative*, since it determines which of two branches of the manifold will be computed.

The Compute Manifolds window should now look like Figure 7.48(b). You can give the manifold a name, e.g., `unstable` (Figure 7.48(b), see `Manifold Name`). If no name is given, a default name will be chosen.

Press **Compute Unstable Manifold**. A window will appear that displays the output of the algorithm; close that window when the computing stops and go back to the Compute Manifolds window (Figure 7.48).

We will now compute the stable manifold; go to `function` and select `Stable Manifold`. We give this manifold the name `stable`. Press **Compute Stable Manifold**.

We have computed an unstable and a stable manifold of the same fixed point at given parameter values. Close the Compute Manifolds window and return to the Starter window.

In the Starter window, go to the `Unstable Manifold` section and press the **select** button. A Data Browser window pops up and lists the available unstable manifolds; we only have one available.[10] We select the manifold by double-clicking on the name in the list. Repeat the same process for `Stable Manifold` in the Starter window.

7.3.2 Loading the computed manifolds into the workspace

Often the computed manifolds are used only to find connections and will not be permanently stored. However, it is possible to load them in the MATLAB workspace to keep them for further usage. The following MATLAB session demonstrates how this can be done.[11]

```
>> cli=CommandLineInterface;
>> manifolds=cli.getManifolds

manifolds =
  struct with fields
      stable: [1x1 Manifold]
    unstable: [1x1 Manifold]
         all: [1x4 struct]
```

The fields `stable` and `unstable` correspond to the manifolds that have been selected in the Starter window. The field `all` contains all manifolds computed so far. To inspect the content of a particular curve, we proceed as follows.

```
>> M=manifolds.all(1).manifold
M =
  Manifold with properties:
    points: [2x4000 double]
    arclen: 25.4053
      optM: [1x1 struct]
    man_ds: [1x1 struct]
      name: 'UnstableManifold_1'
```

At the end of this session the unstable manifold is contained in the $[2 \times 4000]$ array `M.points`. This method allows the user to plot, use and store the manifold in different ways.

[10] Here you can plot the computed manifold by pressing the corresponding **Plot xxx** button.

[11] You can skip this section if you are not interested in such low-level functionality.

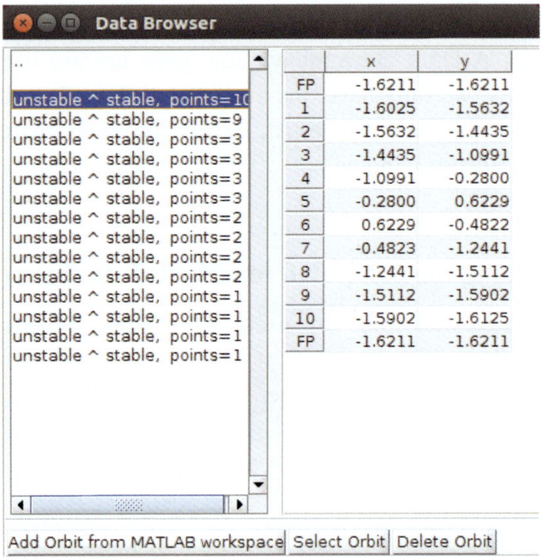

Figure 7.49 The intersection between the unstable and stable manifolds has produced multiple candidates for a connecting orbit. Select the connecting orbit by double-clicking.

7.3.3 Finding intersections of stable and unstable manifolds

Once a stable and an unstable manifold are selected, the **Compute Intersections** button becomes available. We produce a list of intersections between the two manifolds by pressing this button. The end of the computations is announced by the message `Intersections computed between unstable and stable manifolds.` Afterwards, press the **Select** button above the **Plot** button to select such an intersection as a candidate connecting orbit for continuation. A browser window pops up that lists the available connecting orbits produced by the intersection (Figure 7.49). Select the largest intersection (with ten points) by double-clicking.

The unstable manifold, stable manifold and connecting orbit can be plotted by pressing the **Plot** button (Figure 7.51). The MatcontM menu in the upper-right corner can be used to fine-tune the produced image, while the MATLAB `Print` or `Edit` menus allow us to print or edit it (Figure 7.52).

7.3.4 Continuation of homoclinic orbit in one parameter

Once the connecting orbit has been selected, we are ready for continuation. Select a as the free parameter. The Starter window should now look like the

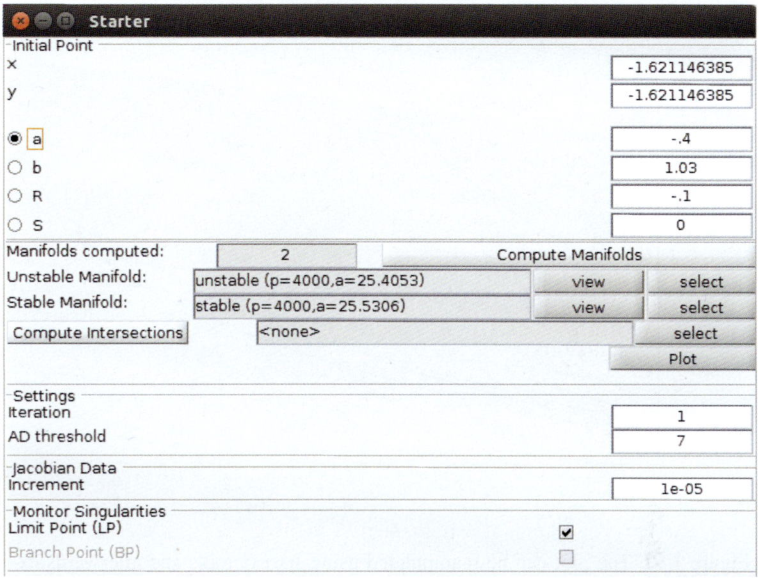

Figure 7.50 The Starter window after configuring a homoclinic continuation.

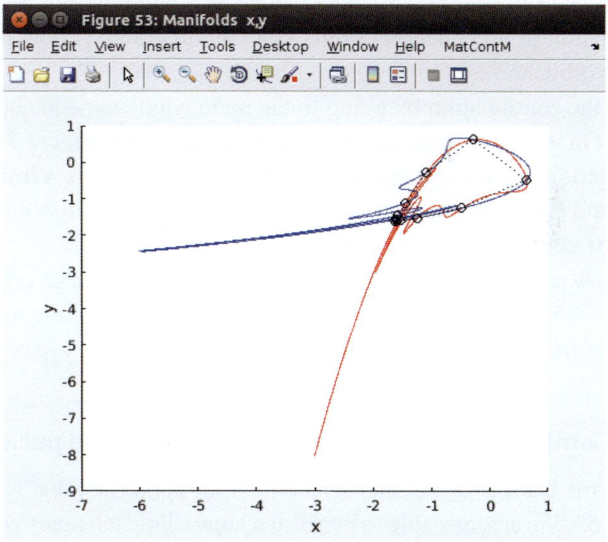

Figure 7.51 The **Plot** button produces an image of the stable and unstable manifold and the connecting orbit.

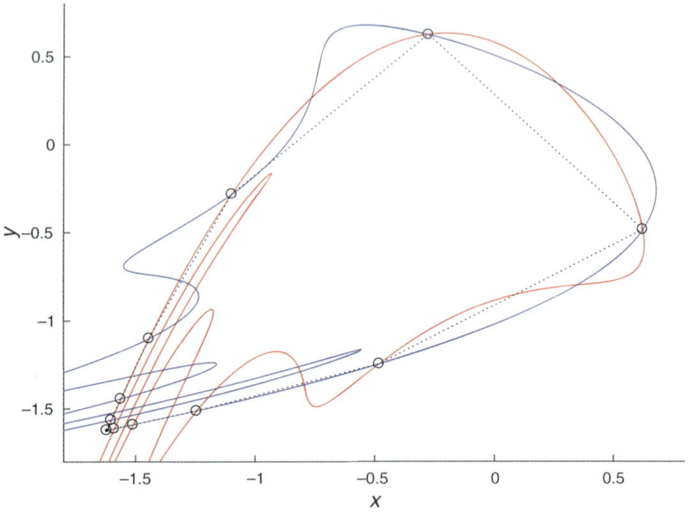

Figure 7.52 The plot can be manipulated using MATLAB tools and afterwards exported to an image file.

window depicted in Figure 7.50. The values of x and y do not play a role in the continuation of a connecting orbit. The "initial point" is now the selected homoclinic orbit.

We start the continuation by going to the main window menu and selecting: **Compute| Forward**. We find one *limit point* on the curve (Figure 7.53), where the computations should be resumed. When finished, press **View Curve** to open the Data Browser and then press **View CurveData** to view the numerical output of the continuation algorithm.

Notice how each point of the continuation curve represents an orbit. The last rows of the x matrix are reserved for parameters. Now the last row contains values of the free parameter a.

7.3.5 Continuation of homoclinic tangency in two parameters

Go back to the Data Browser and select the limit point (LP) that was detected in Figure 7.53. We are now able to perform a homoclinic tangency continuation starting from the detected limit point.

We need to select two free parameters in the Starter window, parameters a and b. The homoclinic connecting orbit associated with the limit point is automatically selected and is named LP_HO-points (Figure 7.55). When in the Starter window after selecting the LP_HO point (Figure 7.55), we can compute

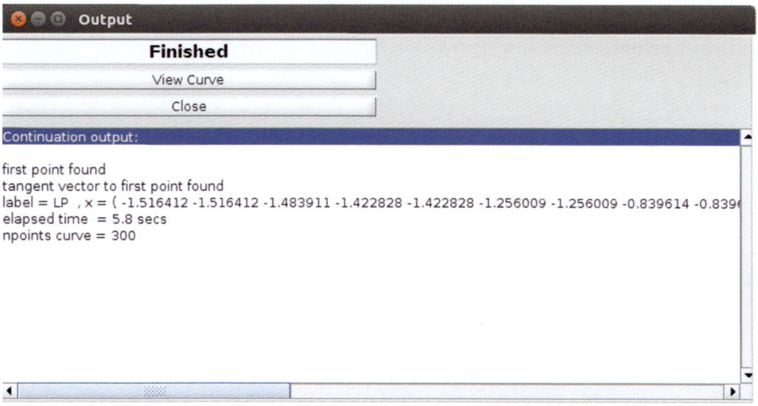

Figure 7.53 Output of the homoclinic connection continuation.

the stable and unstable manifold by pressing the **Compute Manifolds** button. We can use the same settings as in Figure 7.48(b), but use other names for the manifolds. The default value for the fixed point for computing the manifolds is the fixed point associated with `LP_HO-points`. Compute the stable and unstable manifold and select them in the Starter window, after deleting the previously computed `stable` and `unstable` manifolds. Press the **Plot** button to generate the figure. We do not need to compute an intersection because the orbit is already provided by `LP_HO-points`. The corresponding connecting orbit is represented in Figure 7.54 by black circles and black dashed lines. The stable and unstable manifold of the saddle fixed point are also visualized.

We advise setting `MaxNumPoints` to 80 in the Continuer window to reduce the computation time. We are ready to start the homoclinic tangency continuation by selecting **Compute|Forward**. Next, select **Compute|Backward**.

Open a Plot2D window to visualize the computations in the parameter plane. Use the configuration as in Figure 7.56. Select **MatContM|Redraw diagram** in the plot window. The result is shown in Figure 7.57. Notice that the tangency curve (red) forms a wedge. This wedge actually consists of two tangency curves (upper and lower) meeting at a sharp point at the left side. The continuation process accidentally hopped onto the other tangency curve when it encountered the sharp point on the left. This does not always happen, especially when stricter continuation thresholds are selected.

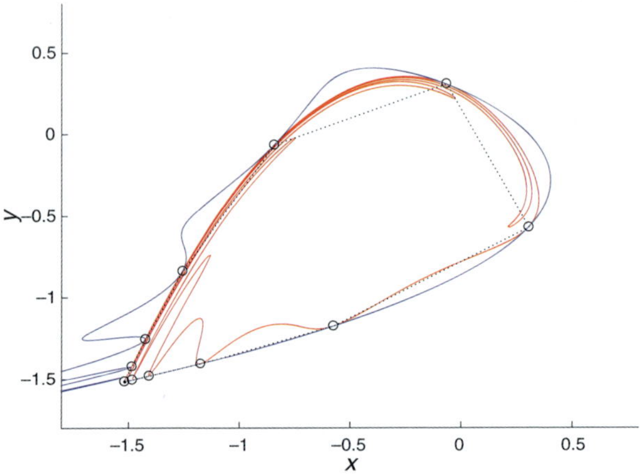

Figure 7.54 Homoclinic tangency.

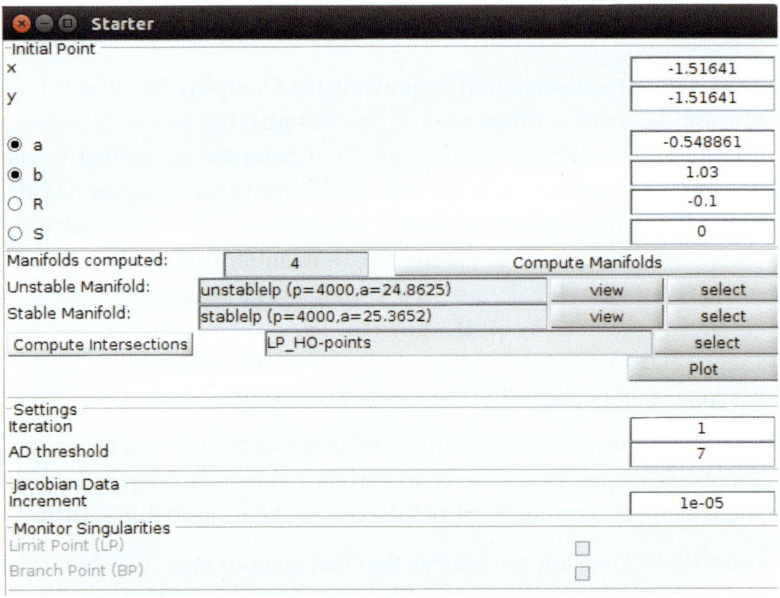

Figure 7.55 Starter window for the homoclinic tangency continuation.

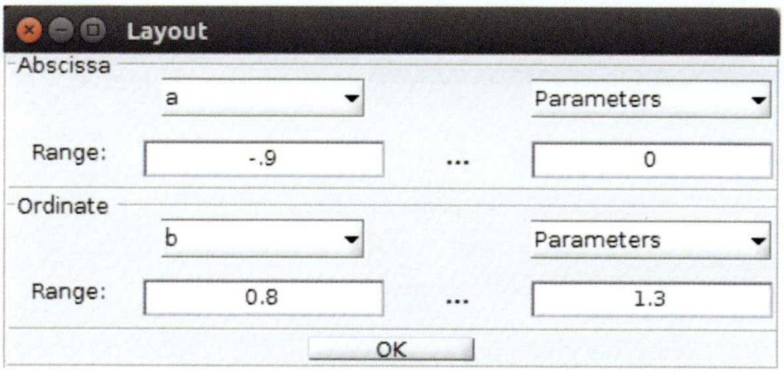

Figure 7.56 Layout options.

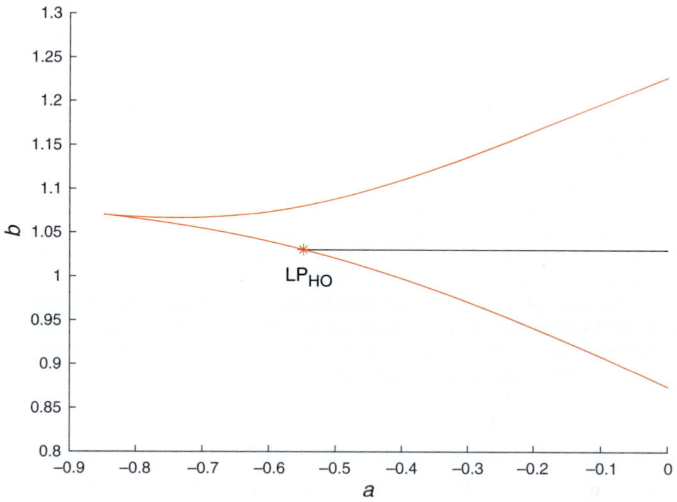

Figure 7.57 Homoclinic tangency curve with two branches meeting at a cusp.

7.3.6 Continuation of a heteroclinic orbit obtained through the MATLAB workspace

We will now compute a heteroclinic bifurcation curve using a connecting orbit that was obtained outside MatcontM.[12] For example, enter in the MATLAB command window:[13]

[12] You can skip this section if you are not interested in such low-level functionality.
[13] You can copy/paste this into MATLAB if you are reading this in a digital format.

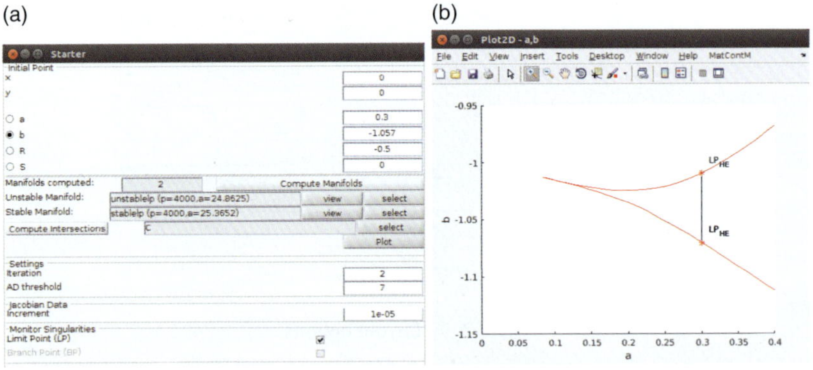

Figure 7.58 (a) Starter window of a heteroclinic connection continuation. (b) Output of heteroclinic tangency continuation.

```
C=[ 0.46661702380495,0.51950188769804 ,0.48317182408537, ...
    0.47311690109635 ,0.6123,0.841195,1.22990435,0.6419816, ...
        0.1347307,-0.14345697, -0.33379938281268,-0.38062126993587, ...
   -0.40424194997464,-0.41621339121235,-0.42230324914752, ...
   -0.42861702380495;0.46661702380495,0.37639405084666, ...
    0.43915841488454, 0.45696220552142, 0.2067,-0.276064, ...
   -1.3327006,-1.0930203,-0.7998431,-0.6233856,-0.49516248970159, ...
   -0.46255049612568,-0.44591635361118,-0.43743735402657, ...
   -0.43311132511221,-0.42861702380495];
```

The orbit is stored as a matrix under the variable name C and is now available in the MATLAB workspace. We will now enter this orbit into MatcontM. First, make sure you have selected a `Connecting Orbit` as the initial point type (Figure 7.45) and select the `HE-Curve` (Figure 7.46). Now open the Starter window and press the **select** button above the **Plot** button. You will now see a list of the available connecting orbits for selection. You can add C by clicking on the button **Add Orbit from MATLAB workspace** below.

Enter the variable name C; the orbit now appears in the list and is available for selection. After selecting the orbit, the name will appear in the Starter window. For this example, make sure that `Iteration` is set to 2 and b is selected as the free parameter. Set the values of parameter a to 0.3, b to -1.057, R to -0.5 and S to 0.

The Starter window should now look like Figure 7.58. Proceed by selecting **Compute| Backward**. You should detect at least one limit point. You can then select that point for a heteroclinic tangency continuation.

7.3.7 Additional problems

A. **Shear map**
Consider the following planar map

$$\Psi = \psi \circ \varphi,$$

where

$$\varphi : \begin{pmatrix} x \\ y \end{pmatrix} \mapsto \begin{pmatrix} \lambda^u x \\ \lambda^s y \end{pmatrix}$$

and

$$\psi : \begin{pmatrix} x \\ y \end{pmatrix} \mapsto \begin{pmatrix} x - cf(x+y) \\ y + cf(x+y) \end{pmatrix}$$

with

$$f(z) = \begin{cases} 0 & \text{for } z \le 1, \\ (z-1)^2 & \text{for } z > 1. \end{cases}$$

Fix $\lambda^s = 0.4$, $\lambda^u = 2.0$ in Ψ and compute the stable and unstable invariant manifolds of the origin for $c = 0.5, 0.75, 0.9$ and 1.0. Find the critical parameter value c^* corresponding to a homoclinic tangency.

B. **McMillan Map**
Consider the following planar map

$$\begin{pmatrix} x \\ y \end{pmatrix} \mapsto \begin{pmatrix} y \\ -x + \dfrac{2\mu y}{1+y^2} \end{pmatrix}, \tag{7.5}$$

where $\mu > 1$, and its perturbation

$$\begin{pmatrix} x \\ y \end{pmatrix} \mapsto \begin{pmatrix} y \\ -x + \dfrac{2\mu y}{1+y^2} + \varepsilon(\beta x + \gamma y) \end{pmatrix}, \tag{7.6}$$

where $\varepsilon > 0$ and (β, γ) are parameters. Study stable and unstable invariant manifolds of the saddle fixed point $(x, y) = (0, 0)$ of the maps, as well as intersections of these manifolds.

1. Study the map (7.5):

   a. Prove that the set of points $(x, y) \in \mathbb{R}^2$ satisfying

   $$x^2y^2 + x^2 + y^2 - 2\mu xy = C$$

   is invariant with respect to map (7.5) for any real constant C. Illustrate this by numerical simulations in MatcontM.

b. Prove that the origin $(x, y) = (0, 0)$ is a saddle fixed point of (7.5) and compute its stable and unstable invariant manifolds to verify that they form a "figure-of-eight." This is a very degenerate case.

2. Study the map (7.6) for $\epsilon = 0.05$ and $\gamma = 1.9$ when $(\beta, \mu) \in [-0.5, 2.0] \times [1, 4.5]$ using MatcontM:

   a. For $(\beta, \mu) = (0.1, 2.0)$, compute and plot branches of the stable and unstable invariant manifolds of the saddle $(0, 0)$ that emanate from it into the positive quadrant. *Warning*: These manifolds eventually pass through other quadrants of the (x, y)-plane.
   b. Locate with MatcontM intersection points of the invariant manifolds.
   Hint: You have to accurately compute rather long branches of the manifolds to ensure that they intersect sufficiently many times near the saddle. For that, you may need to tune parameters of the algorithm.
   c. Continue the obtained approximation of the primary homoclinic orbit to the saddle with respect to parameter β and detect two limit points. Verify that at the corresponding parameter values the stable and the unstable invariant manifolds of $(0, 0)$ are (almost) tangent.
   d. Continue the found LPs with respect to parameters (β, μ) and describe the domain of existence of Poincaré homoclinic structure. What happens as $(\beta, \mu) \to (0, 1)$?

C. **Euler scheme for Lorenz system**
Compute in MatcontM the one-dimensional unstable invariant manifold of the saddle fixed point $(0, 0, 0)$ in the explicit *Euler scheme* for the Lorenz system

$$\begin{pmatrix} x \\ y \\ z \end{pmatrix} \mapsto \begin{pmatrix} x + h\sigma(y - x) \\ y + h(rx - y - xz) \\ z + h(xy - bz) \end{pmatrix},$$

with $\sigma = 10, b = 8/3, r = 8.37$ and $h = 0.1$.

7.4 Tutorial 4: computation of Lyapunov exponents

In this tutorial we compute Lyapunov exponents for three different examples. We start with the planar Hénon Map. It is one of the most studied maps with a chaotic attractor. Through the geometric construction of Smale's horseshoe of stretching and folding back, this example exhibits chaotic dynamics. Next we consider the well-known logistic map as it has the classical route to chaos via successive PD bifurcations. Finally, we look at an economic model for asset prices (Gaunersdorfer, Hommes, and Wagener, 2008), as we can monitor some bifurcations of invariant curves. While Lyapunov exponents are useful indicators, one should be careful with their interpretation. While a positive largest exponent suggests chaos, it does not substitute a proof of chaotic dynamics. Similarly a zero Lyapunov exponent suggests an invariant curve but the numerical computation may not converge to zero quickly. So this has to be interpreted with care.[14]

7.4.1 Example 1: Hénon Map

For this example we use the standard *Hénon Map*, i.e., map (7.4) introduced in Section 7.3 but now with $R = S = 0$. We will compute the Lyapunov exponents for a single set of parameter values. We will explain how various options in the algorithm affect speed and accuracy.

Preparation and input

Load the map via **System|Systembrowser**. Set **Type|Initial Point|Point** and select the computation of the Lyapunov exponents via **Type|Curve|Lyapunov Exponents (QR-method)**. In the Starter window we set $x = 0.2, y = 0.3$ and $a = 1.4$, $b = -0.3$ and $R = S = 0$. Also in the Starter window, we choose the settings

```
Lyapunov steps                  100000
normsteps                           10
report every x normalizations     2000
transient iterations             10000
```

Now press **Compute|Forward**. The Output window appears, where every 20 000 (=10*2000) steps an intermediate result is shown.

Results

In the MATLAB workspace we now find a new variable `lyapunovExponents`, an array with two numbers. These are the computed Lyapunov exponents and we have $\lambda_1 = 0.4205$ and $\lambda_2 = -1.6245$. If we increase the number of steps, our estimate will eventually converge, independent of the initial point. The first iterates may still just be transients, before they reach the attractor. Therefore, it may be advantageous to start the computation only after some number of transient steps such that only the expansion/contraction near the attractor is sampled.

In Figure 7.59(b) we have plotted the ongoing estimates for the Lyapunov

[14] MATCONTM is available at http://sourceforge.net/projects/matcont/files/matcontm/. This tutorial has been tested on MATLAB R2017b with `matcontm5p4`.

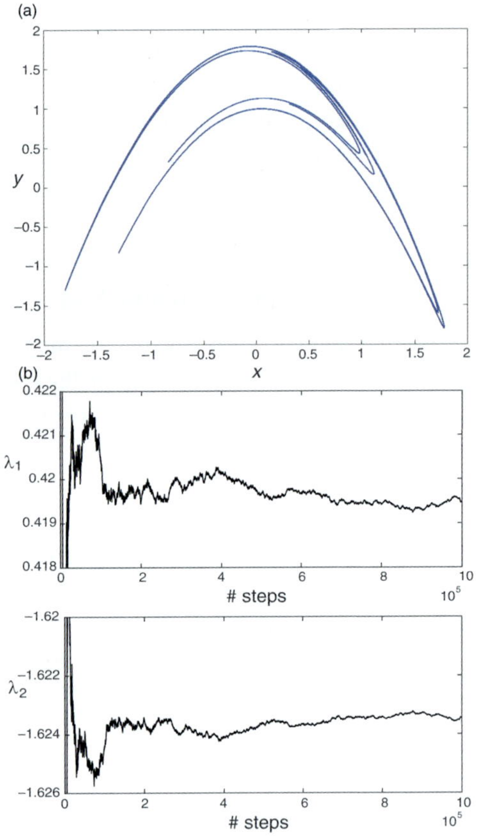

Figure 7.59 (a) Phase plot of the Hénon attractor. (b) The ongoing estimates of the two Lyapunov exponents shown separately as the orders of magnitude differ.

exponents.[15] One may observe the fluctuations which are still visible toward high step numbers. It is clear that a more accurate result requires many steps. As a sort of error we take the difference of the minimum and maximum of the ongoing estimates during the second half of the computation. Then we get $\lambda_1 = 0.420 \pm 0.002$ and $\lambda_2 = -1.625 \pm 0.002$. There is a clear trade-off between speed and accuracy. Setting the number of Lyapunov steps=1e+07 ($= 10^7$), we find $\lambda_1 = 0.4190 \pm 0.0002$ and $\lambda_2 = -1.6230 \pm 0.0002$.

The algorithm determines the expansion/contraction in every direction. If the Lyapunov vectors become completely aligned, then the algorithm for the most contractive direction becomes less reliable. Therefore the vectors need

[15] These results are not immediately available to the user.

to be orthonormalized. This is computationally expensive and so this is not done at every step. To avoid numerical overflow, however, the number of steps for orthonormalization cannot be set too high. In this example, `normsteps = 20` would be too high as $\exp(\lambda_1 - \lambda_2)^{20} \approx 6e{+}17$. After 20 steps, the second Lyapunov vector will be numerically the same as the first and any numerical result would be erratic.

7.4.2 Example 2: Logistic Map

For the Logistic Map we want to show how the exponents can be computed for a range of parameter values. That is, as the attractor changes we can monitor how the exponents change.

Preparation and input

The logistic map is given by

$$F_1 : x \mapsto ax(1 - x).$$

We create a new system with the model name `logistic` with coordinate x and parameter a (see Figure 7.60(a)). We set $x = 0.1$ as the initial condition as otherwise we stay in the origin forever. For the parameters we choose a concatenated array of values to monitor how the single Lyapunov exponent λ changes as the parameter a varies. We take non-equidistant steps to speed up the computation. We set $a = [1 : 0.05 : 2.99, 3 : 0.005 : 4]$ (see Figure 7.60(a)), and set a as an active parameter by clicking the button. Next press **Compute|Forward**. This takes a while, and then the results are written to the workspace into a structure `lyapunovExponents` with two fields, the parameter and the exponents. Finally, we plot our results as follows

```
>> figure;
>> a = lyapunovExponents.a;
>> plot(a,lyapunovExponents.exponents,a,0*a);
>> xlabel('a'); ylabel('\lambda')
```

Results

If all went well, you now have Figure 7.61(a). We have also plotted another familiar graph for this map, which is the coordinate x found by simply iterating the map (see Figure 7.61(b)). Until $a = 1$, the origin $x = 0$ is the only fixed point and all orbits converge to it. At $a = 1$ we encounter a branching point where a positive fixed point $x^* = (a - 1)/a$ appears. At $a = 2$, the multiplier of the fixed point x^* equals 0. This is visible as the Lyapunov exponent λ has

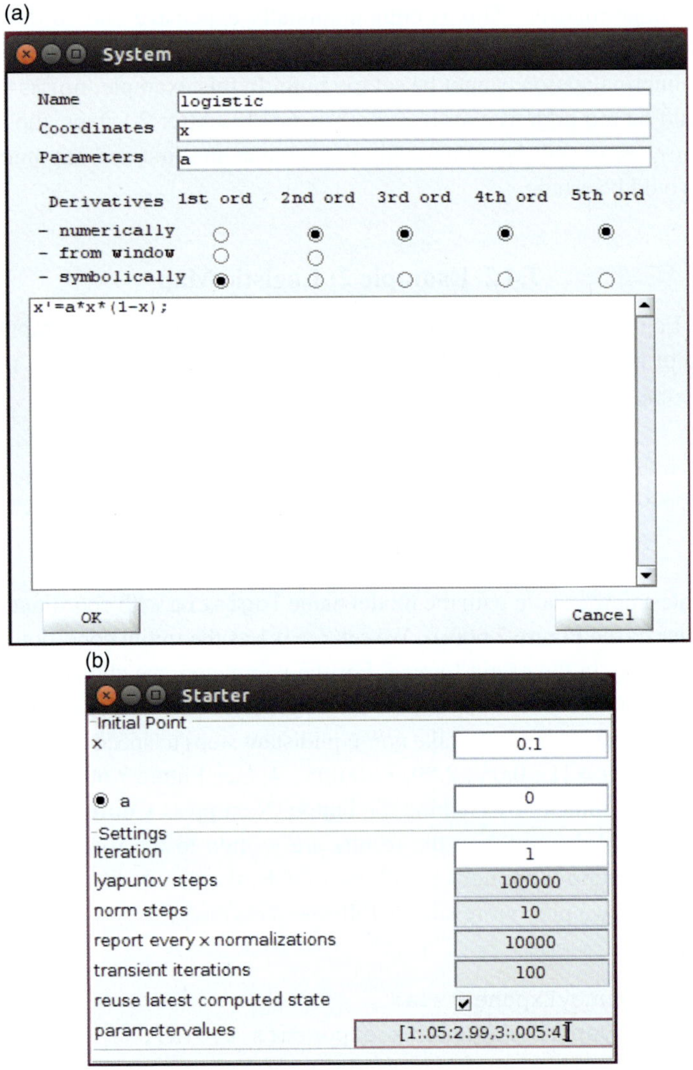

Figure 7.60 (a) The system input for the Logistic Map. (b) The settings for the computation with active parameter a (button ticked). The array for the parameter values is automatically expanded.

a vertical asymptote at $a = 2$. This phenomenon is referred to as superstability in work on "shrimps" (Vitolo, Glendinning, and Gallas, 2011a). At $a = 3$, the first period-doubling occurs which is visible as the Lyapunov exponent λ increases to zero, and next decreases again. The increase may be interpreted

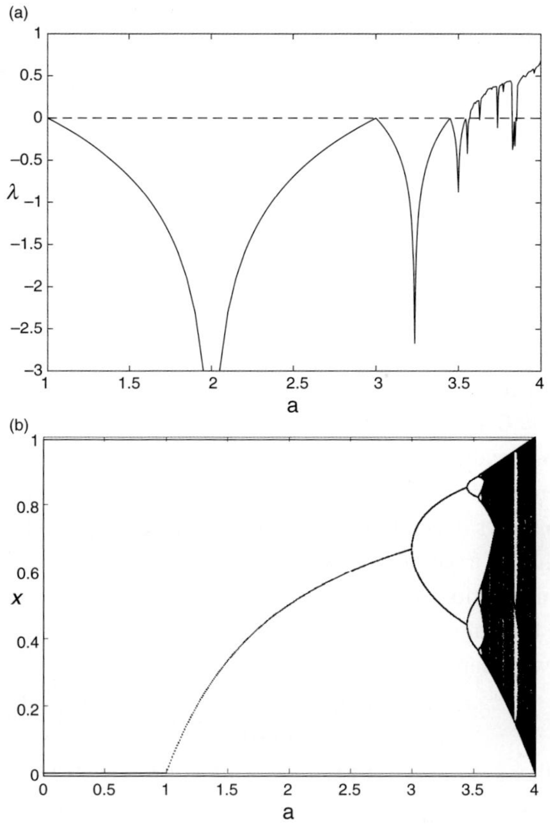

Figure 7.61 (a) The Lyapunov exponent λ as a function of the parameter a. (b) The set of x-values visited by the attractor.

that in the corresponding direction the attractor becomes less attracting until it loses stability. The next period-doubling occurs at $a = 1 + \sqrt{6} \approx 3.45$ where the exponent increases to zero and decreases again. This repeats until $a \approx 3.57$, where chaos sets in. Here we have positive Lyapunov exponents alternating with regions where we have stable cycles.

7.4.3 Example 3: resonances and quasi-periodic bifurcations

In this example we consider the dynamics of an economic model of volatility in asset prices (see Gaunersdorfer, Hommes, and Wagener (2008)). In memoryless form and with reduced parameter space the model may be written as

$$F: \begin{pmatrix} x_1 \\ x_2 \\ x_3 \\ x_4 \end{pmatrix} \mapsto \begin{pmatrix} 1 - n(x))vx_1 + n(x)(x_1 + g(x_1 - x_2)) \\ x_1 \\ x_2 \\ x_3 \end{pmatrix},$$

where

$$n(x) = \frac{e^{-x_1^2} e^{-bu_2}}{e^{-bu_1} + e^{-bu_2}},$$

with

$$u_1 = (x_1 - vx_3)^2 \text{ and } u_1 = (x_1 - x_3 - g(x_3 - x_4))^2.$$

The origin is a fixed point of the model that exhibits an NS bifurcation for $g = 2R$. For $v \approx 0.45$ the NS bifurcation is degenerate. Here, a quasi-periodic saddle-node bifurcation leads to a stable large "outer" invariant curve. We focus on the computation and explaining the computational results and do not interpret the dynamics in economic terms. The idea here is to follow the invariant curve as we change a parameter. We encounter resonances and a quasi-periodic saddle-node bifurcation, where the invariant curve ceases to exist. We want to show how the Lyapunov exponents may be classified to demarcate bifurcations that are otherwise hard to find.

Input

We specify the system `volatility` with coordinates `x1,x2,x3,x4` and parameters `g,v,R,b` as in Figure 7.62(a). Next we select **Type|Initial Point|Point** and **Type|Curve|Compute Lyapunov Exponents (QR-method)**. In the Starter window we set the initial condition `x1=3,x2=3,x3=1,x4=1` and parameters

```
g=2.0, v=0.8, R=1.01, b=8.
```

With these settings we start in the bistable region on the large invariant curve. Next we set g as the active parameter, `normsteps=5`, `lyapunov steps=1e+06` and `g=2 : −0.0005 : 1.7` (see Figure 7.62(b)). Next select **Compute|Forward**. This may take a while.

Results

Using the same plot commands as for the logistic map, but now with g instead of a, we obtain Figure 7.63(b). As $\lambda_{3,4}$ have a different order of magnitude we also show a plot with the range restricted to $\lambda_{1,2}$. As we decrease from $g = 2.0$ to $g \approx g_c := 1.72633$, we see that $\lambda_1 \approx 0$ most often. This indicates a closed invariant curve. We could use a threshold $|\lambda_1| < 10^{-4}$ for classification in this

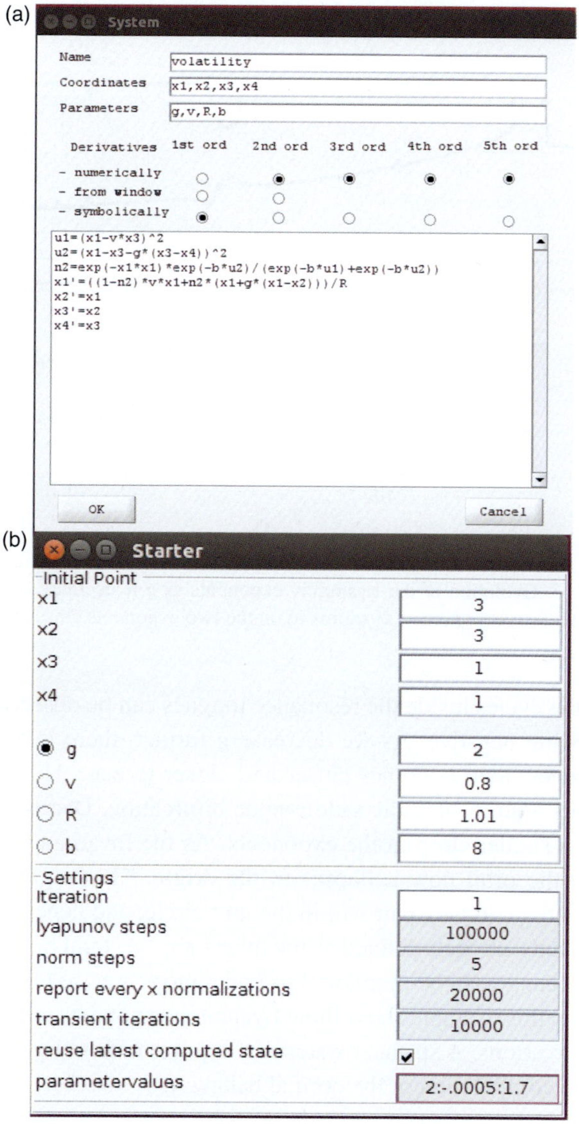

Figure 7.62 (a) The system input for the map. (b) Settings for the computation of Lyapunov exponents.

case, while all other exponents are more negative here. When a resonance is encountered, the dynamics on the invariant curve reduces to a cycle with a high period. In the figure we see resonances of period 17, 18, 19, and 20. The

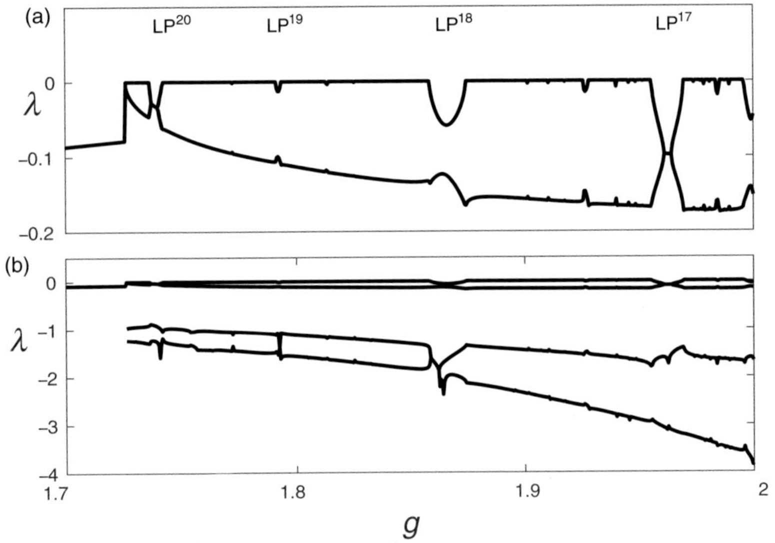

Figure 7.63 Evolution of the Lyapunov exponents as g is decreased from 2.0 to 1.7. (b) All four exponents, (a) zooms in on the two exponents close to zero.

stability of the cycles inside the resonance tongues can be observed by all four exponents being negative. As we decrease g further, there is a critical value g_c where we see that λ_2 comes closer and closer to zero. Here the invariant curve exhibits a quasi-periodic saddle-node bifurcation. Decreasing g further, we observe a sudden drop in the exponents. As the invariant curve is lost as an attractor, the orbit now collapses to the origin. The origin has two zero multipliers and a complex pair within the unit circle, and hence only two Lyapunov exponents are well-defined as the others are $-\infty$ (and hence not drawn). The sudden change can be monitored to find a more precise bifurcation value. Summarizing, this example shows how Lyapunov exponents may hint at quasi-periodic bifurcations. A stronger statement requires a more rigorous treatment involving the computation of the normal behavior.

7.4.4 Additional problems

A. **Pecora Economic Model**

Consider the following economic model (Pecora, 2018)

$$\begin{pmatrix} x \\ q \end{pmatrix} \mapsto \begin{pmatrix} q \\ q + \gamma q(a - c - b(1 + \ln(wq + (1 - w)x))) \end{pmatrix},$$

with $a = 4, b = 3, c = 2$ and w, γ are free parameters. There is a fixed point $x^* = q^* = e^{\frac{a-c-b}{b}}$. Fixing $0 < w < 1$ and increasing γ, one encounters either an NS or a PD bifurcation. For $w = 0.4, 0.6, 0.8$, compute Lyapunov exponents to study how the fixed point loses stability and how complex dynamics appears.

B. **Lotka–Volterra Map**

Consider the following three-dimensional Lotka–Volterra Map (Bischi and Tramontana, 2010)

$$\begin{pmatrix} x \\ y \\ z \end{pmatrix} \mapsto \begin{pmatrix} x(e - x + ay + bz) \\ y(e + bx - y + az) \\ z(e + ax + by - z) \end{pmatrix},$$

with $e = 2$ and $b = -0.5$. Compute Lyapunov exponents for $a = 0.55$ to $a = 0.68$ and use those to identify various bifurcations. Specifically, consider the first exponent around $a \approx 0.57$ and the second Lyapunov exponent near $a \approx 0.658$ and $a \approx 0.667$. Add simulations to illustrate the differences in dynamics.

Part Three

Applications

Applications make systematic theory more attractive. In this Part we will deal with a variety of examples.

We start with a chapter on the Generalized Hénon Map (GHM). This is a real planar map appearing in studies of diffeomorphisms with codim 2 generalized homoclinic and heteroclinic tangencies. One case deals with a saddle fixed point with a non-transversal homoclinic orbit and has saddle value near 1; this case is excluded by earlier works (Gavrilov and Shilnikov (1972); Palis and Takens (1993)). The transition through 1 of this saddle value implies several bifurcation scenarios. Thus this map serves as a good (test) example to apply the results of Chapters 3 and 4.

Then we go to a higher dimensional example from adaptive control where the center manifold has dimension strictly less than the number of state variables at the codim 2 bifurcation points. Therefore the results of Chapter 4 are necessary to analyze the scenarios.

Next we turn to an economic model of a duopoly. We present a thorough analysis of all codim 1 and 2 bifurcations as well as a numerical study to explore chaos. Finally we study bifurcations of periodic orbits in an epidemic model with periodic forcing. The model has several codim 2 points and we check their non-degeneracy numerically. This example shows how to implement such a stroboscopic map into MatcontM.

Some of the calculations can still be done by hand and/or symbolically, with maple (2014) or mathematica® (Wolfram Research, Inc., 2017). For others we rely on numerical methods as presented in Chapters 5 and 6.

8
The Generalized Hénon Map

8.1 Introduction

The real planar map

$$\begin{pmatrix} X \\ Y \end{pmatrix} \mapsto \begin{pmatrix} 1 + Y - aX^2 \\ bX \end{pmatrix} \tag{8.1}$$

was first introduced by Hénon (1976) as a planar diffeomorphism that mimics essential stretching and folding properties of the Poincaré map of the Lorenz system and has a strange attractor. It was also mentioned by Hénon (1976) that any quadratic planar map with constant Jacobian can be put into the canonical form (8.1) by a linear coordinate transformation. Making another linear transformation, we can write (8.1) as

$$\begin{pmatrix} x \\ y \end{pmatrix} \mapsto \begin{pmatrix} y \\ \alpha - \beta x - y^2 \end{pmatrix}, \tag{8.2}$$

which is called the *standard Hénon Map* in what follows. Since the late 1970s, the standard Hénon Map served as an important but artificial example to illustrate many analytical results and numerical techniques of dynamical systems theory.

It is remarkable that the map (8.2) was earlier derived by Gavrilov and Shilnikov (1972,1973) as the principal part of the Poincaré maps near the non-transverse homoclinic orbit to a saddle fixed point, where its stable and unstable invariant manifolds have quadratic tangency (see Section 2.2.1 and Palis and Takens (1993)). This phenomenon has codimension 1 and is called the *homoclinic tangency*. Denote by λ and γ the eigenvalues of the saddle fixed point, so that $0 < |\lambda| < 1 < |\gamma|$. If the *saddle quantity* $\sigma = |\lambda\gamma| < 1$, a one-parameter unfolding of such a singularity leads to the standard Hénon Map (8.2) with $\beta_k = O(\sigma^{-k})$, where k enumerates shrinking definition strips of the Poincaré maps near the critical homoclinic orbit. Thus, the appearing Hénon maps in

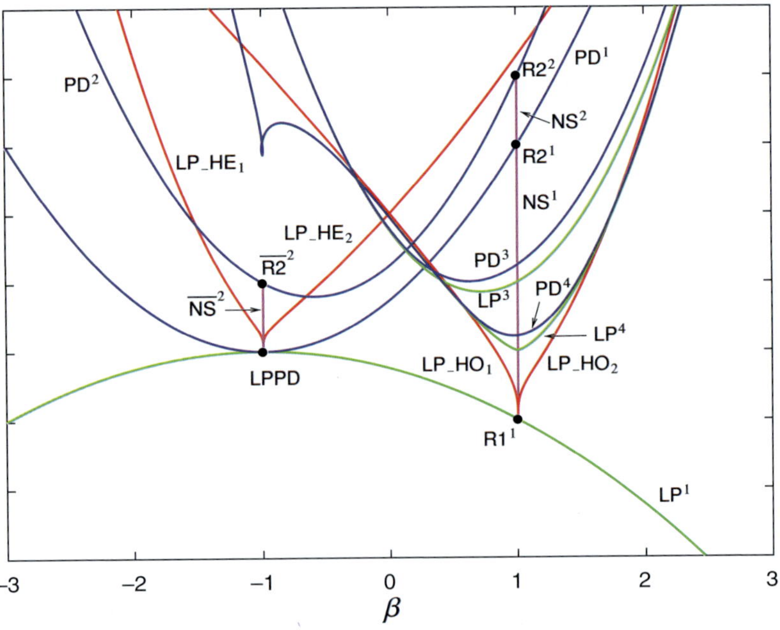

Figure 8.1 Principal bifurcation curves of the standard Hénon Map (8.2).

this case have small Jacobians: $\beta_k \to 0$ as $k \to \infty$. If $\sigma > 1$, a similar result holds for the inverse Poincaré maps.

The standard Hénon map and related problems giving rise to Smale's horseshoe have been the object of many studies. An overview of references can be found, for example, in Mira (1997).

Many details on the bifurcation diagram of (8.2) in the (β, α)-plane have been obtained by Mira (1987). In Figure 8.1, principal bifurcation curves of (8.2) recomputed with MatcontM are shown, which constitute a small part of the bifurcation set described in Mira (1987). A "backbone" of the bifurcation diagram is formed by three curves: LP^1 corresponding to the existence of a fixed point with eigenvalue 1; PD^1 corresponding to the existence of a fixed point with eigenvalue -1; and NS^1, where (8.2) has a fixed point with eigenvalues $e^{\pm i\theta}, 0 < \theta < \pi$. Crossing the curves LP^1 or PD^1 for $\beta \neq \pm 1$ results in non-degenerate fold or flip (period-doubling) bifurcations, respectively. In contrast, a bifurcation at NS^1 is degenerate, since for $|\beta| = 1$, the map (8.2) is area-preserving (conservative). At intersections of these curves, fixed points with double linear degeneracy exist: with eigenvalues ± 1 (point LPPD), with a double eigenvalue 1 (point $\mathsf{R1}^1$); and with a double eigenvalue -1 (point $\mathsf{R2}^1$). These points are origins of other curves in the parameter plane corresponding to local and global bifurcations in (8.2). From LPPD and $\mathsf{R2}^1$ the bifurcation lines $\overline{\mathsf{NS}}^2$ and NS^2 start, respectively, where the standard Hénon Map has a period-2 cycle with eigenvalues $e^{\pm i\theta}, 0 < \theta < \pi$. From LPPD, $\mathsf{R1}^1$ and $\mathsf{R2}^1$, other curves emanate, where (8.2) exhibits global bifurcations. Some of these curves are also shown in Figure 8.1. These are curves $\mathsf{LP\_HO}_{1,2}$, where the stable and unstable invariant manifolds of a saddle fixed point are tangent (homoclinic tangency), and $\mathsf{LP\_HE}_{1,2}$, where a tangency of a stable manifold of one saddle with an unstable manifold of another saddle occurs. Notice that curves $\mathsf{LP\_HO}_1$ and $\mathsf{LP\_HE}_2$ intersect at point $(\beta, \alpha) = (0, 2)$ on the α-axis, along which the standard Hénon Map is non-invertible and reduces to the scalar uni-modular map $y \mapsto \alpha - y^2$. For $\alpha = 2$ this scalar map indeed has an orbit connecting two fixed points and another orbit that is homoclinic to one of these fixed points. According to Mira (1987), the standard Hénon Map (8.2) has an infinite number of bifurcation curves corresponding to fold (LP^k) and flip (PD^k) bifurcations of k-cycles with period $k > 1$. Some of these are depicted in Figure 8.1. Several higher degeneracies on these curves were also reported and analyzed by Mira (1987).

Due to its conservative nature for $|\beta| = 1$, the bifurcation diagram of the standard Hénon Map (8.2) exhibits a number of highly degenerate features. This prevents applying Mira's results in homoclinic studies, where the standard Hénon Map only *approximates* the actual Poincaré map.

In this section, we study an extension of (8.2), namely the map

$$\begin{pmatrix} x \\ y \end{pmatrix} \mapsto \begin{pmatrix} y \\ \alpha - \beta x - y^2 + Rxy + S y^3 \end{pmatrix}, \tag{8.3}$$

where R and S are constants. This map is called the *Generalized Hénon Map* (GHM). For $R = S = 0$ it reduces to (8.2). Of course, one can add many different terms to the standard Hénon Map to remove its degeneracy. Our motivation to study this particular extension of (8.2) is that it appears in the bifurcation analysis of non-transversal homoclinic orbits and heteroclinic cycles of both codim 1 and 2.

We focus on the following two codim 2 bifurcations of maps with homoclinic tangencies, which will be described in more detail in Section 8.2:

(1) the critical diffeomorphism in $\mathbb{R}^2$ has a neutral saddle ($\sigma = 1$) with a quadratic homoclinic tangency (see Gonchenko and Gonchenko (2000, 2004) and Gonchenko (2002) the analogous case in $\mathbb{R}^3$ is considered by Gonchenko and Ovsyannikov (2005));

(2) the critical diffeomorphism in $\mathbb{R}^3$ has a saddle with a generalized homoclinic tangency (i.e., the unstable manifold of the saddle has a quadratic tangency to its stable manifold but is non-transversal to leaves of the strong stable foliation in the stable manifold at the homoclinic points) (see (Gonchenko, Gonchenko, and Tatjer, 2002c)).

There are other global bifurcations of generic maps, where GHM appears naturally. If a diffeomorphism in $\mathbb{R}^2$ has two saddle fixed points O_1 and O_2 connected by two heteroclinic orbits, one of which is non-transversal (codim 1), then GHM appears as a rescaled first-return map when $(\sigma_1 - 1)(\sigma_2 - 1) < 0$, where σ_i is the saddle quantity of O_i (Gonchenko, Shilnikov, and Stenkin, 2002b). If a diffeomorphism in $\mathbb{R}^3$ has a codim 1 homoclinic tangency to a saddle-focus fixed point with eigenvalues $\nu_{1,2} = \lambda e^{\pm i\varphi}, \nu_3 = \gamma$, where $0 < \lambda < 1 < \gamma$, $0 < \varphi < \pi$, and $\lambda^2\gamma < 1$, then GHM appears when $\lambda\gamma > 1$ (Gonchenko, Shilnikov, and Turaev, 2002a).

GHM seems to play an important role in other homoclinic studies. From Champneys, Härterich, and Sandstede (1996) it follows that the Poincaré map near a non-transverse homoclinic orbit to a saddle–saddle equilibrium in a three-dimensional ODE is a smooth extension of the standard Hénon Map (8.2) that can be reduced to GHM with $S = 0$ and small β and R. It is also known that GHM with $|\beta| = 1$ and $R = 0$ appears in the analysis of two-dimensional area-preserving diffeomorphisms with homoclinic tangencies (Gonchenko and Shilnikov, 2003), and three-dimensional divergence-free ODEs with a homoclinic orbit to a saddle-focus equilibrium (Biragov, 1987). It seems that the appearance of GHM as a rescaled first-return map can be expected in other cases of homoclinic and heteroclinic tangencies, when the so-called "effective dimension of the problem" (Turaev, 1996) can change.

This chapter is organized as follows. In Section 8.2 we describe in detail two codim 2 homoclinic bifurcations, where the GHM appears as a rescaled Poincaré map near the homoclinic orbit. Section 8.3 is devoted to analytical and numerical study of the bifurcation diagram of (8.3) in case $S = 0$, and then in the general case $S \neq 0$. In Section 8.4 we discuss the correspondence between the bifurcations of the GHM and those of the original diffeomorphisms with the codim 2 homoclinic tangencies. Some open problems are discussed in Section 8.5.

8.2 Homoclinic bifurcations and GHM

As mentioned in the previous section, here we formulate two problems with homoclinic tangencies (Gonchenko and Gonchenko, 2000, 2004) leading to the GHM. Namely, we consider (1) a two-dimensional diffeomorphism with a homoclinic tangency of a neutral saddle; and (2) a three-dimensional diffeomorphism with a generalized homoclinic tangency. In both cases the analysis leads to a two-parameter family $f_\mu, \mu = (\mu_1, \mu_2) \in \mathbb{R}^2$ of diffeomorphisms close to the diffeomorphism f_0. The first parameter μ_1 in both cases is the splitting parameter, i.e., it measures the displacement of the stable and unstable manifolds. The second parameter μ_2 is the deviation of the saddle quantity from 1 in the first case, and a characterization of the geometry in the second case.

We are interested in orbits located entirely in a small neighborhood U of the critical homoclinic orbit. This neighborhood consists of a small neighborhood U_0 of the saddle fixed point and a finite number of small neighborhoods of points of the homoclinic orbit outside of U_0. We call a periodic orbit (cycle) *p-round* if it lies entirely in U and has exactly p points within each of the small neighborhoods located outside U_0. Bifurcations of single-round orbits (1-round orbits) are studied in the following with the help of first-return (or Poincaré) maps.

8.2.1 Homoclinic tangency of a neutral saddle in $\mathbb{R}^2$

Here we consider bifurcations in a generic two-parameter family f_μ of planar diffeomorphisms satisfying at $\mu = 0$ the following conditions:

A. f_0 has a saddle fixed point O with eigenvalues γ, λ, such that $0 < |\lambda| < 1 < |\gamma|$;
B. the saddle quantity $\sigma \equiv |\lambda\gamma| = 1$;
C. the invariant manifolds $W^u(O)$ and $W^s(O)$ have a quadratic tangency at points of a homoclinic orbit Γ (see Figure 8.2).

8.2.2 Domains of definition

As usual in homoclinic studies, the Poincaré map is constructed as the composition of two maps: "local" – defined in U_0; and "global" – defined along the part of the homoclinic orbit in $U \setminus U_0$. In other words, single-round periodic orbits of period $k + n_0$ for all $k > \bar{k}$ (the meaning of $\bar{k}$ will be explained later) are fixed points of $f_\mu^{k+n_0} \equiv f_\mu^{n_0} \circ f_\mu^k$. In this construction, we will use a general representation of the global map $T_1 = f_\mu^{n_0}$ (based only on the geometry of tangency) and a special form of the local map $T_0^k = f_\mu^k$, in which nonlinear terms are asymptotically small for $k \to \infty$.

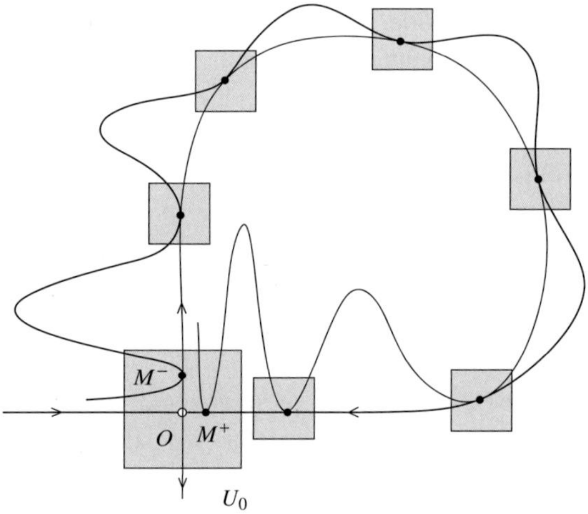

Figure 8.2 Tangency of the stable and the unstable invariant manifolds.

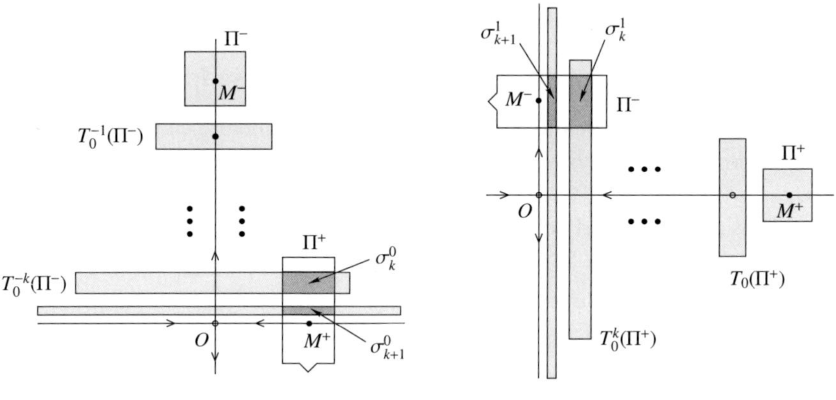

Figure 8.3 Domains of definition: $\sigma_k^1 = f_0^k(\sigma_k^0)$.

Let us consider two points of the critical homoclinic orbit Γ in U_0: M^+ on a local stable manifold and M^- on a local unstable manifold.[1] It is obvious that there is an integer n_0 such that $M^+ = f_0^{n_0}(M^-)$.

Next we can choose two small neighborhoods: Π^+ (of point M^+) and Π^- (of point M^-). Consider the forward images of Π^+ under f_0. As is shown in Figure 8.3, there is some $\bar{k}$, such that for all $k \geq \bar{k}$ there is a "good" intersection

[1] In this neighborhood there are infinitely many such points – we choose any two of them.

of $f_0^k(\Pi^+)$ and Π^-. "Good" means that $f_0^k(\Pi^+)\backslash\Pi^-$ consists of two components. Denote $f_0^k(\Pi^+)\cap\Pi^-$ by σ_k^1. We can also iterate Π^- under f_0^{-1} to obtain domains $\sigma_k^0 = f_0^{-k}(\Pi^-) \cap \Pi^+$, which are the pre-images of σ_k^1, i.e., $\sigma_k^1 = f_0^k(\sigma_k^0)$. It is easy to see that the Poincaré map for the single-round $(k + n_0)$-periodic orbit is defined only in σ_k^0. Indeed, only σ_k^0 may consist of orbits which end up in Π^- after k iterations.[2]

Now we can define $T_1 \circ T_0^k = f_\mu^{k+n_0}$ as the first-return map in such a strip for all μ with sufficiently small $\|\mu\|$. For details we refer to Gonchenko and Gonchenko (2000); Gonchenko, Gonchenko, and Tatjer (2002c). Here we give only a representation of the global map T_1, because it is used below

$$\begin{aligned}\bar{x}_0 - x^+ &\equiv ax_1 + b(y_1 - y^-) + e_{20}x_1^2 + e_{11}x_1(y_1 - y^-) + e_{02}(y_1 - y^-)^2 + \cdots,\\ \bar{y}_0 &\equiv \mu_1 + cx_1 + d(y_1 - y^-)^2 + f_{20}x_1^2 + f_{11}x_1(y_1 - y^-) + f_{30}x_1^3\\ &\quad + f_{21}x_1^2(y_1 - y^-) + f_{12}x_1(y_1 - y^-)^2 + f_{03}(y_1 - y^-)^3 + \cdots .\end{aligned} \tag{8.4}$$

Here all coefficients are smooth functions of μ; moreover, $bcd \neq 0$. Note that the definition of the global map depends on n_0.

8.2.3 Rescaling results

Here we present only the results of the rescaling performed by Gonchenko and Gonchenko (2000). The Poincaré map for any $k > \bar{k}$ can be reduced to the map

$$\begin{cases} \bar{X} = Y, \\ \bar{Y} = \alpha - \beta X - Y^2 + R\lambda^k XY + S\lambda^k Y^3 + o(\lambda^k), \end{cases} \tag{8.5}$$

where λ is the stable eigenvalue of the saddle, α, β, X, Y cover all finite values as $k \to \infty$, and R, S depend on the coefficients of the global map (8.4), namely

$$R = 2a - \frac{b}{d}f_{11} - 2\frac{c}{d}e_{02}, \quad S = -\frac{bc}{d^2}f_{03}. \tag{8.6}$$

There is the following correspondence between (α, β) and (μ_1, μ_2)

$$\begin{aligned}\alpha &= -d\gamma^{2k}[\mu_1 - \gamma^{-k}(y^- + \dots) + c\lambda^k(x^+ + \dots)],\\ \beta &= -bc(1 + \mu_2)^k(1 + \dots).\end{aligned} \tag{8.7}$$

[2] If we consider 2-round periodic orbits the situation would be different, because then we should think about $i + j + 2n_0$-periodic orbits of type (i, j). The 2-round periodic orbit of type (i, j) has i points near the saddle, then n_0 points near the "global" part of the homoclinic orbit, then j points near the saddle, and n_0 points near the homoclinic orbit. In this case, the domain of definition consists of two strips, σ_i^0 and σ_j^0.

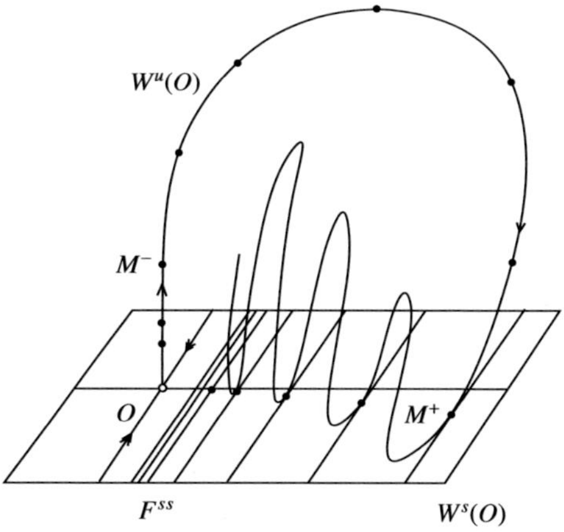

Figure 8.4 Generalized homoclinic tangency.

8.2.4 Generalized homoclinic tangency in $\mathbb{R}^3$

Here we consider a generic two-parameter family f_μ of three-dimensional diffeomorphisms satisfying at $\mu = 0$ the following conditions:

A. f_0 has a saddle fixed point O with eigenvalues $\lambda_1, \lambda_2, \gamma$, such that $0 < |\lambda_2| < |\lambda_1| < 1 < |\gamma|$;
B. $|\lambda_1 \gamma| > 1, \ |\lambda_2 \gamma| < 1$;
C. the invariant manifolds $W^u(O)$ and $W^s(O)$ have a quadratic tangency at points of a homoclinic orbit Γ (see Figure 8.4);
D. $T_1(P^{ue}(M^-))$ is transversal to W^s_{loc} at M^+ but is tangent to $F^{ss}(M^+)$ at M^+ where M^+, M^- are some points on local stable and local unstable manifolds respectively, P^{ue} is the tangent plane to the extended unstable manifold at the point M^-, T^1 is the global map from a neighborhood of M^- to M^+, $F^{ss}(M^+)$ is a leaf of the strong stable foliation containing M^+.[3]

8.2.5 Domains of definition

As in the planar case, we can choose two points on the homoclinic orbit in a sufficiently small neighborhood U_0 of the saddle fixed point: M^+ on the local stable manifold, and M^- on the local unstable manifold. We also choose sufficiently small neighborhoods Π^+ and Π^- of points M^+ and M^-. We write $\bar{k}$ for the minimal k such that $f_0^k(\Pi^+)$ and Π^- have a "good" intersection. And we

[3] In this section we consider only one of the two cases treated by Gonchenko, Gonchenko, and Tatjer (2002c) (case I). Case II is similar.

denote $\sigma_k^1 = f_0^k(\Pi^+) \cap \Pi^-$ and $\sigma_k^0 = f_0^{-k}(\Pi^-) \cap \Pi^+$. The picture is similar to Figure 8.3 but in three dimensions (with an additional contraction direction).

Also we need a representation for the global map T_1:

$$\begin{aligned} \bar{x}_1 - x_1^+ &= a_{11}x_1 + a_{12}x_2 + \mu_2(y - y^-) + e_{02}(y - y^-)^2 + \cdots , \\ \bar{x}_2 - x_2^+ &= a_{21}x_1 + a_{22}x_2 + b_2(y - y^-) + \cdots , \\ \bar{y} &= \mu_1 + c_1x_1 + c_2x_2 + d(y - y^-)^2 + \cdots . \end{aligned} \tag{8.8}$$

Here all coefficients are smooth functions of μ; moreover, $c_1 d \neq 0$. As in the planar case, the definition of the global map depends on the choice of points $M^\pm$.

8.2.6 Rescaling results

For a family f_μ of diffeomorphisms of $\mathbb{R}^3$ close to the diffeomorphism f_0 with the generalized homoclinic tangency, the following result is valid (Gonchenko, Gonchenko, and Tatjer, 2002c). The first-return map for $|\lambda_1\gamma| > 1$, $|\lambda_2\gamma| < 1$ and for any $k > \bar{k}$ can be reduced on some invariant two-dimensional manifold to the map

$$\begin{cases} \bar{X} = Y, \\ \bar{Y} = \alpha - \beta X - Y^2 + R\lambda_1^k XY + o(\lambda_1^k), \end{cases} \tag{8.9}$$

where

$$R = 2\left(a_{11} - \frac{e_{02}c_1}{d}\right) \tag{8.10}$$

and

$$\begin{aligned} \alpha &= -d\gamma^{2k}[\mu_1 - \gamma^{-k}(y^- + \ldots) + \lambda^k(c_1x^+ + \ldots)], \\ \beta &= -c_1(\mu_2 + \ldots)(\lambda_1\gamma)^k. \end{aligned} \tag{8.11}$$

Note that in this case only the quadratic extra term appears (the term with y^3 exists but has order of $O(\gamma^{-k}) = o(\lambda^k)$ since $|\lambda_1\gamma| > 1$).

When $|\lambda_1\gamma| > 1$ and $|\lambda_2\gamma| > 1$, another extension of the standard Hénon Map appears (Gonchenko, Gonchenko, and Tatjer, 2002c).

8.3 Bifurcation diagrams of GHM

In this section we study the GHM

$$\begin{pmatrix} x \\ y \end{pmatrix} \mapsto F_{\alpha,\beta} = \begin{pmatrix} y \\ \alpha - \beta x - y^2 + Rxy + Sy^3 \end{pmatrix}, \tag{8.12}$$

where R and S are (not necessarily small) constants. Then we consider the limit case when R and S are substituted by $R\lambda^k$ and $S\lambda^k$, respectively. Since $|\lambda| < 1$, these sequences tend to zero as $k \to \infty$. For the figures the representative values $R = \pm 0.5$ and $R = \pm 0.1$ have been used.

8.3.1 Codim 1 and 2 bifurcations of fixed points

This subsection is organized as follows. First we consider (8.12) with $S = 0$. We derive the bifurcation curves for local codim 1 bifurcations and verify their non-degeneracy conditions. At the codim 2 points we compute the critical normal form coefficients. Thus we prove the non-degeneracy of these bifurcations for $R \neq 0$. Having this knowledge, we consider $S \neq 0$ and obtain similar results.

8.3.2 The Quadratic Extension ($S = 0$)

Proposition 8.1 *The GHM* (8.12) *with* $S = 0$ *has the following codim 1 bifurcations of fixed points*:

> ***Fold:*** *If* $R \neq 1$, *there is a non-degenerate fold bifurcation for* $\alpha = \frac{(\beta+1)^2}{4(R-1)}$ *with the critical fixed point at* $x = y = \frac{\beta+1}{2(R-1)}$.
> ***Flip:*** *If* $R \neq -1, 2$, *there is a non-degenerate flip bifurcation for* $\alpha = \frac{1}{4}(\beta + 1)^2(3 - R)$ *with the critical fixed point at* $x = y = \frac{\beta+1}{2}$.
> ***Neimark–Sacker:*** *If* $R \neq 0, 1, 2$, *there is an* NS *bifurcation for*
>
> $$\alpha = \frac{(\beta - 1)(\beta - 1 + 2R)}{R^2}$$
>
> *with the critical fixed point at* $x = y = \frac{\beta-1}{R}$. *It is defined for* $\beta \in [\frac{2-3R}{2-R}, \frac{2+R}{2-R}]$ *when* $R \in (0, 2)$ *and* $\beta \in [\frac{2+R}{2-R}, \frac{2-3R}{2-R}]$ *when* $R \notin [0, 2]$. *The bifurcation is non-degenerate away from strong resonances.*

Proof We write DF for the Jacobian matrix of F with respect to (x, y). These curves follow easily from the fixed point equation, i.e., $F_{\alpha,\beta}(x) - x = 0$, together with a condition on the multipliers. These conditions are given by the following equations: $\det(DF - I) = 0$ (fold), $\det(DF + I) = 0$ (flip) and $\det(DF) - 1 = 0$ (NS). We have to exclude neutral saddles for the NS bifurcation and therefore restrict to the interval as given above.

Then we have to check that these codim 1 bifurcations are non-degenerate. We use the formulas for the normal form coefficients from Section 2.1, formulas (2.7), (2.12) and (2.18). The bilinear form corresponding to the GHM with

$S = 0$ is

$$B(x,y) = \begin{pmatrix} 0 \\ R(x_1y_2 + y_1x_2) - 2x_2y_2 \end{pmatrix}, \tag{8.13}$$

while the vectors

$$q_{fold} = \begin{pmatrix} 1 \\ 1 \end{pmatrix}, \quad p_{fold} = \begin{pmatrix} -\beta + \frac{(\beta+1)R}{2(R-1)} \\ 1 \end{pmatrix}, \tag{8.14}$$

$$q_{flip} = \begin{pmatrix} -1 \\ 1 \end{pmatrix}, \quad p_{flip} = \begin{pmatrix} \frac{1}{2}(\beta(2-R) - R) \\ 1 \end{pmatrix}, \tag{8.15}$$

$$q_{ns} = \begin{pmatrix} \gamma \\ 1 \end{pmatrix}, \quad p_{ns} = \begin{pmatrix} -\gamma \\ 1 \end{pmatrix}, \tag{8.16}$$

where $\gamma = \frac{(\beta-1)(R-2)+\sqrt{(\beta-1)^2(R-2)^2-4R^2}}{2R}$ and satisfy the conditions:

$$\begin{array}{lll} (DF)q_{fold} = q_{fold}, & (DF)^T p_{fold} = p_{fold}, & \langle p_{fold}, q_{fold}\rangle \neq 0, \\ (DF)q_{flip} = -q_{flip}, & (DF)^T p_{flip} = -p_{flip}, & \langle p_{flip}, q_{flip}\rangle \neq 0, \\ (DF)q_{ns} = e^{i\theta} q_{ns}, & (DF)^T p_{ns} = e^{-i\theta} p_{ns}, & \langle p_{ns}, q_{ns}\rangle \neq 0. \end{array}$$

We do not give intermediate calculations, but directly present the normal form coefficients

$$b_0 = \frac{4(1-R)^2}{\beta(2-R) - (2-3R)}, \tag{8.17}$$

$$c_0 = \frac{4(1+R)}{(\beta(2-R) - (2+R))(\beta+1)(R-2)}, \tag{8.18}$$

$$L_1 = \frac{(1-R)R^2}{2(\beta(2-R) - (2-3R))}. \tag{8.19}$$

□

From the expressions above we see that for $R = -1, 0, 1, 2$ the map is more degenerate.

Proposition 8.2 *The GHM* (8.12) *with* $S = 0, R \neq 0$, *has the following codim 2 bifurcations of fixed points*:

Fold–flip *There is a fold–flip bifurcation of the fixed point* $x = y = 0$ *for* $(\alpha, \beta) = (0, -1)$. *The normal form coefficients are* $a(0) = \frac{1}{2}(1-R), b(0) = \frac{1}{4}(1+R), c(0) = \frac{1}{4}(1-R), d(0) = -\frac{1}{8}(5+3R)$.

Resonance 1:1 *There is a resonance 1:1 at* $(\alpha, \beta) = \left(\frac{4(-1+R)}{(2-R)^2}, \frac{2-3R}{2-R}\right)$ *for the fixed point* $x = y = \frac{\beta-1}{R}$. *The normal form coefficient is* $s = \text{sign}((1-R)R)$.

Resonance 1:2 *There is a resonance 1:2 at* $(\alpha,\beta) = \left(\frac{4(3-R)}{(2-R)^2}, \frac{2+R}{2-R}\right)$ *for the fixed point* $x = y = \frac{\beta-1}{R}$. *The normal form coefficients are* $c_0 = -(1+R)$ *and* $d_0 = (1-R)R/4$.
Resonance 1:3 *There is a resonance 1:3 at* $(\alpha,\beta) = \left(\frac{5-2R}{(2-R)^2}, \frac{2}{2-R}\right)$ *for the fixed point* $x = y = \frac{\beta-1}{R}$. *The normal form coefficient is* $a_2(0) = \frac{(1-R)\left(R-i\sqrt{3}(2+R)\right)}{4(1+R+R^2)}$.
Resonance 1:4 *There is a resonance 1:4 for the fixed point* $x = y = 0$ *at* $(\alpha,\beta) = (0,1)$. *The normal form coefficient is* $A_0 = \frac{R(1-R)+i(-2+3R+R^2)}{\sqrt{(1+R^2)(4+4R+2R^2)}}$.

Proof We calculate common points of the three bifurcation curves and find the first three codim 2 bifurcations. It is easy to see that on the NS bifurcation curve we encounter the strong resonances 1:3 and 1:4, while we move from the resonance 1:1 point to the 1:2 point. Solving for the eigenvalues, the parameter values for these bifurcations can be computed explicitly. Next we determine the normal form coefficients to verify non-degeneracy. We use the approach from Chapter 4. Throughout the proof we have $B(u,v) = (0, -2u_2v_2 + R(u_1v_2 + u_2v_1))$ and $C(u,v,w) = (0,0)$.

Fold–flip. Using the formulas from Section 4.2 with the critical eigenvectors

$$q_{fold} = 2p_{fold} = -\begin{pmatrix} 1 \\ 1 \end{pmatrix}, \quad q_{flip} = 2p_{flip} = \begin{pmatrix} 1 \\ -1 \end{pmatrix}, \tag{8.20}$$

we find for the truncated critical normal form

$$\begin{pmatrix} x_1 \\ x_2 \end{pmatrix} \mapsto \begin{pmatrix} x_1 + a(0)x_1^2 + b(0)x_2^2 + c(0)x_1^3 + d(0)x_1x_2^2 \\ -x_2 + x_1x_2 \end{pmatrix} \tag{8.21}$$

the following coefficients

$$a(0) = \frac{1}{2}(1-R), \quad b(0) = \frac{1}{2}(1+R), \quad c(0) = \frac{1}{4}(1-R), \quad d(0) = -\frac{1}{4}(5+3R).$$

We see that, indeed, depending on R, the following cases occur.

	sign $a(0)$	sign $b(0)$	
$R < -1$	$+$	$-$	case 3
$-1 < R < 1$	$+$	$+$	case 1
$R > 1$	$-$	$+$	case 2

Next we calculate the critical coefficient determining the non-degeneracy of the LPPD point (see Section 3.4.1) to be $c_{NS} = \frac{1}{4}R^2(1-R)$, meaning that closed invariant curves coming from the NS bifurcation of the period-2 cycle will be unstable for case 1 and stable for case 2. Note that we excluded $R = 0$ and $R = 1$. For $(\alpha,\beta,R) = (0,-1,0)$ we actually have a codim 3 point, since the coefficient c_{NS} is then equal to zero.

Resonance 1:1. We shift the critical fixed point to the origin:

$$\begin{pmatrix} x \\ y \end{pmatrix} \mapsto \begin{pmatrix} y \\ -x + 2y + Rxy - y^2 \end{pmatrix}. \tag{8.22}$$

(Generalized) eigenvectors satisfying (4.32) are

$$q_0 = \frac{1}{2}\sqrt{2}(1,1), \; q_1 = \frac{1}{4}\sqrt{2}(-1,1)$$

and

$$p_0 = \sqrt{2}(-1,1), \; p_1 = \frac{1}{2}\sqrt{2}(1,1).$$

Evaluation of (4.36) yields $a_0 = \frac{1}{2}\sqrt{2}(R-1)$ and $b_0 = \sqrt{2}(\frac{1}{2}R - 1)$ so that $s = 2a(b-2a) = R(1-R)$, which defines the stability of the bifurcating closed invariant curve.

Resonance 1:2. The shifted map can be written as

$$\begin{pmatrix} x \\ y \end{pmatrix} \mapsto \begin{pmatrix} y \\ -x - 2y + Rxy - y^2 \end{pmatrix}. \tag{8.23}$$

(Generalized) eigenvectors satisfying (4.32) are given by

$$q_0 = \frac{1}{2}\sqrt{2}(1,-1), \; q_1 = \frac{1}{4}\sqrt{2}(1,1)$$

and

$$p_0 = \sqrt{2}(1,1), \; p_1 = \frac{1}{2}\sqrt{2}(1,-1).$$

Evaluating (4.44) and (4.45) with these vectors we find $c_0 = -\frac{1}{4}(1+R)$ and $d_0 = \frac{1}{8}(2+R)(3+R)$. For non-degeneracy we need $3c_0 + d_0 = \frac{1}{8}R(R-1) \neq 0$.

Resonance 1:3. Shifting the fixed point to the origin the critical map can be written as

$$\begin{pmatrix} x \\ y \end{pmatrix} \mapsto \begin{pmatrix} y \\ x - y - y^2 + Rxy \end{pmatrix}. \tag{8.24}$$

To evaluate (4.49) and (4.50), we introduce

$$q = \frac{1}{4}(2\sqrt{2}, \sqrt{2} - \sqrt{6}i) \text{ and } p = (\sqrt{2} - \frac{i}{\sqrt{6}}, \sqrt{\frac{2}{3}}i).$$

We then find $b_0 = \frac{\sqrt{2}}{24}(\sqrt{3}i - 3)(1 + i\sqrt{3} + 2R)$ and $c_0 = \frac{R+2}{12}(i\sqrt{3} - 2R + 1)$. So that $a_2(0) = \Re\left(\frac{\bar{\mu}c_0}{|b_0^2|} - 1\right) = \frac{R(1-R)}{2(1+R+R^2)}$.

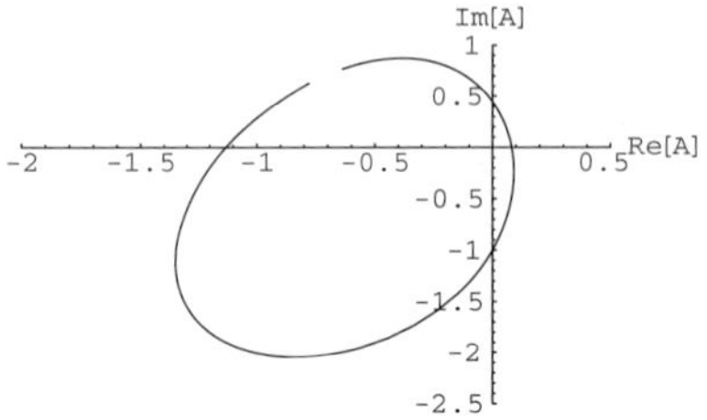

Figure 8.5 The complex values of A_0 as a function of R. In the limit $R \to \pm\infty$ the graph "closes."

Resonance 1:4. We apply the same procedure as before with $\mu = e^{i\pi/2}$. The critical map (no shift) is given by

$$\begin{pmatrix} x \\ y \end{pmatrix} \mapsto \begin{pmatrix} y \\ -x + Rxy - y^2 \end{pmatrix}. \tag{8.25}$$

To determine the normal form coefficients we can simply use the vectors $q = p = \frac{1}{2}\sqrt{2}(1, i)$. Evaluating (4.55) and (4.56), we obtain

$$c_0 = \frac{1}{8}\left(2 - 3R - R^2 + i(R - R^2)\right), \quad d_0 = \frac{1}{8}\left(-2 - R + R^2 + i(3R + R^2)\right).$$

We use these to compute the normal form coefficient for the R4 bifurcation theorem (3.29)

$$A_0 = -\frac{ic_0}{|d_0|} = \frac{R(1 - R) + i(R^2 + 3R - 2)}{\sqrt{(1 + R^2)(2R^2 + 4R + 4)}}.$$

We plot A_0 as a parametric function of R in Figure 8.5. □

8.3.3 The cubic extension ($S \neq 0$)

Proposition 8.3 *The GHM* (8.12) *with* $R, S \neq 0$ *has the following codim 1 bifurcations of fixed points*:

Fold. *There is a fold bifurcation at*

$$\alpha = \frac{2(1 - R)^3 + 9(\beta + 1)(1 - R)S \pm 2\left((R - 1)^2 + 3(\beta + 1)S\right)^{3/2}}{27S^2}$$

of the fixed point

$$x = y = \frac{(1 - R) \pm \sqrt{(R - 1)^2 + 3(\beta + 1)S}}{3S}.$$

The fold normal form coefficient is given by

$$c_{fold} = \frac{-6S\sqrt{(1-R)^2+3(\beta+1)S}}{(3(\beta-1)S - R(1-R)) \pm R\sqrt{(1-R)^2+3(\beta+1)S}}. \tag{8.26}$$

Flip. *There is a flip bifurcation of the fixed point*

$$x = y = \frac{1 \pm \sqrt{1-3(\beta+1)S}}{3S}$$

at $\alpha =$

$$\frac{2(1-3R)+9(\beta+1)(R+1)S \pm 2(1-3R+6(\beta+1)S)\sqrt{1-3(\beta+1)S}}{27S^2}.$$

The normal form coefficient is given by $c_{flip} =$

$$\frac{3S^2\left(-3+R-3R^2+6(\beta-1)S \pm 7R\sqrt{1-3(\beta+1)S}\right)}{\left(3(\beta-1)S + R(-1 \mp \sqrt{1-3(\beta+1)S})\right)\left(3(\beta+1)S + R(-1 \mp \sqrt{1-3(\beta+1)S})\right)}. \tag{8.27}$$

Neimark–Sacker. *There is an* NS *bifurcation of the fixed point* $x = y = \frac{\beta-1}{R}$ *at*

$$\alpha = \frac{(\beta-1)R(\beta-1+2R)-(\beta-1)^3S}{R^3}$$

for which we have

$$c_{NS} = \frac{R^2((1-R)R-3S(\beta-1)}{(\beta(2-R)-(2-3R))R-3(\beta-1)^2S}. \tag{8.28}$$

Proof The bifurcation curves follow from the same equations as in the previous proof. We should note that, unlike in the quadratic case, here the fold and flip curves consist of two branches that are connected. To prove the nondegeneracy we have to find the critical eigenvectors and higher-order derivatives as before. We show computations for one branch only.

$$B(x,y) = \begin{pmatrix} 0 \\ R(x_1y_2+y_1x_2) - 2x_2y_2\left(R+\sqrt{(1-R)^2+3(\beta+1)S}\right) \end{pmatrix}, \tag{8.29}$$

$$q_{fold} = \begin{pmatrix} 1 \\ 1 \end{pmatrix}, \qquad p_{fold} = \begin{pmatrix} -\beta + \frac{R\left(-1+R+\sqrt{(1-R)^2+3(\beta+1)S}\right)}{3S} \\ 1 \end{pmatrix}. \tag{8.30}$$

With these definitions we calculate $c_{fold} = \langle p, B(q,q)\rangle / \langle p, q\rangle$ as given in the proposition on one branch. On the other it is similar. We proceed with the flip bifurcation.

$$B(x,y) = \begin{pmatrix} 0 \\ R(x_1y_2 + y_1x_2) - 2x_2y_2\sqrt{1-3(\beta+1)S} \end{pmatrix},$$

$$C(x,y,z) = \begin{pmatrix} 0 \\ 6S\,x_2y_2z_2 \end{pmatrix}, \tag{8.31}$$

$$q_{flip} = \begin{pmatrix} -1 \\ 1 \end{pmatrix}, \qquad p_{flip} = \begin{pmatrix} \beta + \frac{R\left(-1+\sqrt{1-3(\beta+1)S}\right)}{3S} \\ 1 \end{pmatrix},$$

we calculate $c_{flip} = \langle p, C(q,q,q) - 3B(q,(A-I)^{-1}B(q,q))\rangle / 6\langle p,q\rangle$ and find the expression on one branch.

For the NS bifurcation we omit the intermediate calculations, but the coefficient c_{NS} is obtained in the same way. □

The above codim 1 bifurcations are degenerate in several points. We treat them in two propositions. The cusp, generalized flip and generalized NS are collected together, since they are determined by vanishing of the corresponding normal form coefficient. The fold–flip and the strong resonances fall into another group, since an extra condition on the multipliers is imposed.

Proposition 8.4 *The GHM* (8.12) *with* $R, S \neq 0$ *has the following codim 2 bifurcations of fixed points*:

Cusp. *There is a cusp bifurcation of the fixed point* $x = y = \frac{1-R}{3S}$ *at* $\alpha = \frac{(R-1)^3}{27S^2}$ *and* $\beta = -1 - \frac{(R-1)^2}{3S}$.

Generalized period-doubling. *With* α, x, y *defined in the previous Proposition, the flip bifurcation is degenerate at*

$$\beta = -1 + \frac{1}{24S}\left(12 - 4R - 37R^2 \pm 7R\sqrt{-8+8R+25R^2}\right).$$

Chenciner. *The* NS *bifurcation of the previous proposition is degenerate at* $\alpha = \frac{(R-1)(R^2+R-2-18S)}{27S^2}$, $\beta = 1 + \frac{R(1-R)}{3S}$, *while the fixed point satisfies* $x = y = \frac{1-R}{3S}$.

Proof This follows easily from the expressions for the normal form coefficients at the codim 1 bifurcations. □

Proposition 8.5 *The GHM* (8.12) *with* $R, S \neq 0$ *has the following codim 2 bifurcations of fixed points*:

Fold–flip. *There is a fold-flip bifurcation of the fixed point* $x = y = 0$ *at* $\alpha = 0$ *and* $\beta = -1$. *There is another fold–flip bifurcation for* $x = y = \frac{2-R}{3S}$ *and* $\alpha = \frac{(R+1)(R-2)^2}{27S^2}$, $\beta = -1 - \frac{R(R-2)}{3S}$.

Resonance 1:1. *For* $\beta = 1 - \frac{R}{6S}\left(R - 2 \pm \sqrt{(R-2)^2 + 24S}\right)$ *there is a resonance 1:1.*

Resonance 1:2. *For* $\beta = 1 - \frac{R}{6S}\left(R - 2 \pm \sqrt{(R-2)^2 - 24S}\right)$ *there is a resonance 1:2.*

Resonance 1:3. *For* $\beta = 1 + \frac{R}{6S}\left(R - 2 \pm \sqrt{(R-2)^2 - 12S}\right)$ *there is a resonance 1:3.*

Resonance 1:4. *There is a resonance 1:4 at* $\beta = 1$ *and* $\beta = 1 - \frac{R(R-2)}{3S}$.

It is possible that either the 1:1 or the 1:2 resonance does not exist. In the first case even the 1:3 resonance may not be present.

Proof The eigenvalues can easily be computed along the codim 1 curves. Imposing an extra condition on them one finds the reported values for β on the corresponding codim 1 curve. The non-degeneracy of the codim 2 points can also be checked. □

8.3.4 Branch switching in quadratic GHM

Let us also illustrate the developed techniques for branch switching and their implementation in MatcontM, as in this map all planar bifurcations where our switching formulas apply, occur. Let us start with the 1:2 resonance and the fold–flip. For these cases we can apply the algorithms analytically, i.e., with $S = 0$ and R as a parameter. We note that $q = (1, -1)$ in both cases is an eigenvector of the Jacobian matrix corresponding to eigenvalue -1.

For the 1:2 resonance we have $(x_1, x_2) = (0, 0)$ and $(\alpha_0, \beta_0) = \left(\frac{4(3-R)}{(2-R)^2}, \frac{2+R}{2-R}\right)$. Recall that the critical center manifold reduction yields

$$c_0 = \frac{-(1+R)}{2}, \quad d_0 = \frac{1}{4}(6 + 5R + R^2).$$

Applying the algorithm from Section 4.3.2, we find

$$v_{10} = \begin{pmatrix} \frac{-2}{2-R} \\ -1 \end{pmatrix}, \quad v_{01} = \begin{pmatrix} \frac{-4(3-R)}{(2-R)^2} \\ -\frac{-2}{2-R} \end{pmatrix}, \quad \tilde{p} = \begin{pmatrix} 1 \\ -\frac{2+R+R^2}{2(1+R)} \end{pmatrix} \varepsilon, \tag{8.32}$$

so that our prediction for the NS bifurcation curve of the period-2 cycle is

$$(x, y) = \left(\frac{2}{2-R}, \frac{2}{2-R}\right) + \sqrt{\frac{2\varepsilon}{1+R}}\, q,$$

$$(\alpha,\beta) = \left(\frac{4(3-R)}{(2-R)^2}, \frac{2+R}{2-R}\right) + \left(\frac{2(4+3R^2-R^3)}{(1+R)(2-R)^2}, \frac{2R^2}{(1+R)(2-R)}\right)\varepsilon.$$

The fold–flip bifurcation occurs for $(x_1, x_2) = (0,0)$ at $(\alpha_0, \beta_0) = (0,-1)$. The critical center manifold reduction yields

$$a = \frac{1}{2}(1-R), \quad b = \frac{1}{2}(1+R), \quad c_2 = -\frac{1}{4}(1-R), \quad c_4 = \frac{1}{4}(1+R)^2. \tag{8.33}$$

Then applying the algorithm from Section 4.3.2, we find

$$v_{10} = \begin{pmatrix} -2 \\ \frac{R^2}{2-R} \end{pmatrix}, \quad v_{01} = \begin{pmatrix} 0 \\ \frac{-2}{2-R} \end{pmatrix}, \quad \tilde{p} = \begin{pmatrix} -1 \\ \frac{(1-R)R^2}{2(1+R)} \end{pmatrix}\varepsilon. \tag{8.34}$$

Therefore, our prediction for the NS bifurcation curve of the period-2 cycle emanating here is

$$(x,y) = (0,0) + \sqrt{\frac{2\varepsilon}{1+R}}q, \quad (\alpha,\beta) = (0,-1) + \left(2, \frac{-2R^2}{(1+R)(2-R)}\right)\varepsilon.$$

Let us compare the predictions with the exact expressions for these curves. Consider the following set

$$\alpha = (1+\beta)(\beta - 1 - R + R^2)/R^2, \tag{8.35}$$

$$x_1 = \frac{R(\beta+1) + \sqrt{(R-2)(\beta+1)(2+R-\beta(2-R))}}{2R}, \tag{8.36}$$

$$x_2 = \frac{R(\beta+1) - \sqrt{(R-2)(\beta+1)(2+R-\beta(2-R))}}{2R}, \tag{8.37}$$

when

$$\frac{1}{4}(3-R)(\beta+1)^2 \leq \alpha \leq \frac{5+4R+R^2+2\beta(3+R)+\beta^2(5+2R-R^2)}{2(2+2R+R^2)}.$$

It consists of two different pieces, where an NS bifurcation of a cycle of period-2 occurs. If we take $0 \leq \tilde{\varepsilon} \ll 1$ and consider the linear approximations of (8.35) near $\beta = -1 - \frac{R^2}{2-R}\tilde{\varepsilon}$ and $\beta = \left(\frac{2+R}{2-R}\right) + \frac{R^2}{2-R}\tilde{\varepsilon}$, we find for the 1:2 resonance

$$(\alpha,\beta) = \left(\frac{4(3-R)}{(2-R)^2}, \frac{2+R}{2-R}\right) + \left(\frac{4+3R^2-R^3}{(2-R)^2}, \frac{R^2}{2-R}\right)\tilde{\varepsilon} + O(\tilde{\varepsilon}^2),$$

$$(x,y) = \left(\frac{2}{2-R}, \frac{2}{2-R}\right) + \sqrt{\tilde{\varepsilon}}q + O(\tilde{\varepsilon})$$

and for the fold–flip bifurcation

$$(\alpha,\beta) = (0,-1) + \left(1+R, \frac{-R^2}{2-R}\right)\tilde{\varepsilon} + O(\tilde{\varepsilon}^2),$$

$$(x,y) = (0,0) - \sqrt{\tilde{\varepsilon}}q + O(\tilde{\varepsilon}).$$

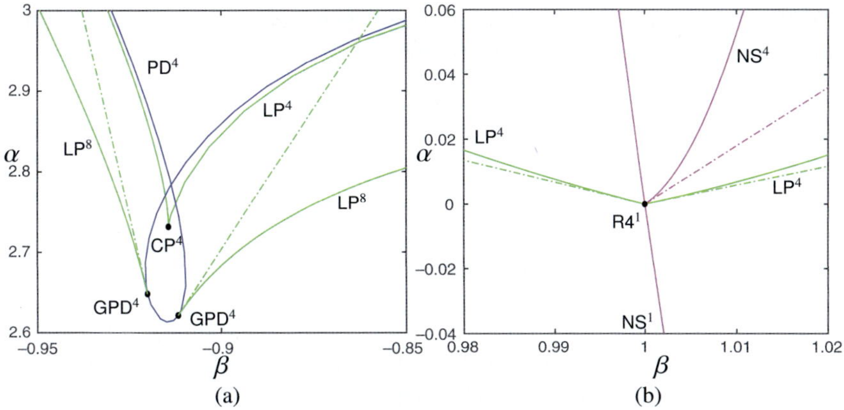

Figure 8.6 The solid curves have been computed using continuation. The dashed lines are the predictions from the switching algorithms.

So up to positive factor our results coincide up to first order in ε.

For the other two cases a numerical approach is more illuminating. We use $R = -0.1$. There is a PD curve of period-4 cycles along which there are two generalized period-doublings near $(\alpha, \beta) = (2.7, -0.92)$. We find a fold curve and a PD curve of cycles of period 4. Then we produce the approximations to the fold curves of the eighth iterates in the generalized PD points and from these continue the fold curves of cycles of period 8. In Figure 8.6(a) we show the continuation results and also the approximation curves. Next we consider the 1:4 resonance. For this specific value of R, all additional local branches emerge from the 1:4 resonance. In Figure 8.6(b) we show the continuation results and also the approximate curves in the parameter plane.

8.3.5 Small coefficients

As we saw in Section 8.2, the Poincaré map for both codim 2 homoclinic tangency cases can be reduced to the map

$$\begin{cases} \bar{x} = y, \\ \bar{y} = \alpha - \beta x - y^2 + R\lambda^k xy + S\lambda^k y^3 + o(\lambda^k), \end{cases} \tag{8.38}$$

where $|\lambda| < 1$ is the stable eigenvalue of the saddle, α, β, x, y cover all finite values as $k \to \infty$. The closer we are to the neutral saddle, the higher k we have to consider. Since R and S scale with λ^k, the case when k is large corresponds to exponentially small (in k) coefficients R and S. It follows from the previous text that the codim 1 bifurcations are non-degenerate away from the codim 2 bifurcations. If we use the parametric form of the NS bifurcation curve

$$\alpha = \cos^2\psi - 2\cos\psi, \quad \beta = 1 - R\cos\psi, \quad 0 < \psi < \pi, \tag{8.39}$$

we find

$$c_{NS} = \frac{R}{4\cos^2\left(\frac{\psi}{2}\right)}\lambda^k + o(\lambda^k). \tag{8.40}$$

Thus, for small R the first Lyapunov coefficient, for the stability of the closed invariant curve emerging from the NS bifurcation, has the same sign as R.

The codim 2 bifurcations which occur only if $S \neq 0$, are at infinity and therefore inaccessible. For the codim 2 bifurcation we give the constant and linear terms in λ^k.

Fold–flip. The normal form coefficients are

$$a(0) = 1 - R\lambda^k + o(\lambda^k),\ b(0) = 1 + R\lambda^k + o(\lambda^k),$$

$$c(0) = \frac{3}{2}(1 - R\lambda^k) + 3S\lambda^k + o(\lambda^k), \quad d(0) = -\frac{1}{2}(5 + 3R\lambda^k) + 3S\lambda^k + o(\lambda^k).$$

The critical coefficient c_{NS} at the fold–flip point is[4]

$$c_{NS} = (R^2 + 2RS + 8s_{21})\lambda^{2k} + o(\lambda^{2k}).$$

Resonance 1:1. The normal form coefficient s has the same sign as $R\lambda^k$.

Resonance 1:2. The normal form coefficients are $c_0 = -2 + 2(R - 4S)\lambda^k + o(\lambda^k)$ and $d_0 = \frac{1}{2}R\lambda^k + o(\lambda^k)$.

Resonance 1:3. The non-degeneracy is given by the real part of the normal form coefficient $a_2(0) = -\frac{i}{\sqrt{3}} + \frac{1}{6}(R + i(R - 2S))\lambda^k + o(\lambda^k)$.

Resonance 1:4. The normal form coefficient A determining the bifurcation sequence is given by $A_0 = -i + \left((\frac{1}{2} + 2i)R - 4iS\right)\lambda^k + o(\lambda^k)$.

We see immediately that when $R \to 0$, the bifurcations of the strong resonances become degenerate. For the fold–flip it follows that the Lyapunov coefficient for the NS bifurcation of period-2 cycles becomes degenerate if we do not consider $O(\lambda^{2k})$-terms.

[4] We took every possible term of the order of λ^{2k} in the second equation of (8.38). The coefficient s_{21} in front of the x^2y-term was the only relevant coefficient.

8.3.6 Description of bifurcations

Here we describe the bifurcation diagram of the quadratic GHM for both positive and negative R. The bifurcations of the standard Hénon Map remain as a backbone (compare Figure 8.1 with Figures 8.7 and 8.8). However, from the previous considerations, it follows that the bifurcations are non-degenerate for $R \neq 0$ and that the cubic term can be neglected if S is small. We describe the local codim 2 bifurcations of low period of the GHM, since these mostly organize the parameter plane. These points are also the origins of some global bifurcations, which are then addressed. Below the LP^1 curve there are no fixed points, and the dynamics are not interesting. On this curve there are two codim 2 points.

First we have the fold–flip point LPPD, whose unfolding for small R corresponds to case 1 of Section 3.4.1. Near this bifurcation an unstable invariant closed curve appears and is destroyed. To show what happens, we take $\alpha = 0.3$ and vary β (see Figures 8.9 and 8.10). From (1) to (3) we see the stable and unstable manifolds of the two saddle fixed points getting closer until they intersect. Furthermore we see that a resonance tongue of period 12 is present

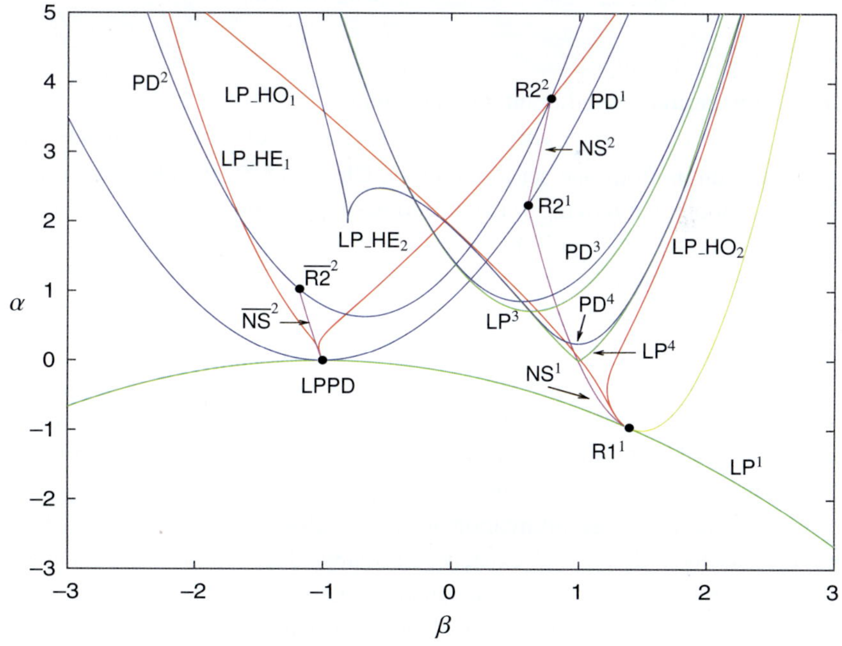

Figure 8.7 Bifurcation diagram for the GHM for $R = -0.5$.

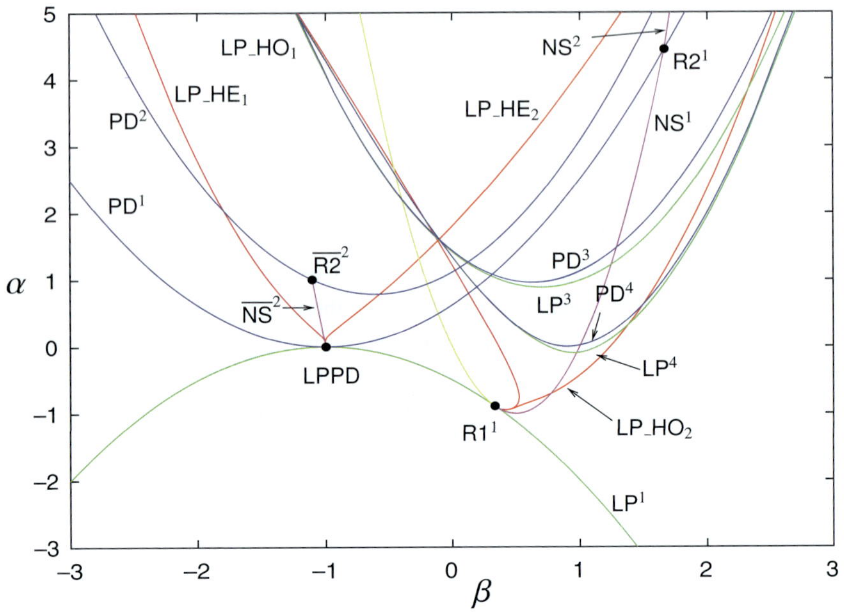

Figure 8.8 Bifurcation diagram for the GHM for $R = 0.5$.

around the period-2 fixed point. Going from (3) to (4), this orbit gains stability and a closed invariant curve is born. It consists of two disjoint sets and an iterate jumps from one set to the other. Close to $\overline{\mathsf{NS}}^2$ the closed invariant curve is smooth, but moving to (5) it becomes bigger and loses its smoothness. Eventually it "merges" with the stable manifold of the period-12 orbit and is destroyed. Then from (5) to (6) the period-12 cycles disappear through a fold bifurcation and the manifolds move away from each other.

The second codim 2 point $\mathsf{R1}^1$ on the LP^1 curve is a 1:1 resonance. From this point a Neimark–Sacker curve NS^1 emerges. The corresponding closed invariant curve is stable for negative R and unstable for positive R. The structure is completed with resonance tongues emerging from NS^1 and curves of homoclinic tangencies (see Figure 8.11).

The Neimark–Sacker bifurcation generates closed invariant curves; here we discuss their domain of existence. Possible destruction scenarios for a closed invariant curve inside the resonance tongue were studied theoretically by Afraimovich and Shilnikov (1983); Aronson et al. (1982), and Broer, Simó, and Tatjer (1998). We consider here the lowest weak resonance, i.e., of order 5, which is located inside the homoclinic wedge of the 1:1 resonance. The bifur-

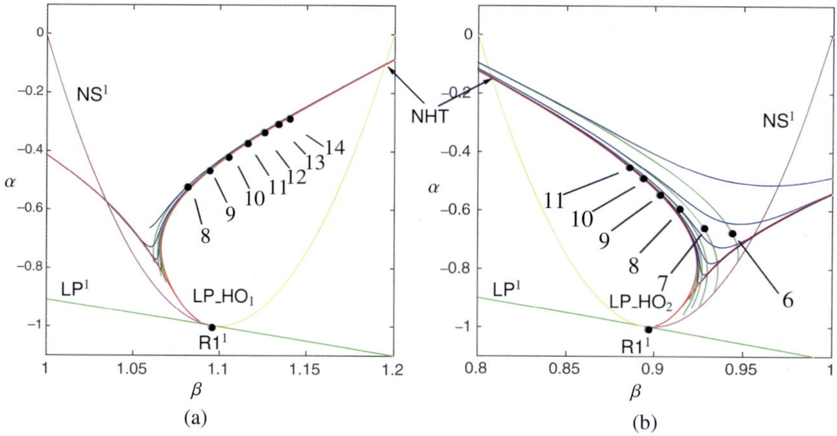

Figure 8.16 Resonance 1:1 points near a homoclinic tangency for (a) $R = -0.1$, (b) $R = 0.1$, indicated by dots. Resonance 1:2 points are not shown since these lie very close to the 1:1 points. The NS curves connecting 1:1 and 1:2 resonances are not visible for the same reason.

1:3 resonance is encountered. Apart from a period-3 cycle with a homoclinic connection, nothing special occurs. Then finally we meet the 1:2 resonance. For small R, this involves only one of the two possible cases, namely the one from which the Neimark–Sacker curve NS^2 for the period-2 cycle emanates. The behavior of the generated invariant circle is determined by the sign of R. Continuing further we observe a cascade of non-degenerate 1:2 resonances $\mathsf{R2}^k$ located on the flip curves and connected by Neimark–Sacker bifurcation segments (in Figure 8.7 only $\mathsf{R2}^1$ and $\mathsf{R2}^2$ are visible).

We now give more details on global bifurcations. Above the PD curve there always exists a transversal heteroclinic orbit connecting the two period-1 saddles. The other invariant manifolds of these saddles can also intersect or become tangent. The wedges of the corresponding heteroclinic and homoclinic tangencies (delimited by $\mathsf{LP\_HE}_{1,2}$ and $\mathsf{LP\_HO}_{1,2}$) form boundaries of parameter regions where nontrivial hyperbolic sets exist. The boundary curves het_2 and hom_1 intersect, giving rise to a double (homoclinic/heteroclinic) tangency (DHT) point (see Figures 8.14 and 8.15). We start in region (1), where there are two saddle fixed points and two transversal heteroclinic orbits which are always present. Going to region (2), the stable manifold of x_1 becomes tangent to the unstable manifolds of this point and two transversal homoclinic orbits are born. From (1) to (3) the unstable manifolds of x_2 touch the stable manifold of x_1 and two heteroclinics are born. In region (4), all six transversal orbits are present.

Finally, let us point out that there exists a homoclinic tangency to a neutral saddle in the GHM (see points NHT in Figure 8.16). This means that GHM, which itself was derived to study the homoclinic tangency to a neutral saddle, also exhibits this codim 2 bifurcation. It implies the existence of a fractal bifurcation set near this singularity in the parameter plane. In particular, we see that the fold and flip curves accumulate on the homoclinic tangency curve, while strong 1:1 resonances (slowly) approach the corresponding NHT points.

8.4 Interpretation

Here we discuss the correspondence between the obtained bifurcation diagrams of the GHM and two bifurcations of diffeomorphisms with codim 2 homoclinic tangencies.

8.4.1 Homoclinic tangency with a neutral saddle in the plane

We begin with a note on invertibility. The standard Hénon Map has the line $\beta = 0$ in the parameter plane, which corresponds to zero Jacobian (i.e., the map is noninvertible). The GHM has no such line in the parameter plane, but in the phase plane it always has a line of noninvertibility for all α, β. This line $y = \frac{\beta}{R\lambda^k} = \frac{-bc}{R}\gamma^k$ shifts to infinity as $k \to \infty$. The noninvertibility effects are, therefore, inaccessible (not observable) in the original map f_μ near the codim 2 point (when k is sufficiently large), which is consistent with the fact that f_μ is a diffeomorphism.

The correspondence between parameters (α, β) and (μ_1, μ_2) is given by (8.7) from Section 8.2. From the second equation in (8.7) it follows that in this case μ_2 has the asymptotic form

$$\mu_2 = \left(-\frac{\beta}{bc}\right)^{\frac{1}{k}} - 1 \tag{8.41}$$

for $k \to \infty$. Thus, in the (μ_2, μ_1)-plane we can see only one half of the (α, β)-plane of the GHM, depending on the sign of bc, which is determined by the global map: $bc < 0$ if the global map is orientation-preserving, $bc > 0$ if the global map is orientation-reversing.

8.4.2 Interpretation for fixed k

The relationships (8.7) imply that bifurcations which occur in GHM at finite values of parameters (α, β) are present in the original map (even infinitely many times – for every $k \geq \bar{k}$) and that the corresponding parameter values (μ_1, μ_2) tend to the point $(0, 0)$ as $k \to \infty$. Moreover, here we have infinitely many regions in the (μ_1, μ_2)-plane, where the bifurcation diagram of the GHM appears in the rescaled coordinates. All these regions accumulate on the line $\mu_1 = 0$.

For all $k \geq \bar{k}$ we have the same (rescaled) picture – a "half" of the bifurcation diagram of the GHM that is described in the previous section. Note that in the considered case the stability type is the same for fixed points of GHM as for the corresponding orbits in the original planar map. A fixed point of GHM corresponds to a single-round periodic orbit with period $k + n_0$ of the original map. A cycle of period m corresponds to a cycle of period $m(k + n_0)$, more precisely to a m-round orbit of period $m(k + n_0)$, which exactly m times did k iterations near the saddle (from strip σ_k^0 to σ_k^1) and n_0 iterations along the "global" part of homoclinic trajectory (from strip σ_k^1 to σ_k^0). A closed invariant curve in GHM corresponds to a closed invariant curve of the $k+n_0$ iterate of the original map (in general, it has $k+n_0$ disjoint components). The parameters α, β on the curves of homoclinic and heteroclinic tangencies in GHM correspond to the tangencies of some one-dimensional manifolds of $k+n_0$-periodic orbits.

8.4.3 Interpretation for all periods

If the original diffeomorphism f_μ is orientation-preserving, then $\lambda\gamma > 0$ and the global map T_1 also preserves the orientation. In this case we have the accumulation of the fold bifurcation curves LP_k^1, the flip bifurcation curves PD_k^1 and the Neimark–Sacker bifurcation curves NS_k^1, as well as domains of existence of closed invariant curves, to the line $\mu_1 = 0$ (Gavrilov and Shilnikov, 1973). We denote the strips between the fold and flip curves by D_k. These strips also accumulate on the line $\mu_1 = 0$. Moreover, we also have the accumulation of curves of homoclinic tangencies and curves of heteroclinic tangencies of the invariant manifolds of single-round periodic orbits of saddle type.

When $\gamma > 0$, the accumulation on the line $\mu_1 = 0$ is monotone (see Figure 8.17(a)).

When $\gamma < 0$, the accumulation is non-monotone, so that the curves are located above or below of the line $\mu_1 = 0$ depending on the parity of k (see Figure 8.17(b)).[5]

If the original diffeomorphism f_μ is orientation-reversing, then $\lambda\gamma < 0$ and the global map T_1 can either preserve or reverse the orientation, depending on

[5] Note that strips in parameter space between LP_k^1 and PD_k^1 with different k can intersect when $c > 0$. This means that, for some values of parameters (in the intersection of D_i and D_j), two single-round periodic orbits of different period can coexist. Moreover, these coexisting orbits can have different types of stability.

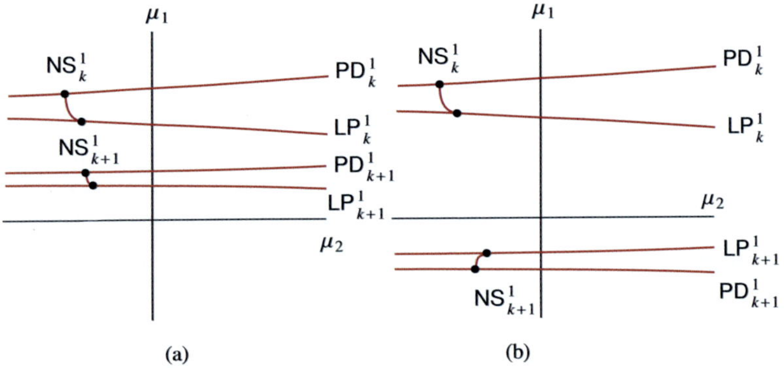

Figure 8.17 Bifurcation curves near the planar neutral tangency when the global map is orientation-preserving.

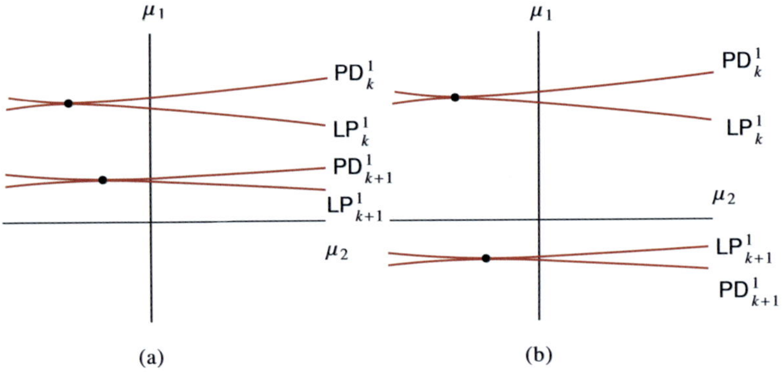

Figure 8.18 Bifurcation curves near a planar neutral tangency when the global map is orientation-reversing.

the parity of n_0. If T_1 preserves the orientation, we get, as described above, the accumulation of the bifurcation curves LP_k^1 and PD_k^1 connected by NS_k^1. If T_1 reverses the orientation, we will see no curves NS_k^1 but, instead, the accumulation of the curves LP_k^1 and PD_k^1 touching at the fold–flip points, as $k \to \infty$ (see Figure 8.18). As above, this accumulation is monotone (Figure 8.18(a)) or non-monotone (Figure 8.18(b)), depending on the sign of γ. Therefore, the whole bifurcation diagram in the case when f_μ reverses the orientation is a superposition of those sketched in Figures 8.17 and 8.18.

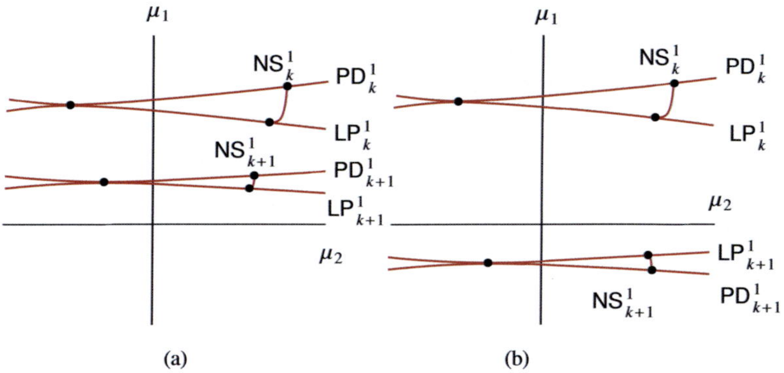

Figure 8.19 Accumulation of bifurcation curves in the three-dimensional case: (a) $\gamma > 0$; (b) $\gamma < 0$.

8.4.4 Generalized homoclinic tangency in $\mathbb{R}^3$

The correspondence between parameters (α,β) and (μ_1,μ_2) is given in this case by (8.11) from Section 8.2. Contrary to the planar case, the whole (α,β)-plane is projected to (μ_1,μ_2) in this case. Indeed, from the second equation in (8.11) follows the linear asymptotic correspondence between μ_2 and β in this case, namely

$$\mu_2 = -\frac{\beta}{c_1(\lambda_1\gamma)^k} \tag{8.42}$$

for $k \to \infty$.

Recall that in this case the GHM is obtained via the reduction to a stable two-dimensional invariant center manifold. Therefore a stable point (invariant curve) in GHM corresponds to a stable orbit (invariant curve) of thethree-dimensional diffeomorphism, but a completely unstable point (unstable invariant curve) in GHM corresponds to a saddle orbit (saddle invariant curve) in the original diffeomorphism. Further interpretation is similar to the case of the planar neutral tangency. In particular, the bifurcation curves and regions accumulate on $\mu_1 = 0$ (see Figure 8.19).

8.5 Discussion

In this chapter we gave a rather detailed characterization of the bifurcation structure of the GHM (8.3). This allows us to establish the following facts about the codim 2 homoclinic tangencies.

1. The non-degeneracy of codim 1 and 2 bifurcations of fixed points in (8.3), including all strong resonances, is verified analytically by computing the corresponding normal form coefficients.
2. Accumulation of fold and flip bifurcation curves on heteroclinic tangencies is observed numerically.
3. Cascades of non-degenerate 1:2 resonances of k-cycles are found in the GHM.
4. It is shown that the GHM itself has a homoclinic tangency of a neutral saddle.
5. The above bifurcation phenomena are interpreted in terms of the original maps with the codim 2 homoclinic tangencies. In particular, infinite cascades of homoclinic tangencies of neutral saddles are predicted.

In the following we provide some additional comments on these issues.

The accumulation of fold and flip bifurcation curves on the homoclinic tangency curve was first proved by Gavrilov and Shilnikov (1972). It is not guaranteed that all fold curves originating from NS^1 at weak resonance points have to approach the homoclinic tangency curve. However, the fold curves in GHM (originating from NS^1) indeed look as if they all approach the corresponding homoclinic tangency curves $\mathsf{LP\_HO}_{1,2}$. It seems that the $\frac{1}{q}$-resonant periodic orbits are exactly the *single-round* orbits of period q, which exist near homoclinic tangency (see Section 8.2 for more details). It would be very interesting to explain the correspondence between 2-round (and more round) orbits near the homoclinic tangency and orbits of the original map. There are some topological arguments which do not allow us to associate directly $\frac{2}{q}$-resonant orbits with 2-round periodic orbits.

Our numerical analysis demonstrated that the double homoclinic tangency point DHT moves to the half-plane $\beta < 0$ for both positive and negative small values of $R \neq 0$ (see Figure 8.14(b)). This implies that such codim 2 global bifurcations will not be present near the homoclinic tangency to a neutral saddle, if the considered planar diffeomorphism preserves the orientation. Topological reasons for this effect are not clear.

We conclude this discussion with numerical evidence (Figure 8.16) that the resonances 1:1 and 1:2 indeed accumulate to NHT as sketched in Figure 8.17(a). Note that we see only one-half of the parameter plane. Figure 8.16 shows bifurcation curves for orbits with relatively high period together with strong resonances. Figure 8.16 shows that the 1:1 resonance points with periods $\geq$ 10 lie practically either on $\mathsf{LP\_HO}_1$ (a) or on $\mathsf{LP\_HO}_2$ (b). However, according to Section 8.4, these resonance points must converge to the NHT point. This is not very evident from the figure, since even the 1:1

points of period 14 (Figure 8.16 (a)) or period 11 (Figure 8.16 (b)) are rather far from the NHT points. Fortunately, this is not a contradiction with the theory. Indeed, from (8.7) in Section 8.2 it follows that the 1:1 resonance point of period-k lies, on the parameter plane, at a distance of order of $O(\gamma^{-k})$ from the curve of the homoclinic tangency, where γ is the unstable multiplier of the saddle, while the distance from this 1:1 point to the NHT point is only of order $O(\frac{1}{k})$. To see this, recall that (8.41) implies that the 1:1 resonance k-point has the coordinate $\mu_2 \sim (\frac{1}{-bc})^{\frac{1}{k}} - 1 \sim \frac{1}{k}\ln(\frac{1}{-bc})$. Thus, such points are exponentially close to the corresponding homoclinic tangency curve but only polynomially close to the NHT point. These two very different asymptotics are clearly visible in Figure 8.16.

9

Adaptive Control Map

Consider a model for controlling a single-input/single-output plant (see (Golden and Ydstie, 1988; Frouzakis, Adomaitis, and Kevrekidis, 1991) for references). Regulating the output to a constant by low-order feedback leads to the following map $G\colon \mathbb{R}^3 \mapsto \mathbb{R}^3$

$$G\colon \begin{pmatrix} x \\ y \\ z \end{pmatrix} \mapsto \begin{pmatrix} y \\ bx + k + zy \\ z - \dfrac{ky}{c + y^2}(bx + k + zy - 1) \end{pmatrix}, \tag{9.1}$$

where k and b are considered as main parameters. The coefficient b measures the mismatch between the reference and the real model. The coefficient k represents an error in the assumption on how strongly the output variable is controlled. Choosing $b = 0, k = 1$ implies no errors. The coefficient c comes from the projection algorithm. It is positive to avoid division by zero. Since the possible bifurcation sequences are determined by the critical normal form coefficients, we may also choose the constant c to prevent too complex behavior.

9.1 Local bifurcations

There are two codim 1 bifurcation curves of fixed points with several codim 2 points. In Frouzakis, Adomaitis, and Kevrekidis (1991) the system was studied numerically near strong resonance points. We can compute the second- and third-order derivatives of this map symbolically (see Section 9.3) and thus compute the critical normal form coefficients for the strong resonances analytically and study their dependence on the parameter c, which is typically fixed at $c = 1/10$.

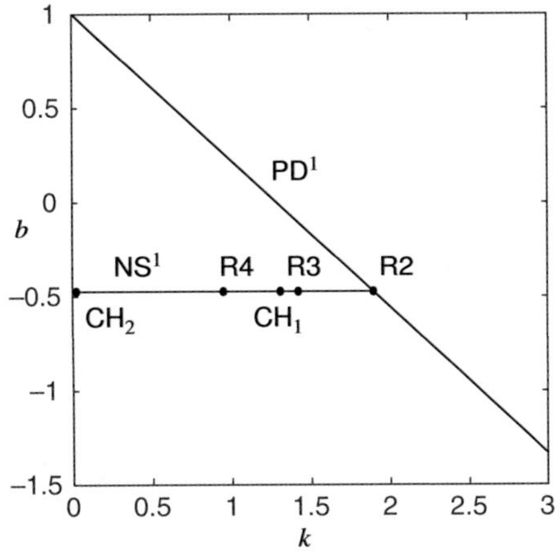

Figure 9.1 Local bifurcations of (9.1) for $c = 1/10$: PD^1 – flip (period-doubling), NS^1 – Neimark–Sacker bifurcation of the fixed point. Labels Rx denote the strong resonances; $\mathsf{CH}_{1,2}$ are the Chenciner points. The codim 2 points are studied in this section.

This map has one fixed point given by $(x, y, z) = (1, 1, 1 - b - k)$. Local bifurcation analysis reveals two codim 1 bifurcation curves shown in Figure 9.1. For

$$\mathsf{PD}^1: \quad b_f = 1 - \left(\frac{1}{2} + \frac{1}{4(c+1)}\right) k$$

the fixed point undergoes a period-doubling, while if

$$\mathsf{NS}^1: \quad b_{NS} = -\left(\frac{c+1}{c+2}\right)$$

and $k^2 + 4bk < 0$ an invariant curve emerges from the fixed point via the NS bifurcation.

Starting from $k = 0$ on the NS curve and tracing it until we meet the PD curve, we encounter strong resonances. These occur for the following values of k:

$$\mathsf{R4}: \quad k = \frac{2(c+1)}{c+2}, \qquad \mathsf{R3}: \quad k = \frac{3(c+1)}{c+2}, \qquad \mathsf{R2}: \quad k = \frac{4(c+1)}{c+2}. \tag{9.2}$$

As is observed by Frouzakis, Adomaitis, and Kevrekidis (1991), at $(k, c) \approx (1.308, 0.1)$ the NS bifurcation is degenerate. A second degenerate point on the curve was overlooked there.

Next we compute the critical normal form coefficients using expressions for the relevant eigenvectors and the derivatives of (9.1) up to and including the fifth order, as given in Section 9.3.

- **Resonance 1:2.** A straightforward computation using (4.44) and (4.45) yields

$$c_0 = \frac{4(2+c)^2(1-3c-2c^2-2c^3)}{3(1+c)^2(18+24c+11c^2+2c^3)}, \tag{9.3}$$

$$d_0 = \frac{4(2+c)^2(-12-7c+6c^2+37c^3+32c^4+8c^5)}{(6+4c+c^2)(3+5c+2c^2)^2}. \tag{9.4}$$

For small c up to $c \approx 0.617$ the coefficient d_0 is negative, while c_0 is positive up to $c \approx 0.271$.
- **Resonance 1:3.** Now we use the formula (4.50) to find

$$a_2(0) = \frac{(2+c)(-3-4c+6c^2+33c^3+18c^4)}{6(1+3c^2)(7+9c+3c^2)}. \tag{9.5}$$

Since $c > 0$ but small, this means that the invariant closed curve existing close to the 1:3 bifurcation is stable.
- **Resonance 1:4.** This bifurcation has the most complicated bifurcation sequences. Using (4.55) and (4.56), we get

$$A_0 = \frac{(2+c-4c^2+3c^3+2c^4) - i(1+36c+40c^2+17c^3+4c^4)}{\sqrt{(5+6c+2c^2)(1-10c+83c^2+12c^3+4c^4+8c^5+2c^6)}} \operatorname{sign}(3+2c). \tag{9.6}$$

Consider now the case with small positive c. The real part of A_0 is always positive. If we start with $c = 0$ we find $|A_0| = 1$, and thus we will encounter the bifurcation sequences IV(a), III(a), III, V, V(a), VI and VIII (here the labeling from Section 3.3.4 is used). Actually, for small negative c we are within the unit circle in the A_0-plane, and the dynamics is simple.
- **Chenciner bifurcation.** For the NS bifurcation we derive using (2.18) the following expression for the first Lyapunov coefficient:

$$\begin{aligned} L_1 =& (b(1+b)^2(1+2b)(4+4b+3b^2) \\ &+ (4+21b+50b^2+102b^3+92b^4-7b^5)k \\ &+ (14b+72b^2+79b^3-3b^4)k^2 + (1+3b)(3+5b)k^3) \\ &/(2b^2(1+b)((1+b)^2+k)(b(1+b)^2+4bk+k^2))). \end{aligned}$$

For $c = 1/10$, we have $b_{NS} = -11/21$ and L_1 vanishes at two points, namely: CH_1 with $k \approx 1.30064$ and CH_2 with $k \approx 0.0214$ (see Figure 9.1). The first point was reported by Frouzakis, Adomaitis, and Kevrekidis (1991) but not analyzed; the second point was not mentioned.

For $c = 0.1$, $b \approx -0.52381$, and $k \approx 1.30064$, we find $e^{-i\theta}c_1 \approx 0.37721i$. So the first Lyapunov coefficient vanishes at this point. We compute the derivatives up to fifth order (see Section 9.3) and then solve recursively for h_{jk}. This gives

$$
\begin{aligned}
h_{20} &\approx (-1.4588 - 1.5723i, 0.5516 + 2.0727i, 1.6970 - 1.7342i),\\
h_{11} &\approx (-0.7019, -0.7019, 1.1871),\\
h_{30} &\approx (-4.7566 - 2.9282i, -5.3569 + 1.5822i, 4.2621 - 4.3465i),\\
h_{21} &\approx (1.5723 - 1.7334i, 1.9215 + 1.1523i, 3.1354 - 2.1245i),\\
h_{31} &\approx (-6.4616 + 18.7151i, 9.7893 - 12.2545i, 1.9851 - 3.2360i),\\
h_{22} &\approx (6.1653, 6.1653, 29.1252).
\end{aligned}
$$

Using these values for the second Lyapunov coefficient (4.28), we find $L_2 \approx -17.5164$. The same procedure can be carried out for the second point, yielding $L_2 \approx 3423$. This implies that both Chenciner bifurcations in (9.1) are non-degenerate.

9.2 Numerical continuation

From a control point of view, coexistence of the fixed point together with other attractors is undesirable. In the original paper (Frouzakis, Adomaitis, and Kevrekidis, 1991) numerical continuation of fixed points and computation of one-dimensional stable and unstable invariant manifolds were used. The coexistence of global stable attractors together with the fixed point and tangencies of stable and unstable manifolds were then deduced from the phase portraits. The authors pointed out that it was difficult to characterize the bifurcations near the codim 2 points. However, first, we computed the critical normal form coefficients symbolically and thus were able to check that these are indeed non-degenerate codim 2 bifurcations. Second, we were able to verify the hypothesis that some global bifurcations occurred. Since these are present in the normal form for certain values of the critical normal form coefficients, by continuity they should also be observed away from the resonances. The analysis shows as well that a specific choice of $c > 0$ may lead to more desirable bifurcation sequences.

In Frouzakis, Adomaitis, and Kevrekidis (1991), the authors also pay attention to a period-5 resonance bubble. The resonance tongue starts at the NS bifurcation. Initially it corresponds to an unstable closed invariant curve. Increasing b, the tongues meet a quasi-periodic bifurcation saddle-node bifurcation that emerges from the Chenciner point at $k = 0.0214$. Here the tongue bends and then corresponds to a stable invariant curve. In Frouzakis, Adomaitis, and Kevrekidis (1991) the bifurcation diagram contains a zone with several question marks as the details became complicated. In Figure 9.2 we show the full horn starting from the NS bifurcation and the bubble. The bubble is slightly

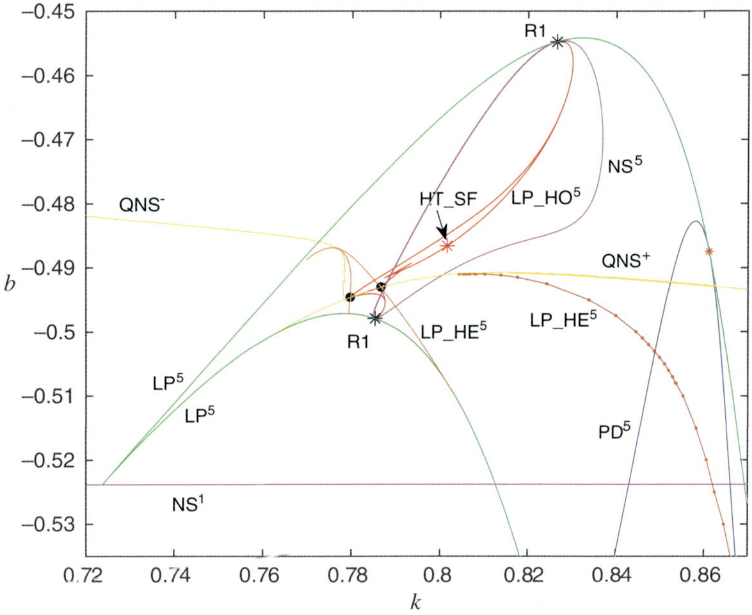

Figure 9.2 The full resonance horn. The dark magenta curve (straight horizontal line) is an NS^1 curve, all other curves are bifurcation curves of the fifth iterate. Green = LP^5 curves, magenta solid = NS^5 curve, magenta dashed = neutral saddle fixed points. The magenta curves touch the green curves in R1 points. Red = homoclinic tangency curves $\mathsf{LP\_HO}^5$, rooted in the R1 points. Orange and yellow =heteroclinic tangencies. Blue = PD^5.

deformed with respect to our discussion in Section 3.2. Note we have $q = 5$ so that analysis formally does not apply. Still, quite some structure is preserved. Indeed, the close-up in Figure 9.3 resolves the bubble in quite some detail. In addition, one may note the presence of HT_SF and HT_NS bifurcation points on the LP_HO curve. These are codim 2 homoclinic bifurcations highlighting the use of MatcontM.

9.3 Derivatives for the adaptive control map

Here we give the expressions of the eigenvectors on the NS curve along which all four codim 2 bifurcations occur. We fix $c = -\left(\frac{2b+1}{b+1}\right)$ rather than b, since this simplifies the formulas. The equivalent expressions can easily be computed.

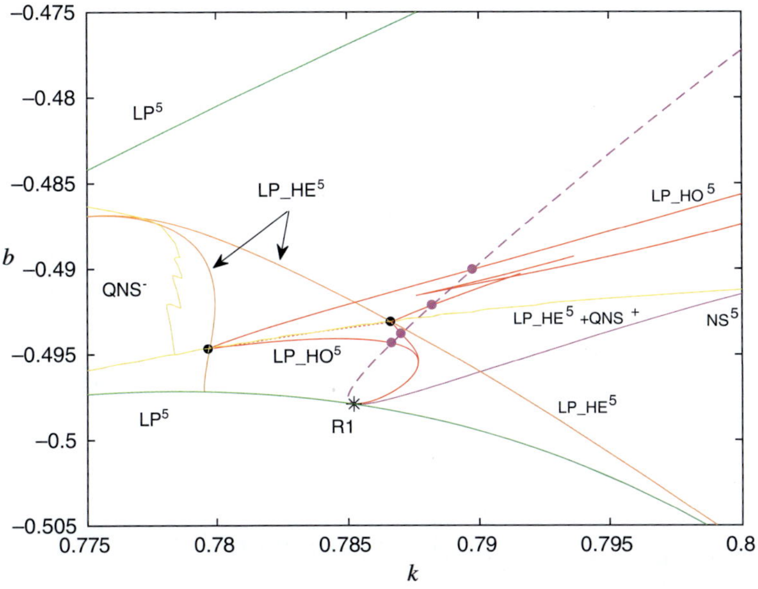

Figure 9.3 Zoom of Figure 9.2. The four magenta dots on the dashed magenta curve are HT_NS bifurcation points. Note that the form of one of the LP_HO curves suggests the nearby presence of a swallowtail bifurcation of homoclinic orbits.

The Jacobian matrix at the fixed point is

$$A = \begin{pmatrix} 0 & 1 & 0 \\ b & 1 - b - k & 1 \\ k(1+b) & \dfrac{(k(1+b)(1-b-k)}{b} & 1 + k + \dfrac{k}{b} \end{pmatrix}$$

with eigenvalues

$$\lambda_1 = \frac{2b + k - \sqrt{k(4b+k)}}{2b}, \quad \lambda_2 = \frac{2b + k + \sqrt{k(4b+k)}}{2b}, \quad \lambda_3 = -b.$$

If $k^2 + 4bk < 0$, then the eigenvectors corresponding to λ_1 are

$$q = \left(-\frac{k(3b+k) + (b+k)\sqrt{k(4b+k)}}{(2kb(b+1)}, \frac{-k + \sqrt{k(4b+k)}}{2k(b+1)}, 1 \right),$$

$$p = \frac{1}{\alpha}\left(-\frac{1}{2}\left(k + \sqrt{k(4b+k)}\right), -\frac{1}{2b}\left(k(2b+1) + \sqrt{k(4b+k)}\right), 1 \right),$$

where

$$\alpha = \frac{(1+b+k)(k+4b) - (1+3b+k)\sqrt{k^2+4b}}{2b(1+b)}.$$

Then the derivative tensors are given by

$$B(p,q) = \begin{pmatrix} 0 \\ p_2q_3 + p_3q_2 \\ \frac{(1+b)(2+3b)k}{k}(p_1q_2 + p_2q_1) + \frac{2(1+b)(2+3b)k(1-b-k)}{b^2}p_2q_2 \\ +\frac{(1+b)(1+2b)k}{b^2}(p_2q_3 + p_3q_2) \end{pmatrix}.$$

It is easy to see that for the third- and higher-order tensors only the third component will be nonzero. We write

$$C(p,q,r) = \begin{pmatrix} 0 \\ 0 \\ \sum_{i,j,k=1}^{3} C_{ijk}p_iq_jr_k \end{pmatrix},$$

where at least two indices of C_{ijk} are equal to 2, otherwise the coefficient is zero. We have

$$C_{122} = C_{212} = C_{221} = \frac{2(1+b)^2(4+7b)k}{b^2},$$

$$C_{222} = \frac{6(1+b)^2(4+7b)k(1-b-k)}{b^3},$$

$$C_{322} = C_{232} = C_{223} = \frac{2(1+b)^2(1+2b)(4+5b)k}{b^3}.$$

In a similar manner we find

$$D(p,q,r,s) = \begin{pmatrix} 0 \\ 0 \\ \sum_{i,j,k,l=1}^{3} D_{ijkl}p_iq_jr_ks_l \end{pmatrix},$$

where at least three indices of D_{ijkl} are equal to 2, otherwise the coefficient is zero. We have

$$D_{1222} = D_{2122} = D_{2212} = D_{2221} = \frac{6(1+b)^2(8+24b+17b^2)k}{b^3},$$

$$D_{2222} = \frac{24(1+b)^2(8+24b+17b^2)k(1-b-k)}{b^4},$$

$$D_{3222} = D_{2322} = D_{2232} = D_{2223} = \frac{24(1+b)^2(1+2b)(2+3b)k}{b^4}.$$

And finally the fifth-order tensors are given by

$$E(p,q,r,s,t) = \begin{pmatrix} 0 \\ 0 \\ \sum_{i,j,k,l,m=1}^{3} E_{ijklm} p_i q_j r_k s_l t_m \end{pmatrix},$$

where

$$\begin{aligned} E_{12222} &= E_{21222} = E_{22122} = E_{22212} = \\ E_{22221} &= \frac{24(1+b)^3(16+52b+41b^2)k}{b^4}, \\ E_{22222} &= \frac{120(1+b)^3(16+52b+41b^2)k(1-b-k)}{b^5}, \\ E_{32222} &= E_{23222} = E_{22322} = E_{22232} = \\ E_{22223} &= \frac{24(1+b)^2(1+2b)(16+44b+29b^2)k}{b^5} \end{aligned}$$

and all other coefficients are zero.

10
Duopoly Model of Kopel

In this chapter we consider a two-dimensional map first introduced by M. Kopel that describes the competition of two companies in a duopoly market. It is based on a joint work with Matthias Aengenheyster.

10.1 Description of the model

Kopel's duopoly model is a well-known model describing the dynamics of two companies competing on a common market, the underlying mathematics of which were first described by Cournot (1838). The firms produce goods x and y, respectively, being equivalent and interchangeable from the point of view of consumers. At each discrete time k the firms find their optimal, i.e., profit-maximizing, production for the next time $k+1$ using a reaction function on the competitor's expected production, determined using adaptive expectations. This leads to the recurrent relation

$$\begin{pmatrix} x_{k+1} \\ y_{k+1} \end{pmatrix} = \begin{pmatrix} (1-\rho)x_k + \rho\mu y_k(1-y_k) \\ (1-\rho)y_k + \rho\mu x_k(1-x_k) \end{pmatrix} \tag{10.1}$$

or, equivalently, to iterating the map

$$\begin{pmatrix} x \\ y \end{pmatrix} \mapsto f(x,y) = \begin{pmatrix} (1-\rho)x + \rho\mu y(1-y) \\ (1-\rho)y + \rho\mu x(1-x) \end{pmatrix}, \tag{10.2}$$

where essentially the future production is a weighted mean of the optimal production computed from the competitor (weighted with the adjustment coefficient ρ) and the own current production (weighted with $(1-\rho)$). For a full derivation and economic justification see Bischi and Kopel (2001). The economic interpretation implies that $0 \le \rho \le 1$. As in the previous studies, we will ignore this restriction for most of the analysis but keep in mind that only those

ρs are economically relevant. We use homogeneous players with the same values for μ, ρ. In this model, fixed points correspond to the player's Nash equilibria. It is therefore interesting to study (co)existence, stability and bifurcations of these fixed points as functions of the parameters as they describe the behavior of firms following this simple model.

The Kopel system has been investigated by several authors, starting with Kopel (1996). Govaerts and Khoshsiar Ghaziani (2008) analytically computed fixed points and their stability regions as well as normal form coefficients of period-doubling (PD) bifurcations, and used MatcontM for numerical stability analysis of 2-,3- and 4-cycles. Agiza (1999) investigated chaos and its control in the Kopel system, while Bischi and Kopel (2001) and Anderson, Myran, and White (2005) looked at multistability and complex basins of attraction.

In this chapter we extend the previous analysis by analytical computation of the normal form coefficient of the Neimark–Sacker (NS) bifurcation of the fixed point. We also analytically locate and characterize a number of codim 2 bifurcations via computing the normal form coefficients. We then use the bifurcation software MatcontM to find bifurcation curves of the fixed points up to the 32nd iterate and extend the previous study of chaos in the Kopel system by the computation of Lyapunov exponents as a function of both parameters μ, ρ, enabling us to identify key regions in the parameter space. The main feature of our approach is a combination of analytic and numerical techniques.

10.2 Fixed points and codim 1 bifurcations

Fixed points, PD and NS bifurcations of the first iterate have been analyzed previously (Agiza, 1999; Govaerts and Khoshsiar Ghaziani, 2008). Here we only briefly summarize the fixed points and their linear stability regions determined by the Jacobian matrix $A = Df$, i.e.,

$$A = \begin{pmatrix} 1-\rho & (1-2y)\mu\rho \\ (1-2x)\mu\rho & 1-\rho \end{pmatrix}, \tag{10.3}$$

in Table 10.1 and Figure 10.1, while presenting codim 1 bifurcation curves in Table 10.2 and Figure 10.2.

Note that the PD points $3a$ and $3b$ are actually one continuous PD curve coinciding at $(\mu, \rho) = (1 + \sqrt{5}, 2)$. Note that the lines $\mu = 1$ and $\mu = 3$ correspond to branch points (BP) (for a complete treatment we refer to Govaerts and Khoshsiar Ghaziani (2008)). At $\mu = 1$ the origin turns unstable while E_2 crosses into the positive quadrant and turns stable. At $\mu = 3$ E_2 loses stability to the pair of fixed points E_3, E_4. No further intersections between the fixed

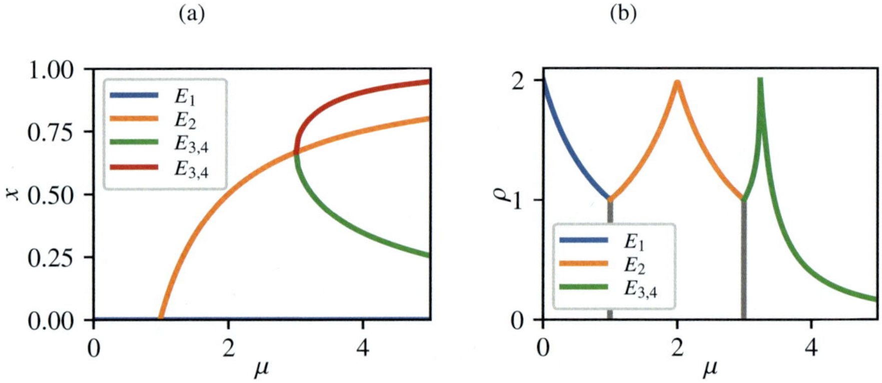

Figure 10.1 (a) Fixed points as functions of μ. (b) Stability regions of fixed points.

Table 10.1 Fixed points and stability regions.

Point	x	y	μ	ρ
E_1	0	0	$\lvert\mu\rvert < 1$	$\rho \le \frac{2}{1+\lvert\mu\rvert}$
E_2	$\frac{\mu-1}{\mu}$	$\frac{\mu-1}{\mu}$	$\lvert\mu - 2\rvert < 1$	$\rho \le \frac{2}{1+\lvert\mu-2\rvert}$
E_3,E_4	$\frac{1+\mu\mp\sqrt{-3-2\mu+\mu^2}}{2\mu}$	$\frac{1+\mu\pm\sqrt{-3-2\mu+\mu^2}}{2\mu}$	$3 < \mu < 1+\sqrt{5}$	$0 < \rho < \frac{2\left(1-\sqrt{4+2\mu-\mu^2}\right)}{(\mu-3)(\mu+1)}$
			$1+\sqrt{5} < \mu$	$\rho < \frac{2}{(\mu-3)(\mu+1)}$

Table 10.2 Codim 1 bifurcations.

Point	ρ	μ
PD_{1a}	$\frac{2}{1-\mu}$	$0 < \mu < 1$
PD_{1b}	$\frac{2}{1+\mu}$	$0 < \mu$
PD_{2a}	$\frac{2}{3-\mu}$	$0 < \mu < 3$
PD_{2b}	$\frac{2}{-1+\mu}$	$1 < \mu$
PD_{3a}	$\frac{2\left(1-\sqrt{4+2\mu-\mu^2}\right)}{-3-2\mu+\mu^2}$	$3 < \mu < 1+\sqrt{5}$
PD_{3b}	$\frac{2\left(1+\sqrt{4+2\mu-\mu^2}\right)}{-3-2\mu+\mu^2}$	$3 < \mu < 1+\sqrt{5}$
NS	$\frac{2}{-3-2\mu+\mu^2}$	$\mu > 1+\sqrt{5}$
BP_1	ρ	1
BP_3	ρ	3

points exist. In particular, there are no limit points of the first iterate as branch and limit points are mutually exclusive. The simplicity of both branch points can be verified using formulas from Section 4.3.1.

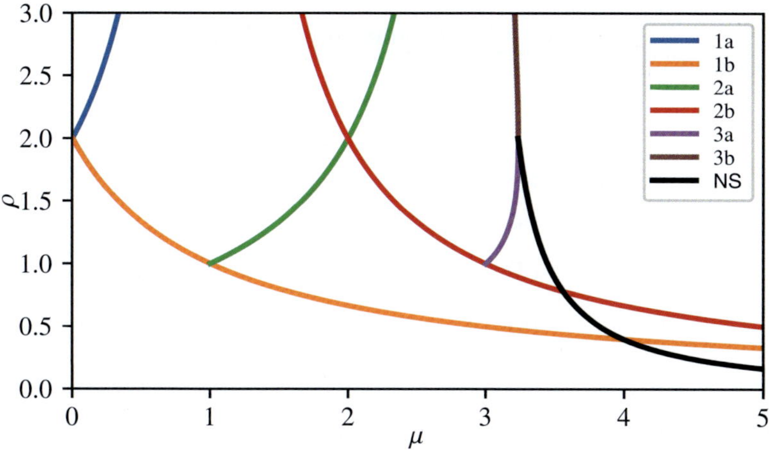

Figure 10.2 PD and NS curves in the (μ, ρ)-plane.

10.3 Normal forms of codim 1 bifurcations

10.3.1 Normal forms for PD

We compute the normal form coefficient c_0 for each PD point using (2.12), i.e.

$$c_0 = \frac{1}{6}\langle p, C(q,q,q)\rangle - \frac{1}{2}\langle p, B(q,(A - I_n)^{-1}B(q,q))\rangle \tag{10.4}$$

with the quadratic and cubic multilinear forms B and C, identity matrix I_n and inner product $\langle p, q\rangle = p^T q$. We first compute the eigenvectors p_s, q_s such that

$$Aq_s = -q_s, \quad A^T p_s = -p_s. \tag{10.5}$$

Then we normalize these vectors to fulfill

$$\langle q, q\rangle = \langle p, q\rangle = 1 \tag{10.6}$$

by setting

$$q = \frac{q_s}{\sqrt{\langle q_s, q_s\rangle}}, \tag{10.7}$$

$$p = \frac{p_s}{\langle q, p_s\rangle} \tag{10.8}$$

and then use these vectors in (10.4). The expressions for p, q for all six PD curves can be found in Govaerts and Khoshsiar Ghaziani (2008). The computations result in the coefficient c_0 (Table 10.3). A supercritical (subcritical)

Table 10.3 PD normal form coefficients.

Point	c_0
PD_{1a}	$\frac{2\mu^2}{(\mu-1)^2}$
PD_{1b}	$\frac{2\mu^2}{\mu^2-1}$
PD_{2a}	$\frac{2\mu^2}{3-4\mu+\mu^2}$
PD_{2b}	$\frac{2\mu^2}{(\mu-1)^2}$
PD_{3a}	$-\frac{8\mu^2}{3+2\mu-\mu^2}\left(1-\frac{1}{\sqrt{4+2\mu-\mu^2}}\right)$
PD_{3b}	$-\frac{8\mu^2}{3+2\mu-\mu^2}\left(1+\frac{1}{\sqrt{4+2\mu-\mu^2}}\right)$

PD bifurcation is characterized by a positive (negative) normal form coefficient c_0. It therefore follows directly from Table 10.3 and the validity regions in Table 10.2 that PD bifurcations at curves $1a$ and $2b$ are always supercritical and at $2a$ is subcritical. $1b$ is subcritical for $\mu < 1$. Hence, PD curves bounding the stability regions in ρ correspond to subcritical bifurcation for $\mu < 2$ (so no stable objects exist for larger ρ), while for $2 \leq \mu \leq 3$ the boundary ($2b$) corresponds to the supercritical bifurcation, prompting the existence of 2- and eventually higher cycles. We will return to this in the numerical section.

10.3.2 Normal form for NS

The eigenvalues of the Jacobian matrix A along the NS curve are given by

$$\lambda_{1,2} = u \pm iv \tag{10.9}$$

with

$$u = \frac{-5 - 2\mu + \mu^2}{-3 - 2\mu + \mu^2}, \tag{10.10}$$

$$v = \frac{2\sqrt{-4 - 2\mu + \mu^2}}{-3 - 2\mu + \mu^2}. \tag{10.11}$$

First, we compute the eigenvectors p_s, q_s satisfying

$$Aq_s = e^{i\theta_0} q_s, \quad A^T p_s = e^{-i\theta_0} p_s \tag{10.12}$$

with $e^{i\theta_0} = u + iv, e^{-i\theta_0} = u - iv$. Then we define

$$q = \frac{q_s}{\sqrt{\langle q_s, q_s \rangle}}, \tag{10.13}$$

$$p = \frac{p_s}{\langle q, p_s \rangle}, \tag{10.14}$$

where now $\langle p,q\rangle = \bar{p}^T q$, so that $\langle q,q\rangle = \langle p,q\rangle = 1$. We arrive at

$$p = \frac{1}{\sqrt{2}}\sqrt{\frac{-3-2\mu+\mu^2+\sqrt{-3-2\mu+\mu^2}}{-4-2\mu+\mu^2}}\left(i\frac{(-1+\sqrt{-3-2\mu+\mu^2})}{\sqrt{2}\sqrt{-4-2\mu+\mu^2}},1\right),$$

$$q = \frac{1}{\sqrt{1+\frac{(1+\sqrt{-3-2\mu+\mu^2})^2}{-4-2\mu+\mu^2}}}\left(i\frac{1+\sqrt{-3-2\mu+\mu^2}}{\sqrt{-4-2\mu+\mu^2}},1\right),$$

Then we compute the first Lyapunov coefficient $L_1 = \Re(e^{-i\theta_0}c_1^0)$, where, according to (2.18),

$$c_1^0 = \frac{1}{2}\langle p, C(q,q,\bar{q}) + B(\bar{q},(e^{2i\theta_0}I_n - A)^{-1}B(q,q)) + 2B(q,(I_n - A)^{-1}B(q,\bar{q}))\rangle,$$

and arrive at the following compact expression

$$L_1 = -\frac{4(\mu-2)(\mu-1)^2\mu^3}{(\mu-3)^4(\mu+1)^4}. \tag{10.15}$$

Supercriticality of the NS point was already suggested from numerical investigation by Govaerts and Khoshsiar Ghaziani (2008) and we are now able to prove this hypothesis analytically. A supercritical (subcritical) NS bifurcation is characterized by a negative (positive) normal form coefficient L_1. The condition $L_1 < 0$ follows directly from (10.15) as $\mu > 1 + \sqrt{5} \approx 3.2361$ and hence $\mu^3 > 0$ and $\mu - 2 > 0$. Therefore the fraction is positive for all relevant values of μ and we have $c_0 < 0$, proving supercriticality.

10.4 Codim 2 bifurcations

We use our analysis of codim 1 bifurcations to locate codim 2 bifurcations on the codim 1 curves.

10.4.1 LPPD

An LPPD (limit point period doubling, also called fold–flip) point is characterized by its multipliers being $\lambda_1 = 1$ (LP) and $\lambda_2 = -1$ (PD). This obviously implies that such points lie on PD curves. We therefore compute the eigenvalues of the Jacobian matrix along the PD curves and find the expressions for λ_1, as shown in Table 10.4.

We attempt to solve $\lambda_1 = 1$ with each of these expressions for μ and find two points at $(\mu,\rho) = (1,1)$ and $(3,1)$. They lie at the intersections of the PD curves $(1b, 2a)$ and $(2b, 3a)$ and the branch points at $\mu = 1, \mu = 3$. Therefore

Table 10.4 Multipliers along the PD curves.

	λ_1
PD_{1a}	$\frac{1+3\mu}{\mu-1}$
PD_{1b}	$\frac{-1+3\mu}{1+\mu}$
PD_{2a}	$\frac{-5+3\mu}{-3+\mu}$
PD_{2b}	$\frac{-7+3\mu}{-1+\mu}$
PD_{3a}	$\frac{-13-6\mu+3\mu^2+4\sqrt{4+2\mu-\mu^2}}{-3-2\mu+\mu^2}$
PD_{3b}	$\frac{13+6\mu-3\mu^2+4\sqrt{4+2\mu-\mu^2}}{3+2\mu-\mu^2}$

these are in fact BPPD (branch point period doubling) points, similar to what Khoshsiar Ghaziani, Govaerts, and Sonck (2012) find in a different system, and in contrast to what Govaerts and Khoshsiar Ghaziani (2008) state from a numerical investigation of this same Kopel system. The identification as BPPD directly results from our finding that there are no limit points of fixed points in the system. In the numerical section we will see that MatcontM identifies these points as LPPD, as BPPD are not generic on PD curves. Of course, MatcontM is not able to compute reasonable LP curves from them. Note that both BPPD are located at $\rho = 1$, that is on the boundary of the region of economic interest. With our expressions for the PD normal form coefficients, it is clear that PD curve $1b$ changes from sub- to supercritical at $(\mu, \rho) = (1, 1)$. Interestingly, no such behavior is observed for $(\mu, \rho) = (3, 1)$.

10.4.2 Strong resonances

Further, we investigate if strong resonances can be found. Strong 1:2 resonance points R2 lie on PD curves and require $\lambda_{1,2} = -1$, which we can solve from Table 10.4. We find an R2 point at $(\mu, \rho) = (1 + \sqrt{5}, 2)$ where the NS curve intersects the PD curve belonging to fixed points 3 and 4. At that point, the PD bifurcation switches from sub- ($\rho < 2$, $3a$) to supercritical ($\rho > 2$, $3b$).

For a strong 1:3 resonance point R3 on an NS curve we have the condition

$$\theta_0 = \frac{2\pi}{3}, \quad \mu_{1,2} = \mathrm{e}^{\pm i\theta_0}, \tag{10.16}$$

where $\mathrm{e}^{i\theta} = u + iv$ has been determined previously. So we compute

$$\theta_0 = \frac{2\pi}{3}, \tag{10.17}$$

$$\arctan\left(\frac{v}{u}\right) = \frac{2\pi}{3}, \tag{10.18}$$

$$\frac{v}{u} = \tan\left(\frac{2\pi}{3}\right) = -\sqrt{3}, \tag{10.19}$$

$$\frac{2\sqrt{-4-2\mu+\mu^2}}{-3-2\mu+\mu^2} = -\sqrt{3}, \tag{10.20}$$

which reduces to $\mu = 1 + \frac{4}{\sqrt{3}}$. Hence we have one R3 point at $(\mu, \rho) = \left(1 + \frac{4}{\sqrt{3}}, \frac{3}{2}\right)$.

Strong 1:4 resonance points R4 are found from the condition

$$\theta_0 = \frac{\pi}{2}, \quad \lambda_{1,2} = e^{\pm i\theta_0} = \pm i. \tag{10.21}$$

We cannot use the same procedure as before because $\tan(\theta)$ has a singularity at $\pi/2$, so we use the fact that the multipliers are purely imaginary. We compute the multipliers along the NS curve as

$$\lambda_{1,2} = \frac{-5-2\mu+\mu^2 \pm 2\sqrt{4+2\mu-\mu^2}}{-3-2\mu+\mu^2}. \tag{10.22}$$

For $\mu > 1 + \sqrt{5}$ the square root is imaginary and $\lambda_{1,2}$ are complex conjugates as required for NS. So for the multipliers to be purely imaginary, we have to require that the real part vanishes, that is

$$-5 - 2\mu + \mu^2 = 0, \tag{10.23}$$

which is easily solved for $\mu = 1 + \sqrt{6}$. Hence there is an R4 point at $(\mu, \rho) = \left(1 + \sqrt{6}, 1\right)$.

10.4.3 Further bifurcations

The analytical form of the NS normal form coefficient enables us to prove that there are no Chenciner bifurcations on the NS curve, as this would require

$$\mu_{1,2} = e^{\pm i\theta_0}, \quad L_1 = 0 \tag{10.24}$$

and it is trivial to show that $L_1 < 0 \; \forall \, \mu > 1 + \sqrt{5}$. It is likewise straightforward to show that there are no generalized period doubling (GPD) points as this would require a PD normal form coefficient $c_0 = 0$, which is not possible for $\mu > 0$.

10.5 Codim 2 normal form coefficients

Following the procedures and expressions given in Chapter 4, we proceed for the first time to determine analytically the normal form coefficients of the found codim 2 points, that is the strong resonances and BPPD points.

10.5.1 Resonance 1:2

We start with the R2 point at $(\mu, \rho) = (1 + \sqrt{5}, 2)$ and determine the generalized eigenvectors of the Jacobian matrix as

$$q_0 = \begin{pmatrix} 1 \\ 0 \end{pmatrix}, \quad q_1 = -\frac{1}{4}\begin{pmatrix} \sqrt{15} \\ 1 \end{pmatrix}, \tag{10.25}$$

$$p_0 = \begin{pmatrix} 0 \\ -4 \end{pmatrix}, \quad p_1 = \begin{pmatrix} 1 \\ -\sqrt{15} \end{pmatrix}, \tag{10.26}$$

where we have chosen the additional normalization of $\langle q_0, q_0 \rangle = \langle q_1, q_1 \rangle = 1$ in order to uniquely define the vectors. It is then straightforward to determine the critical normal form coefficients

$$c_0 = 64(3 + \sqrt{5}) \approx 335.11, \tag{10.27}$$

$$d_0 = 128(3 + \sqrt{5}) \approx 670.22 \tag{10.28}$$

in the normal form (3.33). This agrees with the numerical values computed in MatcontM.

10.5.2 Resonance 1:3

For R3 at $(\mu, \rho) = (1 + \frac{4}{\sqrt{3}}, \frac{3}{2})$, we determine the eigenvectors as

$$q = \frac{1}{2\sqrt{2 + \sqrt{3}}}\begin{pmatrix} -2 - \sqrt{3} \\ i \end{pmatrix}, \quad p = \sqrt{2 + \sqrt{3}}\begin{pmatrix} -2 + \sqrt{3} \\ i \end{pmatrix}, \tag{10.29}$$

fulfilling $\langle p, q \rangle = \langle q, q \rangle = 1$. Via the intermediate variables b_0, c_0 (the latter requires the use of a bordered system, as described in Section 4.2) we arrive at the normal form coefficient

$$a_2(0) = \frac{\Re(e^{-2i\pi/3}c_0) - |b_0|^2}{|b_0|^2} = -\frac{4}{3}, \tag{10.30}$$

which appears in Theorem 3.25. This result agrees with the value returned by MatcontM.

10.5.3 Resonance 1:4

We proceed likewise for the R4 point at $(\mu, \rho) = (1 + \sqrt{6}, 1)$, where we determine the eigenvectors as

$$q = \frac{\sqrt{2+\sqrt{2}}}{2}\begin{pmatrix} -1 \\ \frac{i}{1+\sqrt{2}} \end{pmatrix}, \quad p = \frac{\sqrt{2+\sqrt{2}}}{2}\begin{pmatrix} -2+\sqrt{2} \\ i\sqrt{2} \end{pmatrix} \tag{10.31}$$

with $\langle p, q\rangle = \langle q, q\rangle = 1$. Via c_0, d_0 we find the critical coefficient (3.55) for the normal form (3.54),

$$A_0 = -\frac{ic_0}{|d_0|} = -1, \tag{10.32}$$

which agrees with the numerical result obtained from MatcontM. Notice the 1:4 resonance is therefore degenerate.

10.5.4 BPPD

For the two BPPD points we apply the algorithm given in Section 4.2 for an LPPD point. We start with the point at $(\mu, \rho) = (1, 1)$ and find the eigenvectors

$$q_1 = \frac{1}{\sqrt{2}}\begin{pmatrix} -1 \\ -1 \end{pmatrix}, \quad p_1 = \frac{1}{\sqrt{2}}\begin{pmatrix} -1 \\ -1 \end{pmatrix}, \tag{10.33}$$

$$q_2 = \frac{1}{\sqrt{2}}\begin{pmatrix} 1 \\ -1 \end{pmatrix}, \quad p_2 = \frac{1}{\sqrt{2}}\begin{pmatrix} 1 \\ -1 \end{pmatrix}, \tag{10.34}$$

where we have also normalized q as before. We find the intermediate constants $a_1 = \sqrt{2}, e_1 = -\sqrt{2}, b_1 = \sqrt{2}$ and use the bordering technique to determine the vectors h_{20}, h_{11}, h_{02} and constants $c_i, i = 1 \ldots 4$ that all turn out to be equal to zero. We arrive at the normal form coefficients

$$a_0 = -\frac{1}{2}, b_0 = -1, c_0 = 0, d_0 = 0. \tag{10.35}$$

We conclude that this BPPD is non-degenerate as $a_0 b_0 \neq 0$.

In the same manner we proceed for the second BPPD at $(\mu, \rho) = (3, 1)$ with the eigenvectors

$$q_1 = \frac{1}{\sqrt{2}}\begin{pmatrix} -1 \\ 1 \end{pmatrix}, \quad p_1 = \frac{1}{\sqrt{2}}\begin{pmatrix} -1 \\ 1 \end{pmatrix}, \tag{10.36}$$

$$q_2 = \frac{1}{\sqrt{2}}\begin{pmatrix} 1 \\ 1 \end{pmatrix}, \quad p_2 = \frac{1}{\sqrt{2}}\begin{pmatrix} 1 \\ 1 \end{pmatrix}. \tag{10.37}$$

However, here it turns out that already $a_1 = e_1 = b_1 = 0$, while $c_1 = c_2 = -c_3 = -c_4 = 27$. Hence we arrive at $b_0 = 0$ and conclude that this BPPD is more degenerate.

10.6 Numerical analysis using MatcontM

10.6.1 Higher-order bifurcation curves

Using the numerical continuation software MatcontM, we extend the analysis to bifurcations of higher iterates. We focus on the region to the right of the R4 point, where we discover rich behavior. It also partially extends to $\rho \leq 1$ and is therefore also interesting economically.

In Figure 10.3(a) we see on the left the BPPD point at $(\mu, \rho) = (3, 1)$ and the stability region of E_3, E_4 to the right of it. We find a large number of bifurcation curves of higher iterates. The vast majority of numerically determined curves refer to bifurcations of the fixed points E_3, E_4, and the prominent role of R4 is clear. In addition we also find PD curves of E_2 – for these, $x = y$ holds. Shown as dashed lines are PD^2, PD^4, PD^8 and PD^{16}.[1] They stem from a period-doubling cascade of the PD curve $2b$ and were also identified by Govaerts and Khoshsiar Ghaziani (2008). These bifurcations lead to k-cycles and eventually chaos (shown in the following). The blue PD curve crossing the NS^1 curve at R4 is a PD^2 curve, the red curve to its right an NS^4. Note how in particular the pairs of PD and NS curves of the same iterate lead to regions of multistability of period-1, -2, -4 and other cycles. In order to investigate the economically interesting lower region we zoom into the region right of R4 in the panel (b) and Figure 10.4.

At the R4 point two LP^4 originate (red and gray, colors refer to Figure 10.4). They both intersect the same PD^4 curve (green), extending between them in a "butterfly" shape, at two LPPD points. Therefore there exists a 4-cycle in the area enclosed by the three curves. It is most interesting that a kind of two-dimensional period-doubling cascade can be observed. A family of LP curves of various iterates extends from close to the center of the butterfly to the lower right. In consecutive LPPD points they are touched by PD curves of the eighth (light blue), sixteenth (red) and thirty-second (violet) iterate, where the distances progressively decrease (bottom-right panel in Figure 10.3). Note that the cascade takes place for $\rho < 1$. Economically speaking, when the other firm's profit generates large positive externalities (large μ) and is weighted

[1] From now on superscripts refer to the period of the iterate, i.e., PD^2 is a period-doubling of the second iterate.

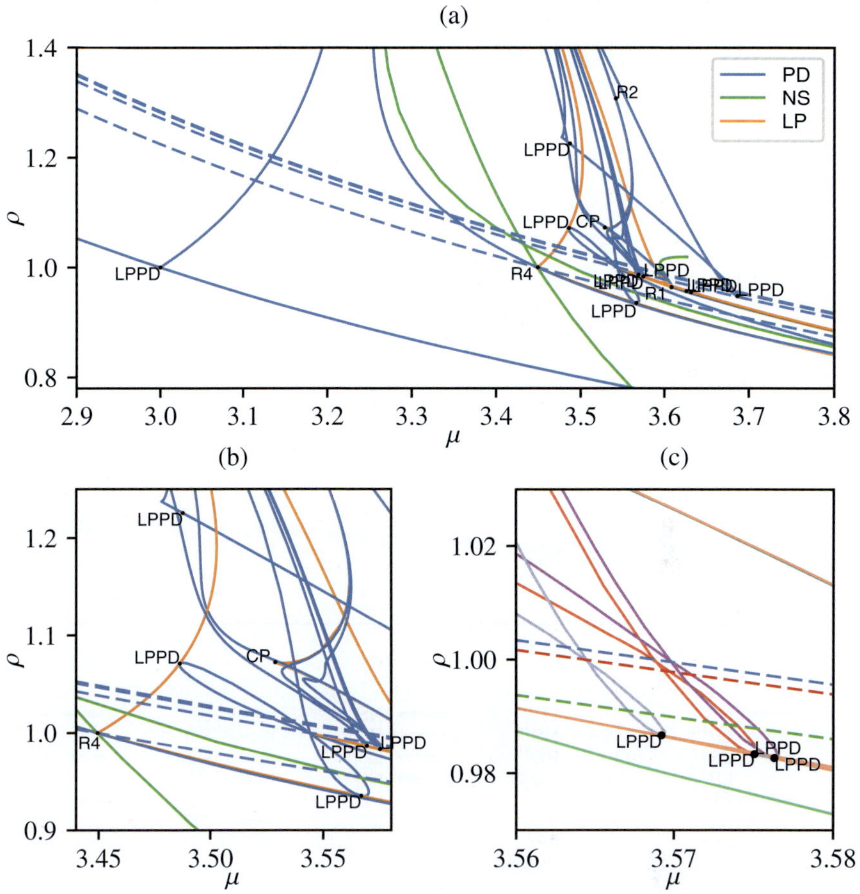

Figure 10.3 Codim 1 curves and codim 2 points determined with MatcontM. PD and BP (blue), NS (green) and LP (orange) of various k-cycles. Solid lines give analytically found curves and bifurcations of E_3. Bifurcations of E_2 (i.e., $x = y$) shown as dashed lines. (a) All found curves for $(\mu, \rho) \in [2.9, 3.8] \times [0.78, 1.4]$. The focus is on the stability region of $E_{3,4}$ between the upward sloping PD curve $3a$ (left) and the downward sloping NS curve (center) with the R4 point. (b) Zoom into the region right of R4. (c) Zoom into the lower right region of the bottom-left plot. Here colors are – for easier identification – the same as in Figure 10.4.

heavily (ρ close to 1) the dynamics of the duopoly may change drastically for small parameter changes. The higher PD curves form rather complicated shapes, though none shows a butterfly-like shape of PD^4. Several curves show strong resonances and further LPPD points. Notable is an LPPD point of PD^{16}

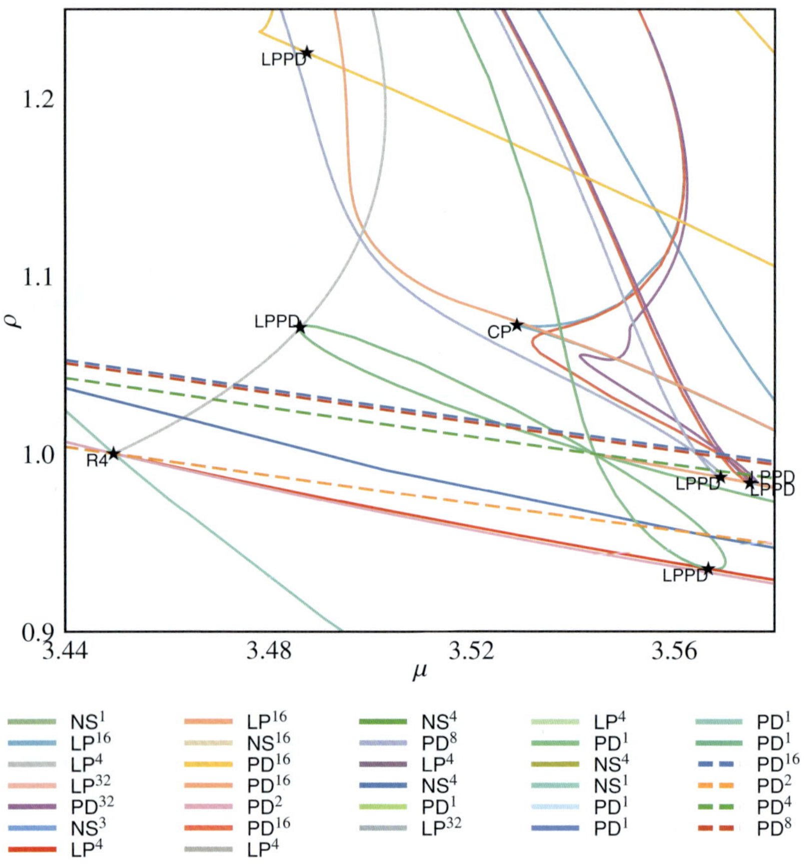

Figure 10.4 Codim 1 curves and codim 2 points determined with MatcontM. Coloring is the same as in the panel (c) of Figure 10.3. Dashed lines show PD curves of E_2.

(red) at $(\mu, \rho) = (3.54199, 1.30731)$. The LP^{16} curve (cyan) that originates there has a cusp point at $(\mu, \rho) = (3.60936, 0.96426)$. It is clear that the regions shown contain a very complex pattern of coexisting stable and unstable cycles of various iterates.

10.6.2 Chaos in the Kopel Map

Due to the limitations in understanding highly complex bifurcation diagrams, we now take a step back and investigate the dynamics more globally. The presence of chaos is suggested, for instance, by the observed period-doubling cascades. Chaos was previously studied in this map, among others, by Agiza (1999). We compute the largest and smallest Lyapunov exponents $\Lambda_{max}, \Lambda_{min}$ as a function of both parameters.[2] Due to possible coexistence of attractors we do so for 200 initial conditions (x_i, y_i), randomly drawn from a uniform distribution between 0 and 1 per (μ, ρ)-parameter point and average the results. This procedure at the same time allows us to get a numerical estimate of the basin size, i.e., the number of initial conditions that converge to an attractor. We first show results for the parameter space $(\mu, \rho) \in [1, 5] \times [0, 3]$ (Figure 10.5) and then move to our primary region of interest.

In the top panel of Figure 10.5 we clearly identify the stability regions of fixed points $E_2, E_{3,4}$ and note how they are delimited by the identified bifurcation curves of the first iterate. We clearly identify super- and subcritical period-doublings and note the region with $\Lambda_{max} \approx 0$ to the right of NS^1, indicative of invariant curves and as expected from a supercritical NS bifurcation. In panel (b) we limit the dynamics to the diagonal $x = y$ and thereby include only bifurcations of E_2, while in panel (c) all orbits with $x = y$ are excluded. Comparing panel (a) with (b) and (c) we can single out regions being dominated by either of the two bifurcation sequences. There are essentially two regions of chaotic behavior in the system, one above the PD curve 2b of E_2, likely driven by the period-doubling cascade mentioned earlier and one to the right of the NS curve where E_3, E_4 lose stability via NS or PD. Interestingly, note how it is possible to yield dynamics with $x \neq y$ for $\mu < 3$ via the period-doubling cascade of E_2, as already found by Govaerts and Khoshsiar Ghaziani (2008). This "tongue" in panel (c) does therefore not stem from $E_{3,4}$.

We now again move to the region to the right of R4 where we observe a lot of interesting behavior in the Lyapunov charts (Figure 10.6). Generally, chaos tends to be more likely and stronger for larger ρ. Somewhat surprisingly, most found bifurcation curves are not very successful in identifying "interesting" patterns in the Lyapunov chart. Exceptions are, for instance, the LPPD point at the intersection of the upper LP^4 and PD^4 curves at the boundary of a large stable window (left panel of 10.6) and the family of LP curves starting at the center of the butterfly that run parallel to the approximate boundary of the stable region. However, there remain a great many features that one would expect to be explainable in terms of bifurcation diagrams, such as the strong stability point at $(\mu, \rho) = (3.50, 1.00)$ and the stability windows in the chaotic

[2] We follow an algorithm on http://csc.ucdavis.edu/~chaos/courses/nlp/ by Jim Crutchfield and implemented in Python. Consequently, all exponents here are given in base-2 logs. Apart from this scaling factor, the results obtained are identical within numerical accuracy to those obtained in MatcontM.

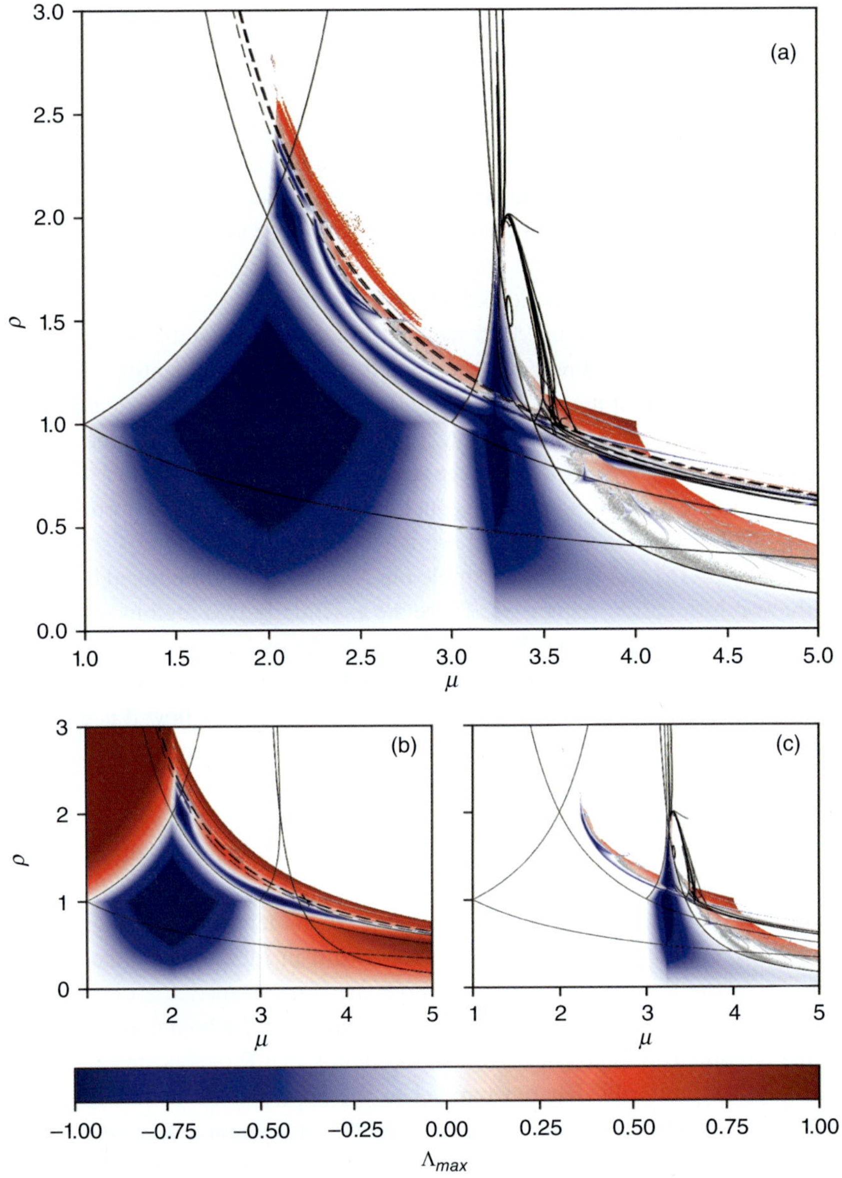

Figure 10.5 Λ_{max} for $\mu, \rho \in [1,5] \times [0,3]$ with bifurcation curves from before. (a) All attractors. (b) Attractors with $x = y$. (c) Attractors with $x \neq y$. Solid lines: codim 1 curves of 1-cycles and higher cycle of E_3. Dashed lines: codim 1 curves of higher cycles of E_2.

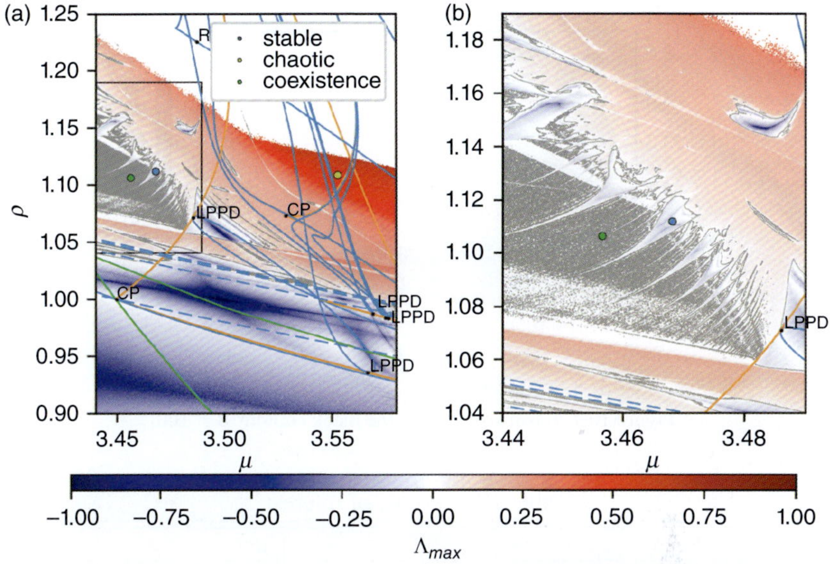

Figure 10.6 Λ_{max}. Codim 1 curves labeled as before: PD (blue), NS (green), LP (orange). Zoomed region (b) is shown as the black rectangle in (a). The marked points are discussed in the text.

regime. Also it would be very interesting to explain what exactly happens at the sharp boundary where all attractors lose stability. However, this problem remains so far unsolved. To the left of the mentioned LPPD point we see stable windows in the chaotic region that repeat in an almost fractal way to the lower right. We consider some points here as case studies in the next subsection.

The scan for Lyapunov exponents also gives us a good indication of the basin size B, defined as the percentage of initial conditions that converge to some attractor and do not diverge to infinity (Figure 10.7).

This quantity appears quite independent of the details of chaos. For $\rho < 1$ no orbits go to infinity. However, above a certain line B rapidly increases with ρ so that initial conditions in the upper part of the $E_{3,4}$ stability region have a high chance of diverging to infinity. Besides the economic reasoning, this further justifies paying less attention to regions with ρ substantially larger than 1.

10.6.3 Bistability and coexistence

In the previous section we have shown rather complicated Lyapunov charts. Parts of their behavior may be due to the coexistence of several attractors with

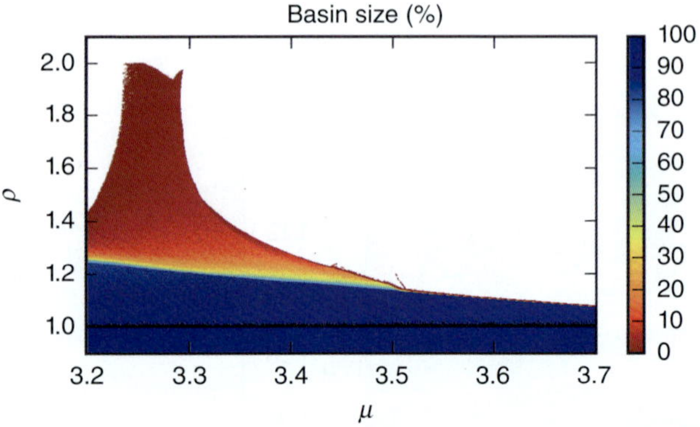

Figure 10.7 Basin size. Below the black contour $B = 100$.

Table 10.5 Points of interest.

	Point	μ	ρ
1	Fully chaotic region	3.55298	1.10839
2	Large ρ, small basin size close to blow-up	3.24	1.81
3	Shaded region in between stable windows	3.45653	1.1064
4	Stable window	3.46824	1.11197

different Λ_{max}. Therefore, we investigate four points in more detail that are given in Table 10.5 and Figure 10.6.

Stable orbits in chaotic domain. There are regions in the (μ, ρ)-space where almost all initial conditions are chaotic. Choose, for instance, $(\mu, \rho) = (3.55298,\ 1.10839)$ where the $\Lambda_{max} > 0$ for almost any initial condition $x_0, y_0 \in [0, 1] \times [0, 1]$. We look at the isolated points with $\Lambda_{max} < 0$ in detail and find that eventually all of these points converge to a stable 3-cycle on $x = y$ (Figure 10.8). The transient toward the cycle occurs over a highly complex object that corresponds to the attractor traced by the aperiodic orbits for the same parameter values. Hence we observe coexistence between a 3-cycle related to E_2 and a symmetric chaotic attractor.

Region close to blow-up. When increasing ρ the attractors become less and less stable in the sense that a much larger fraction of orbits disappears to infinity. As an example, consider $\mu, \rho = 3.24, 1.81$ where all Λ_{max} cluster in a very narrow range around $\Lambda_{max} = -0.243$ but most (>99.8%) of the initial conditions diverge, resulting in a particular basin structure (Figure 10.9). Here the only attractors are the 1-cycle E_3 and E_4.

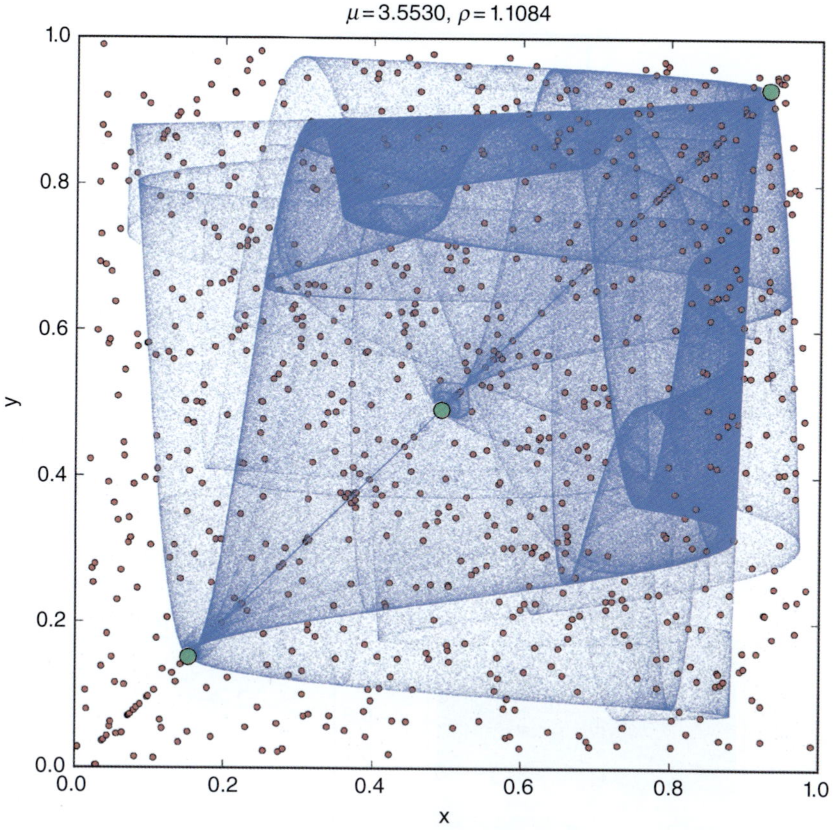

Figure 10.8 Stable cycles. Red: initial conditions with $\Lambda_{max} < 0$; blue: orbits tracing the attractor; green: stable 3-cycle.

Coexistence of stability and chaos. In the shaded regions (Figure 10.6) the structure of Λ_{max} in the x, y-plane changes markedly. There appears a two-dimensional fractal structure where all initial conditions with $\Lambda_{max} > 0$ converge either to (a chaotic attractor on) the diagonal $x = y$ if $x_0 = y_0$ or to a closed invariant curve around E_3, E_4. The initial conditions with $\Lambda_{max} < 0$ also converge to the invariant curves. We expect eventually a convergence to an n-cycle. However, since Λ_{max} is only very slightly negative in these cases, convergence was not observed within thousands of iterations.

Windows of stability. When μ, ρ are slightly varied such that (μ, ρ) enters a window of stability (left panel in Figure 10.6), the invariant curve turns into an

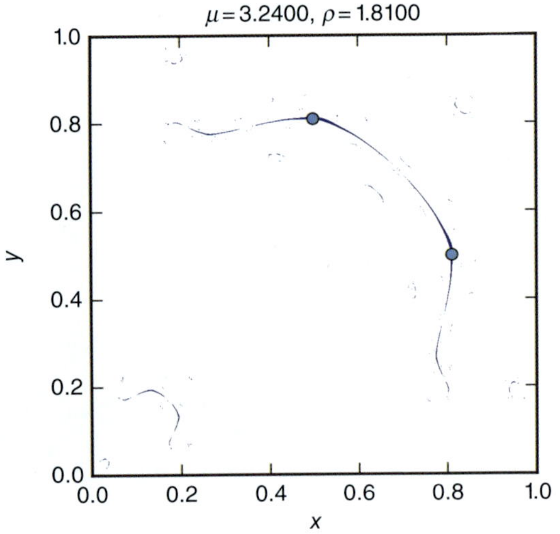

Figure 10.9 Λ_{max}. Blue: $\Lambda_{max} < 0$. E_3, E_4 indicated by blue dots.

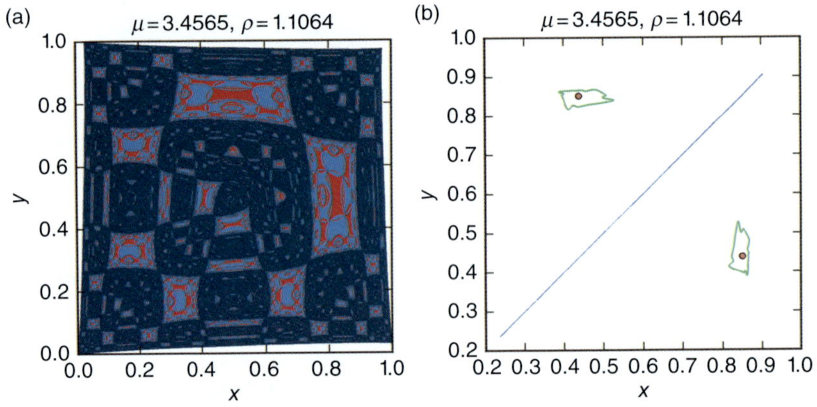

Figure 10.10 Coexistence. (a) Λ_{max}. Gray lines give $\Lambda_{max} = 0$. Blue: $\Lambda_{max} < 0$, red: $\Lambda_{max} > 0$. (b) Attractor on the diagonal (blue) and on the invariant curves (green) around E_3, E_4 (red dots).

n-cycle, in this example a 19-cycle. Note how the shape is qualitatively similar to the invariant curve in Figure 10.10(b). The dynamics are still chaotic on the diagonal, but the attractor only covers parts of the diagonal.

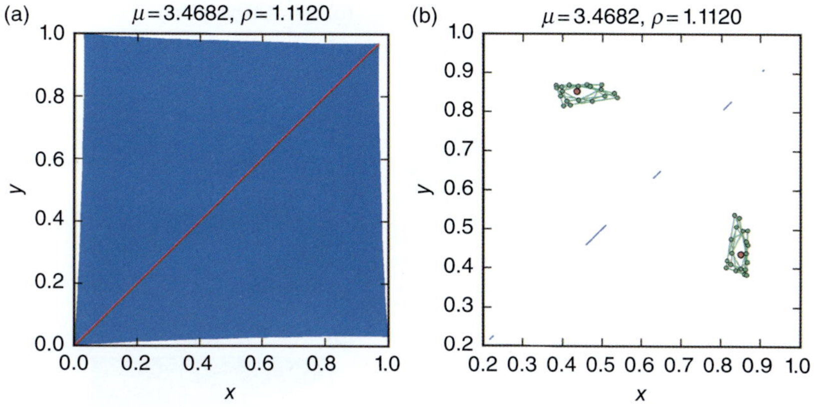

Figure 10.11 Windows of stability. (a) Λ_{max}. Blue: $\Lambda_{max} < 0$, red: $\Lambda_{max} > 0$. (b) Attractor. Chaotic on the diagonal (blue) and 19-cycle (green) around E_3, E_4 (red dots).

10.6.4 Regime identification

Summing up, we can identify four regimes:

1. all orbits are stable and converge to an n-cycle;
2. all orbits are chaotic;
3. both stable n-cycles and chaotic orbits occur;
4. no stable attractors exist and all orbits diverge to infinity.

Figure 10.12 shows these regimes to the right of R4: (a) Chaos exists above the $\mathsf{NS}^{(1)}$ curve and above the E_2 period-doubling cascade. All orbits are only chaotic in a rather small region below the blow-up line. (b) Stability windows exist in the chaotic region. (c) Coexistence is a common feature over large parts of the domain here investigated. This is likely largely due to structures stemming from E_2 and E_3, E_4, respectively, as illustrated. (d) The basin of attraction of stable and chaotic attractors decreases strongly with ρ. Beyond $\rho = 2$ they are virtually non-existent for $\mu > 3$.

It becomes clear that even Lyapunov charts are somewhat limited in their explanatory power as they only describe the "average" behavior at a parameter point and do not *per se* inform on the existence and properties of multiple attractors. As an example we show the attractors for $(\mu, \rho) = (3.5, 1)$, a point with strongly negative Λ_{max}. At this point there is no chaos and several highly stable 4-cycles coexist. They appear from bifurcations of E_2 and $E_{3,4}$.

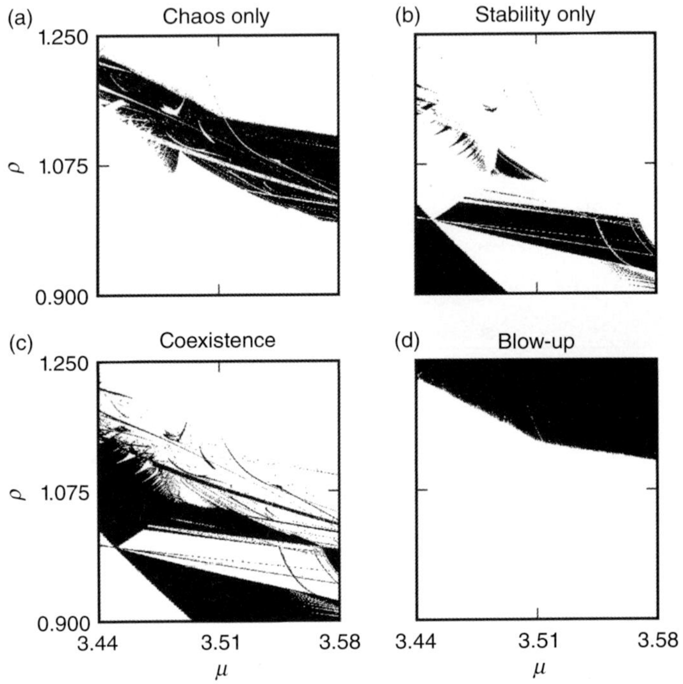

Figure 10.12 Stability properties of attractors in $(\mu, \rho) \in [0, 5] \times [0, 3]$, shown in black.

10.7 Conclusions

In this chapter we contributed to the study of Kopel's Duopoly Model. Partially in an extension of Govaerts and Khoshsiar Ghaziani (2008) we were able to analytically compute the normal form coefficient of the $\mathsf{NS}^{(1)}$ point and prove its unconditional supercriticality. In addition to the normal forms of all $\mathsf{PD}^{(1)}$ points, we also computed for the first time parameter values and normal form coefficients for all codim 2 points of the first iterate analytically. We have shown, similar to Khoshsiar Ghaziani, Govaerts, and Sonck (2012), that the LPPD points found by Govaerts and Khoshsiar Ghaziani (2008) are actually BPPD points. We then used MatcontM to compute a large number of bifurcation curves of higher n-cycles, extending our knowledge of the complex behavior inside and outside the economically interesting region $\rho < 1$. Some features, such as accumulating PD points, already suggested the presence of chaos, which we then extensively investigated by numerically computing Lyapunov exponents of the system as a function of ρ and μ. We were able to iden-

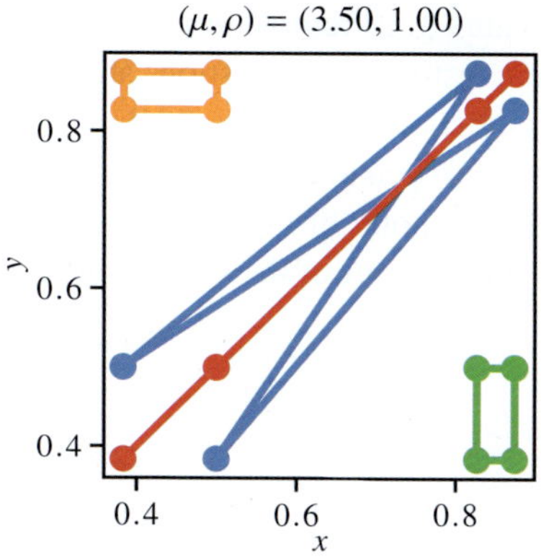

Figure 10.13 Attracting cycles at $(\mu, \rho) = (3.5, 1)$.

tify and study in detail several interesting parameter regions. We were able to show patterns of coexistence of chaos and stable n-cycles and complex-shaped basins of attraction of periodic and chaotic attractors. Our results suggest that the coexistence strongly relates to interacting dynamics of E_2 and E_3, E_4. Extensions of this work could include, among others, the identification of the various borders in the Lyapunov charts that have not yet been attributed to bifurcations of the map.

We believe our results to be useful both theoretically for our understanding of iterated maps and perhaps even from an economic point of view. We have shown the usefulness of the combination of analytical and numerical techniques to the understanding of a system. As already mentioned by Bischi and Kopel (2001), the chosen nonlinear reaction functions introduce multiple Nash equilibria (the fixed points) and the model hence features path dependence. We have demonstrated extensively that the behavior of these coexisting equilibria may be drastically different as there may be coexistence of fixed points, stable cycles and chaotic attractors. For managing an economic system functioning according to such a model it would clearly be important to know parameter values where such complex dynamics occur and if possible steer clear of them, in particular of regions with a large chance of chaotic behavior. Charts such as our Figure 10.12 may then prove useful.

Nevertheless, substantial gaps remain in our knowledge of the Kopel Map. In our view the most interesting open questions are the interpretation of certain borders in the Lyapunov charts and in particular the blow-up line where all attractors are abruptly destroyed and for certain μ the basin shrinks extremely quickly. From an analytical point of view it would moreover be interesting to extend the analysis to bifurcations of the second and higher iterates.

11

The SEIR Epidemic Model

11.1 The model

The SEIR Epidemic Model describes the spread of a non-lethal disease in a large population, which is divided into four classes: susceptible (S), exposed (E), infective (I) and recovered (R). We briefly introduce the model (see Diekmann and Heesterbeek (2000) for details and references). New susceptibles are "born" with the growth rate μ; β is the contact rate between susceptibles and infectives. The exposed become infective with the rate α and the infectives recover with the rate γ. This gives

$$\begin{cases} \dot{S} = \mu - \mu S - \beta S I, \\ \dot{E} = \beta S I - (\mu + \alpha) E, \\ \dot{I} = \alpha E - (\mu + \gamma) I, \end{cases} \tag{11.1}$$

and $R = 1 - S - E - I$. In Kuznetsov and Piccardi (1994), effects of a seasonal variation of the contact rate with other parameters constant

$$\beta = \beta_0(1 + \delta \cos(2\pi t)),$$

were studied numerically by considering the Poincaré map

$$P\colon (S(0), E(0), I(0)) \mapsto (S(1), E(1), I(1)). \tag{11.2}$$

Instead of the original SEIR model (11.1), we used an equivalent system for $s = \ln S, e = \ln E$ and $i = \ln I$. When the variables are small, integrating this equivalent system is numerically more stable. To implement this into MatContM we use the code as in Figure 11.1.

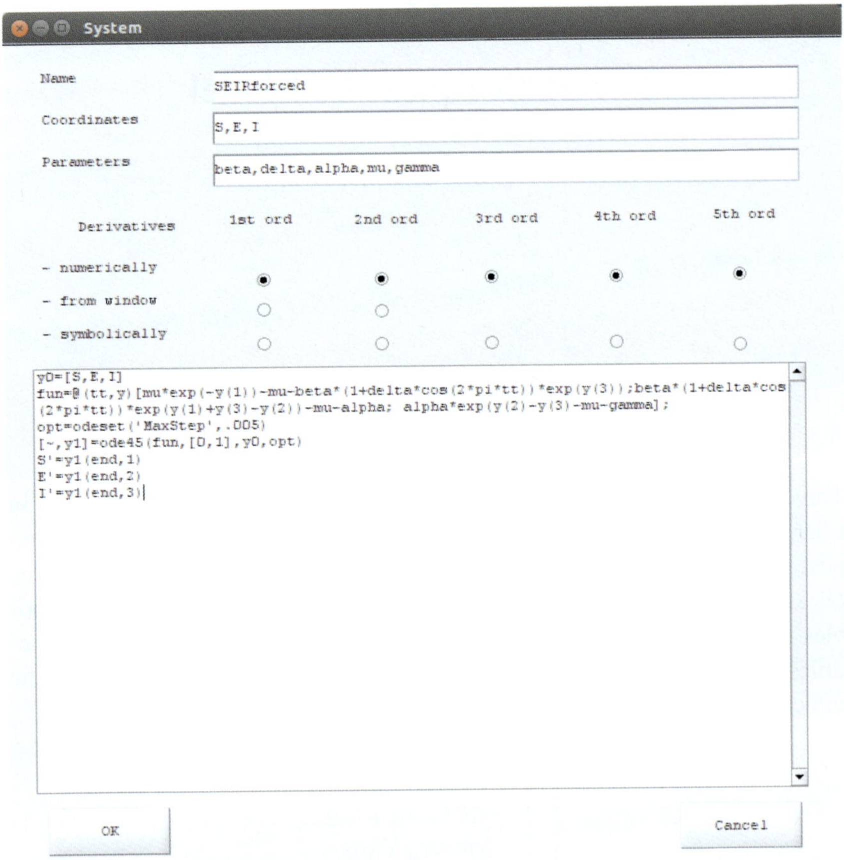

Figure 11.1 Implementing numerical integration over one period of a periodically forced ODE.

11.2 Bifurcation diagram

For measles, the characteristic parameters are $\mu = 0.02$, $\alpha = 35.842$, and $\gamma = 100$. A bifurcation diagram in two parameters (β, the mean value, and δ, the degree of seasonality) was obtained by Kuznetsov and Piccardi (1994). Several codim 2 bifurcations, namely cusps and generalized period-doublings, were found.

We used MatcontM to recompute the codim 1 bifurcation curves LP^k and PD^k of period-k cycles with $k \leq 3$, obtained by Kuznetsov and Piccardi (1994), and to locate the codim 2 points (see Figure 11.2 and Table 11.1). Note that there are two additional cusp points on the LP^3 branch that were overlooked

Table 11.1 The codim 2 points in the periodically forced SEIR model.

Point	Parameters	Coordinates
CP_1	$\delta \approx 0.5327$ $\beta_0 \approx 5928$	$S \approx 0.02229$ $E \approx 0.1887 \times 10^{-7}$ $I \approx 0.5499 \times 10^{-8}$
GPD_1	$\delta \approx 0.03815$ $\beta_0 \approx 2015$	$S \approx 0.05029$ $E \approx 5.3301 \times 10^{-4}$ $I \approx 1.8846 \times 10^{-4}$
GPD_2	$\delta \approx 0.1328$ $\beta_0 \approx 3019$	$S \approx 0.03566$ $E \approx 4.9463 \times 10^{-4}$ $I \approx 1.6794 \times 10^{-4}$
CP_2	$\delta \approx 0.4478$ $\beta_0 \approx 3262$	$S \approx 0.02755$ $E \approx 0.1758 \times 10^{-7}$ $I \approx 0.5868 \times 10^{-8}$
CP_3	$\delta \approx 0.6767$ $\beta_0 \approx 3998$	$S \approx 0.02224$ $E \approx 0.1791 \times 10^{-9}$ $I \approx 0.5748 \times 10^{-10}$

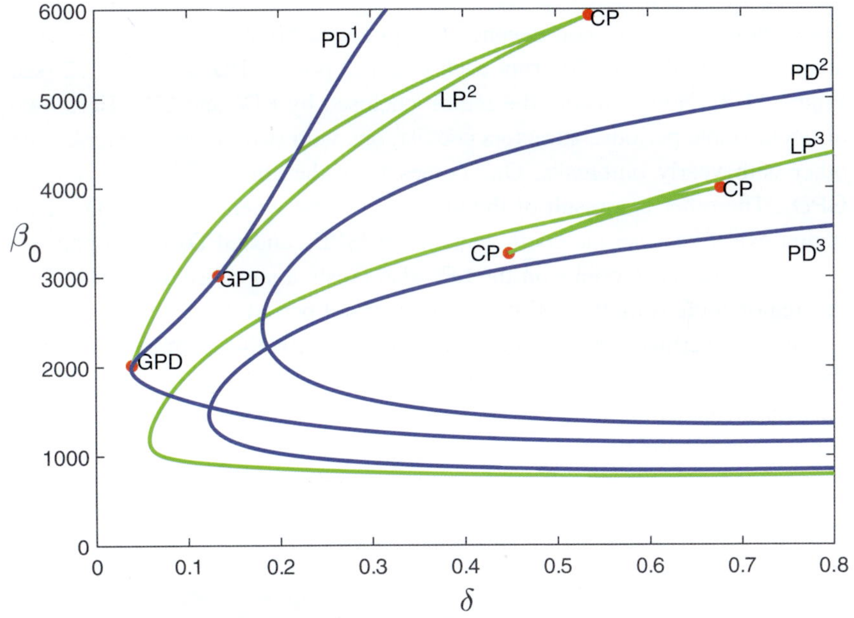

Figure 11.2 Bifurcation diagram of the SEIR model. Labels LP^k (PD^k) denote fold (flip) bifurcation of period-k cycles. Labels CP and GPD corresponds to the cusp and generalized period-doubling points, respectively. These codim 2 points are analyzed in this section.

by Kuznetsov and Piccardi (1994). Automatic differentiation was switched off as it is incompatible with the numerical integration. For some points the maximum step size for integration was lowered for accuracy. Next MatcontM reported the normal form coefficients as computed using finite differences.

For the critical normal form coefficients (2.7) and (4.13) we have found

$$\begin{aligned} \mathsf{CP}_1: &\quad b_0 \approx -6.87 \times 10^{-7}, \quad c_0 \approx -0.224, \\ \mathsf{CP}_2: &\quad b_0 \approx -2.32 \times 10^{-8}, \quad c_0 \approx 0.2680, \\ \mathsf{CP}_3: &\quad b_0 \approx 4.30 \times 10^{-7}, \quad c_0 \approx -0.126. \end{aligned}$$

We see that the cusp points are indeed non-degenerate, so that precisely three cycles of period-2 or -3 exist for nearby parameter values.

For the generalized period-doubling points we have obtained using (2.12) and (4.19) the following values

$$\begin{aligned} \mathsf{GPD}_1: &\quad c_0 \approx 1.34 \times 10^{-6}, \quad d_0 \approx 0.0109, \\ \mathsf{GPD}_2: &\quad c_0 \approx 6.74 \times 10^{-6}, \quad d_0 \approx -0.0044. \end{aligned}$$

Thus, these points are non-degenerate and a fold bifurcation curve LP^2 emanates tangentially to PD^1 from each codim 2 point GPD_k for $k = 1, 2$ (see Figure 11.2). Now consider the region enclosed by PD^1 and LP^2. Here, two different stable period-2 attractors coexist, one with two-yearly outbreaks, the other with yearly outbreaks. One comes from the curve LP^2 originating in GPD_1. The other is a result of the period doubling when crossing PD^1. This region exists because the coefficients d_0 of GPD_1 and GPD_2 have opposing signs. The two cusp points on the LP^3 also create a so-called *swallowtail*. In this region there is multistability of two cycles of period-3.

Since we verified the non-degeneracy numerically, this implies that we indeed deal with codim 2 points. So near the codim 2 points $\mathsf{GPD}_{1,2}$ and CP no other bifurcations exist, and our description of the bifurcations of cycles with period ≤ 4 is complete.

References

Abraham, R., Gardini, L., and Mira, C. 1997. *Chaos in Discrete Dynamical Systems: A Visual Introduction in 2 Dimensions*. Springer-Verlag, New York.

Afraimovich, V. S., and Shilnikov, L. P. 1983. Invariant two-dimensional tori, their breakdown and stochasticity. Pages 3–26 of: *Methods of the Qualitative Theory of Differential Equations*. Gorky State University, Gorky (in Russian).

Agiza, H. N. 1999. On the analysis of stability, bifurcation, chaos and chaos control of Kopel map. *Chaos Solitons Fractals*, **10**(11), 1909–1916.

Al-Hdaibat, B., Govaerts, W., van Kekem, D. L., and Kuznetsov, Yu. A. 2018. Remarks on homoclinic structure in Bogdanov–Takens map. *J. Difference Equ. Appl.*, **24**(4), 575–587.

Allgower, E. L., and Georg, K. 1990. *Numerical Continuation Methods: An Introduction*. Springer-Verlag, Berlin.

Anderson, D. R., Myran, N. G., and White, D.L. 2005. Basins of attraction in a Cournot duopoly model of Kopel. *J. Difference Equ. Appl.*, **11**(10), 879–887.

Andronov, A. A., Leontovich, E. A., Gordon, I. I., and Maier, A. G. 1973. *Theory of Bifurcations of Dynamical Systems on a Plane*. Israel Program for Scientific Translations, Jerusalem.

Arnold, V. I. 1977. Loss of stability of self-induced oscillations near resonance, and versal deformations of equivariant vector fields. *Functional Anal. Appl*, **11**(2), 85–92.

Arnold, V. I. 1983. *Geometrical Methods in the Theory of Ordinary Differential Equations*. Springer-Verlag, New York.

Arnold, V. I., Afraimovich, V. S., Ilyashenko, Yu. S., and Shilnikov, L. P. 1994. *Bifurcation Theory, Dynamical Systems V.* Springer-Verlag, New York.

Aronson, D. G., Chory, M. A., Hall, G. R., and McGehee, R. P. 1982. Bifurcations from an invariant circle for two-parameter families of maps of the plane: a computer-assisted study. *Comm. Math. Phys.*, **83**(3), 303–354.

Arrowsmith, D. K., and Place, C. M. 1990. *An Introduction to Dynamical Systems*. Cambridge University Press, Cambridge.

Back, A., Guckenheimer, J., Myers, M., Wicklin, F., and Worfolk, P. 1992. DsTool: computer assisted exploration of dynamical systems. *Notices Amer. Math. Soc.*, **39**, 303–309.

Baesens, C., and MacKay, R. S. 2007. Resonances for weak coupling of the unfolding of a saddle-node periodic orbit with an oscillator. *Nonlinearity*, **20**(5), 1283–1298.

Baesens, C., Guckenheimer, J., Kim, S., and MacKay, R. S. 1991. Three coupled oscillators: mode-locking, global bifurcations and toroidal chaos. *Phys. D*, **49**(3), 387–475.

Balibrea, F., Oliveira, H. M., and Valverde, J. C. 2017. Topological equivalences for one-parameter bifurcations of maps. *J. Nonlinear Sci.*, **27**(2), 661–685.

Berezovskaya, F. S., and Khibnik, A. I. 1979. On the problem of bifurcations of auto-oscillation near resonance 1:4 (investigation of a model equation). Academy of Sciences of the USSR Scientific Centre of Biological Research, Pushchino (in Russian, with an English summary).

Berezovskaia, F. S., and Khibnik, A. I. 1980. On the bifurcation of separatrices in the problem of stability loss of auto-oscillations near 1:4 resonance. *J. Appl. Math. Mech.*, **44**(5), 938–943 (in Russian).

Beyn, W.-J., and Hüls, T. 2004. Error estimates for approximating non-hyperbolic heteroclinic orbits of maps. *Numer. Math.*, **99**(2), 289–323.

Beyn, W.-J., and Kleinkauf, J.-M. 1997. The numerical computation of homoclinic orbits for maps. *SIAM J. Numer. Anal.*, **34**(3), 1207–1236.

Beyn, W.-J., and Kless, W. 1998. Numerical Taylor expansions of invariant manifolds in large dynamical systems. *Numer. Math.*, **80**(1), 1–38.

Beyn, W.-J., Champneys, A., Doedel, E., Govaerts, W., Kuznetsov, Yu.A., and Sandstede, B. 2002. Numerical continuation, and computation of normal forms. Pages 149–219 of: *Handbook of Dynamical Systems, Vol. 2*. North-Holland, Amsterdam.

Beyn, W.-J., Hüls, T., Kleinkauf, J.-M., and Zou, Y. 2004. Numerical analysis of degenerate connecting orbits for maps. *Internat. J. Bifur. Chaos Appl. Sci. Engrg.*, **14**(10), 3385–3407.

Biragov, V. S. 1987. Bifurcations in a two-parameter family of conservative mappings that are close to the Hénon mapping. Pages 10–24 of: *Methods of the Qualitative Theory of Differential Equations* (in Russian). Gorky State University, Gorky. Translated in *Selecta Math. Soviet*, **9** (1990), 273–282.

Bischi, G. I., and Kopel, M. 2001. Equilibrium selection in a nonlinear duopoly game with adaptive expectations. *J. Econ. Behav. Organ.*, **46**(1), 73–100.

Bischi, G. I. and Tramontana, F. 2010. Three-dimensional discrete-time Lotka–Volterra models with an application to industrial clusters. *Commun. Nonlinear. Sci. Numer. Simul.*, **15**(10), 3000–3014

Bogdanov, R. I. 1975. Versal deformation of a singular point of a vector field on the plane in the case of zero eigenvalues. *Funkcional Anal. i Prilozen.*, **9**(2), 63.

Bogdanov, R. I. 1976a. Bifurcations of a limit cycle of a certain family of vector fields on the plane. *Trudy Sem. Petrovsk.*, **2**, 23–35 (in Russian). Translated in *Selecta Math. Soviet.* **1** (1981), 373–388.

Bogdanov, R. I. 1976b. The versal deformation of a singular point of a vector field on the plane in the case of zero eigenvalues. *Trudy Sem. Petrovsk.*, **2**, 37–65 (in Russian). Translated in *Selecta Math. Soviet.* **1** (1981), 389–421.

Borland, M., Emery, L., Shang, H., and Soliday, R. 2005. *User's Guide for SDDS Toolkit Version 1.30*. `www.aps.anl.gov/asd/oag/software.shtml`

Braaksma, B. L. J., and Broer, H. W. 1987. On a quasiperiodic Hopf bifurcation. *Ann. Inst. H. Poincaré Anal. Non Linéaire*, **4**(2), 115–168.

Brauer, F., and Castillo-Chávez, C. 2001. *Mathematical Models in Population Biology and Epidemiology*. Springer-Verlag, New York.

Broer, H. W., Huitema, G. B., Takens, F., and Braaksma, B. L. J. 1990. Unfoldings and bifurcations of quasi-periodic tori. *Mem. Amer. Math. Soc.*, **83**(421).

Broer, H. W., Roussarie, R., and Simó, C. 1996a. Invariant circles in the Bogdanov–Takens bifurcation for diffeomorphisms. *Ergodic Theory Dynam. Systems*, **16**(6), 1147–1172.

Broer, H. W., Huitema, G. B., and Sevryuk, M. B. 1996b. *Quasi-Periodic Motions in Families of Dynamical Systems Order Amidst Chaos*. Springer-Verlag, Berlin.

Broer, H. W., Simó, C., and Tatjer, J. C. 1998. Towards global models near homoclinic tangencies of dissipative diffeomorphisms. *Nonlinearity*, **11**(3), 667–770.

Broer, H. W., Takens, F., and Wagener, F. O. O. 1999. Integrable and non-integrable deformations of the skew Hopf bifurcation. *Regul. Chaotic Dyn.*, **4**(2), 16–43.

Broer, H. W., Simó, C., and Vitolo, R. 2002. Bifurcations and strange attractors in the Lorenz-84 climate model with seasonal forcing. *Nonlinearity*, **15**(4), 1205–1267.

Broer, H. W., Golubitsky, M., and Vegter, G. 2003. The geometry of resonance tongues: a singularity theory approach. *Nonlinearity*, **16**(4), 1511–1538.

Broer, H. W., Simó, C., and Vitolo, R. 2008a. Hopf saddle-node bifurcation for fixed points of 3D-diffeomorphisms: analysis of a resonance 'bubble'. *Phys. D*, **237**(13), 1773–1799.

Broer, H. W., Simó, C., and Vitolo, R. 2008b. The Hopf–saddle-node bifurcation for fixed points of 3D-diffeomorphisms: the Arnol'd resonance web. *B. Belg. Math. Soc.-Sim.*, **15**(5), 769–787.

Broer, H. W., and Holtman, S. J., Vegter, G., and Vitolo, R. 2011. Dynamics and geometry near resonant bifurcations *Regul. Chaotic Dyn.*, **16**(1–2), 39–50.

Bruschi, C. 2010. *Growing 1D stable and unstable manifolds of n-dimensional maps*. Master's project, Utrecht University.

Carcassès, J. P., and Kawakami, H. 1999. Existence of a cusp point on a fold bifurcation curve and stability of the associated fixed point: case of an n-dimensional map. *Internat. J. Bifur. Chaos Appl. Sci. Engrg.*, **9**(5), 875–894.

Champneys, A. R., Härterich, J., and Sandstede, B. 1996. A non-transverse homoclinic orbit to a saddle-node equilibrium. *Ergodic Theory Dynam. Systems*, **16**(3), 431–450.

Chenciner, A. 1985a. Bifurcations de points fixes elliptiques: I. Courbes invariantes. *Inst. Hautes Études Sci. Publ. Math.*, **61**, 67–127.

Chenciner, A. 1985b. Bifurcations de points fixes elliptiques: II. Orbites periodiques et ensembles de Cantor invariants. *Invent. Math.*, **80**(1), 81–106.

Chenciner, A. 1988. Bifurcations de points fixes elliptiques: III. Orbites périodiques de "petites" périodes et élimination résonnante des couples de courbes invariantes. *Inst. Hautes Études Sci. Publ. Math.*, **66**, 5–91.

Chenciner, A., Gasull, A., and Llibre, J. 1987. Une description complète du portrait de phase d'un modèle d'élimination résonante. *C. R. Acad. Sci. Paris Sér. I Math.*, **305**(13), 623–626.

Cheng, C. Q. 1990. Hopf bifurcations in nonautonomous systems at points of resonance. *Sci. China Ser. A*, **33**(2), 206–219.

Cheng, C. Q., and Sun, Y. S. 1992. Metamorphoses of phase portraits of vector field in the case of symmetry of order 4. *J. Differential Equations*, **95**(1), 130–139.

Chow, S.-N., Li, C. Z., and Wang, D. 1994. *Normal Forms and Bifurcation of Planar Vector Fields*. Cambridge University Press, Cambridge.

Cincotta, P. M., Giordano, C. M., and Simó, C. 2003. Phase space structure of multi-dimensional systems by means of the mean exponential growth factor of nearby orbits. *Phys. D*, **182**(3–4), 151–178.

Coullet, P. H., and Spiegel, E. A. 1983. Amplitude equations for systems with competing instabilities. *SIAM J. Appl. Math.*, **43**(4), 776–821.

Cournot, A. 1838. *Recherches sur les Principes Mathematiques de la Theorie des Richesses*. Irwin, Homewood, IL, 1963 (English translation).

Cushing, J. M., Henson, S. M., and Roeger, L.-I. 2007. Coexistence of competing juvenile–adult structured populations. *J. Biol. Dyn.*, **1**(2), 201–231.

Davydova, N. V. 2004. Old and young: can they coexist? PhD thesis, University of Utrecht.

Demmel, J. W., Dieci, L., and Friedman, M. J. 2000. Computing connecting orbits via an improved algorithm for continuing invariant subspaces. *SIAM J. Sci. Comput.*, **22**(1), 81–94.

Devaney, R. L. 2003. *An Introduction to Chaotic Dynamical Systems*, 2nd ed. Westview Press, Boulder, CO.

Dhooge, A., Govaerts, W., Kuznetsov, Yu. A., Mestrom, W., and Riet, A. M. 2003a. CL_MATCONT: a continuation toolbox in MATLAB. *Proceedings of the 2003 ACM symposium on applied computing*. ACM Press, pp. 161–166.

Dhooge, A., Govaerts, W., and Kuznetsov, Yu. A. 2003b. MATCONT: a MATLAB package for numerical bifurcation analysis of ODEs. *ACM Trans. Math. Software*, **29**(2), 141–164.

Dhooge, A., Govaerts, W., Kuznetsov, Yu. A., Meijer, H. G. E., and Sautois, B. 2008. New features of the software MatCont for bifurcation analysis of dynamical systems. *Math. Comput. Model. Dyn. Syst.*, **14**(2), 147–175.

Dieci, L., and Eirola, T. 1999. On smooth decompositions of matrices. *SIAM J. Matrix Anal. Appl.*, **20**(3), 800–819.

Diekmann, O., and Heesterbeek, J. A. P. 2000. *Mathematical Epidemiology of Infectious Diseases: Model Building, Analysis and Interpretation.* Wiley, Chichester.

Ding, W.-C., Xie, J. H., and Sun, Q. G. 2004. Interaction of Hopf and period doubling bifurcations of a vibro-impact system. *Journal of Sound and Vibration*, **275**(1), 27–45.

Doedel, E. J., Champneys, A. R., Fairgrieve, T. F., Kuznetsov, Yu. A., Sandstede, B., and Wang, X.-J. 1997–2000. *Continuation and Bifurcation Software for Ordinary Differential Equations (with HomCont), User's Guide.* `http://indy.cs.concordia.ca`

Edmunds, J., Cushing, J. M., Costantino, R.F., Henson, S. M., Dennis, B., and Desharnais, R. A. 2003. Park's Tribolium competition experiments: a non-equilibrium species coexistence hypothesis. *J. Anim. Ecol.*, **72**(5), 703–712.

Elphick, C., Tirapegui, E., Brachet, M. E., Coullet, P., and Iooss, G. 1987. A simple global characterization for normal forms of singular vector fields. *Phys. D*, **29**(1–2), 95–127.

England, J. P., Krauskopf, B., and Osinga, H. M. 2004. Computing one-dimensional stable manifolds and stable sets of planar maps without the inverse. *SIAM J. Appl. Dyn. Syst.*, **3**(2), 161–190.

Frouzakis, C. E., Adomaitis, R. A., and Kevrekidis, I. G. 1991. Resonance phenomena in an adaptively-controlled system. *Internat. J. Bifur. Chaos Appl. Sci. Engrg.*, **1**(1), 83–106.

Gaunersdorfer, A., Hommes, C. H. and Wagener, F. O. O. 2008. Bifurcation routes to volatility clustering under evolutionary learning. *J. Econ. Behav. Organ.*, **67**(1), 27–47, Working paper available at `https://ideas.repec.org/p/ams/ndfwpp/00-04.html`.

Gavrilov, N. K., and Shilnikov, L. P. 1972. Three-dimensional dynamical systems that are close to systems with a structurally unstable homoclinic curve: I. *Mat. Sb. (N.S.)*, **88**(130), 475–492.

Gavrilov, N. K., and Shilnikov, L. P. 1973. Three-dimensional dynamical systems that are close to systems with a structurally unstable homoclinic curve: II. *Mat. Sb. (N.S.)*, **90**(132), 139–156.

Gheiner, J. 1994. Codimension-two reflection and nonhyperbolic invariant lines. *Nonlinearity*, **7**(1), 109–184.

Gheiner, J. 1998. Codimension *n* flips. *Ergodic Theory Dynam. Systems*, **18**(5), 1115–1137.

Gheiner, J. 2014. Codimension *n* saddle-nodes. *Nonlinearity*, **27**(9), 2337–2360.

Ginelli, F., Poggi, P., Turchi, A., Chaté, H., Livi, R., and Politi, A. 2007. Characterizing dynamics with covariant Lyapunov vectors. *Phys. Rev. Lett.*, **99**(Sep), 130601.

Golden, M. P., and Ydstie, B. E. 1988. Bifurcation in model reference adaptive control systems. *Systems Control Lett.*, **11**(5), 413–430.

Gonchenko, S. V., and Gonchenko, V. S. 2000. On Andronov–Hopf bifurcations of two-dimensional diffeomorphisms with homoclinic tangencies. Technical report 556. WIAS, Berlin.

Gonchenko, S. V., and Gonchenko, V. S. 2004. On bifurcations of the birth of closed invariant curves in the case of two-dimensional diffeomorphisms with homoclinic tangencies. *Tr. Mat. Inst. Steklova*, **244**, 87–114 (in Russian).

Gonchenko, S. V., and Shilnikov, L. P. 2003. On two-dimensional area-preserving maps with homoclinic tangencies that have infinitely many generic elliptic periodic points. *Zap. Nauchn. Sem. S.-Peterburg. Otdel. Mat. Inst. Steklov. (POMI)*, **300**(8), 155–166, 288–289 (in Russian).

Gonchenko, S. V., Shilnikov, L. P., and Turaev, D. V. 1996. Dynamical phenomena in systems with structurally unstable Poincaré homoclinic orbits. *Chaos*, **6**(1), 15–31.

Gonchenko, S. V., Shilnikov, L. P. and Turaev, D.V. 2002a. *On Dynamical Properties of Diffeomorphisms with Homoclinic Tangencies*. Technical report 795. WIAS, Berlin.

Gonchenko, S. V., Shilnikov, L. P., and Stenkin, O. V. 2002b. On Newhouse regions with infinitely many stable and unstable invariant tori. Pages 80–102 of: *Progress in Nonlinear Science, Vol. 1 (Nizhny Novgorod, 2001)*. RAS, Institute of Applied Physics, Nizhniy Novgorod.

Gonchenko, S. V., Gonchenko, V. S., and Tatjer, J. C. 2002c. Three-dimensional dissipative diffeomorphisms with codimension two homoclinic tangencies and generalized Hénon maps. Pages 63–79 of: *Progress in Nonlinear Science, Vol. 1 (Nizhny Novgorod, 2001)*. RAS, Institute of Applied Physics, Nizhniy Novgorod.

Gonchenko, V. S. 2002. *On bifurcations of Two-Dimensional Diffeomorphisms with a Homoclinic Tangency of Manifolds of a "Neutral" Saddle. Trudy Mat. Inst. Steklova*, **236**, 86–93 (in Russian).

Gonchenko, V. S., and Ovsyannikov, I. I. 2005. On bifurcations of three-dimensional diffeomorphisms with a homoclinic tangency to a "neutral" saddle fixed point. *Journal of Mathematical Sciences*, **128**(2), 2774–2777.

Gonchenko, V. S., Kuznetsov, Yu. A., and Meijer, H. G. E. 2005. Generalized Hénon map and bifurcations of homoclinic tangencies. *SIAM J. Appl. Dyn. Syst.*, **4**(2), 407–436.

Govaerts, W. J. F. 2000. *Numerical Methods for Bifurcations of Dynamical Equilibria*. Society for Industrial and Applied Mathematics (SIAM), Philadelphia, PA.

Govaerts, W., and Khoshsiar Ghaziani, R. 2006. Numerical bifurcation analysis of a nonlinear stage structured cannibalism population model. *J. Difference Equ. Appl.*, **12**(10), 1069–1085.

Govaerts, W., and Khoshsiar Ghaziani, R. 2008. Stable cycles in a Cournot duopoly model of Kopel. *J. Comput. Appl. Math.*, **218**(2), 247–258.

Govaerts, W., Kuznetsov, Yu. A., and Sijnave, B. 1999. Bifurcations of maps in the software package CONTENT. Pages 191–206 of: *Computer Algebra in Scientific Computing – CASC'99 (Munich)*. Springer, Berlin.

Govaerts, W., Khoshsiar Ghaziani, R., Kuznetsov, Yu. A., and Meijer, H. G. E. 2007. Numerical methods for two-parameter local bifurcation analysis of maps. *SIAM J. Sci. Comput.*, **29**(6), 2644–2667.

Gramchev, T. and Walcher, S. 2005, Normal forms of maps: formal and algebraic aspects. *Acta Applicandae Mathematica*, **87**(1), 123–146

Griewank, A. and Walther, A. 2008. *Evaluating Derivatives: Principles and Techniques of Algorithmic Differentiation* 2nd edition. Society for Industrial and Applied Mathematics (SIAM), Philadelphia, PA.

Griewank, A., Juedes, D., and Utke, J. 1996. ADOL-C: a package for the automatic differentiation of algorithms written in C/C++. *ACM Trans. Math. Softw.*, **22**(2), 131–167.

Guckenheimer, J., and Holmes, P. 1990. *Nonlinear Oscillations, Dynamical Systems, and Bifurcations of Vector Fields*. Springer-Verlag, New York. Revised and corrected reprint of the 1983 original.

Guckenheimer, J., and Meloon, B. 2000. Computing periodic orbits and their bifurcations with automatic differentiation. *SIAM J. Sci. Comput.*, **22**(3), 951–985.

Hassard, B. D., Kazarinoff, N. D., and Wan, Y. H. 1981. *Theory and Applications of Hopf Bifurcation*. Cambridge University Press, Cambridge.

Hénon, M. 1976. A two-dimensional mapping with a strange attractor. *Comm. Math. Phys.*, **50**(1), 69–77.

Hirsch, M. W., Pugh, C. C., and Shub, M. 1977. *Invariant Manifolds*. Springer-Verlag, Berlin.

Holmes, P.J., and Rand, D. A. 1977/1978. Bifurcations of the forced van der Pol oscillator. *Quart. Appl. Math.*, **35**(4), 495–509.

Holmes, P., and Whitley, D. 1984a. Bifurcations of one- and two-dimensional maps. *Philos. Trans. Roy. Soc. London Ser. A*, **311**(1515), 43–102.

Holmes, P., and Whitley, D. 1984b. Erratum: "Bifurcations of one- and two-dimensional maps." *Philos. Trans. Roy. Soc. London Ser. A*, **312**(1523), 601.

Horozov, E. I. 1979. Versal deformations of equivariant vector fields for cases of symmetry of order 2 and 3. *Trudy Sem. Petrovsk.*, **5**, 163–192 (in Russian).

Iooss, G. 1979. *Bifurcation of Maps and Applications*. North-Holland Publishing Co., Amsterdam.

Iooss, G. 1988. Global characterization of the normal form for a vector field near a closed orbit. *J. Differential Equations*, **76**(1), 47–76.

Iooss, G., and Adelmeyer, M. 1998 *Topics in Bifurcation Theory and Applications*. 2nd edition. World Scientific Publishing, River Edge, NJ.

Iooss, G., and Los, J. E. 1988. Quasi-genericity of bifurcations to high-dimensional invariant tori for maps. *Comm. Math. Phys.*, **119**(3), 453–500.

Ipsen, M., Hynne, F., and Sörensen, P. G. 1998. Systematic derivation of amplitude equations and normal forms for dynamical systems. *Chaos*, **8**(4), 834–852.

Jorba, A. 2001. Numerical computation of the normal behaviour of invariant curves of n-dimensional maps. *Nonlinearity*, **14**(5), 943–976.

Kamiyama, K., Komuro, M., Endo, T., and Aihara, K. 2014. Classification of bifurcations of quasi-periodic solutions using Lyapunov bundles. *Internat. J. Bifur. Chaos Appl. Sci. Engrg.*, **24**(12), 1430034, 30.

Katok, A., and Hasselblatt, B. 1995 *Introduction to the Modern Theory of Dynamical Systems.* Cambridge University Press, Cambridge.

Kawakami, H. 1981. Algorithme numérique définissant la bifurcation d'un point homocline. *C. R. Acad. Sci. Paris Sér. I Math.*, **293**(8), 401–403.

Keller, H. B. 1977. Numerical solution of bifurcation and nonlinear eigenvalue problems. Pages 359–384 of: *Applications of Bifurcation Theory*. Academic Press, New York.

Khibnik, A. I., Kuznetsov, Yu. A., Levitin, V. V., and Nikolaev, E. V. 1993. Continuation techniques and interactive software for bifurcation analysis of ODEs and iterated maps. *Phys. D*, **62**(1–4), 360–371.

Khoshsiar Ghaziani, R., Govaerts, W., Kuznetsov, Yu. A., and Meijer, H. G. E. 2009. Numerical continuation of connecting orbits of maps in MATLAB. *J. Difference Equ. Appl.*, **15**(8–9), 849–875.

Khoshsiar Ghaziani, R., Govaerts, W., and Sonck, C. 2012. Resonance and bifurcation in a discrete-time predator–prey system with Holling functional response. *Nonlinear Anal. Real World Appl.*, **13**(3), 1451–1465.

Kleinkauf, J.-M. 1998. Numerische Analyse Tangentialer Homokliner Orbits. PhD thesis, Universität Bielefeld.

Komuro, M., Kamiyama, K., Endo, T., and Aihara, K. 2016. Quasi-periodic bifurcations of higher-dimensional tori. *Internat. J. Bifur. Chaos Appl. Sci. Engrg.*, **26**(7), 1630016, 40.

Kopel, M. 1996. Simple and complex adjustment dynamics in Cournot duopoly models. *Chaos, Solitons & Fractals*, **7**(12), 2031–2048.

Krauskopf, B. 1994. The bifurcation set for the 1:4 resonance problem. *Experiment. Math.*, **3**(2), 107–128.

Krauskopf, B. 1995. The extended parameter space of a model for 1:4 resonance. *CWI Quarterly*, **8**(3), 237–255.

Krauskopf, B. 1997. Bifurcations at ∞ in a model for 1:4 resonance. *Ergodic Theory Dynam. Systems*, **17**(4), 899–931.

Krauskopf, B. 2001. Strong resonances and Takens's Utrecht preprint. Pages 89–111 of: *Global Analysis of Dynamical Systems*. Institute of Physics, Bristol.

Krauskopf, B., and Oldeman, B. E. 2004. A planar model system for the saddle-node Hopf bifurcation with global reinjection. *Nonlinearity*, **17**(4), 1119–1151.

Krauskopf, B., and Osinga, H. M. 1998a. Globalizing two-dimensional unstable manifolds of maps. *Internat. J. Bifur. Chaos Appl. Sci. Engrg.*, **8**(3), 483–503.

Krauskopf, B., and Osinga, H. M. 1998b. Growing 1D and quasi-2D unstable manifolds of maps. *J. Comput. Phys.*, **146**(1), 404–419.

Kurakin, L. G. 1989. Semi-invariant form of criteria for stability of fixed points of a mapping in critical cases. *Izv. Severo-Kavkaz. Nauchn. Tsentra Vyssh. Shkoly Estestv. Nauk.*, **1**, 31–35 (in Russian).

Kuznetsov, Yu. A. 1999. Numerical normalization techniques for all codim 2 bifurcations of equilibria in ODE's. *SIAM J. Numer. Anal.*, **36**(4), 1104–1124.

Kuznetsov, Yu A. 2004. *Elements of Applied Bifurcation Theory*. 3rd edition. Springer-Verlag, New York.

Kuznetsov, Yu. A. 2005. Practical computation of normal forms on center manifolds at degenerate Bogdanov–Takens bifurcations. *Internat. J. Bifur. Chaos Appl. Sci. Engrg.*, **15**(11), 3535–3546.

Kuznetsov, Yu. A., and Levitin, V. V. 1995–1997. *CONTENT: A Multiplatform Environment for Analyzing Dynamical Systems*. `www.staff.science.uu.nl/~kouzn101/CONTENT`.

Kuznetsov, Yu. A., and Meijer, H. G. E. 2005. Numerical normal forms for codim 2 bifurcations of fixed points with at most two critical eigenvalues. *SIAM J. Sci. Comput.*, **26**(6), 1932–1954.

Kuznetsov, Yu. A., and Meijer, H. G. E. 2006. Remarks on interacting Neimark–Sacker bifurcations. *J. Difference Equ. Appl.*, **12**(10), 1009–1035.

Kuznetsov, Yu. A., and Piccardi, C. 1994. Bifurcation analysis of periodic SEIR and SIR epidemic models. *J. Math. Biol.*, **32**(2), 109–121.

Kuznetsov, Yu. A., Muratori, S., and Rinaldi, S. 1992. Bifurcations and chaos in a periodic predator–prey model. *Internat. J. Bifur. Chaos Appl. Sci. Engrg.*, **2**(1), 117–128.

Kuznetsov, Yu. A., Meijer, H. G. E., and van Veen, L. 2004. The fold–flip bifurcation. *Internat. J. Bifur. Chaos Appl. Sci. Engrg.*, **14**(7), 2253–2282.

Laskar, J., Froeschlé, C., and Celletti, A. 1992. The measure of chaos by the numerical analysis of the fundamental frequencies: application to the standard mapping. *Phys. D*, **56**(2–3), 253–269.

Leslie, P. H., and Gower, J. C. 1958. The properties of a stochastic model for two competing species. *Biometrika*, **45**, 316–330.

Leslie, P. H., Park, T., and Mertz, D. B. 1968. The effect of varying the initial numbers on the outcome of competition between two *Tribolium* species. *J. Animal Ecology*, **37**, 9–23.

Lindtner, E., Steindl, A., and Troger, H. 1989. Generic one-parameter bifurcations in the motion of a simple robot. *J. Comput. Appl. Math.*, **26**(1- 2), 199–218.

Lorenz, E. N. 1984. Irregularity: a fundamental property of the atmosphere. *Tellus Series A*, **36**, 98–110.

Los, J. E. 1988. Small divisor phenomena in an invariant curve doubling bifurcation on a cylinder. *Ann. Inst. H. Poincaré Anal. Non Linéaire*, **5**(1), 37–95.

Los, J. E. 1989. Non-normally hyperbolic invariant curves for maps in $\mathbf{R}^3$ and doubling bifurcation. *Nonlinearity*, **2**(1), 149–174.

Maynard Smith, J. 1968. *Mathematical Ideas in Biology*. Cambridge University Press, London

MAPLE. 2014. Version 18, www.maplesoft.com.

MATLAB. 2016. Version 2016b, www.mathworks.com.

Melnikov, V. K. 1963. Qualitative description of strong resonance in a nonlinear system. *Dokl. Akad. Nauk SSSR*, **148**, 1257–1260.

Mira, C. 1987. *Chaotic Dynamics*. World Scientific Publishing, Singapore.

Mira, C. 1997. Some historical aspects of nonlinear dynamics: possible trends for the future. *J. Franklin Inst. B*, **334**(5–6), 1075–1113.

Murdock, J. 2003. *Normal Forms and Unfoldings for Local Dynamical Systems*. Springer-Verlag, New York.

Neimark, Ju. I. 1959. Some cases of the dependence of periodic motions on parameters. *Dokl. Akad. Nauk SSSR*, **129**, 736–739.

Neimark, Ju. I. 1972. *The Method of Point Transformations in the Theory of Nonlinear Oscillations*. Nauka, Moscow (in Russian).

Neirynck, N., Govaerts, W., Kuznetsov, Yu. A., and Meijer, H. G. E. 2018. Numerical bifurcation analysis of homoclinic orbits embedded in one-dimensional manifolds of maps. *ACM Trans. Math. Softw.* **44** (3), article no. 25.

Neishtadt, A. I. 1978. Bifurcations of the phase pattern of an equation system arising in the problem of stability loss of self-oscillations close to 1:4 resonance. *J. Appl. Math. Mech.*, 830–840.

Newhouse, S., Palis, J., and Takens, F. 1983. Bifurcations and stability of families of diffeomorphisms. *Inst. Hautes Études Sci. Publ. Math.*, **57**, 5–71.

Nusse, H. E., and Yorke, J. A. 1998. *Dynamics: Numerical Explorations*. 2nd edition. Springer-Verlag, New York.

Paez Chavez, J. 2010 Starting homoclinic tangencies near 1:1 resonances. *Internat. J. Bifur. Chaos Appl. Sci. Engrg*, **20**(10), 3157–3172.

Palis, J., and de Melo, W. 1982. *Geometric Theory of Dynamical Systems*. Springer-Verlag, New York.

Palis, J., and Takens, F. 1993. *Hyperbolicity and Sensitive Chaotic Dynamics at Homoclinic Bifurcations*. Cambridge University Press, Cambridge.

Palmer, T. N. 1995. The influence of north-west Atlantic sea surface temperature: an unplanned experiment. *Weather*, **50**(12), 413–419.

Peckham, B. B., and Kevrekidis, I. G. 1991. Period doubling with higher-order degeneracies. *SIAM J. Math. Anal.*, **22**(6), 1552–1574.

Peckham, B. B., and Kevrekidis, I. G. 2002. Lighting Arnold flames: resonance in doubly forced periodic oscillators. *Nonlinearity*, **15**(2), 405–428.

Pecora, N. 2018. Analysis of 1:4 resonance in a monopoly model with memory. *Chaos, Solitons & Fractals*, **110**, 95–104.

Pryce, J. D., Khoshiar Ghaziani, R., de Witte, V., and Govaerts, W. 2010. Computation of normal form coefficients of cycle bifurcations of maps by algorithmic differentiation. *Math. Comput. Sim.*, **81**(1), 109–119.

Roose, D., and Hlavacek, V. 1985. A direct method for the computation of Hopf bifurcation points. *SIAM J. Appl. Math.*, **45**(6), 879–894.

Rousseau, C. 1990. Codimension 1 and 2 bifurcations of fixed points of diffeomorphisms and periodic solutions of vector fields. *Ann. Sci. Math. Québec*, **13**(2), 55–91.

Sacker, R. J. 1965. A new approach to the perturbation theory of invariant surfaces. *Comm. Pure Appl. Math.*, **18**, 717–732.

Saleh, K. 2005. Organising centres in the semi-global analysis of dynamical systems. PhD thesis, University of Groningen.

Saleh, K., and Wagener, F. O. O. 2010. Semi-global analysis of periodic and quasi-periodic normal-internal k:1 and k:2 resonances. *Nonlinearity*, **23**(9), 2219–2252.

Sanders, J. A., and Verhulst, F. 1985. *Averaging Methods in Nonlinear Dynamical Systems*. Springer-Verlag, New York.

Shilnikov, A., Nicolis, G., and Nicolis, C. 1995. Bifurcation and predictability analysis of a low-order atmospheric circulation model. *Internat. J. Bifur. Chaos Appl. Sci. Engrg.*, **5**(6), 1701–1711.

Shoshitaishvili, A. N. 1975. The bifurcation of the topological type of the singular points of vector fields that depend on parameters. *Trudy Sem. Petrovsk.*, **1**, 279–309 (in Russian).

Simó, C. 1990. On the analytical and numerical approximation of invariant manifolds. Pages 285–329 of: *Les Méthodes Modernes de la Mécanique Céleste [Modern Methods in Celestial Mechanics]*. Available at `www.maia.ub.es/dsg/2004/index.html`.

Simó, C. 2005. On the use of Lyapunov exponents to detect global properties of the dynamics. Pages 631–636 of: *EQUADIFF 2003*. World Scientific Publishing, Hackensack, NJ.

Song, Y. L. 2004. The juvenile/adult Leslie–Gower competition model. Preprint.

Takens, F. 1974. Forced oscillations and bifurcations. Pages 1–59 of: *Applications of global analysis: I.* Mathematical Institute, Utrecht. Reprinted in *Global Analysis of Dynamical Systems*, edited by H.W. Broer, B. Krauskopf and G. Vegter. Institute of Physics, Bristol, 2001.

Takens, F., and Wagener, F. O. O. 2000. Resonances in skew and reducible quasi-periodic Hopf bifurcations. *Nonlinearity*, **13**(2), 377–396.

Turaev, D. 1996. On dimension of non-local bifurcational problems. *Internat. J. Bifur. Chaos Appl. Sci. Engrg.*, **6**(5), 919–948.

van Gils, S. A. 1984. *On a Formula for the Direction of Hopf Bifurcation*. Technical Report TW/225. CWI, Amsterdam.

Vanderbauwhede, A. 1982. *Local Bifurcation and Symmetry*. Pitman (Advanced Publishing Program), Boston, MA.

Vanderbauwhede, A. 1988. Center manifolds and their basic properties: an introduction. *Delft Progr. Rep.*, **12**(1), 57–78.

Vanderbauwhede, A., and van Gils, S. A. 1987. Center manifolds and contractions on a scale of Banach spaces. *J. Funct. Anal.*, **72**(2), 209–224.

Vanhulle, L. 2017. Computational detection of homoclinic and heteroclinic connections for iterated maps (in Dutch). MSc thesis, Ghent University. Available at `https://lib.ugent.be/en/catalog/rug01:002376270?i=0&q=Laura+Vanhulle`.

Veen, L. van. 2003. Baroclinic flow and the Lorenz-84 model. *Internat. J. Bifur. Chaos Appl. Sci. Engrg.*, **13**(8), 2117–2139.

Vitolo, R. 2003. *Bifurcations in 3D diffeomorphisms: a study in experimental mathematics*. PhD thesis, University of Groningen.

Vitolo, R., Broer, H. W., and Simó, C. 2010. Routes to chaos in the Hopf-saddle-node bifurcation for fixed points of 3D-diffeomorphisms. *Nonlinearity*, **23**(8), 1919.

Vitolo, R., Glendinning, P., and Gallas, J. A. C. 2011a. Global structure of periodicity hubs in Lyapunov phase diagrams of dissipative flows. *Phys. Rev. E*, **84**, 016216.

Vitolo, R., Broer, H. W., and Simó, C. 2011b. Quasi-periodic bifurcations of invariant circles in low-dimensional dissipative dynamical systems. *Regul. Chaotic Dyn.*, **16**(1–2), 154–184.

Wan, Y. H. 1978. Bifurcation into invariant tori at points of resonance. *Arch. Rational Mech. Anal.*, **68**(4), 343–357.

Wen, G.-L., and Xu, D. 2004a. Feedback control of Hopf–Hopf interaction bifurcation with development of torus solutions in high-dimensional maps. *Phys. Lett. A*, **321**(1), 24–33.

Wen, G.-L., and Xu, D. 2004b. Implicit criteria of eigenvalue assignment and transversality for bifurcation control in four-dimensional maps. *Internat. J. Bifur. Chaos Appl. Sci. Engrg.*, **14**(10), 3489–3503.

Wen, G.-L., Wang, Q.-G., and Chiu, M.-S. 2006. Delay feedback control for interaction of Hopf and period doubling bifurcations in discrete-time systems. *Internat. J. Bifur. Chaos Appl. Sci. Engrg.*, **16**(1), 101–112.

Wiggins, S. 2003. *Introduction to Applied Nonlinear Dynamical Systems and Chaos*. 2nd edition. Springer-Verlag, New York.

Wolfram Research, Inc. 2017. *Mathematica, Version 11.1*. Wolfram Research, Champaign, IL.

Xie, J., and Ding, W. 2005. Hopf–Hopf bifurcation and invariant torus $\mathbf{T}^2$ of a vibro-impact system. *Internat. J. Non-Linear Mech.*, **40**(4), 531–543.

Yagasaki, K., 1998. Numerical detection and continuation of homoclinic points and their bifurcations for maps and periodically forced systems, *Internat. J. Bifur. Chaos Appl. Sci. Engrg.*, **8**(7), 1617–1627

Zegeling, A. 1993. Equivariant unfoldings in the case of symmetry of order 4. *Serdica*, **19**(1), 71–79.

Zholondek, K. 1983. Versality of a family of symmetric vector fields on the plane. *Mat. Sb. (N.S.)*, **120(162)**(4), 473–499.

Index